REDOX PROTEINS IN SUPERCOMPLEXES AND SIGNALOSOMES

REDOX PROTEINS IN SUPERCOMPLEXES AND SIGNALOSOMES

edited by

Ricardo O. Louro

Instituto de Tecnologia Química e Biológica António Xavier,
Universidade Nova de Lisboa, Portugal

Irene Díaz-Moreno

Instituto de Bioquímica Vegetal y Fotosíntesis, cicCartuja,
Universidad de Sevilla - CSIC

CRC Press is an imprint of the
Taylor & Francis Group, an **informa** business

First published in hardback 2020
First published in paperback 2024

Published 2016 by CRC Press
2385 NW Executive Center Drive, Suite 320, Boca Raton FL 33431

and by CRC Press
4 Park Square, Milton Park, Abingdon, Oxon, OX14 4RN

CRC Press is an imprint of Taylor & Francis Group, LLC

Publisher's Note
The publisher has gone to great lengths to ensure the quality of this reprint but points out that some imperfections in the original copies may be apparent.

ISBN: 978-1-4822-5110-4 (hbk)
ISBN: 978-1-138-89398-6 (pbk)
ISBN: 978-0-429-06949-9 (ebk)

DOI: 10.1201/b19087

Contents

Preface

THIS BOOK COMES AS a product of its time, a time when scientific paradigms about biology are changing as a consequence of evolving methodologies and understanding of the ensuing experimental results. Over the past 10 years, the structural and functional organization of the inner mitochondrial membrane and oxidative phosphorylation have been the focus of intense debate. Within this domain, it is becoming widely accepted that respiratory complexes can be rearranged/reorganized into supercomplexes in response to different stimuli, carbon sources or stress conditions, and serve as a crucial adaptive mechanism regulated by mitochondria. Nowadays, the organization and dynamics of the mitochondrial electron transport chain is the subject of passionate discussion, for which two different models are being considered: the *fluid model*—also known as the *random collision model*—proposes independent diffusional motions for all membrane proteins and redox components; the *solid model* suggests specific interactions between individual respiratory components to form stable assemblies. Thus, the *fluid model* states that each respiratory complex would act as an individual entity, whereas the *solid model* strongly supports the oligomerization in supercomplexes upon the discovery of detergent-based strategies for their isolation and solubilisation.

However, neither model accounts for all the experimental evidence, since the supercomplex assembly is inherently plastic and it evolves during cell lifespan and strongly varies from cell to cell. The apparently opposed models—solid versus fluid—are reconciled in the *plasticity model*, which assumes that the supercomplex organization is not fixed but in equilibrium with randomly dispersed respiratory complexes in living cells under physiological conditions. Indeed, the ensemble of supercomplexes tightly depends on the changes of phospholipid composition of the membrane due to genetic or dietary reasons, the mitochondrial membrane potential, the so-called supercomplex assembly factors, and the phosphorylation state of the protein subunits of the complexes.

Evidence of *respiratory supercomplexes in all life domains* has been published in the past 10 years, demonstrating that the supramolecular organization of respiratory chains is an ancient metabolic strategy widespread in nature. Indeed, respiratory supercomplexes have been described in different eukaryotic organisms, such as algae, fungi, plants, and mammals, and also in bacteria, possibly to overcome the poor compartmentalization of bacterial enzymes. In plants and green algae, there is also evidence of scaffolds between the photosystems and the light-harvesting complexes to form *photosynthetic supercomplexes* inside chloroplasts. Moreover, the photosystem I and the cytochrome b_6f are associated in a supercomplex that is functional in cyclic electron transport. To date, however, we are not aware of electron transport supercomplexes in cyanobacteria, despite the proximity of the photosynthetic apparatus and the respiratory chain inside the thylakoid membrane.

Altogether, these findings suggest that the organization in supramolecular assemblies provides *functional advantages* in the context of the mitochondrial respiratory function and oxidative phosphorylation. Then, CoQ and cytochrome *c channelling* promotes direct transfer of electrons between two adjacent enzymes—from complex I to complex III and from complex III to complex IV, respectively—by successive reduction and re-oxidation of the intermediate—CoQ or cytochrome *c*—without its diffusion in the bulk medium. Evidence for possible channelling comes from the mitochondrial *respirasome* $I_1III_2IV_1$, the largest supercomplex, which represents the minimal unit able to perform complete aerobic respiration from NADH to oxygen. It is also worth mentioning that the respirasome assembly preserves *the structural integrity and activity of complex I*, when bound to complex III. In addition, the supramolecular structure of mitochondrial respiratory complexes act as a factor *limiting reactive oxygen species (ROS) generation*, mainly by complexes I and III in the mitochondria, especially during ageing-dependent decay of mitochondrial function. Therefore, the plastic and dynamic nature of supercomplexes lies at the basis of a control mechanism of the ROS concentration in the cell. Under physiological conditions, ROS are messengers acting through redox modifications in signalling proteins. Under homeostasis, ROS levels are controlled by not only supercomplexes but also *detoxifying cellular signalosomes* existing inside the mitochondria, such as thiorredoxin and cytochrome *c*-signalling networks. In contrast, when mitochondria are malfunctioning, ROS levels are increased, causing pathological effects such as lipid peroxidation, leading to membrane raft formation, protein oxidation including cytochrome *c*, mitochondrial DNA damage, and supercomplex dismantling. Thus, modulation of respiratory supercomplexes arises as a new approach to regulate disease progression.

Yet, controversially, a number of recent studies report that mitochondrial respirasomes may not be the preferential functional forms of the mitochondrial respiratory chain, since the lack of respirasomes in some organisms does not lead to evident pathological phenotypes. Although much remains to be discovered about the function of supercomplexes in biological systems, it is clear, nonetheless, that the countless and ever-growing advances in the field have helped the scientific community overcome the scepticism about the reality of biological supercomplexes. In other words, *respirasomes (supercomplexes) exist because they respire (function).*

Nevertheless, we consider that this novel and widely accepted scheme of the mitochondrial respiratory chain, dynamically organised in supercomplexes even under physiological conditions, deserves to be discussed in a book, so as to explain the state of the art to readers and stimulate the debate. The other aim is one of shedding light on the impact of supercomplex assembly on mitochondrial morphology and its physiology and biogenesis in an effort to understand the molecular mechanisms of pathological situations, including ageing. Without any hint of doubt, answers to these and other questions will certainly come by in the forthcoming years.

Keeping this in mind, this book aims to bring together the most exciting research from leading practitioners in the field. It is impossible to cover all aspects of *supercomplexes and signalosomes* into a single volume, but we sincerely hope that the selection of themes herein presented will help the readers grasp the state of the art in this multidisciplinary field and understand the exciting development and quick progress of current research in biosciences. To further this aim, several of the figures appearing in black and white in the printed book were collected in the CRC Press website for this book (http://www.crcpress.com/product/isbn/9781482251104), from where they are freely downloadable.

We are most grateful not only to all authors who have generously accepted the invitation to contribute to this book with their interesting and cutting-edge chapters, but also to the editorial office of CRC Press for their excellent technical assistance and support.

Ricardo O. Louro
Universidade Nova de Lisboa

Irene Díaz-Moreno
Universidad de Sevilla

Editors

Ricardo O. Louro completed his undergraduate studies in chemistry at Universidade Nova de Lisboa, Portugal, where he continued to obtain a PhD in biochemistry. After two short postdoctoral periods in Stockholm and Leiden, he returned to Portugal.

Dr. Louro's research work focuses on the investigation of the molecular bases for biological electron transfer phenomena relevant for bioenergetic processes and the applications of this knowledge to blue-, red- or white-biotechnology. He has published more than 70 research papers and book chapters on the subjects of biological electron transfer, bioenergetics, and metalloproteins. He has also presented his work at more than 30 international conferences and seminar series organized by university departments.

During his career, Dr. Louro has taught at several universities in Portugal and organized a series of advanced courses on metals in biology, which has attracted applicants from all over the world. He has also organized several international conferences focussing on the interface of chemistry and biology.

Dr. Louro has received numerous awards for his scientific contributions and is a nominated Burgen scholar by the Academia Europaea. He is a member of the Portuguese Biochemical Society, the Society for Biological Inorganic Chemistry, and the International Society for Microbial Electrochemical Technologies. He is currently the leader of the Inorganic Biochemistry and NMR laboratory at Instituto de Tecnologia Química e Biológica António Xavier, Oeiras, Portugal.

Dr. Irene Díaz-Moreno is an associate professor of biochemistry and molecular biology at the Institute of Plant Biochemistry and Photosynthesis—IBVF of the Scientific Research Centre Isla de la Cartuja—cicCartuja, Seville, Spain.

She was awarded a PhD with European mention from the University of Seville, Spain, in 2005. Dr. Irene Díaz-Moreno has worked in collaboration with groups at the Universities of Göteborg (Sweden) and Leiden (The Netherlands) on molecular recognition between metalloproteins involved in electron-transfer processes. She was an EMBO postdoctoral fellow (2006–2008) at the NIMR-MRC in London (UK), working on the regulatory mechanisms of mRNA decay by RNA-binding proteins. In 2010, she won a permanent position at the University of Seville, where she is developing research projects in biointeractomics, as well as on the post-translational regulation of biological macromolecules (www.ibvf.csic.es/en/biointeractomics).

The significance of all her works has been published in high-impact journals such as *Nature Structural and Molecular Biology*, *Nucleic Acids Research*, *Chemistry and Biology*, *Structure*, and *The Journal of Biological Chemistry.*

Contributors

Justine M. Abais
Department of Pharmacology & Toxicology
School of Medicine
Virginia Commonwealth University
Richmond, Virginia

Rebeca Acín-Pérez
Fundación Centro Nacional de Investigaciones Cardiovasculares
Instituto de Carlos III
Madrid, Spain

Jun-Xiang Bao
Department of Pharmacology & Toxicology
School of Medicine
Virginia Commonwealth University
Richmond, Virginia

Sara Cogliati
Fundación Centro Nacional de Investigaciones Cardiovasculares
Instituto de Carlos III
Madrid, Spain

Isabel Cruz-Gallardo
Instituto de Bioquímica Vegetal y Fotosíntesis, cicCartuja
Universidad de Sevilla—(CSIC)
Sevilla, Spain

Miguel A. De la Rosa
Instituto de Bioquímica Vegetal y Fotosíntesis, cicCartuja
Universidad de Sevilla—(CSIC)
Sevilla, Spain

Antonio Díaz-Quintana
Instituto de Bioquímica Vegetal y Fotosíntesis, cicCartuja
Universidad de Sevilla—(CSIC)
Sevilla, Spain

William Dowhan
Department of Biochemistry & Molecular Biology
University of Texas-Houston, Medical School
Houston, Texas

José A. Enriquez
Fundación Centro Nacional de Investigaciones Cardiovasculares
Instituto de Carlos III
Madrid, Spain

Bruno M. Fonseca
Instituto de Tecnologia Química e Biológica António Xavier
Universidade Nova de Lisboa
Oeiras, Portugal

James K. Fredrickson
Pacific Northwest National Laboratory
Richland, Washington

Maria L. Genova
Dipartimento di Scienze Biomediche e Neuromotorie
Alma Mater Studiorum—Università di Bologna
Bologna, Italy

Anna M. Ghelli
Dipartimento di Farmacia e Biotecnologie FABIT
Università di Bologna
Bologna, Italy

Katiuska González-Arzola
Instituto de Bioquímica Vegetal y Fotosíntesis, cicCartuja Research Center
Universidad de Sevilla—Spanish National Research Council
Sevilla, Spain

Emma B. Gutiérrez-Cirlos
Unidad de Biomedicina, Facultad de Estudios Superiores Iztacala
Universidad Nacional Autónoma de México
Mexico City, Mexico

Hidenori Ichijo
Laboratory of Cell Signaling
Graduate School of Pharmaceutical Sciences
The University of Tokyo
Bunkyo, Japan

Kentaro Ifuku
Division of Integrated Life Science
Graduate School of Biostudies
Kyoto University
Kyoto, Japan

Giorgio Lenaz
Dipartimento di Scienze Biomediche e Neuromotorie
Alma Mater Studiorum—Università di Bologna
Bologna, Italy

Tchern Lenn
School of Biological and Chemical Sciences
Queen Mary University of London
London, United Kingdom

Pin-Lan Li
Department of Pharmacology & Toxicology
School of Medicine
Virginia Commonwealth University
Richmond, Virginia

Toshiya Machida
Laboratory of Cell Signaling
Graduate School of Pharmaceutical Sciences
The University of Tokyo
Bunkyo, Japan

Jonathan Martínez-Fábregas
Instituto de Bioquímica Vegetal y Fotosíntesis, cicCartuja Research Center
Universidad de Sevilla—Spanish National Research Council
Sevilla, Spain

Ana M.P. Melo
Instituto de Investigação Científica Tropical
Lisboa, Portugal

Eugenia Mileykovskaya
Department of Biochemistry & Molecular Biology
University of Texas-Houston, Medical School
Houston, Texas

Blas Moreno-Beltrán
Instituto de Bioquímica Vegetal y Fotosíntesis, cicCartuja Research Center
Universidad de Sevilla—Spanish National Research Council
Sevilla, Spain

Conrad W. Mullineaux
School of Biological and Chemical Sciences
Queen Mary University of London
London, United Kingdom

Isao Naguro
Laboratory of Cell Signaling
Graduate School of Pharmaceutical Sciences
The University of Tokyo
Bunkyo, Japan

Catarina M. Paquete
Instituto de Tecnologia Química e Biológica António Xavier
Universidade Nova de Lisboa
Oeiras, Portugal

Kevin M. Rosso
Pacific Northwest National Laboratory
Richland, Washington

Michela Rugolo
Dipartimento di Farmacia e Biotecnologie FABIT
Università di Bologna
Bologna, Italy

Liang Shi
Pacific Northwest National Laboratory
Richland, Washington

Toshiharu Shikanai
Department of Botany
Graduate School of Science
Kyoto University
Kyoto, Japan

Miguel Teixeira
Instituto de Tecnologia Química e Biológica António Xavier
Universidade Nova de Lisboa
Oeiras, Portugal

Ming Tien
The Pennsylvania State University
University Park, Pennsylvania

Valentina C. Tropeano
Dipartimento di Farmacia e Biotecnologie FABIT
Università di Bologna
Bologna, Italy

John M. Zachara
Pacific Northwest National Laboratory
Richland, Washington

Yang Zhang
Department of Pharmacology & Toxicology
School of Medicine
Virginia Commonwealth University
Richmond, Washington

CHAPTER 1

Multi-Electron Transfer in Biological Systems

Catarina M. Paquete, Bruno M. Fonseca and Ricardo O. Louro

CONTENTS

1.1 INTRODUCTION

The central role that redox reactions play in biological systems may be appreciated by the fact that in the systematic nomenclature of enzymes, oxidoreductases are the first class. Understanding the electron-transfer mechanism of redox proteins with multiple centres requires a combination of structural, thermodynamic and kinetic studies. These studies reveal the molecular details of how electrons are received by the protein, distributed among the redox cofactors and subsequently used for catalysis or reduction of downstream partners, while avoiding side reactions that could generate deleterious reactive species. This makes the study of biological electron-transfer phenomena and redox catalysis a multidisciplinary endeavour that requires a combination of know-how in biology, chemistry and physics. In this chapter, a brief introduction is given to the importance and role of multi-electron reactions in biological systems, how they can be studied and key biological processes that rely on such reactions.

1.2 REDOX CENTRES IN BIOLOGICAL SYSTEMS

Living organisms have adapted a multitude of redox-active cofactors for the purpose of establishing long-range electron-transfer pathways and redox catalysis. These include the following: (1) purely inorganic cofactors such as the iron-sulphur (Fe-S) clusters and their hetero-metallic derivatives such as the FeMoCo cluster of nitrogenases (Eady, 1996); (2) organometallic cofactors such as the hemes, chlorins and molybdopterins (Kupitz et al., 2014; Mowat and Chapman, 2005; Schwarz et al., 2009) and (3) purely organic cofactors such as flavins and quinones, and sidechains of amino acids such as tryptophan and tyrosine (Baradaran et al., 2013; Coulson and Yonetani, 1972; Deller et al., 2008). A survey of structures of redox proteins reveals that the physiological roles assigned to these cofactors appear to be biased towards having hemes and Fe-S clusters predominately in redox chains and having flavins, pterins and quinones in catalytic positions (Page et al., 2003). In inorganic and organometallic redox centres three metals stand out due to their abundance or biological importance: *iron*, *copper* and *manganese*.

1.2.1 Iron

Iron is a transition metal with great chemical versatility found in the first row of group eight of the periodic table. It displays several redox-, spin- and coordination states accessible in a biological context. Iron-based cofactors play a key role in biological electron transfer by establishing the backbone of the intramolecular wiring for long-distance conduction—be they hemes, as in the hexadecaheme cytochrome from *Desulfovibrio* (Matias et al., 2002), or Fe-S clusters, as in the peripheral arm of respiratory complex I (Sazanov and Hinchliffe, 2006).

1.2.1.1 Hemes

The heme is one of the most versatile redox cofactors in biology and plays roles in long-range electron transfer, signalling and catalysis (Mayfield et al., 2011). It is formed by a protoporphyrin IX ring coordinated to an iron atom via the four central nitrogens. This forms the basis for *b*-type cytochromes, where the heme is held in the protein structure by the axial ligands of the iron. In *c*-type cytochromes the heme is additionally bound covalently to two cysteines in a characteristic heme-binding sequence —$CX_{2-4}CH$, where the histidine provides one of the axial ligands of the iron (Verissimo and Daldal, 2014). Modifications at the periphery of the heme give rise to other types of hemes and cytochromes such as the heme *a* found in mitochondrial heme-copper oxidases (Papa et al., 1998). The reduction potential of heme proteins spans almost 1 V, from −475 mV (vs. SHE) to +450 mV. It can be modulated by solvent exposure, which leads to lower potentials (Stellwagen, 1978), by the nature of axial ligands with His-Met coordination leading to an increase of ~150 mV in reduction potential versus His-His (Dolla et al., 1994), and smaller effects caused by heme planar distortions (Ma et al., 1998) and electrostatic interactions (Fonseca et al., 2012). The heme propionates play a central role in linking the redox state of the heme with pH, leading to the redox-Bohr effect (Xavier, 1986). In addition to the modulation of the reduction potential, hemes afford coordination number transitions typically between 5 and 6, as in cytochromes *c′* and nitric oxide synthases (Girvan and Munro, 2013). Hemes can undergo spin state transitions from $S = 0$ to $S = 5/2$ (Coletta et al., 1997). The iron in the heme typically presents transitions between +2/+3 redox states, but can also reach +4 and even a state with +4 iron/porphyrin cation in compound I of horseradish peroxidase and cytochromes P450 (Rittle and Green, 2010). All these properties make hemes versatile cofactors capable of performing electron transfer, ligand binding for transport and signalling, as well as oxygen activation for catalysis.

1.2.1.2 Fe-S Clusters

Fe-S clusters are considered the most ancient of redox cofactors in biology. Their structure bears strong similarities to that of Fe-S minerals, suggesting a co-opting of the chemical properties of the minerals for the primordial metabolic processes. In several proteins, Fe-S clusters can be reconstituted by adding iron and sulphur in reducing conditions (Tsibris and Woody, 1970). These clusters cover the widest range of reduction potentials for a single kind of cofactor with a range of over 1 V from approximately −700 mV (vs. SHE) for some ferredoxins to approximately +400 mV for some HiPIPs (Liu et al., 2014). Although they are found mostly in redox chains that establish long-distance electron-transfer pathways, they, and their derivatives, are also found in the catalytic sites of numerous enzymes such as hydrogenases (Peters et al., 1998). Fe-S clusters are very sensitive to oxidative damage and therefore they also play an important role in oxidative stress signalling and regulation (Beinert and Kiley, 1999). Fe-S clusters are found in nature with a wide degree of complexity, from the simple rubredoxin type of clusters with one iron coordinated by four cysteines to the P-cluster of nitrogenases containing 8 irons and 7 sulphurs with a structure that is redox-state dependent (Ribbe et al., 2014). The irons are typically bound to the polypeptide chain by cysteine side-chains but other alternatives exist such as in the 2Fe-2S Rieske-type clusters where two histidines and two cysteines coordinate the iron atoms (Liu et al., 2014). In the clusters with more than one iron atom, each pair of irons is coordinated by two inorganic sulphur atoms. In Fe-S clusters with multiple irons, electronic coupling between the irons allows a multitude of spin states to emerge, from diamagnetic ($S = 0$) up to $S = 9/2$ for 2Fe-2S clusters (Achim et al., 1999). The pH dependence of the reduction potential of some Fe-S clusters such as the 3Fe-4S clusters and some HiPIPs arise from ionizable amino acid sidechains in their vicinity (Camba et al., 2003; Luchinat ct al., 1994). Fe-S clusters can often accommodate other metals such as molybdenum (Mo) in the case of nitrogenases and these hybrid clusters can have important roles in reaction catalysis (see section 1.6.3).

1.2.2 Copper

Copper is unlikely to have been significantly bioavailable in early earth, given the low solubility in environments of low redox potential (Williams and Fraústo da Silva, 1996). It has however emerged as a major player in the modern oxygen-rich atmosphere, where it has the central catalytic role in the terminal oxidase of aerobic respiratory chains and in multi-copper oxidases, responsible for degradation of toxic compounds (Martins et al., 2002). These roles derive from the fact that copper centres have reduction potentials in the positive range versus SHE up to 1 V, as in the multi-copper enzyme ceruloplasmin (Machonkin et al., 1998). Copper proteins can be classified according to the organization of the centres in type 1 and type 2 for mononuclear centres, type 3 and Cu_A for dinuclear centres and Cu_Z for tetranuclear centres. Type 1 and Cu_A centres typically have blue colour due to an intense absorption band at ~600 nm. Proteins containing type 1 and Cu_A centres have a characteristic topology called the cupredoxin fold organized as an 8 strand beta-barrel (Rydén and Hunt, 1993). This makes them suitable for electron-transfer reactions due to the low reorganization energy of this type of protein secondary structure. The structures of the apo-, reduced- and oxidized forms of cupredoxins are virtually identical, showing that this rigid structure effectively pre-defines the coordination geometry of the copper, in accordance with the entatic state hypothesis (Vallee and Williams, 1968). Among the copper centres, Cu_A is unique in biology since it displays a metal-metal bond, allowing for electron delocalization between the two Cu atoms. As a consequence the two Cu(I) atoms in the reduced state become formally mixed-valence Cu (1.5) in the oxidized, also called resting, state (Kroneck et al., 1988).

Type 2, type 3 and Cu_Z centres have catalytic roles in the enzymes where they are incorporated but the major catalytic activity involving copper atoms is performed by the binuclear heme-copper site in aerobic terminal oxidases (Wikström, 2004). Heme-copper oxidases are the final complex in aerobic respiratory chains where an oxygen molecule bound between heme a_3 (in mitochondrial HCO) and CuB is reduced to two water molecules concomitantly with consumption of protons from the inner side of the membrane and the pumping of additional protons to the outside (see Section 1.6.2).

1.2.3 Manganese

Manganese is a relatively abundant element in the crust of the earth that plays a central role in biology. It is the key element in the oxygen-evolving centre (OEC) that performs the light-driven water-splitting reaction (Kern et al., 2012). This process dramatically changed the atmospheric composition of the earth and enabled the emergence of complex life forms. The OEC is located in a large multi-subunit membrane complex called photosystem II (PSII), found in the chloroplasts of plants, algae and in cyanobacteria (see Section 1.6.1).

Manganese is also an important element in the protection from reactive oxygen species. High cellular concentrations of manganese (II) are correlated with resistance to UV radiation by *Deinococcus radiodurans* (Ghosal et al., 2005). Furthermore, manganese is found in the active site of catalases, peroxidases and superoxide dismutases. Some of these proteins contain mononuclear manganese sites whereas others contain dinuclear manganese sites.

In addition to these catalytic activities that arise from the fact that manganese can display three oxidation states that interconvert with potentials relevant for biology, other enzymes use dinuclear centres of manganese for catalysis that is not redox dependent. Arginases are well studied given their key role in the last step of the urea cycle, an essential metabolic pathway in the metabolism of nitrogen compounds in mammals, and in NO synthesis in macrophages (Kanyo et al., 1996).

1.3 ELECTRON TRANSFER THEORY

Biological electron-transfer reactions occur between cofactors within a single protein (intramolecular electron transfer) or between proteins engaged in transient or stable interactions (intermolecular electron transfer). While in intramolecular electron-transfer reactions the redox cofactors are close to each other within a single protein, in intermolecular electron-transfer reactions between transiently bound partners it is necessary that the two redox cofactors within two proteins approach each other (Ubbink, 2009). This usually occurs by diffusion and random collision, and it is necessary to consider several steps in the whole process: formation of an encounter complex, eventual subsequent rearrangements and conformational changes that optimize the coupling between the redox centres (Berg and von Hippel, 1985; Ubbink, 2009), the actual electron-transfer event and, finally, dissociation of the protein-protein complex (Scheme 1.1).

$$A^{ox} + D^{red} \overset{K_1}{\leftrightarrow} \left[A^{ox} - D^{red}\right] \overset{K_x}{\rightarrow} \left[A^{ox} - D^{red}\right]^{*} \overset{k_{ET}}{\rightarrow} \left[A^{red} - D^{ox}\right] \overset{K_2}{\leftrightarrow} A^{red} + D^{ox} \qquad \text{Scheme 1.1}$$

In Scheme 1.1, D is the electron donor and A is the electron acceptor, K_1 is the binding constant of the protein-protein complex, K_2 is the dissociation constant for the products of electron transfer, and k_{ET} is the electron-transfer rate constant. K_x is the rate of protein rearrangement or conformational changes within the complex that may be required to optimize or activate the system for electron transfer (Davidson, 2008). It is clear from this scheme that in an intermolecular electron-transfer process the experimentally observed rate may not be a true electron-transfer kinetic rate. The observed rate constant may contain information on the biological system and reaction mechanisms. For example, structural fluctuations may have a significant impact in the rates of electron transfer, and may modulate the observed rates depending on the dynamic regime for the fluctuations (Skourtis et al., 2010). When the structural fluctuations occur at timescales shorter than the timescale for electron transfer, the electron-transfer kinetics can become gated by these fluctuations (Davidson, 2008). In multi-redox centre enzymes, the distribution of the electrons among the centres, ligand binding or protonation can modify the timescale of the dynamic fluctuations, making certain states of protein reduction more catalytically active than others (Fourmond et al., 2013; Paquete et al., 2014). In these cases, the analysis of protein electron-transfer reactions depends on the assignment of the rate-limiting step (Davidson, 2008). When the rate-limiting step is the electron-transfer reaction the theoretical frame for the parameters that control the kinetics of the reaction is nowadays well established.

1.3.1 Marcus Theory

For electron transfer to occur, sufficient coupling must exist between the electronic orbitals of the reactants. Given that the electron is approximately 600 times lighter than the lightest atom, and that nuclear motion is much slower than electronic motion, in most cases the Franck-Condon principle applies for biological electron-transfer reactions (Marcus et al., 1954). In these conditions the nuclei involved in the reaction do not change position during the electron-transfer event, and the nuclear rearrangement must occur before the actual electron transfer (Marcus and Sutin, 1985). Considering that the reactant and product states behave as ideal harmonic oscillators, the potential energy profile of both reactant and product states can be defined as simple parabolas in a reaction coordinate diagram (Figures 1.1 and 1.2). For electron transfer to occur the total potential energy of the reactants and surrounding medium must be equal to the total potential energy of the products and surrounding medium. This occurs at the intersection of the potential energy surfaces of the reactant and product states and depends on the thermal fluctuations of the reactant states (Figure 1.1). This is denominated the activation energy necessary for the electron transfer to occur:

$$\Delta G^{*} = \frac{\left(\Delta G^{0} + \lambda\right)^{2}}{4\lambda} \tag{1.1}$$

where:

ΔG^0 is determined by the reduction potential difference between the electron donor and the electron acceptor

λ is the reorganization energy, which is the energy required for the electronic transition to occur without nuclear motion (McLendon, 1988). This term is a sum of the inner-sphere reorganization energy (λ_i), which is the energy associated with the change of the intramolecular coordinates of the redox site itself, and the medium reorganization energy (λ_o) that is associated with the change in the surrounding solvent (Krishtalik, 2011)

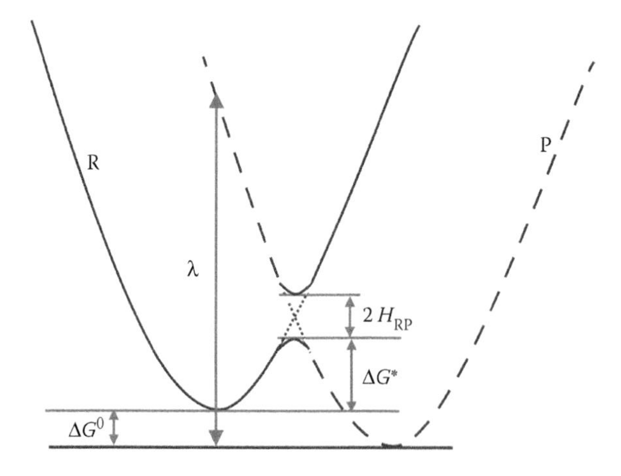

FIGURE 1.1 Schematic representation of reactant and product potential energy curves as a function of reaction coordinates of the reactants and products in electron-transfer reactions.

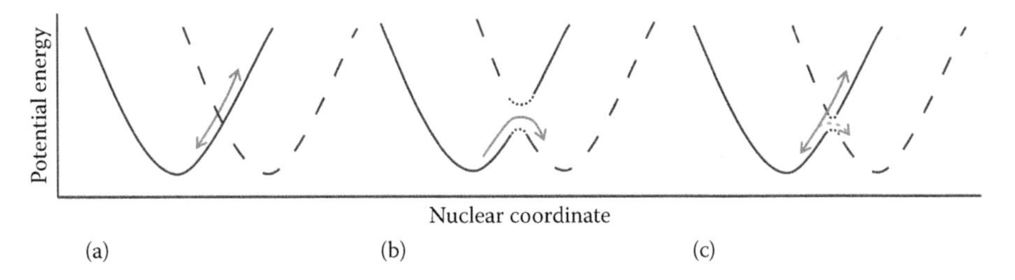

FIGURE 1.2 Schematic representation of reactant and products potential energy curves as a function of reaction coordinates of the reactants and products in electron-transfer reactions: (a) diabatic, (b) adiabatic and (c) nonadiabatic.

At the intersection of the two parabolas the quantum mechanical electronic coupling factor (H_{RP}), which is directly related to the strength of the electronic interaction between the electron donor and the electron acceptor, will determine the fate of the reaction. If H_{RP} is zero, electron transfer will not occur regardless of the energy, and the reaction is *diabatic* (Figure 1.2a). By contrast, in conditions of strong coupling, H_{RP} is larger than k_BT (k_B is the Boltzmann constant and T is the absolute temperature) and the probability of crossover when the activation energy is achieved is unitary (Schrauben et al., 2012). Such a system is called *adiabatic* and corresponds to the classical transition-state theory (Figure 1.2b). In the vast majority of cases, biological electron transfer occurs in conditions of weak electronic coupling where there is a finite probability $1>P>0$ that the electron will transition to the product state. This corresponds to a relatively narrow gap between the reactant and product potential energy profile at the intersection state represented as $2H_{RP}$ (Figure 1.2c), making the reaction *nonadiabatic* (Davidson, 2008).

The rate of the electron transfer in nonadiabatic reactions is described by semi-classical theory of electron transfer (Marcus and Sutin, 1985). It predicts that the efficiency of electron transfer between an electron donor and an electron acceptor depends on: the driving force for the electron transfer, the nuclear reorganization in the electron donor and acceptor and the electronic coupling between the reactants and products at the transition state according to:

$$k_{ET} = \sqrt{\frac{4\pi^3}{h^2\lambda k_B T}} H_{RP}^2 \exp\left[-\frac{\left(\Delta G^0 + \lambda\right)^2}{4\lambda k_B T}\right] \tag{1.2}$$

This theory predicts that the rate constants increase with the driving force. The function reaches a maximum at $-\Delta G^0 = \lambda$, and then it starts to decrease in the so-called inverted region (Marcus and Sutin, 1985). The electronic coupling, H_{RP}, is temperature independent and proportional to $e^{-\beta R}$, where R is the edge-to-edge distance between the electron donor and the electron acceptor (in Å) and β is the decay constant for tunnelling that depends on the intervening medium (Gray and Winkler, 2005). This led to the development of a simple empirical expression that describes the dependence of the tunnelling rate (k_{ET} in s^{-1}) with distance and driving force (Moser et al., 1992). This was later refined to account for variations of β with protein packing in exergonic electron-transfer reactions (Page et al., 1999):

$$\log k_{ET} = 13 - (1.2 - 0.8\rho)(R - 3.6) - 3.1\frac{(\Delta G^0 + \lambda)^2}{\lambda} \tag{1.3}$$

In Equation 1.3, ρ accounts for the packing density of the protein atoms in the path between the redox centres. This value can range from 0 to 1, where 0 corresponds to the interstitial space in the protein structure that is outside the united der Walls atomic radii ($\beta = 2.8$ Å^{-1}), and 1 corresponds to the fully packed medium ($\beta = 0.9$ Å^{-1}) (Page et al., 1999). A modified version of this equation was developed that is applicable also for endergonic electron-transfer reactions (Crofts and Rose, 2007).

These expressions were critical to predict electron-tunnelling rates in many biological systems (Davidson, 2008; Gray and Winkler, 1996, 2003). As a general outcome, the distance between electron donor and electron acceptor is the dominant parameter that controls the kinetic-tunnelling rate of an electron-transfer event. Indeed, single-step electron transfer that is physiologically relevant is unlikely to occur for distances greater than 25 Å, and edge-to-edge distance for neighbouring redox centres rarely goes beyond 14 Å (Page et al., 2003; Winkler and Gray, 2014). For distances that exceed this upper limit, fast electron transfer is achieved by placing several redox centres within the protein in a way that splits the distance between an electron donor and an electron acceptor into a series of short electron-transfer steps (Winkler and Gray, 2014). This is called *multi-step hopping*. However, this structural organization has functional consequences. The low dielectric medium inside proteins provides little electrostatic shielding, and charges in closely packed redox centres may produce strong repulsive electrostatic interactions. This electrostatic repulsion can however be minimized by the concomitant uptake of positive charges together with the electrons, such as protons.

1.4 COOPERATIVE EFFECTS AND COUPLED PHENOMENA

Proteins and enzymes that contain multiple redox active centres are intrinsically (anti) cooperative biological devices owing to the fact that electrostatic repulsion is an unavoidable property of charged particles such as electrons. The intensity of electrostatic interactions between charged particles decays with distance and is modulated by the dielectric properties of the intervening medium. The protein interior typically displays low dielectric permittivity and is spatially anisotropic due to the nature and packing of the amino acid residues. Modelling of protein electrostatic properties is the subject of intense research (Warshel, 2014). For the typical distances found between neighbouring redox cofactors in multi-centre proteins and supercomplexes, the dielectric properties of the protein interior lead to electrostatic interactions of substantial magnitude. In the absence of redox-linked conformational changes, these interactions can be estimated using simple empirical models (Figure 1.3) (Louro et al., 2004). For instance using the Debye-Hückel formalism

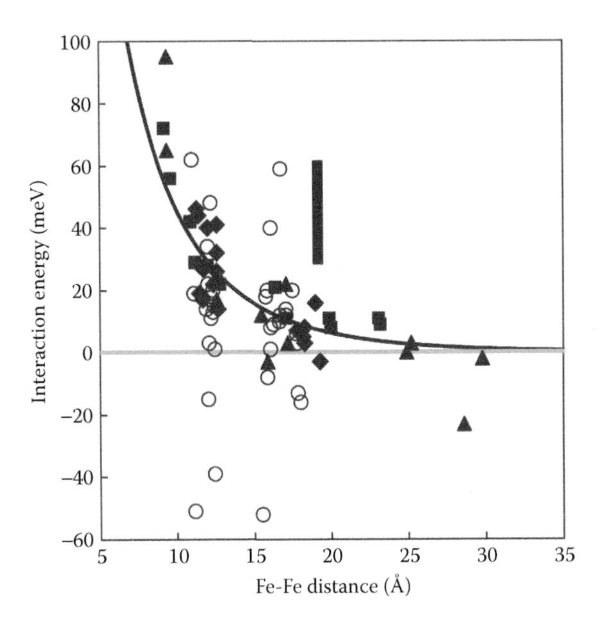

FIGURE 1.3 Distance dependence of the pairwise interactions between hemes in several multi-heme cytochromes. The solid line was obtained using Equation 1.4 considering a dielectric constant of 8.6 and a Debye length of 7.7 Å. (Reproduced from Fonseca, B.M. et al., *FEBS Lett.*, 586, 2012, 504–509. With permission.)

it is possible to estimate these electrostatic interactions using solely the knowledge on the distance between redox cofactors (Fonseca et al., 2012):

$$V_{\mathrm{i}} = k\frac{1}{\varepsilon \cdot r}\exp\left(\frac{-r}{r_{\mathrm{D}}}\right) \tag{1.4}$$

In this equation k stands for the electron charge divided by the vacuum electric permittivity multiplied by 4π (1.44×10^{-9} Vm), r_{D} and ε are empirically determined parameters formally related with the Debye length and the dielectric permittivity at the protein interior, respectively, and r is the distance (in angstrom) between the redox centres.

1.4.1 Cooperative Effects

The twentieth century saw great improvements in the understanding of cooperative phenomena in biological macromolecules (Wyman and Gill, 1990). The observation by Christian Bohr in the early 1900s of sigmoid shape in binding curves of oxygen to haemoglobin, most likely the most studied protein (Brunori, 1999; Weissbluth, 1967), was the original and main target of this interest. In the 1910s Archibald Hill proposed a mathematical formulation for oxygen binding to haemoglobin that fits the experimental data (Hill, 1910):

$$\theta = \frac{[\mathrm{L}]^{n}}{K_{\mathrm{d}} + [\mathrm{L}]^{n}} \tag{1.5}$$

In this equation [L] is the unbound ligand concentration, K_d is the apparent dissociation constant, and n is the Hill coefficient that determines the number of binding ligands. This model had the virtue of simplicity, with only one adjustable parameter, but had very little adherence to physical reality, assuming the simultaneous binding of multiple ligands to the protein given by the Hill coefficient. The good fit to the experimental data for haemoglobins is typically achieved with Hill coefficients between 2 and 3. However, the simultaneous encounter of three or four particles in the correct orientation is an extremely unlikely event, unsuitable to maintain one of the most important physiological processes in vascular organisms. Therefore, the Hill coefficient became a surrogate measurement of cooperativity between ligand-binding sites. Values between 0 and 1 indicate negative cooperativity, where binding of one ligand decreases the affinity of the others. Values between 1 and the number of ligand binding sites indicate positive cooperativity, with binding of one ligand increasing the affinity of the protein for the subsequent ligands. A value of 1 indicates independent binding sites. Gilbert Adair proposed a more realistic model of oxygen binding to haemoglobin in the 1920s, which considered the binding of oxygen to haemoglobin in four sequential steps (Adair, 1925). Each of the four sequential binding events is described by a binding constant that can be adjusted to fit the experimental data:

$$K_{4i} = \frac{\left[Hb_4(O_2)_i\right]}{\left[Hb_4\right]\left[O_2\right]^i} \quad i = 1,2,3,4 \tag{1.6}$$

However, a fit of the haemoglobin oxygen binding data requires that the later binding events are stronger than the earlier. This presented a conundrum. Statistical thermodynamics decreases the apparent affinity as more ligands are bound to the macromolecule because there are less free sites available (Cantor and Schimmel, 1980). Exploration of the reasons for this apparent contradiction had to wait for the knowledge of the structure of haemoglobin. This was achieved by Max Perutz in the 1950s, showing a symmetric protein that changes conformation upon ligand binding (Perutz, 1990).

Symmetry provides a rational for analysing cooperative phenomena using relatively simple models given that it provides a structural justification for degeneracy among different centres. This was the basis for the Monod-Wyman-Changeaux model (Monod et al., 1965) and the Koshland-Némethy-Filmer model (Koshland et al., 1966) of oxygen binding to haemoglobin.

However, the spatial arrangement and the nature of redox cofactors within redox proteins and enzymes with multiple centres are not symmetrical in the vast majority of the cases. Therefore, to achieve a detailed characterization of the redox activity of these proteins, the discrimination of the redox state of the individual cofactors is required.

As shown in Figure 1.4 with a symmetric protein the binding of ligands can be analysed considering a sequential binding of *each of ligand* according to the Adair model. When the protein is not symmetrical, experimental data that discriminate the different sites allow a more detailed analysis. The level of detail depends on the capacity of the experimental data to discriminate the different sites and states that can be assumed by the protein.

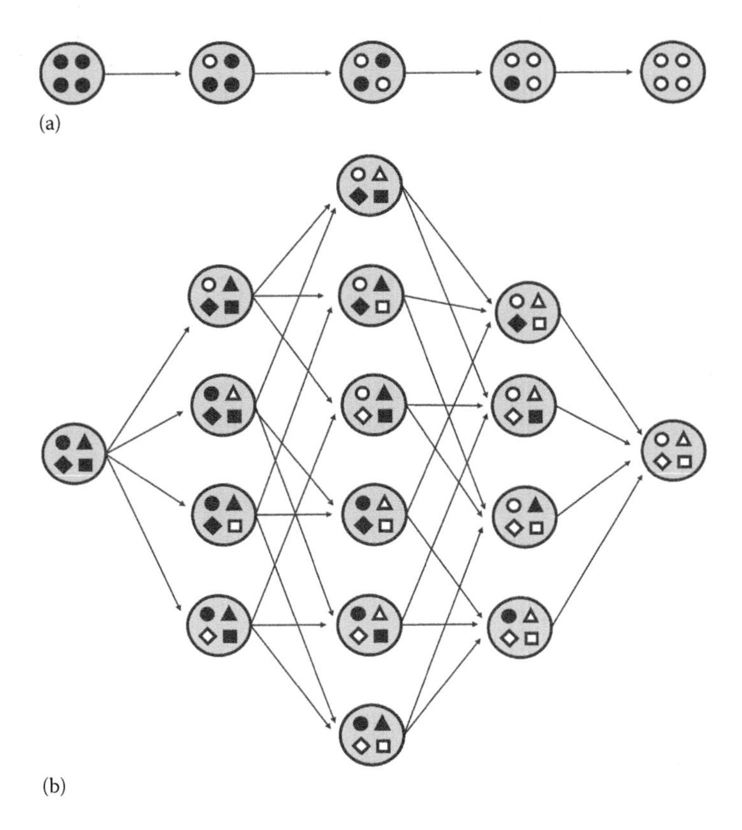

FIGURE 1.4 Diagram of ligand binding to a protein with multiple centres. If a protein is symmetrical, partially bound states with the same number of ligands are structurally degenerate and cannot be discriminated (a). When the protein is not symmetrical and the binding sites can be discriminated, the full statistical thermodynamic distribution must be considered (b). (Palmer, G., and Olson, J.S., Concepts and approaches to understanding of electron transfer processes in enzymes containing multiple redox centres. In: Coughlan, M.P, ed., *Molybdenum and Molybdenum-Containing Enzymes*, Pergamon Press, London, 1980.)

For sufficiently discriminating methods, binding of the ligands to *each individual site* may be defined (Paquete and Louro, 2014). Given that redox cofactors within proteins are so close that the interactions between the different sites cannot be ignored, the number of parameters that need to be experimentally defined increases faster than the number (n) of centres. Considering only pairwise interactions the number of parameters is given by Equation 1.7:

$$n+\frac{n(n-1)}{2} \tag{1.7}$$

Positive cooperativity can arise form factors other than conformational changes of the redox proteins. The chemical nature of the redox cofactors is an important source of cooperativities. Flavins and quinones can exchange two electrons with their partners. Depending on the chemical environment of these molecules the one-electron reduced state may be stabilized and a stepwise transfer of two electrons takes place, as in the case of flavodoxins (Smith et al., 1977). Alternatively, the intermediate one-electron reduced state is not formed and a two-electron step occurs. This situation corresponds formally to a very strong positive cooperativity and is essential in the phenomenon of electron bifurcation (Nitschke and Russell, 2012). In this process, the two electrons are donated to different acceptors, with the first going to an acceptor with a potential higher than the midpoint of the flavin or quinone and the second going to an acceptor with a potential lower than that of the flavin or quinone. This phenomenon is spontaneous as long as the overall change in free energy for the transfer of the two electrons is negative, and is essential for the operation of the bc_1 complex of mitochondrial respiratory chains (Swierczek et al., 2010). It has also recently been identified as essential for the anaerobic metabolism of *Chlostridia* and methanogens (Buckel and Thauer, 2013).

The aforementioned description focused on the interaction between cofactors that bind the same kind of ligands, electrons in this case, and therefore corresponds to homotropic cooperativity. An added layer of complexity derives from the fact that electrons are charged particles and therefore affected by other charged particles such as protons.

1.4.2 Coupled Phenomena

Coupling between protons and electrons is a key feature of proteins involved in bioenergetic processes, and is at the core of biological energy transduction (Williams, 2005). As this occurs between particles of different nature this interaction is described as heterotropic cooperativity. The strength of the interaction defines the separation between the pK_as of the acid-base centre that binds the proton in the oxidized and in the reduced states of the protein. In the pH range between the values of the $pK_{a\,ox}$ and $pK_{a\,red}$ the reduction potential of a redox centre is modulated by pH, giving rise to the well-known redox-Bohr effect (Papa et al., 1979). If a single proton affects the reduction potential of a redox centre, the redox-Bohr effect is described by the following equation:

$$E_B - E_A = \frac{2.3RT}{F}\left(pK_{ox} - pK_{red}\right) \tag{1.8}$$

where:

E_B and E_A are the reduction potential of basic and acidic forms, respectively (Louro et al., 1996)

The pH dependence of the reduction potential depends on the separation between the $pK_{a\,ox}$ and $pK_{a\,red}$ until it reaches a limiting value of RT/nF for separations of more than 3 pH units, with n indicating the number of electrons coupled to proton transfer (Figure 1.5). The consequence is that, when the pH dependence of reduction potentials is measured for only a limited pH range, it may be unclear how many electrons and protons are coupled in the process.

As a consequence of the pH dependence of the reduction potentials, for redox proteins with multiple redox centres, a proper evaluation of the redox interactions requires

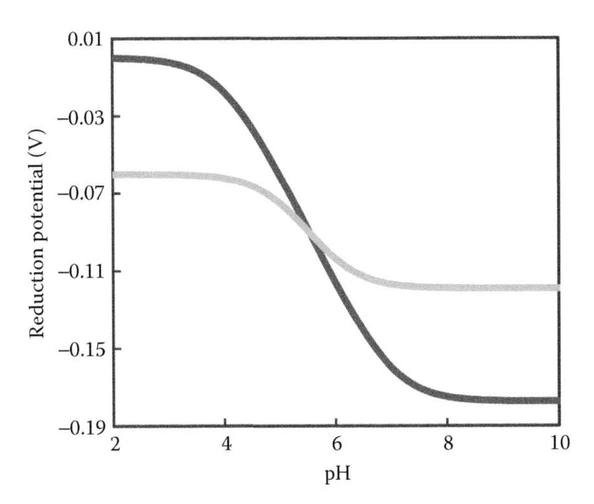

FIGURE 1.5 pH dependence of reduction potential of a single electron for two different separations of the $pK_{a\,ox}$ and $pK_{a\,red}$: black line $pK_{a\,ox} = 4$ and $pK_{a\,red} = 7$; grey line $pK_{a\,ox} = 5$ and $pK_{a\,red} = 6$.

the collection of experimental data at more than one pH value to parse the contributions of homotropic and heterotropic interactions (Paixão et al., 2010). For processes involving multiple electrons and multiple protons, the algebra becomes unpractical and using a formalism based on the free energy of the processes is mathematically more efficient. Each electron- and proton-binding reaction is characterized by a free energy of the reaction, and homotropic and heterotropic cooperativities simply add or subtract from that free energy (Paquete and Louro, 2014). The resulting free energies are then converted back to reduction potentials and pK_as using standard thermodynamic relationships between free energy and equilibrium constants.

The redox-Bohr effect defines a thermodynamic coupling between electrons and protons in the physiological range and corresponds to a square thermodynamic cycle (Figure 1.6):

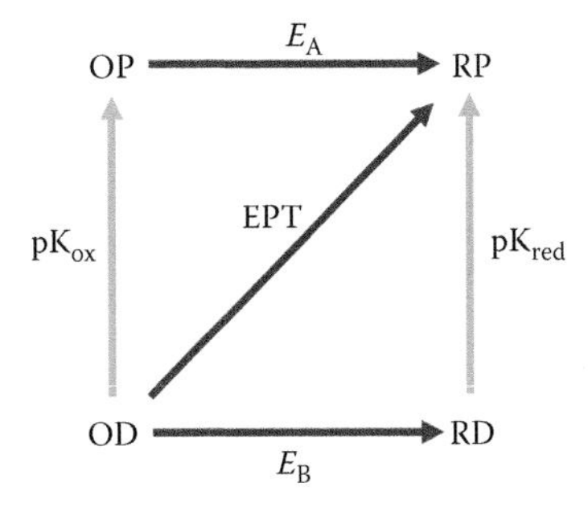

FIGURE 1.6 Square thermodynamic scheme with states labelled as OD (oxidized-deprotonated), OP (oxidized-protonated), RP (reduced-protonated) and RD (reduced-deprotonated). Vertical transitions denote proton transfer and horizontal transitions denote electron transfer. EPT denotes concerted electron-proton transfer.

1.4.3 Proton Coupled Electron Transfer—Concerted and Sequential Modes

The diagram in Figure 1.6 clearly reveals that for thermodynamically coupled systems two different mechanistic pathways can occur: sequential proton and electron transfer or concerted proton/electron transfer. Which of the mechanisms actually takes place in a given protein or enzyme has consequences for the interpretation of bioenergetic phenomena. Although the energetics of concerted and sequential transfer between two states is necessarily the same, the kinetics can be fundamentally different due to different reaction paths. However, the experimental distinction between the two mechanisms can be clouded by the fact that electrons and protons are delocalized quantum-mechanical entities and the timescales of the transfer events may be very short. Therefore, it is considered that concerted electron and proton transfer takes place when an intermediate cannot be observed, and the transfer of coupled electrons and protons occurs in one step (Hammes-Schiffer and Stuchebrukhov, 2010). This transfer can itself occur by two different mechanisms: (1) hydrogen atom transfer (HAT), where both the electron and the proton are transferred between the same donor and acceptor and (2) electron-proton transfer (EPT), where the electron and the proton are transferred between donors and acceptors in different locations (Layfield and Hammes-Schiffer, 2014). These two alternatives are fundamentally different. In the case of HAT a neutral species is transferred, whereas in the case of EPT two charged species are transferred with a global electrostatic effect that depends on whether the electron and proton are moving in the same direction or in different directions.

In the case of sequential transfer, two mechanistic pathways need to be considered: electron transfer first (EP) or proton transfer first (PE). The first of the two steps in a sequential mechanism is thermodynamically uphill followed by the second downhill process (Figure 1.7). When progressing around the thermodynamic cycle of Figure 1.6 the re-establishment of the starting condition requires that the overall change in free energy is zero.

A return to the starting state by a different pathway requires a conformational change or interaction with different partners, but allows the establishment of a *turning wheel*

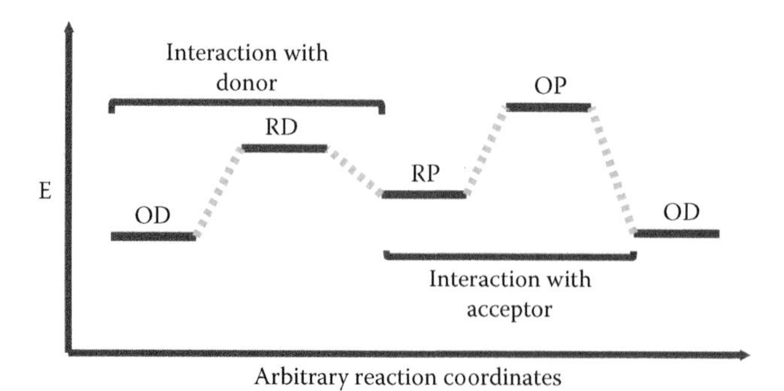

FIGURE 1.7 Free energy diagram for the transition between states in the thermodynamic cycle of Figure 1.6.

(Wyman, 1975), a hallmark of transducing enzymatic devices. The overall rate for each of the EP or PE is given by Equation 1.9:

$$\frac{1}{k} = \frac{1}{k_P} + \frac{1}{k_e} \tag{1.9}$$

One of the sequential pathways may dominate or even both may coexist with different levels of contribution to the overall observed rate. This depends on the lifetime of the intermediate state and the relative rate of the forward and backward reactions (Hammes-Schiffer and Stuchebrukhov, 2010).

From Equation 1.9 it is clear that either electron transfer or proton transfer can be rate limiting to the overall process. Given that biological electron transfer typically takes place in the nonadiabatic regime (Gray and Winkler, 2003), it is expected that the rate constant for proton transfer is larger than that of electron transfer. Therefore, electron transfer will be rate limiting. For the opposite case or in the cases of concerted transfer, the rate of proton transfer will affect the overall rate. In these conditions, replacement of protons for heavier isotopes of the hydrogen nucleus will influence the observed rates.

1.4.4 Kinetic Isotope Effect

The kinetic isotope effect is given by the rate of the reaction involving protons versus the rate of the reaction involving a different hydrogen isotope, typically deuterium (k_H/k_D). Very large H/D kinetic isotope effects (>7) are typically associated with hydrogen tunnelling (Klinman and Kohen, 2013). This is a quantum mechanical phenomenon that depends on the magnitude of overlap of the wave functions in the reactants and in the products state. The extension of the wave function is strongly dependent on the mass of the particle and therefore this overlap is diminished for the heavier hydrogen isotopes.

Kinetic isotope effect is also observed in other cases of coupled electron proton transfer and in sequential transfer where the proton transfer is rate limiting. It will depend on the properties of the biological system such as the donor–acceptor distance, or the experimental conditions such as temperature and solvation (Farver et al., 2001).

1.5 METHODS FOR THE STUDY OF BIOLOGICAL ELECTRON TRANSFER

All of the earlier-described phenomena required the development of experimental methods capable of probing the details of redox reactions. Electrochemical methods feature prominently in the study of biological electron transfer. In recent years, *protein film voltammetry* of redox proteins and enzymes adsorbed at the surface of electrodes has provided important insights on the mechanistic details of electron transfer. The potential is swept up and down the potential scale and the current injected or withdrawn from the electrode is measured. When the potential sweep is conducted at sufficiently slow scan rates the experiment is considered to occur in equilibrium and the anodic and cathodic peaks occur at nearly the same value of applied potential. The peak positions correspond to the reduction or oxidation potential of the centre being measured with a height proportional to the scan rate and to the electrode coverage, and a width that depends on the experimental temperature and the number of electrons involved in the process. As the scan rate is increased, equilibrium is not reached and the anodic and cathodic peaks become broader and change position, giving rise to the *trumpet* plots. These changes provide information on the rate of interfacial electron transfer between the protein and the electrode (Léger and Bertrand, 2008). Moreover, when another chemical process is coupled with electron transfer, such as protonation, distortions in the *trumpet* plot provide information on the rate of the coupled process. Catalytic processes can also be studied by protein film voltammetry (Savéant, 2008). In the presence of substrate, a catalytic current is observed in addition to the noncatalytic current. The shape of these data can be quite complex, requiring the development of mechanistic models to interpret the results but yield insights on the catalytic mechanism of the enzyme (Jones et al., 2003).

In parallel with electrochemical methods, spectroscopic methods hold a prominent place in the characterization of biological electron-transfer processes. The coordination compounds of transition metals that constitute the redox cofactors in biological macromolecules as well as the redox active organic cofactors such as flavins and quinones are spectroscopically active. This has stimulated the use of the entire electromagnetic spectrum to explore the structural and dynamic aspects of electron transfer.

Nuclear magnetic resonance (*NMR*) *spectroscopy* is based on the interactions between electromagnetic radiation in the radio wave frequency and nuclei that are sensitive to an external magnetic field (Keeler, 2002). NMR can be used to provide structural and geometric information on electron-transfer proteins, as well as information on the rates of intra- and intermolecular electron-transfer reactions (Coutinho and Xavier, 1994). In the course of electron-transfer reactions, states with unpaired electrons are often generated. Unpaired electrons are very strong magnetic spins that cause dramatic changes in NMR spectra with respect to chemical shift range and nuclear relaxation rates (Banci et al., 1991; Louro, 2013). These changes in peak position and line width can be used to follow redox changes (Simonneaux and Bondon, 2005). The sensitivity of NMR signals to chemical environment has also been employed to study protein–protein interactions between redox partners (Prudêncio and Ubbink, 2004). The regions at the surface of redox partners responsible for recognition and docking can be identified (Zuiderweg, 2002), and the ensemble population of the encounter complex, which precedes the actual docked complex that is competent for electron transfer, can be characterized (Bashir et al., 2011).

Electronic paramagnetic resonance (*EPR*) *spectroscopy* also relies on an applied magnetic field, but in this case it is used to probe unpaired electrons in redox proteins and enzymes using microwave radiation (Hagen, 2013). Only the cofactors containing transition metals with unpaired electrons, or amino acid radicals can be observed by EPR. The typical rates of electron relaxation in biological redox centres make it necessary to perform the experiments at cryogenic temperatures, often below the boiling point of nitrogen. EPR provides information on molecular geometry, number of paramagnetic centres, structural organization of paramagnetic centres within a protein and information on the electronic structure of multi-nuclear clusters (Saito et al., 2013). Reaction kinetic can also be monitored by EPR by flash freezing the sample at predetermined times to trap reaction intermediates (Pievo et al., 2013).

Many complexes of transition metals display intense absorption bands in the UV-visible spectral regions, making them easy to study by light absorption methods (ultraviolet or visible light), such as *UV-visible spectroscopy*. The absorption bands can be assigned to specific centres in the protein providing information on the composition and number of centres (Petrenko and Neese, 2007). The absorbance (A) of a solution is directly proportional to the concentration (C) and the length of the optical path (b), providing a simple frame for quantifying the number of centres in dilute solutions:

$$A = \varepsilon b C \tag{1.10}$$

In this equation ε is the absorption coefficient ($M^{-1}cm^{-1}$). This simple relationship, known as the Lambert-Beer law, can be used in diverse experimental set-ups to extract mechanistic information on redox reactions, by following their equilibrium concentrations in redox titrations (Leslie Dutton, 1978). The UV-visible bands can also be used to monitor reaction kinetics. Bimolecular reactions can be followed using stopped-flow, a concept proposed by Britton Chance to follow the reaction mechanism of catalases (Chance, 1949). Modern stopped-flow devices ensure complete mixing of the reagents in the millisecond timescale (<3–5 ms), allowing the monitoring of fast multi-electron reactions (Meints et al., 2011).

Unimolecular reactions are typically much faster and can be photo-initiated. This results from the generation of an excited state that is a strong oxidizer (Winkler and Gray, 2014) or by photo-dissociation of a ligand that is blocking the progress of subsequent reactions (Brzezinski and Wilson, 1997). A flash of a laser of the correct wavelength provides the necessary initiation event. *Flash photolysis* provides the means to follow reactions that take place in the timescale of picoseconds or slower (Brindell et al., 2008). Photo-initiated experiments were instrumental to determine the value of β that defines the distance dependence of electron-transfer reactions (Winkler and Gray, 2014). They were also essential for exploring the catalytic cycle of the terminal reductase of the aerobic respiratory chains (Winkler et al., 1995).

At the high energy end of the electromagnetic spectrum, *Mössbauer spectroscopy* probes the nuclear transitions using gamma-ray radiation (Gutlich et al., 2011). Iron is the best target for Mössbauer spectroscopy and, given its prime role in biology, has been widely studied. Mössbauer spectroscopy provides information on the valence state, symmetry of the environment around the iron, number of ligands and covalency of bonds. Significant absorption of radiation with the energy of gamma-rays requires the freezing of the sample due to the recoil effect caused by such high-energy photons. Therefore, Mössbauer spectroscopy is performed at cryogenic temperatures below the boiling point of nitrogen or helium.

Often, spectroscopic and electrochemical methods can be combined to provide more detailed information on reaction mechanisms of redox proteins. Vibrational spectroscopies use specific marker bands assigned to the redox centres to monitor changes in their redox and spin states as well as changes in geometry. Both *Raman* and *infra-red spectroscopies* can be performed in samples placed on the surface of nanostructured metallic electrodes giving rise to signal enhancements up to 10^6, improving the detection limit of redox centres of interest (Ataka and Heberle, 2006; Murgida and Hildebrandt, 2008).

1.6 EXAMPLES OF BIOLOGICALLY IMPORTANT MULTI-ELECTRON REACTIONS

This vast repertoire of methods has been put to use to determine the detailed molecular mechanisms of key multi-centre redox enzymes.

1.6.1 Photosynthetic Water Splitting

The water-plastoquinone photo-oxidoreductase, also known as photosystem II (PSII), is the first protein complex in the light-dependent reactions of oxygenic photosynthesis and is responsible for the production of nearly all of the earth's atmospheric oxygen (Rochaix, 2014; Vinyard et al., 2013; Yano and Yachandra, 2014). PSII is a membrane protein complex located in the thylakoid membranes of cyanobacteria, algae and plants. It is a homodimer, having a total molecular weight of approximately 700 kDa with each monomer containing several different subunits and cofactors. The core of PSII, which consists of a pseudo-symmetric heterodimer of two homologous proteins D1 and D2, contains a singular metallo-oxo cluster. The cluster comprises four manganese ions and one divalent calcium ion, known as the oxygen-evolving complex (OEC) or as the water-splitting complex (Rochaix, 2014; Vinyard et al., 2013; Yano and Yachandra, 2014).

The OEC undergoes a sequential abstraction of four electrons and four protons from two water molecules concomitantly with the absorption of four quanta of energy by PSII (Vinyard et al., 2013). One molecule of oxygen is produced (Cox and Messinger, 2013; Debus, 2014; Yano and Yachandra, 2014). During this process, the OEC sequentially delivers the electrons from the water molecules via a redox-active tyrosine to a distinctive pair of chlorophyll molecules called the P680. The P680 absorbs light at 680 nm and transfers an electron to a nearby pheophytin, which subsequently transfers the electron through a series of other electron carriers to an exchangeable plastoquinone pool and from there to the photosynthetic protein complexes embedded in the thylakoid membrane. The positive charge that is formed on P680 results in a powerful oxidant, the $P680^+$. Each time a photon of light extracts an electron from P680, the oxidized $P680^+$ extracts an electron from the water molecules bound at the OEC catalytic core, reducing it back to P680 and resetting it for a subsequent photosynthetic event (Cox and Messinger, 2013; Debus, 2014; Yano and Yachandra, 2014). The photosynthetic water oxidation reaction is shown in Scheme 1.2:

$$2\ H_2O \rightarrow O_2 + 4\ H^+ + 4e^- \qquad \text{Scheme 1.2}$$

Nearly 50 years have passed since the first experimental evidence showed that oxygen is released through a cyclic reaction of OEC within the PSII (Joliot et al., 1969; Kok et al., 1970). These early experiments demonstrated that if dark-adapted photosynthetic material is exposed to a series of single turnover flashes, oxygen yield is maximum after the third flash and then cyclically after every fourth flash. This became the basis for the S-state cycle model (Figure 1.8) (Kok et al., 1970). This model proposes that the OEC cycles through a series of five intermediate S-states (S_i, $i = 0 - 4$), representing the number of oxidizing equivalents stored on the OEC driven by the energy of the four successive photons

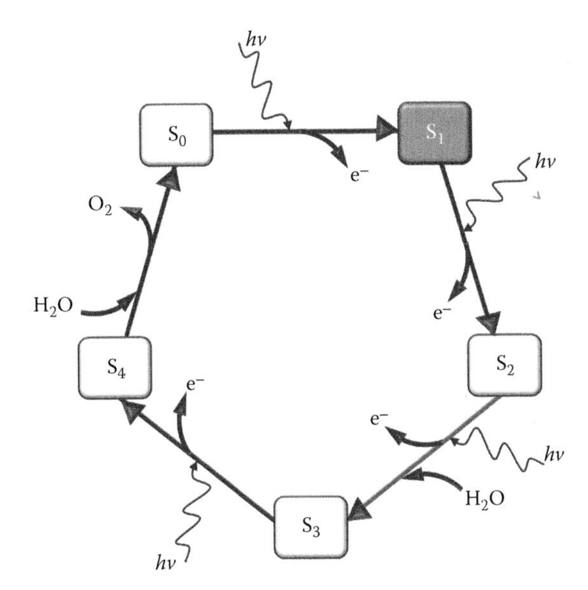

FIGURE 1.8 Scheme of S-state cycle of the Mn_4CaO_5 cluster during water oxidation. The grey box corresponds to the dark-adapted state of PSII (S1 state), while the white boxes correspond to the illuminated states of PSII.

absorbed by PSII. When PSII is dark-adapted, the S_0 state is oxidized by the redox-active tyrosine Z to the S_1 state. Illumination of dark-adapted PSII (S_1 state) with visible light leads to increments in the oxidation state of the OEC by removal of one electron, with the OEC acting like a redox capacitor for the water oxidation reaction until the four oxidizing equivalents are accumulated (S_4 state). A precisely choreographed sequence of proton- and electron-transfer steps occur, with the release of protons preventing the redox potential of the OEC from rising to levels that would stop its subsequent oxidation by the redox-active tyrosine. Recently, a proton-release pattern of 1 (S_0 to S_1)/0 (S_1 to S_2)/1 (S_2 to S_3)/2 (S_3 to S_0) has been proposed (Suzuki et al., 2009). Once four oxidizing equivalents are accumulated in the OEC (S_4 state), a spontaneous reaction occurs that results in the oxidation of water, the release of dioxygen gas and the reformation of the S_0 state and consequently the S_1 state via the redox-active tyrosine. During this cycle, it remains however somewhat unclear how the electron- and proton-transfer reactions are coupled with each other in individual S-state transitions and in which order the electron and protons are released from the OEC (Cox and Messinger, 2013; Vinyard et al., 2013; Yano and Yachandra, 2014).

The structure of the OEC has been the subject of intense study over the past few years, with the first X-ray structure of PSII in its dark stable state (S_1 state) determined to a resolution of 3.8 Å in 2001 (Zouni et al., 2001). This revealed for the first time the architecture of PSII and the overall shape and location of the OEC. Several other structures with slightly higher resolution were obtained subsequently (Ferreira et al., 2004; Guskov et al., 2009) though it was only in 2011 that structural elucidation at near atomic resolution for the PSII and its OEC was achieved with a X-ray structure at 1.9 Å (Umena et al., 2011). The crystallographic structures of PSII complexes showed that the OEC consists of a Mn_4CaO_5 cluster.

However, the structure of this metallo-oxo cluster is still contentious since there is evidence that the manganese atoms are reduced by the high-intensity X-rays used, altering the OEC structure. Nonetheless, crystallography in combination with a variety of other less damaging spectroscopic methods such as extended X-ray absorption fine structure, Fourier transform infrared and EPR (Cox et al., 2013; Debus, 2014; Mamedov et al., 2008; Pushkar et al., 2007) have started to give a fair idea of the structure of the Mn_4CaO_5 cluster. Though this is still an ongoing debate, it is widely accepted that the OEC structure is composed by an irregular heterocubane-like Mn_3CaO_x unit containing a more distant protruding Mn referred to as the dangler Mn. This atom seems to serve as a substrate-binding site for the water molecules in the higher S-states (Kern et al., 2013; Yano and Yachandra, 2014; Yano et al., 2006).

Recently, structural advances have been achieved for the illuminated S-states (S_2 and S_3) using femtosecond X-ray spectroscopy (Kern et al., 2013; Kupitz et al., 2014). These studies show no apparent structural changes taking place between the S_1 and S_2 states (Kern et al., 2013) while in the case of the S_2 to S_3 state transition, PSII undergoes significant changes at the Mn_4CaO_5 core of the OEC. These include an elongation of the metal cluster (increase in the distance between the heterocubane and the dangler Mn), accompanied by changes in the protein environment, which could allow the binding of a water molecule between the more distant dangler Mn and the Mn_3CaO_x heterocubane during the S_2 to S_3 transition (Kupitz et al., 2014). Thought these works show great potential and open new windows in the structural study of other illuminated S-states, the interpretation of these results should be made with caution due to the low resolution ($\approx$5 Å) obtained.

Also, the issue of the Mn oxidation states along the S-state cycle is an ongoing debate among the scientific community, with different combinations of Mn oxidation states for each intermediate S-state ranging from +2 to +5 (Vinyard et al., 2013; Yano and Yachandra, 2014). Nonetheless, a recent resonant inelastic X-ray scattering study has emphasized that the assignment of formal Mn oxidation states alone is not sufficient for understanding the complex nature of the electronic structure in multi-nuclear clusters such as the OEC. This is because the electrons are strongly delocalized, making the assignment of oxidation states to each individual Mn ion highly challenging (Glatzel et al., 2013). Furthermore, it is likely that ligands surrounding the cluster are intimately involved in the delocalization of the electron density as the cluster is being oxidized and in modulating its redox chemistry.

In order to unravel the water-splitting mechanism at atomic detail, the resolution must be further improved, and the nature of conformational changes and electron delocalization in the cluster must be elucidated. This understanding of the natural water oxidation reaction at an atomic level could lead to the development of artificial photosynthetic devices and ultimately the development of renewable carbon-neutral energy sources.

1.6.2 Respiratory Oxygen Reduction

One of the largest superfamilies of aerobic terminal oxidases is the heme copper oxidases that catalyse the final step of the aerobic respiratory chains. Cytochrome *c* oxidases (C*c*O) belong to the class A of this family (Pereira et al., 2001), being among the best-studied heme-copper oxidases. They are located in the mitochondrial inner membrane in eukaryotes or in the cytoplasmic membrane of bacteria. C*c*O is responsible for the four-electron reduction of molecular oxygen to water. This process is coupled to proton pumping across the membrane, contributing to the generation of the proton gradient that powers the production of ATP by ATP-synthase (Scheme 1.3) (Babcock and Wikström, 1992; Papa et al., 1979):

$$O_2 + 4e^- + 8H^+ (\text{N-side}) \rightarrow 2H_2O + 4H^+ (\text{P-side}) \qquad \text{Scheme 1.3}$$

In each catalytic cycle, four *chemical* protons are taken up by C*c*O from the N-side of the membrane (mitochondrial matrix or cytoplasmic side in bacteria), to combine with oxygen to form water, and four additional *pumped* protons are transferred from the N-side to the P-side of the membrane (inter-membrane space in mitochondria or periplasmic space in bacteria) (Figure 1.9) (Wikström, 1977).

In mammals, C*c*O exists as a homodimer, where a single monomer is made of 13 different subunits, while in bacteria the number of subunits is usually less than four (Iwata et al., 1995; Richter and Ludwig, 2009; Tsukihara et al., 1996). A single monomeric functional unit of the enzyme contains four redox-active metallic centres: a Cu_A, a heme *a*, and the binuclear centre (BNC) that consists of heme a_3 and another copper site (Cu_B) (Figure 1.9).

The C*c*O reaction begins with the reduction of the Cu_A site by cytochrome *c* at the P-side. The electron is transferred from Cu_A to heme *a*, and then to the BNC. The fully oxidized BNC,

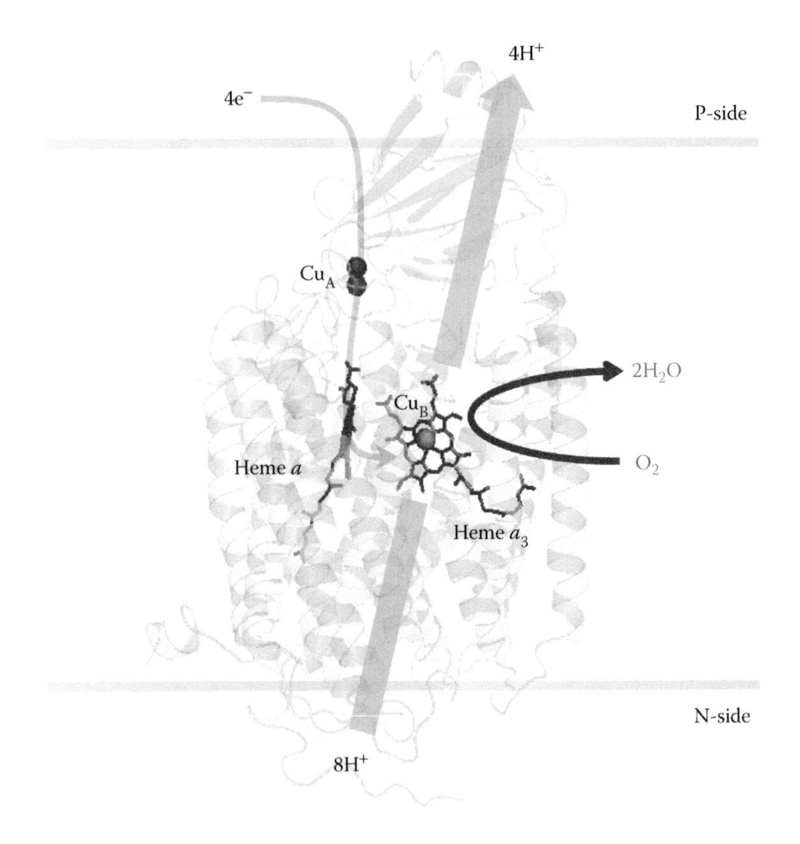

FIGURE 1.9 Crystal structure of the catalytic core formed by subunits I and II of the bovine C*c*O (PDB 2OCC). (Yoshikawa, S. et al., *Science*, 280, 1998, 1723–1729.) The arrows indicate the electron and proton pathways.

O state is reduced by two successive one-electron transfer steps, forming the one-electron reduced *E* state and the two-electron reduced *R* state. These $O \rightarrow E \rightarrow R$ steps are known as the reductive steps of the catalytic cycle. In the *R* state, molecular oxygen binds to heme a_3 to form the *A* state, and if further electrons do not become available from heme *a*, then *A* rapidly converts to the relatively stable peroxy intermediate P_M state. In this state an extra electron is necessary for the formation of the ferryl state, which is given by a tyrosine coordinated to one of the Cu_B histidine ligands (Kaila et al., 2010). In contrast, if heme *a* is already reduced when *A* is formed, then the necessary electron comes from heme *a* and the intermediate P_R state is produced. The classical ferryl intermediate *F* state can be formed either by the reduction of P_M state with an electron from heme *a* together with the uptake of a proton, or by uptake a proton from the N-side by the P_R intermediate state. The transfer of another electron and the uptake of more protons lead to the regeneration of state *O* (Figure 1.10). The $P \rightarrow F \rightarrow O$ steps are known as the oxidative part of the reaction cycle of C*c*O.

The molecular mechanism of proton pumping of C*c*O that involves a complex choreography of electron and proton transfer steps remains a subject of intense debate. It was proposed that the electron transfer to the BNC, which promotes the binding of oxygen,

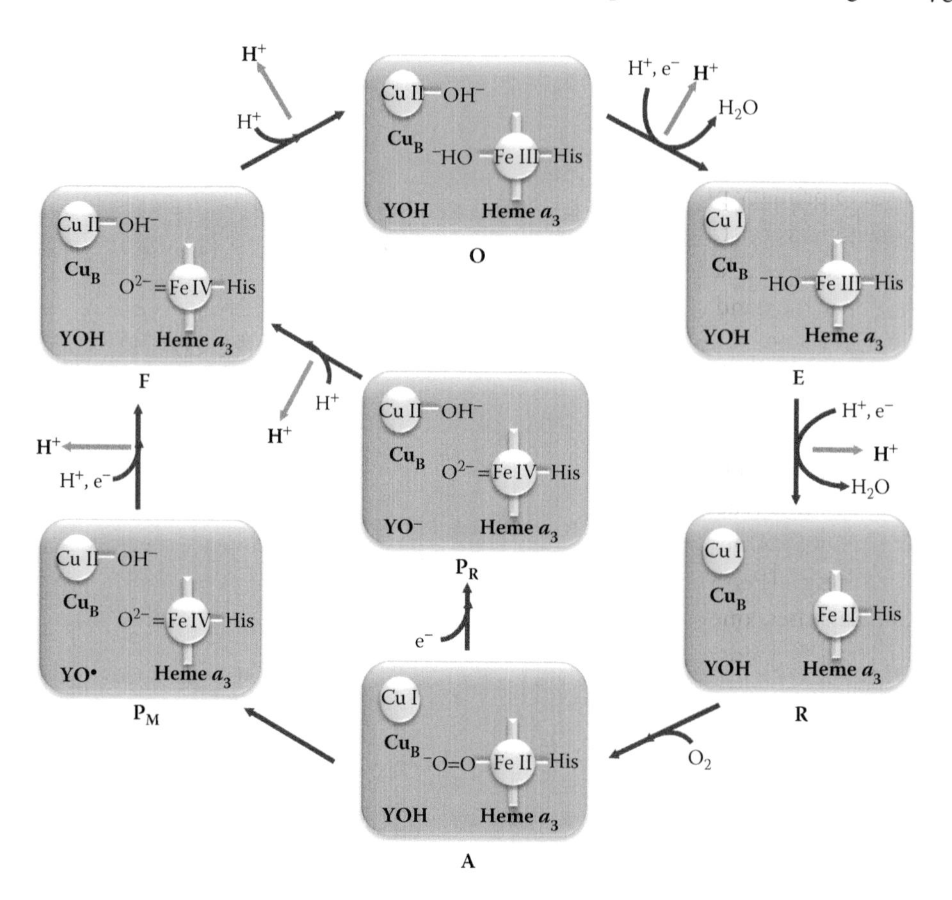

FIGURE 1.10 Diagram of the catalytic cycle of C*c*O. Within the shaded box is represented the BNC with Cu_B and heme a_3 redox centres, as well as the tyrosine that acts as the proton donor to the active site in the P_R state.

must be coordinated with the uptake and delivery of chemical protons in order to ensure the directional turnover of the proton activation cycle (Xavier, 2004). Nevertheless, though many different mechanisms have been proposed, the nature of the gates that separate *chemical* protons (consumed in water formation) from the *pumped* protons has not yet been elucidated. Based on the X-ray structures of C*c*Os, three possible proton transfer pathways (D-, K- and H-pathways) were proposed (Iwata et al., 1995; Tsukihara et al., 2003). It is believed that the K channel provides one or both *chemical* protons required in the reductive steps of the catalytic cycle ($O \rightarrow R \rightarrow E$), while the D-channel delivers the other protons to the BNC and conducts all four *pumped* protons from the N-side to the P-side (Rich and Maréchal, 2013). Both K- and D-channel are conserved between mitochondrial and bacterial C*c*Os. Nevertheless, another proton pathway was identified in the bovine mitochondrial C*c*O. It was proposed that this pathway, denominated as H-pathway, only translocates *pumped* protons from the N- to the P-side of the membrane (Tsukihara et al., 2003).

1.6.3 Biological Nitrogen Cycle

Biologically accessible nitrogen is essential for the existence of life (Capone et al., 2006), being a key element for the construction of many key biomolecules such as proteins and nucleic acids. Although it is the most abundant component of the atmosphere, nitrogen is largely inaccessible in this form to the majority of organisms. Only when nitrogen is converted from N_2 into ammonia (NH_3), a process known as nitrogen fixation, does it become accessible to primary producers, such as plants (Hoffman et al., 2014; Kim and Rees, 1994). Dinitrogen is a very stable compound, with two nitrogen atoms joined by a triple covalent bond that requires a large amount of energy to break down. The whole process requires a total of 8 electrons and at least 16 ATP molecules (Scheme 1.4):

$$N_2 + 8H^+ + 8e^- + 16ATP \rightarrow 2NH_3 + H_2 + 16ADP + 16\ P_i \qquad \text{Scheme 1.4}$$

Only a limited group of Prokaryotes is capable of carrying out this highly demanding process and all of them contain a two-component metalloenzyme complex known as nitrogenase that catalyses this reaction (Hoffman et al., 2014; Hu and Ribbe, 2013; Peters et al., 1995; Rees et al., 2005). One of the components, known as the Fe-protein, is a homodimeric protein containing a [Fe_4-S_4] cluster at the interface between the two identical subunits, and two ATP-binding sites, one on each subunit. The other component is a $\alpha_2\beta_2$ heterotetrameric protein with a functional unit established via each $\alpha\beta$ dimer, with each dimer containing two metal cofactors. One of these cofactors is a distinctive [Fe_8-S_7] cluster called the P-cluster. The other cofactor, denominated FeMo-cofactor (Fe_7-S_9-Mo-homocitrate), contains molybdenum (Mo) as the heterometal and the organic acid R-homocitrate, which is coordinated by its C-2 carboxyl and hydroxyl groups to the Mo atom (Figure 1.11) (Hoffman et al., 2013; Hu and Ribbe, 2013; Rubio and Ludden, 2008). Although there is evidence for participation of the inorganic portion of FeMo-cofactor in substrate binding and catalysis (Barney et al., 2004), the exact role of homocitrate in catalysis is still unknown though its replacement by other organic acid drastically changes substrate specificity and results in impaired N_2 reduction (Rubio and Ludden, 2008). In some organisms, alternative heterometal clusters

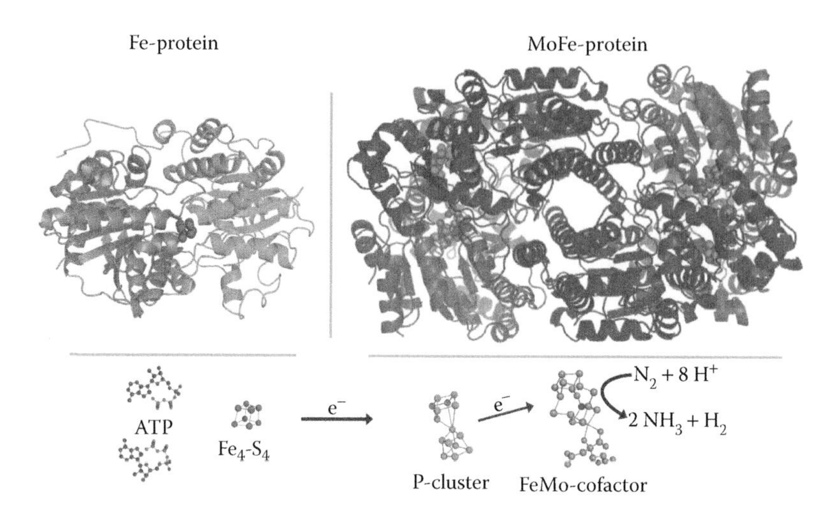

FIGURE 1.11 **(See colour insert.)** Mo-Nitrogenase with cofactors. The top shows ribbon diagrams of the homodimeric Fe-protein component (right) and the $\alpha_2\beta_2$ heterodimeric MoFe-protein component (left). The two subunits of the Fe-protein are shown in different shades of green, while the [Fe_4-S_4] cluster is represented in red. The α- and β-subunits of the MoFe-protein are shown in pink and blue, respectively, while the P-cluster and the FeMo-cofactor are represented in green and orange, respectively. Representations were made with PyMol, using the nitrogenase structure from *Azotobacter vinelandii* (PDB code 1M1Y). Represented below are the ATP molecules and the metal clusters involved in the catalysis process.

containing vanadium (V) or iron (Fe) instead of Mo also exist and are produced only under certain conditions such as particularly low Mo levels (Robson et al., 1986).

In this protein, catalysis is initiated when the reduced MgATP-bound form of the Fe-protein transiently associates with the MoFe-protein, prompting the intermolecular electron transfer and hydrolysis of the two ATP molecules. This binding process induces large conformational changes within these proteins (Danyal et al., 2011; Hoffman et al., 2013; Hu and Ribbe, 2013). The heterocomplex undergoes several cycles of association and dissociation, transferring with each cycle one electron from the Fe-protein to the MoFe-protein, until sufficient electrons have been accumulated to enable substrate binding and reduction. Kinetic studies show that the FeMo-cofactor must accumulate four electrons before the binding of N_2 to the active site (Lukoyanov et al., 2007; Yang et al., 2013). The P-cluster, which is located near the docking interface, acts as an intermediary for electron transfer from the Fe-protein's [Fe_4-S_4] cluster to the MoFe-active site (Figure 1.11).

An important feature of nitrogenase is that this enzyme is designed to bind N_2 and conduct multiple rounds of reduction and protonation without the release of partially reduced intermediates, converting N_2 quantitatively to two NH_3 molecules. A proposed macroscopic model, designated as the *deficit-spending* model, reflects the creation of an electron deficit at the P-cluster and its subsequent backfilling, describing the likely electron-transfer events that occur within this enzyme (Danyal et al., 2011; Hoffman et al., 2014; Mayweather et al., 2012). This model assumes that all electrons passed from the Fe-protein to the MoFe-protein must accumulate in the FeMo-cofactor or one of its bound-activated intermediate states (Figure 1.12).

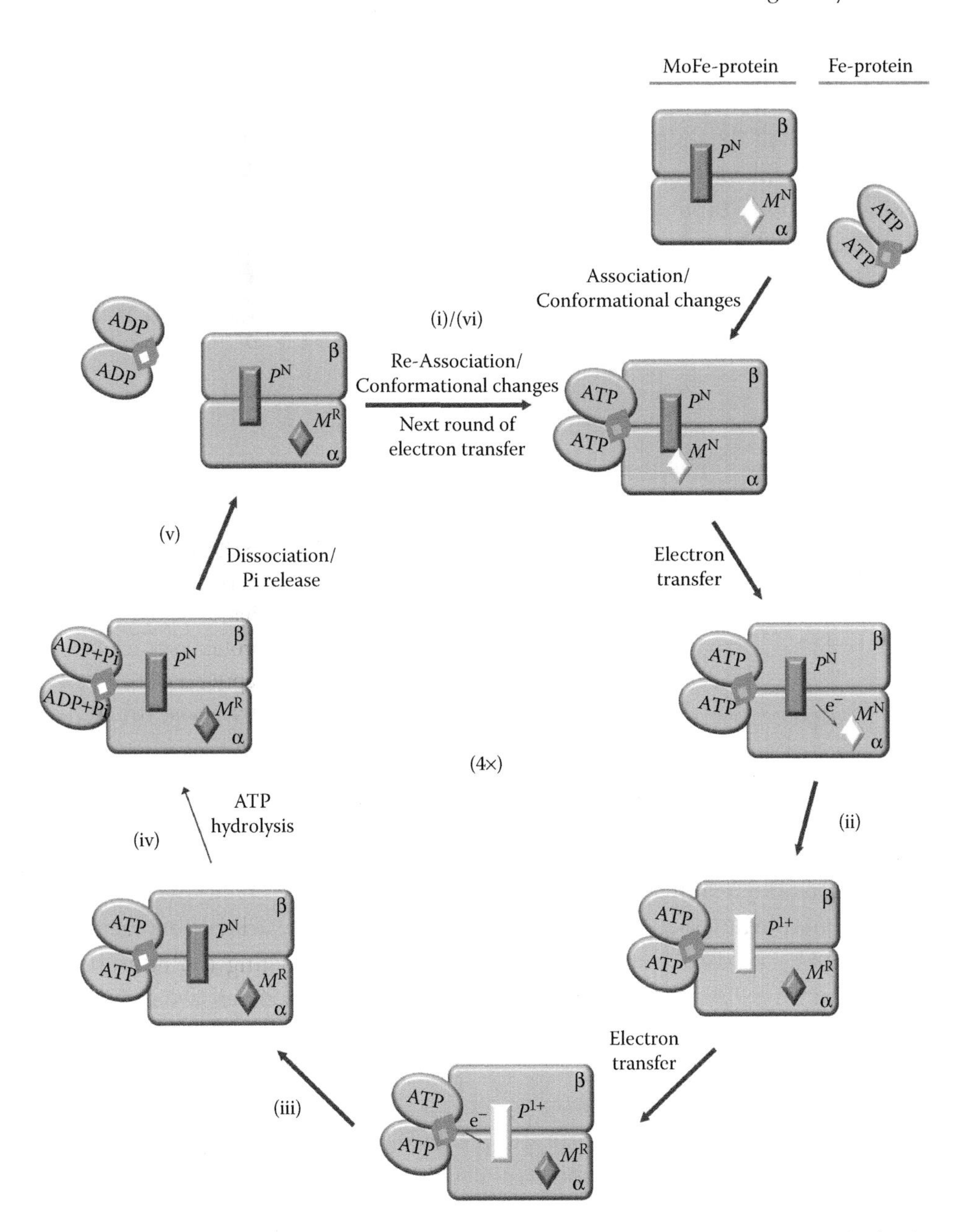

FIGURE 1.12 Nitrogenase catalytic cycle. This cycle consists of the following steps: (i) the formation of a complex between the ATP-bound Fe-protein and the MoFe-protein; (ii) the intramolecular electron transfer from the resting P-cluster (P^N) to the resting FeMo-cofactor (M^N), resulting in a reduced FeMo-cofactor (M^R) and an oxidized P-cluster (P^{1+}); (iii) the Fe-protein transfers an electron to the P^{1+} cluster, resulting in its reduction back to the P^N state and the consequent oxidation of the Fe-protein; (iv) ATP hydrolysis in the Fe-protein; (v) the dissociation of the complex following the release of phosphate (P_i) and (vi) the re-reduction of the Fe protein and the replacement of ADP with ATP, restarting the cycle for a subsequent round of electron-transfer events to accumulate electrons on the FeMo-cofactor. The reduced FeMo-cofactor (M^R) must accumulate 4 electrons before the binding of N_2 to the active site occurs. A change in the colour of the cofactors, from grey to white, corresponds to the loss of an electron via electron transfer to a neighbouring cofactor.

Currently, the knowledge on how nitrogenase accumulates electrons in the FeMo-cofactor comes from the characterization of trapped intermediate states during turnover with protons, N_2 (N≡N), diazene (HN=NH) and hydrazine (H_2N—NH_2). Recently, the characterization of the proton turnover state has provided insights into the question of how the FeMo-cofactor accumulates electrons and also into the mechanism behind N_2 reduction (Hoffman et al., 2013; Lukoyanov et al., 2007; Seefeldt et al., 2012). A *reductive elimination* model has also been proposed, suggesting that nitrogenase catalyses N_2 reduction by the sequential addition of electrons and protons to a FeMo-cofactor-bound N_2 (Hoffman et al., 2014; Seefeldt et al., 2013; Yang et al., 2013).

$$\mathrm{N \equiv N \rightarrow HN = NH\ (diazene) \rightarrow H_2N - NH_2\ (hydrazine) \rightarrow 2\ NH_3\ (ammonia)} \quad \text{Scheme 1.5}$$

The binding of N_2 to the four-electron reduced state of the FeMo-cofactor involves the loss of two metal-bound hydrides as H_2, with electron and proton addition to the bound N_2 resulting in a metal-bound diazene. The reduction of the bound diazene by addition of protons and electrons consequently yields a hydrazine-bound state and, ultimately results in the release of two NH_3 (Scheme 1.5). However, how these electrons and protons are added to the FeMo-cofactor-bound N_2 remains to be unravelled.

Overall, despite significant advances in recent years, several aspects of the biological nitrogen fixation cycle still remain obscure, hampering the ambition of replacing the venerable Haber-Bosch process with a biologically inspired process of lower environmental footprint.

1.7 CONCLUSION

As seen from the three examples in the previous section, despite the numerous theoretical and experimental advances in our capability to characterize and describe biological multi-electron-transfer processes, many aspects of these processes remain to be understood.

In the context of redox proteins in supercomplexes and signalosomes, two particular challenges need to be met: determine how the intermolecular redox reactions are coordinated with intramolecular redox reactions and how redox reactions are coordinated with other chemical events such as proton transfer or substrate binding and release. These processes occur at different timescales and are constrained by different environmental cues. And yet, they must take place at the right rate and at the right time to avoid the dissipation of the signal or of energy, and avoid the presence of side reactions that generate dangerous reactive species.

REFERENCES

Achim, C., Bominaar, E.L., Meyer, J. et al. Observation and interpretation of temperature-dependent valence delocalization in the [2Fe–2S] + Cluster of a Ferredoxin from Clostridium pasteurianum. *J. Am. Chem. Soc.* 121(15) (1999): 3704–14.

Adair, G.S. The oxygen dissociation curve of hemoglobin. *J. Biol. Chem.* 63 (1925): 529–45.

Ataka, K., and Heberle, J. Use of surface enhanced infrared absorption spectroscopy (SEIRA) to probe the functionality of a protein monolayer. *Biopolymers* 82(4) (2006): 415–9.

Babcock, G.T., and Wikström, M. Oxygen activation and the conservation of energy in cell respiration. *Nature* 356(6367) (1992): 301–9.

Banci, L., Bertini, I., and Luchinat, C. *Nuclear and Electron Relaxation*. New York: VCH (1991).

Baradaran, R., Berrisford, J.M., Minhas, G.S. et al. Crystal structure of the entire respiratory complex I. *Nature* 494(7438) (2013): 443–448.

Barney, B.M., Igarashi, R.Y., Dos Santos, P.C. et al. Substrate interaction at an iron-sulfur face of the FeMo-cofactor during nitrogenase catalysis. *J. Biol. Chem.* 279(51) (2004): 53621–4.

Bashir, Q., Scanu, S., and Ubbink, M. Dynamics in electron transfer protein complexes. *FEBS J.* 278(9) (2011): 1391–400.

Beinert, H., and Kiley, P.J. Fe-S proteins in sensing and regulatory functions. *Curr. Opin. Chem. Biol.* 3(2) (1999): 152–7.

Berg, O.G., and von Hippel, P.H. Diffusion-controlled macromolecular interactions. *Ann. Rev. Biophys. Biophys. Chem.* 14 (1985): 131–60.

Brindell, M., Stawoska, I., Orzeł, L. et al. Application of high pressure laser flash photolysis in studies on selected hemoprotein reactions. *Biochim. Biophys. Acta* 1784(11) (2008): 1481–92.

Brunori, M. Hemoglobin is an honorary enzyme. *Trends Biochem. Sci.* 24(4) (1999): 158–61.

Brzezinski, P., and Wilson, M.T. Photochemical electron injection into redox-active proteins. *Proc. Natl. Acad. Sci. U. S. A.* 94(12) (1997): 6176–9.

Buckel, W., and Thauer, R.K. Energy conservation via electron bifurcating ferredoxin reduction and proton/Na(+) translocating ferredoxin oxidation. *Biochim. Biophys. Acta* 1827(2) (2013): 94–113.

Camba, R., Jung, Y.-S., Hunsicker-Wang, L.M. et al. Mechanisms of redox-coupled proton transfer in proteins: role of the proximal proline in reactions of the [3Fe-4S] cluster in Azotobacter vinelandii ferredoxin I. *Biochemistry* 42(36) (2003): 10589–99.

Cantor, C.R., and Schimmel, P.R. *Part III: The Behavior of Biological Macromolecules. Biophysical Chemistry.* San Francisco, CA: W.H. Freeman (1980).

Capone, D.G., Popa, R., Flood, B. et al. Geochemistry. Follow the nitrogen. *Science* 312(5774) (2006): 708–9.

Chance, B. The primary and secondary compounds of catalase and methyl or ethyl hydrogen peroxide. II. Kinetics and activity. *J. Biol. Chem.* 179(1949): 1341–69.

Coletta, M., Costa, H., De Sanctis, G. et al. pH dependence of structural and functional properties of oxidized cytochrome c″ from Methylophilus methylotrophus. *J. Biol. Chem.* 272(40) (1997): 24800–4.

Coulson, A.F., and Yonetani, T. Oxidation of cytochrome c peroxidase with hydrogen peroxide: Identification of the 'endogenous donor'. *Biochem. Biophys. Res. Commun.* 49(2) (1972): 391–8.

Coutinho, I., and Xavier, A.V. Tetraheme cytochromes. *Methods Enzymol.* 243 (1994): 119–40.

Cox, N., and Messinger, J. Reflections on substrate water and dioxygen formation. *Biochim. Biophys. Acta* 1827(8–9) (2013): 1020–30.

Cox, N., Pantazis, D.A., Neese, F. et al. Biological water oxidation. *Acc. Chem. Res.* 46(7) (2013): 1588–96.

Crofts, A.R., and Rose, S. Marcus treatment of endergonic reactions: A commentary. *Biochim. Biophys. Acta* 1767(10) (2007): 1228–32.

Danyal, K., Dean, D.R., Hoffman, B.M. et al. Electron transfer within nitrogenase: Evidence for a deficit-spending mechanism. *Biochemistry* 50(43) (2011): 9255–63.

Davidson, V.L. Protein control of true, gated and coupled electron transfer reactions. *Acc. Chem. Res.* 41(6) (2008): 730–8.

Debus, R.J. FTIR studies of metal ligands, networks of hydrogen bonds, and water molecules near the active site Mn4CaO5 cluster in photosystem II. *Biochim. Biophys. Acta* 1847(1) (2015): 19–34.

Deller, S., Macheroux, P., and Sollner, S. Flavin-dependent quinone reductases. *Cell. Mol. Life Sci.* 65(1) (2008): 141–60.

Dolla, A., Florens, L., Bianco, P. et al. Characterization and oxidoreduction properties of cytochrome c3 after heme axial ligand replacements. *J. Biol. Chem.* 269(9) (1994): 6340–6.

Dutton, P.L. Redox potentiometry: Determination of midpoint potentials of oxidation-reduction components of biological electron-transfer systems. *Methods Enzymol.* 54 (1978): 411–435.

Eady, R.R. Structure–function relationships of alternative nitrogenases. *Chem. Rev.* American Chemical Society 96(7) (1996): 3013–3030.

Farver, O., Zhang, J., Chi, Q. et al. Deuterium isotope effect on the intramolecular electron transfer in Pseudomonas aeruginosa azurin. *Proc. Natl. Acad. Sci. U. S. A.* 98(8) (2001): 4426–30.

Ferreira, K.N., Iverson, T.M., Maghlaoui, K. et al. Architecture of the photosynthetic oxygen-evolving center. *Science* 303(5665) (2004): 1831–8.

Fonseca, B.M., Paquete, C.M., Salgueiro, C.A. et al. The role of intramolecular interactions in the functional control of multiheme cytochromes c. *FEBS Lett.* 586(5) (2012): 504–9.

Fourmond, V., Baffert, C., Sybirna, K. et al. Steady-state catalytic wave-shapes for 2-electron reversible electrocatalysts and enzymes. *J. Am. Chem. Soc.* 135(10) (2013): 3926–38.

Ghosal, D., Omelchenko, M.V., Gaidamakova, E.K. et al. How radiation kills cells: Survival of Deinococcus radiodurans and Shewanella oneidensis under oxidative stress. *FEMS Microbiol. Rev.* 29(2) (2005): 361–75.

Girvan, H.M., and Munro, A.W. Heme sensor proteins. *J. Biol. Chem.* 288(19) (2013): 13194–203.

Glatzel, P., Schroeder, H., Pushkar, Y. et al. Electronic structural changes of Mn in the oxygen-evolving complex of photosystem II during the catalytic cycle. *Inorg. Chem.* 52(10) (2013): 5642–4.

Gray, H.B., and Winkler, J.R. Electron transfer in proteins. *Annu. Rev. Biochem.* 65 (1996): 537–61.

Gray, H.B., and Winkler, J.R. Electron tunneling through proteins. *Q. Rev. Biophys.* 36(3) (2003): 341–72.

Gray, H.B., and Winkler, J.R. Long-range electron transfer. *Proc. Natl. Acad. Sci. U. S. A.* 102(10) (2005): 3534–9.

Guskov, A., Kern, J., Gabdulkhakov, A. et al. Cyanobacterial photosystem II at 2.9 A resolution and the role of quinones, lipids, channels and chloride. *Nat. Struct. Mol. Biol.* 16(3) (2009): 334–42.

Gutlich, P., Eckhard, B., and Trautwein, A.X. *Mössbauer Spectroscopy and Transition Metal Chemistry.* Springer, Berlin (2011).

Hagen, W.R. EPR spectroscopy. In: Crichton, R.R., and Louro, R.O. (eds.), *Practical Approaches to Biological Inorganic Chemistry.* Elsevier, Amsterdam (2013), pp. 53–76.

Hammes-Schiffer, S., and Stuchebrukhov, A.A. Theory of coupled electron and proton transfer reactions. *Chem. Rev.* 110(12) (2010): 6939–60.

Hill, A.V. The possible effects of the aggregation of the molecules of huemoglobin on its dissociation curves. *J. Physiol.* 40 (1910): iv–vii.

Hoffman, B.M., Lukoyanov, D., Dean, D.R. et al. Nitrogenase: A draft mechanism. *Acc. Chem. Res.* 46(2) (2013): 587–95.

Hoffman, B.M., Lukoyanov, D., Yang, Z.-Y. et al. Mechanism of nitrogen fixation by nitrogenase: the next stage. *Chem. Rev.* 114(8) (2014): 4041–62.

Hu, Y., and Ribbe, M.W. Nitrogenase assembly. *Biochim. Biophys. Acta* 1827(8–9) (2013): 1112–22.

Iwata, S., Ostermeier, C., Ludwig, B. et al. Structure at 2.8 A resolution of cytochrome c oxidase from Paracoccus denitrificans. *Nature* 376(6542) (1995): 660–9.

Joliot, P., Barbieri, G., and Chabaud, R. Un Nouveau Modele des Centres Photochimiques du Systeme II. *Photochem. Photobiol.* 10(5) (1969): 309–29.

Jones, A.K., Lamle, S.E., Pershad, H.R. et al. Enzyme electrokinetics: Electrochemical studies of the anaerobic interconversions between active and inactive states of Allochromatium vinosum [NiFe]-hydrogenase. *J. Am. Chem. Soc.* 125(28) (2003): 8505–14.

Kaila, V.R.I., Verkhovsky, M.I., and Wikström, M. Proton-coupled electron transfer in cytochrome oxidase. *Chem. Rev.* 110(12) (2010): 7062–81.

Kanyo, Z.F., Scolnick, L.R., Ash, D.E. et al. Structure of a unique binuclear manganese cluster in arginase. *Nature* 383(6600) (1996): 554–7.

Keeler, J. *Understanding NMR Spectroscopy.* Wiley, Chichester (2002).

Kern, J., Alonso-Mori, R., Hellmich, J. et al. Room temperature femtosecond X-ray diffraction of photosystem II microcrystals. *Proc. Natl. Acad. Sci. U. S. A.* 109(25) (2012): 9721–6.

Kern, J., Alonso-Mori, R., Tran, R. et al. Simultaneous femtosecond X-ray spectroscopy and diffraction of photosystem II at room temperature. *Science* 340(6131) (2013): 491–5.

Kim, J., and Rees, D.C. Nitrogenase and biological nitrogen fixation. *Biochemistry* 33(2) (1994): 389–97.

Klinman, J.P., and Kohen, A. Hydrogen tunneling links protein dynamics to enzyme catalysis. *Annu. Rev. Biochem.* 82 (2013): 471–96.

Kok, B., Forbush, B., and McGloin, M. Cooperation of charges in photosynthetic O2 evolution—I. A linear four step mechanism. *Photochem. Photobiol.* 11(6) (1970): 457–75.

Koshland, D.E., Némethy, G., and Filmer, D. Comparison of experimental binding data and theoretical models in proteins containing subunits. *Biochemistry* 5(1) (1966): 365–85.

Krishtalik, L.I. The medium reorganization energy for the charge transfer reactions in proteins. *Biochim. Biophys. Acta* 1807(11) (2011): 1444–56.

Kroneck, P.M., Antholine, W.A., Riester, J. et al. The cupric site in nitrous oxide reductase contains a mixed-valence [Cu(II),Cu(I)] binuclear center: A multifrequency electron paramagnetic resonance investigation. *FEBS Lett.* 242(1) (1988): 70–4.

Kupitz, C., Basu, S., Grotjohann, I. et al. Serial time-resolved crystallography of photosystem II using a femtosecond X-ray laser. *Nature* 513(7517) (2014): 261–5.

Layfield, J.P., and Hammes-Schiffer, S. Hydrogen tunneling in enzymes and biomimetic models. *Chem. Rev.* 114(7) (2014): 3466–94.

Léger, C., and Bertrand, P. Direct electrochemistry of redox enzymes as a tool for mechanistic studies. *Chem. Rev.* 108(7) (2008): 2379–438.

Liu, J., Chakraborty, S., Hosseinzadeh, P. et al. Metalloproteins containing cytochrome, iron-sulfur, or copper redox centers. *Chem. Rev.* 114(8) (2014): 4366–469.

Louro, R.O. Introduction to biomolecular NMR and metals. In: Chrichton, R.R., and Louro, R.O. (eds.), *Practical Approaches to Biological Inorganic Chemistry*. Elsevier, Amsterdam (2013), pp. 77–107.

Louro, R.O., Catarino, T., Paquete, C.M. et al. Distance dependence of interactions between charged centres in proteins with common structural features. *FEBS Lett.* 576(1–2) (2004): 77–80.

Louro, R.O., Catarino, T., Salgueiro, C.A. et al. Redox-Bohr effect in the tetrahaem cytochrome c3 from Desulfovibrio vulgaris: a model for energy transduction mechanisms. *J. Biol. Inorg. Chem.* 450 (1996): 34–38.

Luchinat, C., Capozzi, F., Borsari, M. et al. Influence of surface charges on redox properties in high potential iron-sulfur proteins. *Biochem. Biophys. Res. Commun.* 203(1) (1994): 436–42.

Lukoyanov, D., Barney, B.M., Dean, D.R. et al. Connecting nitrogenase intermediates with the kinetic scheme for N2 reduction by a relaxation protocol and identification of the N2 binding state. *Proc. Natl. Acad. Sci. U. S. A.* 104(5) (2007): 1451–5.

Ma, J.G., Zhang, J., Franco, R. et al. The structural origin of nonplanar heme distortions in tetraheme ferricytochromes c3. *Biochemistry* 37(36) (1998): 12431–42.

Machonkin, T.E., Zhang, H.H., Hedman, B. et al. Spectroscopic and magnetic studies of human ceruloplasmin: Identification of a redox-inactive reduced Type 1 copper site. *Biochemistry* 37(26) (1998): 9570–8.

Mamedov, F., Danielsson, R., Gadjieva, R. et al. EPR characterization of photosystem II from different domains of the thylakoid membrane. *Biochemistry* 47(12) (2008): 3883–91.

Marcus, R.A., and Sutin, N. Electron transfers in chemistry and biology. *Biochim. Biophys. Acta* 811 (1985): 265–322.

Marcus, R.J., Zwolinski, B.J., and Eyring, H. The electron tunnelling hypothesis for electron exchange reactions. *J. Phys. Chem.* American Chemical Society 58(5) (1954): 432–7.

Martins, L.O., Soares, C.M., Pereira, M.M. et al. Molecular and biochemical characterization of a highly stable bacterial laccase that occurs as a structural component of the Bacillus subtilis endospore coat. *J. Biol. Chem.* 277(21) (2002): 18849–59.

Matias, P.M., Coelho, A.V, Valente, F.M.A. et al. Sulfate respiration in Desulfovibrio vulgaris Hildenborough. Structure of the 16-heme cytochrome c HmcA AT 2.5-A resolution and a view of its role in transmembrane electron transfer. *J. Biol. Chem.* 277(49) (2002): 47907–16.

Mayfield, J.A., Dehner, C.A., and DuBois, J.L. Recent advances in bacterial heme protein biochemistry. *Curr. Opin. Chem. Biol.* 15(2) (2011): 260–6.

Mayweather, D., Danyal, K., Dean, D.R. et al. Temperature invariance of the nitrogenase electron transfer mechanism. *Biochemistry* 51(42) (2012): 8391–8.

McLendon, G. Long-distance electron transfer in proteins and model systems. *Acc. Chem. Res.* 21(4) (1988): 160–7.

Meints, C.E., Gustafsson, F.S., Scrutton, N.S. et al. Tryptophan 697 modulates hydride and interflavin electron transfer in human methionine synthase reductase. *Biochemistry* 50(51) (2011): 11131–42.

Monod, J., Wyman, J., and Changeux, J.P. On the nature of allosteric transitions: A plausible model. *J. Mol. Biol.* 12 (1965): 88–118.

Moser, C.C., Keske, J.M., Warncke, K. et al. Nature of biological electron transfer. *Nature* 355(6363) (1992): 796–802.

Mowat, C.G., and Chapman, S.K. Multi-heme cytochromes—new structures, new chemistry. *Dalt. Trans.* 21 (2005): 3381–9.

Murgida, D.H., and Hildebrandt, P. Disentangling interfacial redox processes of proteins by SERR spectroscopy. *Chem. Soc. Rev.* 37(5) (2008): 937–45.

Nitschke, W., and Russell, M.J. Redox bifurcations: Mechanisms and importance to life now, and at its origin—A widespread means of energy conversion in biology unfolds .... *Bioessays* 34(2) (2012): 106–9.

Page, C.C., Moser, C.C., Chen, X. et al. Natural engineering principles of electron tunnelling in biological oxidation-reduction. *Nature* 402(6757) (1999): 47–52.

Page, C.C., Moser, C.C., and Dutton, P.L. Mechanism for electron transfer within and between proteins. *Curr. Opin. Chem. Biol.* 7(5) (2003): 551–6.

Paixão, V.B., Vis, H., and Turner, D.L. Redox linked conformational changes in cytochrome c3 from Desulfovibrio desulfuricans ATCC 27774. *Biochemistry* 49(44) (2010): 9620–9.

Palmer, G., and Olson, J.S. Concepts and approaches to understanding of electron transfer processes in enzymes containing multiple redox centres. In: Coughlan, M.P. (ed.), *Molybdenum and Molybdenum-Containing Enzymes*. London: Pergamon Press (1980).

Papa, S., Capitanio, N., Villani, G. et al. Cooperative coupling and role of heme a in the proton pump of heme-copper oxidases. *Biochimie* 80(10) (1998): 821–36.

Papa, S., Guerrieri, F., and Izzo, G. Redox Bohr-effects in the cytochrome system of mitochondria. *FEBS Lett.* 105(2) (1979): 213–6.

Paquete, C.M., and Louro, R.O. Unveiling the details of electron transfer in multicenter redox proteins. *Acc. Chem. Res.* 47(1) (2014): 56–65.

Paquete, C.M., Saraiva, I.H., and Louro, R.O. Redox tuning of the catalytic activity of soluble fumarate reductases from Shewanella. *Biochim. Biophys. Acta* 1837(6) (2014): 717–25.

Pereira, M.M., Santana, M., and Teixeira, M. A novel scenario for the evolution of haem-copper oxygen reductases. *Biochim. Biophys. Acta* 1505(2–3) (2001): 185–208.

Perutz, M. *Mechanisms of Cooperativity and Allosteric Regulation in Proteins.* Cambridge University Press, Cambridge (1990).

Peters, J.W., Fisher, K., and Dean, D.R. Nitrogenase structure and function: A biochemical-genetic perspective. *Annu. Rev. Microbiol.* 49 (1995): 335–66.

Peters, J.W., Lanzilotta, W.N., Lemon, B.J. et al. X-ray crystal structure of the Fe-only hydrogenase (CpI) from Clostridium pasteurianum to 1.8 angstrom resolution. *Science* 282(5395) (1998): 1853–8.

Petrenko, T., and Neese, F. Analysis and prediction of absorption band shapes, fluorescence band shapes, resonance Raman intensities, and excitation profiles using the time-dependent theory of electronic spectroscopy. *J. Chem. Phys.* 127(16) (2007): 164319.

Pievo, R., Angerstein, B., Fielding, A.J. et al. A rapid freeze-quench setup for multi-frequency EPR spectroscopy of enzymatic reactions. *ChemPhysChem* 14(18) (2013): 4094–101.

Prudêncio, M., and Ubbink, M. Transient complexes of redox proteins: Structural and dynamic details from NMR studies. *J. Mol. Recognit.* 17(6) (2004): 524–39.

Pushkar, Y., Yano, J., Glatzel, P. et al. Structure and orientation of the Mn4Ca cluster in plant photosystem II membranes studied by polarized range-extended x-ray absorption spectroscopy. *J. Biol. Chem.* 282(10) (2007): 7198–208.

Rees, D.C., Akif Tezcan, F., Haynes, C.A. et al. Structural basis of biological nitrogen fixation. *Philos. Trans. A. Math. Phys. Eng. Sci.* 363(1829) (2005): 971–84.

Ribbe, M.W., Hu, Y., Hodgson, K.O. et al. Biosynthesis of nitrogenase metalloclusters. *Chem. Rev.* 114(8) (2014): 4063–80.

Rich, P.R., and Maréchal, A. Functions of the hydrophilic channels in protonmotive cytochrome c oxidase. *J. R. Soc. Interface* 10(86) (2013): 20130183.

Richter, O.-M.H., and Ludwig, B. Electron transfer and energy transduction in the terminal part of the respiratory chain—lessons from bacterial model systems. *Biochim. Biophys. Acta* 1787(6) (2009): 626–34.

Rittle, J., and Green, M.T. Cytochrome P450 compound I: Capture, characterization, and C-H bond activation kinetics. *Science* 330(6006) (2010): 933–7.

Robson, R.L., Eady, R.R., Richardson, T.H. et al. The alternative nitrogenase of Azotobacter chroococcum is a vanadium enzyme. *Nature* 322(6077) (1986): 388–90.

Rochaix, J.-D. Regulation and dynamics of the light-harvesting system. *Annu. Rev. Plant Biol.* 65 (2014): 287–309.

Rubio, L.M., and Ludden, P.W. Biosynthesis of the iron-molybdenum cofactor of nitrogenase. *Annu. Rev. Microbiol.* 62 (2008): 93–111.

Rydén, L.G., and Hunt, L.T. Evolution of protein complexity: The blue copper-containing oxidases and related proteins. *J. Mol. Evol.* 36(1) (1993): 41–66.

Saito, K., Rutherford, A.W., and Ishikita, H. Mechanism of tyrosine D oxidation in photosystem II. *Proc. Natl. Acad. Sci. U. S. A.* 110(19) (2013): 7690–5.

Savéant, J.-M. Molecular catalysis of electrochemical reactions. Mechanistic aspects. *Chem. Rev.* 108(7) (2008): 2348–78.

Sazanov, L.A., and Hinchliffe, P. Structure of the hydrophilic domain of respiratory complex I from Thermus thermophilus. *Science* 311(5766) (2006): 1430–6.

Schrauben, J.N., Cattaneo, M., Day, T.C. et al. Multiple-site concerted proton-electron transfer reactions of hydrogen-bonded phenols are nonadiabatic and well described by semiclassical Marcus theory. *J. Am. Chem. Soc.* 134(40) (2012): 16635–45.

Schwarz, G., Mendel, R.R., and Ribbe, MW. Molybdenum cofactors, enzymes and pathways. *Nature* 460(7257) (2009): 839–47.

Seefeldt, L.C., Hoffman, B.M., and Dean, D.R. Electron transfer in nitrogenase catalysis. *Curr. Opin. Chem. Biol.* 16(1–2) (2012): 19–25.

Seefeldt, L.C., Yang, Z.-Y., Duval, S. et al. Nitrogenase reduction of carbon-containing compounds. *Biochim. Biophys. Acta* 1827(8–9) (2013): 1102–11.

Simonneaux, G., and Bondon, A. Mechanism of electron transfer in heme proteins and models: The NMR approach. *Chem. Rev.* 105(6) (2005): 2627–46.

Skourtis, S.S., Waldeck, D.H., and Beratan, D.N. Fluctuations in biological and bioinspired electron-transfer reactions. *Annu. Rev. Phys. Chem.* 61 (2010): 461–85.

Smith, W.W., Burnett, R.M., Darling, G.D. et al. Structure of the semiquinone form of flavodoxin from Clostridum MP. Extension of 1.8 A resolution and some comparisons with the oxidized state. *J. Mol. Biol.* 117(1) (1977): 195–225.

Stellwagen, E. Haem exposure as the determinate of oxidation-reduction potential of haem proteins. *Nature* 275(5675) (1978): 73–4.

Suzuki, H., Sugiura, M., and Noguchi, T. Monitoring proton release during photosynthetic water oxidation in photosystem II by means of isotope-edited infrared spectroscopy. *J. Am. Chem. Soc.* 131(22) (2009): 7849–57.

Swierczek, M., Cieluch, E., Sarewicz, M. et al. An electronic bus bar lies in the core of cytochrome bc1. *Science* 329(5990) (2010): 451–4.

Tsibris, J.C.M., and Woody, R.W. Structural studies of iron-sulfur proteins. *Coord. Chem. Rev.* 5(4) (1970): 417–58.

Tsukihara, T., Aoyama, H., Yamashita, E. et al. The whole structure of the 13-subunit oxidized cytochrome c oxidase at 2.8 A. *Science* 272(5265) (1996): 1136–44.

Tsukihara, T., Shimokata, K., Katayama, Y. et al. The low-spin heme of cytochrome c oxidase as the driving element of the proton-pumping process. *Proc. Natl. Acad. Sci. U. S. A.* 100(26) (2003): 15304–9.

Ubbink, M. The courtship of proteins: Understanding the encounter complex. *FEBS Lett.* 583(7) (2009): 1060–6.

Umena, Y., Kawakami, K., Shen, J.-R. et al. Crystal structure of oxygen-evolving photosystem II at a resolution of 1.9 Å. *Nature* 473(7345) (2011): 55–60.

Vallee, B.L., and Williams, R.J. Metalloenzymes: The entatic nature of their active sites. *Proc. Natl. Acad. Sci. U. S. A.* 59(2) (1968): 498–505.

Verissimo, A.F., and Daldal, F. Cytochrome c biogenesis system I: An intricate process catalyzed by a maturase supercomplex? *Biochim. Biophys. Acta* 1837(7) (2014): 989–98.

Vinyard, D.J., Ananyev, G.M., and Dismukes, G.C. Photosystem II: The reaction center of oxygenic photosynthesis. *Annu. Rev. Biochem.* 82 (2013): 577–606.

Warshel, A. Multiscale modeling of biological functions: From enzymes to molecular machines (nobel lecture). *Angew. Chem. Int. Ed. Engl.* 53(38) (2014): 10020–31.

Weissbluth, M. The physics of hemoglobin. In: *Structure and Bonding.* Springer, Berlin (1967), pp. 1–125.

Wikström, M. Proton pump coupled to cytochrome c oxidase in mitochondria. *Nature* 266(5599) (1977): 271–3.

Wikström, M. Cytochrome c oxidase: 25 years of the elusive proton pump. *Biochim. Biophys. Acta* 1655(1–3) (2004): 241–7.

Williams, R.J.P. Molecular and thermodynamic bioenergetics. *Biochem. Soc. Trans.* 33(Pt 4) (2005): 825–8.

Williams, R.J.P., and Fraústo da Silva, J.J.R. *The Natural Selection of the Chemical Elements—The Environment and Life's Chemistry.* New York: Oxford University Press (1996).

Winkler, J.R., and Gray, H.B. Electron flow through metalloproteins. *Chem. Rev.* 114(7) (2014): 3369–80.

Winkler, J.R., Malmström, B.G., and Gray, H.B. Rapid electron injection into multisite metalloproteins: Intramolecular electron transfer in cytochrome oxidase. *Biophys. Chem.* 54(3) (1995): 199–209.

Wyman, J. The turning wheel: a study in steady states. *Proc. Natl. Acad. Sci. U. S. A.* 72(10) (1975): 3983–7.

Wyman, J., and Gill, S.J. *Binding and Linkage: Functional Chemistry of Biological Macromolecules* (Kelly, A., ed.). University Science Books, Mill Valey, CA (1990).

Xavier, A.V. Energy transduction coupling mechanisms in multiredox center proteins. *J. Inorg. Biochem.* 28(2–3) (1986): 239–43.

Xavier, A.V. Thermodynamic and choreographic constraints for energy transduction by cytochrome c oxidase. *Biochim. Biophys. Acta* 1658(1–2) (2004): 23–30.

Yang, Z.-Y., Khadka, N., Lukoyanov, D. et al. On reversible H2 loss upon N2 binding to FeMo-cofactor of nitrogenase. *Proc. Natl. Acad. Sci. U. S. A.* 110(41) (2013): 16327–32.

Yano, J., Kern, J., Sauer, K. et al. Where water is oxidized to dioxygen: Structure of the photosynthetic Mn4Ca cluster. *Science* 314(5800) (2006): 821–5.

Yano, J., and Yachandra, V. Mn4Ca cluster in photosynthesis: Where and how water is oxidized to dioxygen. *Chem. Rev.* 114(8) (2014): 4175–205.

Yoshikawa, S., Shinzawa-Itoh, K., Nakashima, R. et al. Redox-coupled crystal structural changes in bovine heart cytochrome c oxidase. *Science* 280(5370) (1998): 1723–9.
Zouni, A., Witt, H.T., Kern, J. et al. Crystal structure of photosystem II from Synechococcus elongatus at 3.8 A resolution. *Nature* 409(6821) (2001): 739–43.
Zuiderweg, E.R.P. Current topics mapping protein-protein interactions in solution by NMR spectroscopy. *Biochemistry* 41(1) (2002): 1–7.

CHAPTER 2

Diversity of Interactions in Redox Systems

From Short- to Long-Lived Complexes

Antonio Díaz-Quintana, Isabel Cruz-Gallardo, Miguel A. De la Rosa and Irene Díaz-Moreno

CONTENTS

2.1 BINDING FEATURES OF SHORT- AND LONG-LIVED COMPLEXES

Protein interaction networks modulate the function of living cells. For instance, specific bimolecular interactions control important cellular events such as the regulation of metabolism and the triggering of processes that promote cell life or cell death. The physiological functions of these bimolecular interactions are determined by the latter's strength and duration.

The binding affinity, or interaction strength, of two biomolecules that interact reversibly is defined by the equilibrium dissociation constant, K_D. Assuming that the simplest bimolecular association reactions are reversible and follow second-order kinetics (2.1), the K_D value is given as the ratio between the dissociation and association rate constants, k_{off} and k_{on}, respectively (2.2). Values for k_{on} and k_{off} are usually expressed in $M^{-1}\ s^{-1}$ and s^{-1}, respectively (Janin, 2000; Kastritis and Bonvin, 2013).

$$[A]+[B] \underset{k_{off}}{\overset{k_{on}}{\rightleftarrows}} [AB] \tag{2.1}$$

$$K_D = \frac{[A][B]}{AB} = \frac{k_{off}}{k_{on}} \tag{2.2}$$

In Equations 2.1 and 2.2, [A] and [B] represent the respective concentrations of free biomolecules A and B, while [AB] represents the concentration of the complex they form. The length of bimolecular interactions is denoted by complex lifetime, which is usually expressed in s and can be obtained from $\ln 2/k_{off}$.

Long-lived complexes involve interactions between biomolecules with high affinity and specificity and exhibit lifetimes ranging from minutes to days. K_D values typically associated with such complexes are expressed in nM, with low dissociation rates ($k_{off} < 1\ s^{-1}$). By contrast, short-lived complexes, with lifetimes best expressed with values from μs to ms, are governed by transient interactions with low affinity. This sort of complex displays high K_D values expressed within the range of μM and mM and high k_{off} values ($k_{off} \geq 10\ s^{-1}$) (Figure 2.1).

Short-lived complexes ensure the fast turnover demanded in many biological functions, yet specificity for their partner(s) must be guaranteed. Thus, molecular interactions within any short-lived complex must be able to balance high dissociation rates with high binding specificity. Among the physiological processes for which such transient and reversible interactions required are electron transfer (ET) reactions, signalling cascade triggering and propagation and the regulation of mRNA metabolism (Bashir et al., 2011; Díaz-Moreno and De la Rosa, 2011; Perkins et al., 2010).

Complexes involved in ET reactions are excellent examples of short-lived interactions in which soluble proteins exchange electrons with large membrane-protein complexes in photosynthesis and cellular respiration via transient contacts. In addition to

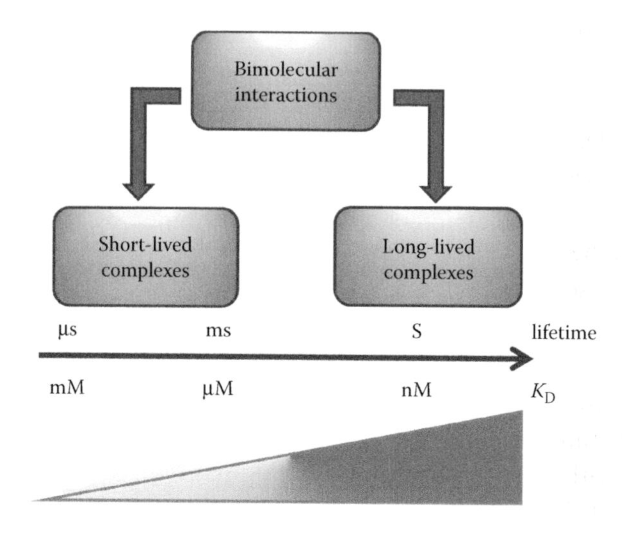

FIGURE 2.1 Bimolecular interactions give rise to short- and long-lived complexes. The binding affinities of such complexes are usually defined by dissociation constant (K_D) and lifetime. The triangle represents how binding affinities decrease from long-lived complexes (grey) to short-lived ones (white).

a high turnover, a certain *promiscuity* in the electron carriers is needed, since their surfaces are often optimized to recognize and interact with multiple redox partners (Schreiber and Keating, 2011). Thus, to transport an electron from a donor protein to a remote acceptor one, carrier must specifically recognize the two partners and dissociate quickly from them (Bashir et al., 2011). ET pathways from protein redox centres to their surfaces are optimized to be as short as possible. However, the surface complementarity between the partners and the interface size are relatively low when compared to long-lived complexes. As a result, redox centres are usually close to a hydrophobic patch where the interacting partners bind (Williams et al., 1995). Thus, complex dissociation can be facilitated in certain redox proteins, such as cupredoxins (De Rienzo et al., 2000), by polar amino acids that surround the hydrophobic region and mediate the entry of water molecules at the interface prior to the disruption of the protein-protein complex. This formation of transient protein complexes is thought to be a two-step process in which the final complex is preceded by an intermediate state, or a so-called encounter complex.

2.2 TWO-STEP ASSOCIATION

The ubiquity of ET reactions among proteins in living cells and their apparent simplicity make them ideal for the development of theoretical protein-protein interaction models and their posterior experimental evaluation. The most simplistic process in a redox reaction involving two partners consists of three principal steps—the bimolecular complex is first formed, ET then takes place and, finally, the complex dissociates (2.3).

$$A_{OX} + B_{RED} \underset{k_{off}}{\overset{k_{on}}{\rightleftharpoons}} A_{OX} \cdot B_{RED} \underset{k'_{ET}}{\overset{k_{ET}}{\rightleftharpoons}} A_{RED} \cdot B_{OX} \underset{k'_{on}}{\overset{k'_{off}}{\rightleftharpoons}} A_{RED} + B_{OX} \tag{2.3}$$

In Equation 2.3, k_{ET} and k'_{ET} are the kinetic rates of the forward and reverse ET reactions, respectively, within the complex. Once the electrical charge is transferred from the donor to the acceptor, the constants for dissociation and association of the partners, k'_{off} and k'_{on}, respectively, may differ from those prior to the ET step.

For the ET step, a prior activation of the complex is required, which may be rate limiting. The ET rate is affected by internal rearrangements, the thermodynamic driving force and the overlap between the electron wave functions of the donor and the acceptor (Marcus and Sutin, 1985; Page, Moser, and Dutton, 2003). As most of the phenomena involved in activation, such as nuclear vibrations, occur within the dead time of the experimental set-ups considered in this chapter, they have therefore been omitted from the following discussion.

In Equation 2.3, the overall ET reaction between the proteins A and B is second order. If association is the limiting step, we should experimentally observe a single-exponential decay of reactants, with a reaction rate depending on donor and acceptor concentrations. Kinetic analyses with UV-visible (UV-vis) absorbance spectroscopy do not yield direct information about the reaction coordinates and experiments must be designed carefully to obtain reliable data. Such data were usually analysed according to kinetic models developed in the early stages of research in the field (Tollin et al., 1986). Furthermore, the experiments should be performed under pseudo-first-order conditions that simplify the analysis. Therefore, if a large excess of oxidant is present with respect to donor concentrations, the observed reaction rate constant k_{obs} for a single exponential kinetic trace will be as follows:

$$k_{obs} = \frac{k_{on} \cdot k_{ET} \cdot [A_{OX}]}{k_{on} \cdot [A_{OX}] + k_{ET} + k_{off}} \quad (2.4)$$

A high temporal resolution is usually achieved by the rapid activation of the reaction, thereby minimizing the dead time for data acquisition that is required. For this purpose, laser-flash set-ups are ideal either to activate light-sensitive redox systems, such as photosystem I (PSI) (Mathis and Setif, 1981), or to trigger redox reactions by means of photosensitizers, such as flavins (Meyer et al., 1983, 1993b; Navarro et al., 1991, 1995; Qin and Kostic, 1992; Rodríguez-Roldán et al., 2008).

According to the two-step kinetic model (Tollin et al., 1986), second-order plots of the reaction rates obtained under pseudo-first-order conditions and low protein concentrations are linear and with two extreme regimes. These plots display the dependency of the reaction rates observed on the concentration of one of the reactants. If the ET rate, k_{ET}, is much higher than k_{off}, the slope of the plot will approach the value of k_{on}. This mechanism was named collisional (Hervás et al., 1995) or collision-limited because the reaction rate is limited by the contact between the two partners. On the other hand, when k_{ET} is much lower than k_{off}, the slope of the plot corresponds to the ratio between k_{ET} and K_D. In this case, a fraction of the electron donor–acceptor complexes is non-productive because dissociation takes place before any electron may be transferred.

The initial, second-order binding step limits the rate of exponential decay at low protein concentrations. According to *in vitro* observations of the reaction rates at different ionic strength values, long-range electrostatics bear on the association between the two partners, which is, for instance, the rate-limiting step in electron donation to PSI in cyanobacteria, such as *Synechocystis* (Hervás et al., 1994, 1995, 1996).

The above formalisms fitted well to the kinetic data for self-exchange reactions (Dixon et al., 1989), as well as for reactions between proteins and redox ligands (Jackman et al., 1987; Meyer et al., 1983; Navarro et al., 1995). It also seems to apply well to the ET reactions involving flavodoxin (Casaus et al., 2002). However, a more complex model may need to be applied to many ET reactions between physiological partners. For instance, collision-limited models mismatched the early kinetic measurements of P700 reduction by plastocyanin (Pc), which showed a biexponential decay with half-lives of 12 and 160 μs (Bottin and Mathis, 1985). They interpreted their results as a clue of the presence of two binding sites in PSI, one of them being productive whereas the other being not, so the donor needed to migrate from one site to another to be able to donate electrons. The fast and slow phases of the kinetic traces (time evolution of the reactive species) fitted to first- and second-order reactions, respectively. Moreover, double-flash experiments suggested that a conformational change—also called rearrangement—precedes the ET step itself. An alternative explanation for these multiphasic kinetics is the obstruction of the active site by an oxidized donor molecule meanwhile it remains bound to the acceptor (see below in this Section and Section 2.4) (Drepper et al., 1996). However, cross-linking experiments provided further clues about the donor–acceptor complex requiring a certain degree of freedom to arrange itself in a productive conformation. Indeed, Pc was shown to be unable to receive electrons from cytochrome *f* (C*f*) when the two proteins are chemically cross-linked (Qin and Kostic, 1993). Analyses using cytochrome c_6 (Cc_6) as donor (Díaz et al., 1994b; Hervás et al., 1992, 1994; Medina et al., 1993; Molina-Heredia et al., 1998) revealed similar behaviour. According to the two-step reaction model mentioned earlier, the initial phase can be attributed to direct ET in a donor–acceptor complex formed before the reaction is triggered, while the second phase was proposed as corresponding to the binding-limited step (Tollin et al., 1986). Thus, the rate constant observed for the fast phase was independent of the viscosity of the media, whereas amplitude decreased when viscosity increased (Hervás et al., 1996). In some cases, the second-order plot of the reaction rate for the slow phase was hyperbolic. According to the two-step reaction mechanism, ET might be the rate-limiting step at saturating donor concentrations. However, the reaction rate observed was clearly lower than that reckoned for the fast phase (Hervás et al., 1995), in agreement with the conformational rearrangement hypothesis. According to it, the reaction comprises at least three steps besides product dissociation: binding, complex rearrangement and intramolecular ET. The dissociation of the reaction products was not taken into account as it cannot be observed in the kinetic traces obtained by single-flash experiments. However, as pointed out before, it might be considered to interpret double-flash experiments (see below in this Section and Section 2.4).

The rearrangement or conformational reorganization concept means that a population of the complexes formed by donors and acceptors is nevertheless unable to transfer electrons (De la Rosa et al., 2002; Hervás et al., 1996) and needs to change their relative orientation or suffer a change of conformation to become a productive complex. However, in another subset of the complexes, the electron is in fact transferred before dissociation takes place, thus giving rise to the fast kinetic phase:

$$A_{OX} + B_{RED} \underset{k_{off}}{\overset{k_{on}}{\rightleftharpoons}} \left(\left[A_{OX} \cdot B_{RED} \right]^{\dashv} \underset{k'_{R}}{\overset{k_{R}}{\rightleftharpoons}} \left[A_{OX} \cdot B_{RED} \right]^{\vdash} \right)$$
$$\underset{k'_{ET}}{\overset{k_{ET}}{\rightleftharpoons}} A_{RED} \cdot B_{OX} \underset{k'_{on}}{\overset{k'_{off}}{\rightleftharpoons}} A_{RED} + B_{OX} \tag{2.5}$$

In Equation 2.5, k_R and k'_R are the rate constants for slow rearrangements of the complex to form a productive conformation ($\vdash$) from an unproductive one ($\dashv$) and vice versa. This rearrangement may act as a reaction bottleneck under certain experimental conditions. In this case, the term *rearrangement* means any event really occurring within the initial complex that makes it capable for ET, without denoting any specific phenomena exclusively.

The relative amplitude, f_{FF}, of the fast kinetic phase was assumed to correspond to the fraction of complexes rearranged when the reaction is triggered (Hervás et al., 1995):

$$f_{FF} = \frac{K_R \cdot [A]}{K_D + (1 + K_R) \cdot [A]} \tag{2.6}$$

According to this equation, K_R is the equilibrium constant for rearrangement, which is equal to the ratio of k_R to k'_R. This population of productive complexes increases with viscosity, in accordance with the fast phase amplitude changes observed in the experimental kinetic traces. In each case, reaction partner properties and experimental conditions determine which of the three reaction steps (binding, rearrangement or intracomplex ET) is rate limiting (Díaz-Quintana et al., 2003).

A general equation was developed by Kostić and co-workers to fit multi-exponential models under pseudo-first-order conditions (Olesen et al., 1999), unsuccessfully attempting to discriminate between the rearrangement hypothesis (Bottin and Mathis, 1985; Hervás et al., 1995) and that of active-site obstruction by the reaction product (Drepper et al., 1996). In the case of P700 reduction by its soluble donors, the matrix form of the differential equation for the rearrangement model is:

$$\dot{c}(t) = \begin{bmatrix} -k_{\mathrm{on}} & k_{\mathrm{off}} & 0 \\ k_{\mathrm{on}} & -k_{\mathrm{off}} - k_{\mathrm{R}} & k'_{\mathrm{R}} \\ 0 & k_{\mathrm{R}} & -k'_{\mathrm{R}} - k_{\mathrm{ET}} \end{bmatrix} \cdot c(t) \tag{2.7}$$

Here, $c(t)$ is a column vector whose elements are the relative fractions of the distinct A species (P700). Diagonalization of the matrix and solution of the resulting eigenvalue equation yields a triexponential (*very fast*, *fast* and *slow*) expression for the function to fit the transient signal recorded $S(t)$:

$$S(t) = a_{\mathrm{vf}} e^{-k_{\mathrm{vf}} t} + a_{\mathrm{f}} e^{-k_{\mathrm{f}} t} + a_{\mathrm{s}} e^{-k_{\mathrm{s}} t} \tag{2.8}$$

The subscripts vf, f and s represent distinct kinetic phases: very fast, fast and slow, respectively. Accordingly, a and k indicate the respective amplitudes and the observed kinetic rates for each phase of the decay curve, and t stands for time.

The rearrangement process corresponding to the three-step mechanism is shared by many biological systems besides redox enzymes. In fact, a two-stage association was previously observed for bacteriophage T4 fibre attachment (Bloomfield and Prager, 1979) and DNA binding through the processing of enzymes like the *lac* repressor (Berg et al., 1981). In both cases, binding is described by a coupled diffusion model where a protein binds non-specifically to its partner's surface—forming the so-called encounter complex—and then diffuses along the partner surface (i.e. slides on it) until it either finds an active conformation or dissociates altogether from its partner (Figure 2.2). Gliding on the partner surface confines the pathway towards the productive complex conformation to a subspace of low and fractal dimensionality.

At this point, it is worth noting that the population of complexes capable of ET reactions may correspond to either a single conformation or a subset of an ensemble of conformers. Moreover, the need for a well-defined conformation for efficient ET depends on how the electron wave function spreads from the cofactor (Marcus and Sutin, 1985; Page et al., 2003).

Expressed in another way, the greater the distance spanned by frontier orbitals, the less defined the electron donor–acceptor complex must be. For instance, the electron

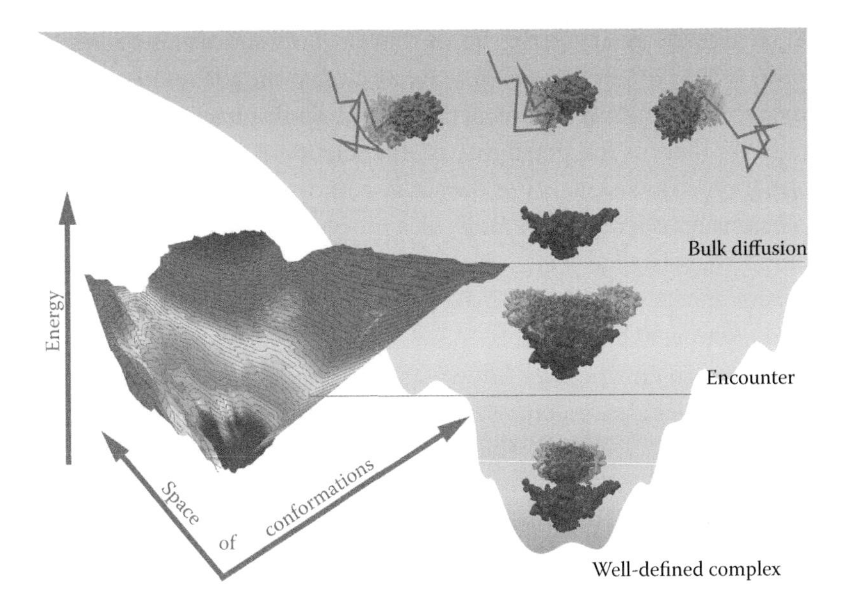

FIGURE 2.2 **(See colour insert.)** Energetic landscape of a transient complex. *Left:* Contour levels correspond to the energy of each point of the 2D projection of the coordinate space. The representation is coloured from low energy level (blue) to high (grey) of each contour. *Right:* Funnel cross-section showing the three major processes involved in the binding event. The proteins participating in the complex are represented by low-resolution surfaces. Partner 1 (purple) remains at a fixed position, whereas partner 2 (magenta) is represented as a mobile partner.

transferred in Pc is much more confined to its single copper atom centre than is the electron in a cytochrome heme porphyrin. This may result in a different degree of conformational restraints for ET. However, as opposed to the three-dimensional diffusion, the rate of diffusion-limited association on a protein surface is barely dependent on partner size, thereby enhancing effectiveness or the active conformation search (Berg and von Hippel, 1985). Thus, the theoretical average time τ required for a molecule with diffusion constant D to reach a small patch on the partner surface with radius ρ, corresponding to the active (or productive) conformation, in the middle of a space with radius R (assuming $R >> \rho$) for unrestrained three-, two- and one-dimensional searches are, respectively:

$$\tau_{3D} = \left(\frac{R^2}{3 \cdot D}\right) \cdot \frac{R}{\rho} \tag{2.9}$$

$$\tau_{2D} = \left(\frac{R^2}{2 \cdot D}\right) \cdot \ln\left(\frac{R}{\rho}\right) \tag{2.10}$$

$$\tau_{1D} = \left(\frac{R^2}{3 \cdot D}\right) \tag{2.11}$$

For instance, NMR experiments have shown that Cc_6 only slides along a limited region on the surface of *Cf* (Díaz-Moreno et al., 2014). According to the above-mentioned arguments, the achievement of a productive conformation is practically guaranteed during the lifetime of the complex. As there is the possibility of dissociation prior to the achievement of the target conformation, dimensionality of the active conformation search may increase again. However, collisions between protein molecules in solution are inelastic and long-lived, so they allow the partners to slide over each other and set their redox centres into a productive apposition. These effects may be enhanced if the target protein is oligomeric, like nitrate reductase (De la Rosa et al., 1981).

A related concept in enzymology known as substrate channelling has been applied to explain how small molecules undergo sequential transformations without diffusing to the bulk phase (Brown et al., 1996). A characteristic paradigm for this phenomenon is the ET flavoprotein from *Methylophilus methylotrophus*. This protein establishes a dual interaction with one of its partners, wherein one of the two flavoprotein domains acts as an anchor for its partner, while that containing the redox centre samples a large range of orientations (Leys et al., 2003).

The variability of binding modes in complexes involving homologous partners has been confirmed in a different timescale by NMR chemical-shift perturbation (CSP) analysis. As reviewed in an earlier work (Prudêncio and Ubbink, 2004), CSPs can be used to calculate equilibrium-binding constants. Chemical shift maps for structure surfaces are indicative of how static or dynamic a complex is, with well-defined complexes demonstrating large CSPs over a specific patch of the protein surface (Crowley et al., 2002), whereas weak and highly dynamic interactions yield a set of small CSPs spread over the whole protein surface (Díaz-Moreno et al., 2005a).

For instance, chemical shifts induced by *Cf* binding and their distribution on the surface of Pc prove to be strongly dependent on the nature of the acceptor (Díaz-Moreno et al., 2005a). In fact, the most populated conformation of the complex changes depending on the organism (Crowley et al., 2001; Díaz-Moreno et al., 2005a; Lange et al., 2005). Notably, in photosynthesis, the surface properties of the soluble electron carrier determine the binding modes, rather than those of the complex membranes as shown in the CSP mapping (Díaz-Moreno et al., 2005b,c) and other kinetic data (Hervás et al., 2005).

Further variability in the data regarding redox processes arises from the fact that the same redox protein may have several different physiological partners, and the corresponding complexes may differ in their dynamic properties. This diversity may correspond to the distinct contexts of the various interactions. Such is the case of cytochrome *c* (C*c*), which plays a role in oxidative phosphorylation and cell signalling (Díaz-Moreno et al., 2011a; García-Heredia et al., 2008; Martínez-Fábregas et al., 2013, 2014a,b). C*c* demonstrates a relatively high affinity towards the cytochrome bc_1 (Cbc_1) complex, and co-crystallizes with it at an ionic strength of 120 mM (Lange and Hunte, 2002). X-ray diffraction data show a single conformation. However, at low ionic strength, cytochrome c_1 (Cc_1) is able to bind to two molecules of C*c* (Moreno-Beltrán et al., 2014). Brownian dynamics (BD) suggests that C*c* is able to slide on the surface of Cc_1. According to this hypothesis, the second site—even if specific to *Cc*—is a local energy-minimum stage along the whole search pathway that remains undetected. *Cc* is also able to form a well-defined complex with cytochrome *c* peroxidase (*Cc*P) (Worrall et al., 2001). However, the non-physiological interaction between bovine cytochrome b_5 (Cb_5) and *Cc* leads to an ensemble of oriented conformations without a dominant one (Hom et al., 2000). Further, *Cc* is a target of several posttranslational modifications such as nitration (Díaz-Moreno et al., 2011b; García-Heredia et al., 2010, 2012; Rodríguez-Roldán et al., 2008) and phosphorylation (García-Heredia et al., 2011) that affect the conformation of the heme moiety (Ly et al., 2012) and its functionality as an electron carrier.

Encounter complexes cannot be analysed by X-ray diffraction or standard NMR methodologies like diamagnetic or paramagnetic effect CSPs, residual dipolar couplings or nuclear Overhauser effect measurements, as they are not sensitive to populations of less than 10% of the molecules. They can be studied, however, using paramagnetic relaxation enhancement (PRE) methodologies. The first encounter complex analysed by the PRE method was that formed between the amino-terminal domain of enzyme I and the phosphocarrier protein HPr (Tang et al., 2006). This complex is as weak as many electron donor–acceptor ones. Notably, the surface buried in the encounter complex conformations is 10 times smaller than that in the stereospecific complex, indicating the encounter as having been stabilized to form the well-defined complex by means of hydrophobic forces.

PRE has also been used to analyse ET complexes—namely, those formed by *Cc* and *Cc*P (Volkov et al., 2006, 2010), Pc and *Cf* (Scanu et al., 2013a,b) and *Nostoc* Cc_6 and *Cf* (Díaz-Moreno et al., 2014). Notably, the dominant, well-defined population in the first complex accounts for more than 70% of the species, while being a clear minority in the last case, in agreement with the poor pseudocontact shift profiles observed for this complex (Díaz-Moreno et al., 2005d).

In the Pc-*Cf* complex from *Nostoc* (Scanu et al., 2013a,b), the model resulting from PRE analyses indicates that electrostatic forces orient the Pc molecules, whereas hydrophobic interactions serve both to further stabilize the encounter complex while, at the same time, allowing the copper protein to slide along the *Cf* surface. According to BD simulations (Gross and Rosenberg, 2006), this type of preorientation is common in Pc-*Cf* complexes in all cyanobacteria. As cyanobacterial Pc shows asymmetry of the charge distribution (Figure 2.3), the hydrophobic patch must lie at the region of positive potential that is attracted by negatively charged *Cf*.

An interesting point for the hypothesis of the sliding motion of Pc on the surface of *Cf* is due, in part, to a constant hydrophobic interaction is that the former requires the entrance of water molecules from one side of the interface and the exclusion of water from the other. It is possible that the balance between repulsive interactions (mainly electrostatic repulsion from the back side of Pc) and attractive ones (hydrophobic and local electrostatic forces) results in the protein's temporary loss of hydrophobic interactions on the surface, permitting water molecules to enter, followed by the recovery of these surface interactions once the protein has slid. This is more likely to happen in Cc_6, as its smaller size and spherical shape brings its negatively charged patch closer to *Cf*. In the case of the Pc-*Cf* complex

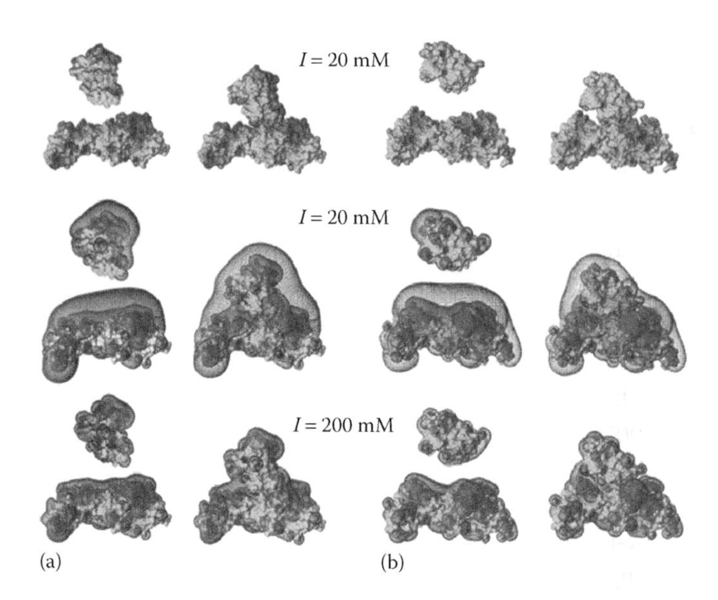

FIGURE 2.3 **(See colour insert.)** Electrostatics of complexes between plastocyanin (Pc) and cytochrome *f* (C*f*) from the cyanobacteria (a) *Phormidium* and (b) *Nostoc*. The electrostatic potentials of Pc and C*f* are represented for both the free proteins and the complex composed by them. *Upper:* Electrostatic surface potential (ESP) calculated with DelPhi at an ionic strength of 20 mM with full scale ranging from $-5\ k_BT/e$ (blue) to $+\ 5\ k_BT/e$ (grey). *Middle* and *lower:* Isopotential surfaces at $\pm 1\ k_BT/e$ (mesh) and $\pm 2.5\ k_BT/e$ (solid) superimposed to the ESP representations at ionic strength of 20 mM or 200 mM. Blue and red surfaces correspond to positive and negative potential values, respectively.

from *Phormidium*, and as deduced from molecular dynamics (MD) calculations, due to the electrostatic strain involving the back part of Pc, the protein is shown to swing (Díaz-Moreno et al., 2009), thereby allowing a partial gap at the interface, which acts as a pivot for the motion. Following this, water molecules may enter from one side and exit through the other during every complete oscillation of Pc, keeping the hydrophobic interactions at the core of the interface.

The diversity in the NMR data matches the variability in the behaviour of the reaction kinetics. For the interaction between Cc_6 and C*f* from *Nostoc*, PRE profiles indicate a major encounter population in which electrostatic forces promote the establishment of hydrophobic forces, but do not take part in the activation barrier. This coincides with the finding that complexes involving *Nostoc* Cc_6 do not show the enthalpy-entropy compensation in activation energy characteristic of other ones and exhibits a rate constant for electron donation to PSI that is independent of temperature (Hervás et al., 1996). Indeed, Cc_6 uses the same interaction surface with its two physiological partners, C*f* and PSI, the complex with the photosynthetic reaction centre being highly dynamic (Díaz-Moreno et al., 2005c, 2014). On the other hand, the *Nostoc* Pc-PSI complex demonstrates strong enthalpy-entropy compensation (Hervás et al., 1996) and the *Nostoc* Pc-C*f* ensemble shows substantial interplay between electrostatic and hydrophobic interactions (Scanu et al., 2013b).

2.3 STRUCTURAL CHANGES IN PARTNERS WITHIN THE COMPLEX

Most of the theoretical background needed to analyse the structural features of the binding phenomena is based on the assumption that proteins act as rigid bodies in most ET complexes. In fact, most NMR data in this kind of interaction yield precise information (Prudêncio and Ubbink, 2004). Many of the structural models are based on rigid-body dynamics or Monte-Carlo methods. In fact, BD simulations, a rigid-body approach, are generally in accord with experimentally calculated reaction rates (De Rienzo et al., 2001, 2002; Gross, 2004, 2007; Gross and Pearson, 2003; Haddadian and Gross, 2005), thereby explaining many of the effects ionic strength and mutations have on the reaction kinetics. Recently, it has been shown that different Pc conformations lead to different BD simulation results (Gross, 2007). Thus, it is important to verify whether the productive binding involves any kind of structural perturbation in any of the partners or their redox centres. Again, the answer depends on the system. For instance, the structure of C*c* is altered when it binds to cytochrome *c* oxidase (C*c*O) (Döpner et al., 1999; Hildebrandt et al., 1993; Sampson and Alleyne, 2001). In this case, we may describe the event as a structural rearrangement, rather than as the diffusion-limited reorientation found in photosynthetic complexes.

In addition, subtle changes occur in other complexes. For instance, in Pc, the copper centre changes geometry during its interaction with either C*f* or PSI (Díaz-Moreno et al., 2006a,c). In the interaction with the former partner, a significant geometry change occurs in the coordination sphere. In addition, shifts in charge distribution within the copper site are observed, depending on the target to which the copper protein binds. In contrast to the effects on the copper protein, the heme moiety of C*f* remains unperturbed, despite one of the axial ligands of the iron atom being located at the core of the interaction surface (Díaz-Moreno et al., 2006c). Moreover, MD showed the regions of the partners that are rich in charged residues to be rather flexible and mobile, despite the general behaviour of the two partners as rigid bodies (Díaz-Moreno et al., 2009). Notably, a conformational change is observed in the C-loop of Pc throughout the computations, explaining both the structural changes in the copper site and the discrepancy between the rigid-body, NMR-based structure and previous site-directed mutagenesis analyses.

Unlike C*f*, which is membrane-anchored, the soluble cytochromes acting as electron carriers undergo changes that affect the electronic configuration of the heme group and/or the protein matrix. In photosynthesis, Cc_6 undergoes small changes in heme moiety (Díaz-Moreno et al., 2006b) and, most likely, in the interaction between helices I and IV (Díaz-Moreno et al., 2005c) when binding to its partners. These changes are consistent with previous observations of a decrease in the difference in redox potential of the electron donor and acceptor within the complex. For instance, the redox potential of cytochrome c_2 (Cc_2) shows an increase of 20 mV upon binding to the photosynthetic reaction centre (Drepper et al., 1997). The structures of *Rhodobacter* Cc_2 that are either free (PDB codes *1cxc* and *2cxb*) or bound to the photosynthetic reaction centre (PDB codes *1l9b* and *1l9j*) (Axelrod and Okamura, 2005) show barely any difference, save an approximately

9° rotation in the axial imidazole ligand. Notably, a perturbation of an NMR signal of the equivalent imidazole was indeed detected for the interaction between Cc_6 and PSI (Díaz-Moreno et al., 2006b).

As regards the complexes in the mitochondrial electron transport chain, *Cc* undergoes a decrease of 40 mV upon binding to *Cc*O. The structural changes that may lead to these changes are presented in Section 2.1, as they also affect complex dissociation. Additionally, MD analyses conducted on the interaction between *Cc* and Complex III in the mitochondrial membrane show a significant structural change in the two partners, although the small soluble carrier is principally effected (Kokhan et al., 2010). The MD results display new salt bridges that were not observable in the X-ray diffraction structure. Notably, this complex shows a K_D value of 0.35 μM (Moreno-Beltrán et al., 2014), which is 10 to 100 times smaller than the values observed for photosynthetic complexes.

2.4 DISSOCIATION AND SITE OBSTRUCTION

Dissociation is another event of great import in redox reactions present in biological processes. From double-flash excitation and site-directed mutagenesis, the dissociation of oxidized Pc was found to limit the demand for electron donor turnover in photosynthesis (Drepper et al., 1996; Kuhlgert et al., 2012). This is because the oxidized donor molecule dissociates slowly enough to transiently obstruct the diffusion of a new reduced donor towards the active site. Notably, this may explain the multiple exponentials observed in the kinetic traces where no diffusional rearrangement has been introduced into the kinetic models (Olesen et al., 1999). Here, both hypotheses, site obstruction by product and complex rearrangement, might be taken into account in a holistic model of the biologically relevant reaction, as they are in no way mutually exclusive events. In the particular case of Cc_6, for instance, in order to ensure efficient turnover by avoiding competition between the two redox forms of Cc_6 for the functional site of the partner, differences in partner affinity for these two forms must exist. This may be due to some structural change in the partners induced by the reaction or simply to some change in electrostatics. In fact, dissociation may be encouraged by electric charge transfer during the reaction (the donor gains a positive charge upon ET), as the electrostatic forces between the partners are affected. For instance, it has been shown that two conserved acidic residues at helices near the donor binding site of PSI mediate a mismatch in the complex that follows the reduction of P700 (Sun et al., 1999). In fact, the mutation of these two residues increases the lifetime of the complex, but has harmful effects *in vivo* (Kuhlgert et al., 2012). This is consistent with the *gliding* dissociation pathway hypothesis (De March et al., 2014), which suggests that the *Cc* glides along the surface of Cc_1 in their functional binding instead of a direct take-off from the Cc_1 surface. In the case of the Pc-*Cf* complex in *Phormidium*, continuum electrostatics calculations at physiological ionic strength values have shown the complex formed to be electrostatically strained, thereby promoting dissociation following the transfer of the charge (Díaz-Moreno et al., 2009).

Besides the electrostatic changes that are concomitant to ET, dissociation is favoured when the structure of the partner is perturbed along the complex formation. For example, intramolecular conformation changes in *Cc* when binding to *Cc*O (Döpner et al., 1999;

Hildebrandt et al., 1993; Sampson and Alleyne, 2001) promote intra-complex ET (Döpner et al., 1999). Again, as the resulting conformation is less stable in its reduced form than in its oxidized state, dissociation is promoted following ET, as suggested by theoretical analyses (Ashe et al., 2012). Similarly, MD analyses of the *Cc*-*Cc*$_1$ complex shows that the structure of the soluble heme protein changes upon binding (Kokhan et al., 2010).

2.5 NATURE OF INTERACTIONS AND COMPLEX INTERFACE

The two principal interactions leading to the formation of the productive ET complex are mainly electrostatic and hydrophobic. Van der Waals forces at the interaction surface are satisfied by water molecules upon the dissociation of the complex, so their net contribution to complex equilibrium may be small. They nevertheless contribute to the enthalpy-entropy compensation, with the low surface complementarity between the partners being in part responsible for their transient nature. The available kinetic data suggest that the weightiness of the distinct interactions changes from prokaryotes to eukaryotes. If orthologous proteins are considered, their diffusion properties may not differ substantially, due to their nearly identical sizes and shapes. However, the binding and ET rates of orthologous proteins with their distinct physiological partners usually differ from one partner to another. The diffusion limit depends on the frictional forces between the reaction partners and the rate at which the water molecules are released from the interaction surface. Hence, the rate constant at zero ionic strength, k_0, cannot be larger than the diffusion limit for the rate constant (Eigen, 1954). At infinite ionic strength, when electric charges are fully screened, the reaction rates are diffusion-limited. However, at lower ionic strength values, when electric charges are able to interact, the rate constant for any given reaction depends exponentially on a factor that is proportional to the ratio between electrostatic potential and thermal energy, k_BT, where k_B is the Boltzmann constant and T temperature (Alberty and Hammes, 1958; Watkins et al., 1994).

$$k(I) = k_{\text{Diff}} + e^{V_{\text{TOT}}(I)/k_B T} \tag{2.12}$$

In Equation 2.12, and as indicated by Tollin and collaborators (Watkins et al., 1994), k_{Diff} is the rate constant of a diffusion-limited reaction at infinite ionic strength and $k(I)$ and $V_{\text{TOT}}(I)$ are the rate constant and total electrostatic potential energy, respectively, at ionic strength I. Here it is important to note that, as k_{Diff} is the sum of two terms that are much smaller and proportional to thermal energy, a purely electrostatic interaction may result in kinetic rates at high ionic strength that are practically independent of temperature. The presence of other phenomena—such as solvation effects—which are temperature dependent may change this tendency.

Protein-protein interfaces are not accessible to solvent in the majority of complexes (Lo Conte et al., 1999). Indeed, the stability of the complex is related to the surface area that becomes buried in the interface (Nooren and Thornton, 2003a,b). Nevertheless, the strength of the hydrophobic interactions is not enough to discriminate between long-lived and transient complexes (Mintseris and Weng, 2003). As regards transient redox complexes, water exclusion may be critical even in cases where the complex interface is relatively small (roughly between 5 and 20 nm^2). In fact, the ET requires the exclusion of mobile water molecules from the donor–acceptor interface as it minimizes the reorganization energy according to Marcus's theory (Marcus and Sutin, 1985). This is why the so-called rearrangement step in the overall reaction mechanism may involve hydrophobic interactions between the two partners (Hervás et al., 1996). In this context, it is not surprising that mutations affecting hydrophobic residues at the interface strongly impair electron donation to PSI (Díaz-Quintana et al., 2002; Sommer et al., 2002, 2004). On the other hand, tightly packed, frozen water molecules can participate in the ET pathway (van Amsterdam et al., 2002).

During the 1980s and 1990s, extensive studies were conducted and results obtained regarding ionic strength dependency by distinct groups. A key for their interpretation was Tollin and co-workers' *parallel plate* model of electrostatic interactions (Watkins et al., 1994), a simplification of the full formalism developed by the same authors to interpret the data and in the absence of structural data for the interface. According to this model, the productive complex occupies an area that is isolated from the solvent, and the dominant contribution to the electrostatic interaction involves mostly the charged residues at the edge of the interface. Hydrophobic effects on the binding energy are not explicitly considered in the model, but included in the fitted rate constant value at infinite ionic strength.

Figure 2.4 represents the interaction surfaces for *Cf*-Pc complexes in *Nostoc* and *Phormidium*, with each residue at the interaction surfaces coloured according to the hydropathy index (Kyte and Doolittle, 1982). Interfaces consist of a neutral core surrounded by several ionizable functional groups or residues at the edge of the interface. This neutral core may be comprised of hydrophobic and neutral polar residues at different ratios, thereby modulating binding affinity towards the corresponding reaction partner. The relevance of the charged residues at the edge of the interface is revealed by several facts. First, many site-directed mutagenesis studies show poor correlations between net charge changes and the effects of mutations on reaction kinetics (Hart et al., 2003). Nevertheless, BD analyses of the interaction between *Cf* and Pc mutants show a clear correlation between the number of collisions between the partners and the net charge of the mutants (Gross and Rosenberg, 2006). This is not contradictory with Watkins's statements if one considers that the wild-type Pc under study has the same net charge (−2) as *Cf*. Thus, Pc might spend a substantial part of its path far from its partner due to the electrostatic repulsion between them. According to Gauss's law, one protein would sense the net charge of the other along this time interval. Second, mutations of residues at the interface edge impair the exclusion of water molecules, thereby making the release of water a rate-limiting step. This is reflected by the fact that the mutations of charged residues at the interface edge show a substantial enthalpy-entropy compensation in the activation energy for the observed reaction rate, while mutations of remote charged residues show enthalpy-entropy compensations that are insignificant according to Sharp's criteria (Sharp, 2001). A good example of how mutations act using electrostatic potential is provided in De la Rosa and co-workers, showing that a charge reversal may have no effect if it is screened by surrounding residues (De la Cerda et al., 1999) or, on the contrary, may have a large impact on the ET reaction rate

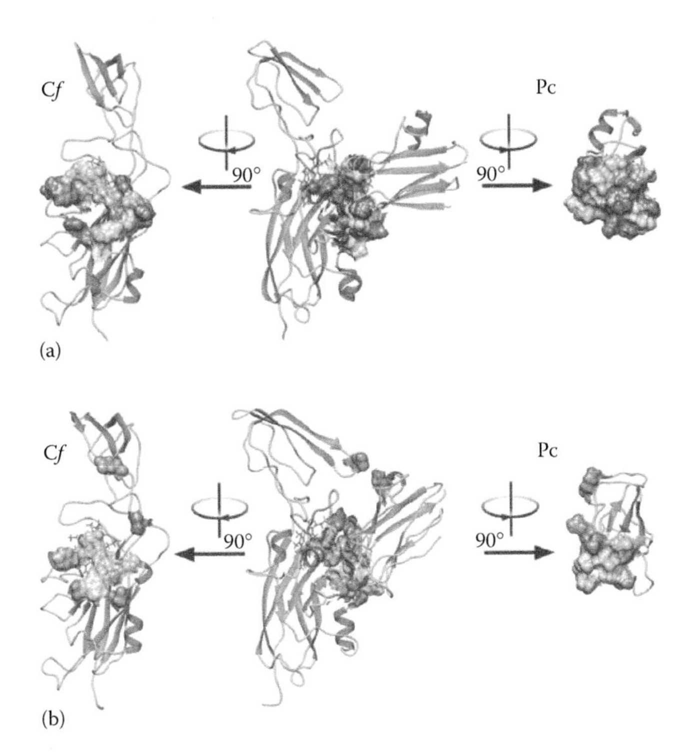

FIGURE 2.4 **(See colour insert.)** Interfaces of complexes between plastocyanin (Pc) and cytochrome *f* (*Cf*) from the cyanobacteria (a) *Phormidium* and (b) *Nostoc*. Surface residues of each partner closer than 4 Å to the surface of the other partner were chosen for the representations. The surfaces are coloured according to the Doodlittle scale of hydrophobicity from −4.5 (green) to +4.5 (brown). The Pc-*Cf* complex is in the middle, with free Pc at the right and free *Cf* at the left following a 90° rotation to better show the residues at the interface.

if affecting electrostatic potential at the edge of the interaction surface (De la Cerda et al., 1999; Díaz-Quintana et al., 2003; Molina-Heredia et al., 1999).

As shown in Figure 2.5 and by the intensity of the symbol colours therein, for electron donors and PSI, the mutations modifying charges at a site distant from the interaction surface of the former cause small activation energy changes for the redox reaction. In addition, enthalpy-entropy compensation can be calculated with a slope at around the average temperature of the experiments, most of which being irrelevant according to Sharp's criteria. As mutations, when taken independently, affect the electrostatic potential near the interaction surface, they deviate from this line. If the effect is mainly due to electrostatics, enthalpy is the major contribution for the changes in activation energy. Some of these mutations may impair the hydrophobic interactions or affect the dynamics within the complex, deviating as well from the first set of mutations with mainly entropic changes in the activation energy. This is the case of the R88E mutant in *Synechocystis* Pc and the V25E substitution in *Nostoc* Cc_6—whose effects are mainly entropic despite the fact that the mutations are charge modifications (Figure 2.5)—due to the location of the mutated residues within the binding interface.

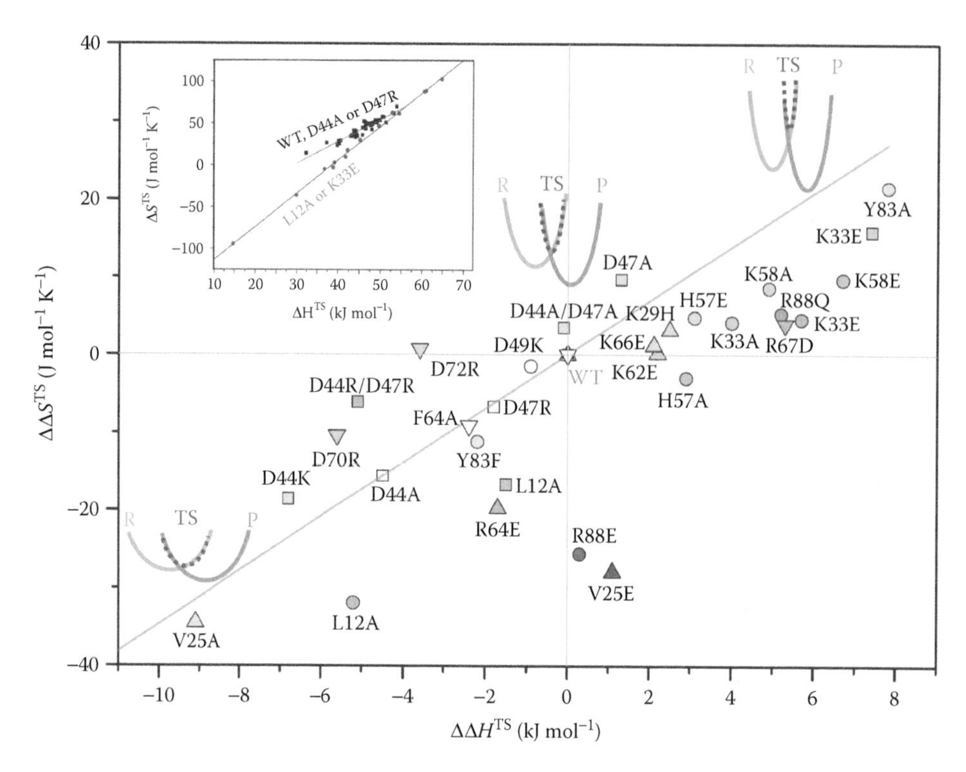

FIGURE 2.5 Perturbation of activation parameters for electron donation to PSI by site-directed mutants of plastocyanin (Pc) and cytochrome *c*6 (C*c*6) from the cyanobacteria *Synechocystis* and *Nostoc*. The mutants of Pc and Cc_6 from *Synechocystis* are represented by squares and upside-down triangles, respectively, whereas the corresponding copper and heme proteins from *Nostoc* are represented by circles and right-side-up triangles, respectively. All symbols are coloured in grey scale according to the absolute value of the perturbation each mutant causes in the apparent activation free energy for the second exponential kinetic phase for ET to PSI (Data from De la Cerda, B. et al., *Biochemistry*, 36, 10125–10130, 1997; Data from De la Cerda, B. et al., *J. Biol. Chem.*, 274, 13292–13297, 1999; Data from Molina-Heredia, F.P. et al., *J. Biol. Chem.*, 274, 33565–33570, 1999; Data from Molina-Heredia, F.P. et al., *J. Biol. Chem.*, 276, 601–605, 2001; Data from Díaz-Quintana, A. et al., *Photosynth. Res.*, 72, 223–230, 2002.), with colour scale ranging from 0 (white) to ±10 kJ mol^{-1} (dark grey). Three reaction coordinate diagrams are included along an isoergonic line to illustrate how enthalpy-entropy compensation relates to the binding landscape. Ordinate represents energy and abscissa corresponds to a projection of molecular coordinates. For simplicity, a harmonic relationship is supposed. Reactant (R) and product (P) coordinates are represented by grey continuous lines. Dark grey discontinuous lines stand for the transition state (TS) well. *Insert:* Enthalpy-entropy compensation observed in two sets of mutants of *Synechocystis* Pc by changing pH or ionic strength conditions. Black dots represent data from wild-type and charge-replacement mutants at residues distant from the interaction surface. (Data from De la Cerda, B. et al., *Biochemistry*, 36, 10125–10130, 1997.) Grey dots represent data from the mutants L12A and K33E. (Data from Díaz-Quintana, A. et al., *Photosynth. Res.*, 72, 223–230, 2002.), lying at either the core or the edge of the interaction surface.

Upon consideration of the complete set of mutations, the same slope is shown, despite the mutations being of a different sort and affecting different kinds of interactions. Consequently, they do not meet Sharp's criteria for consideration as *extra-thermodynamic* or *significant* compensation. In fact, the requirement that changes in free energy be at least three times the thermal energy is not fulfilled (Sharp, 2001). This is so due to the fact that all experimental data regarding these mutations were recorded in the presence of divalent cations (De la Cerda et al., 1999; Molina-Heredia et al., 1999), which specifically dampen the effect of electrostatic mutations near the interface by acting as ion bridges between negative charges. When these cations are absent or titrated into the sample, the changes in thermodynamic parameters become relevant in the case of mutations at the interface and its edge. For instance, changes in salt concentration and pH produce significant changes in the activation free energy (ca. $3k_BT$) and large enthalpy-entropy correlations with slopes deviating from that corresponding to compensation temperature, as opposed to mutations far from the interface edge (Díaz-Quintana et al., 2002), as can be seen in the Figure 2.5 insert.

According to literature (Piguet, 2011), enthalpy-entropy compensation for binding in solution is achieved under two conditions. First, the desolvation entropy (DDS_{desolv}) may remain constant across the experimental series. However, this does not explain how mutations affecting the buried hydrophobic surface are aligned along isoperturbation lines with amino acid substitutions affecting the charges at the edge of the interaction surface.

The second case involves the interplay between desolvation and other phenomena, such as dynamic disorder within the complex due to the breaking of an intermolecular salt bridge or simply affecting the short-range electrostatics. This seems to be the case for the data in the Figure 2.5 insert. When changing buffer conditions, the K33E substitution at the edge of the interface shows the same slope as the L12A mutation at the core of the interface, as well as a wider interval of enthalpies and entropies than the wild-type form and mutations at remote site 2.

How does the relationship between enthalpy and entropy linked to the energetic landscape? Figure 2.5 gives an idea of how this landscape may look at three different positions on an isoergonic line. Assuming a transition between two minima represented by harmonic functions and all wells as having the same volume, a downwards movement along the isoergonic line (lower left-hand corner of Figure 2.5) indicates Hooke's force constants acting in the wells are lower, so the wells broaden. The activation energy (saddle point) is low, so less thermal energy is required to place reactants at the transition state. However, as frequency is inversely proportional to the square root of the force constant, the probability of finding the reactants at the saddle point is much lower. In other words, when the reactants already have the thermal energy equivalent to activation enthalpy, the probability of reaching the transition state is smaller, as reactants and products sample a greater number of coordinates compared to those available at the saddle point. Furthermore, as the wells are shallow, the reactants can abandon this state more easily for an nonreactive state. In the upper right-hand corner of Figure 2.5, wells are deep and the energy barrier is high compared to the minima, so reactants need enough thermal energy to arrive to the transition state. However, the stronger force constant causes the reactants to arrive to the transition state with a higher probability when thermal energy and activation enthalpy are equal. In summary, a larger degree of freedom in the reactants (and products) leads to a lower probability of finding the saddle point and, therefore, more negative entropy of activation. On the contrary, placing reactants in a deep well makes finding the saddle point easier, but greater thermal energy is required.

One of the most controversial points regarding electrostatic interactions driving the binding between the partners is related to the role of the dipole interactions (Van Leeuwen, 1983). In fact, many redox proteins demonstrate a clearly asymmetric charge distribution, leading to a net dipole moment. The example in Figure 2.5 shows Pc molecules to be dipoles, a property that allows them to slide along the negative surface of *Cf*. Whether the copper protein slides or sticks to its partner's surface depends on the balance between the attractive and repulsive forces at either end of the copper protein.

As indicated earlier, the charges surrounding the interaction surface are essential for the complex to be productive, while charges at more remote distances are less important, especially at physiological ionic strength. This requires the protein dipole moment to either promote the interaction or, simply, not be strongly opposed (Watkins et al., 1994). In fact, many complexes present substantial angles between the electrostatic dipole moments of the partners, though these moments do not oppose each other. Again, the interpretation of results strongly depends on the experimental conditions and the phenomenon being observed. A recent report indicates that, at a millisecond timescale, the interaction between adrenodoxin and its partners is not affected by the magnitude of the dipole moment of adrenodoxin (Hannemann et al., 2009). Notably, there was no change registered to the dissociation constant. Most likely, the electrostatic potential at the interface did not change in accordance to Watkins's statements. Unfortunately, none of the mutants in this work affected the direction of the dipole moment and kinetic analysis was performed at low time resolution. Nevertheless, the relevance of the work from the physiological point of view makes it an interesting system to be analysed by other techniques and at higher time resolution in order to obtain further information on the binding mechanism.

As diffusional behaviour depends on the properties of the two molecules, even two orthologs may act in a different manner when reacting with their partners. BD simulations have been of particular relevance to understand the interaction mechanics of redox metalloproteins and are a useful and reliable tool to predict reaction rates. For instance, computations show that the bulk diffusion of Pc is electrostatically steered towards *Cf* and provides the right overall orientation to establish hydrophobic interactions (De Rienzo et al., 2001; Gross, 2004; Haddadian and Gross, 2005). Again, homologous proteins from different organisms behave in a different manner. For instance, in Pc-*Cf* complexes, the orientation of the two partners and their dynamics (Gross and Rossemberg, 2006) are variable, in agreement with NMR data derived from paramagnetic CSPs (Crowley et al., 2001; Díaz-Moreno et al., 2005a,b; Lange et al., 2005; Ubbink et al., 1998).

In the case of photosynthetic complexes involving two soluble electron carriers, Pc and Cc_6, a thorough analysis on the topic has been performed. In most prokaryotes, kinetic rates change monotonically when ionic strength is increased (Hervás et al., 1996). Eukaryotic complexes, in turn, are characterized by their ET reaction rates' bell-shaped dependencies on ionic strength (De la Rosa et al., 2002). Such a bell-shaped profile indicates that the encounter requires a conformational change to optimize the probability of ET. Indeed, biphasic ionic strength dependences—in which the overall reaction rates show a maximum at low-moderate ionic strength values—indicate that electrostatic interactions must be at least partially damped to allow the complexes to accommodate for the ET step. Biphasic ionic strength dependencies were observed in early analyses of the reactions between *Cc* and *Cc*P (Hazzard et al., 1991), ferredoxin and ferredoxin-NADP reductase (Walker et al., 1991), *Cf* and Pc (Meyer et al., 1993b), PSI and flavodoxin (Medina et al., 1992) and Cb_5 and *Cc* (Meyer et al., 1993a).

A striking issue is the effect of divalent cations on the reaction kinetics. The addition of magnesium chloride, for instance, has a large effect on the kinetics of PSI reduction by its donors (Díaz et al., 1994a; Hervás et al., 1995). When time resolution is not sufficient to resolve fast kinetic phases, the effect of divalent ions at moderate concentrations is easily fitted to the double layer theory (Takabe et al., 1983); however, the fits are poorer at low salt concentrations or when high time-resolution kinetic approaches are used. A purported explanation is that divalent ions are able to act as bridges between charges of the same sign belonging to the two partners. A good example is the K33E mutation in *Synechocystis* Pc. This mutation locates at the edge of the interaction surface of the copper protein and hinders the latter's ability to donate electrons to PSI, probably by disrupting a salt bridge at the edge of the interface. The free activation energy increases with a large enthalpy-entropy compensation that indicates that the mutation impairs the partners to set up the hydrophobic interactions required for ET. All these effects are fully reversed by $MgCl_2$ at very low concentrations, but not when monovalent salts are added (Díaz-Quintana et al., 2002). Extensive analysis of NMR data on ET complexes has led to the conclusion that the long-range nature of electrostatics makes it lead reaction partners to an ensemble of conformations with almost equal energies, thereby resulting in nonspecific complexes (Prudêncio and Ubbink, 2004). Figure 2.3 represents the electrostatic potential around Pc and *Cf*. Charges in *Cf* cause the electrostatic potential to be equally distributed around the entire luminal surface of the heme protein, according to the isopotential surfaces represented. This agrees with PRE-NMR data (Díaz-Moreno et al., 2014; Scanu et al., 2013a,b) at low ionic strength, which shows a delocalization of the soluble electron carrier across the surface of *Cf*. It may be speculated that at higher ionic strengths, the populations of some conformations of the ensemble will decrease beyond what can be detected, or simply may cease to exist.

The perspective from NMR of electrostatics playing an unspecified role in the binding event appears to conflict with kinetics analyses and site-directed mutagenesis studies in which specific charged residues are shown to play an essential role for a specific interaction. For instance, kinetic analyses of ET from *Cf* to site-directed mutants of *Phormidium* Pc showed Arg93 to be an essential residue for the reaction (Hart et al., 2003). Surprisingly, however, this residue did not present any significant CSP during NMR analyses (Crowley et al., 2001). Kinetic analysis was further supported by similar analyses for ET to PSI (Díaz-Quintana et al., 2003). Here, MD solved the apparent contradiction; namely, the long side-chain of the highly conserved arginine can change its conformation to form a salt bridge with Glu165 at the *Cf* surface (Cruz-Gallardo et al., 2012; Díaz-Moreno et al., 2009).

It is important to note here that most NMR analyses are performed at low ionic strength. Under such conditions, the nonspecific long-range interactions are favoured. However, where ionic strength increases, the electrostatic interactions are more localized, as shown in Figure 2.3. Indeed, as the isopotential surfaces become more irregular and the electrostatic interactions become short-range, the electrostatic field displacement arising from the charged regions near active sites may constrain the conformation space. On the other hand, kinetic analyses and site-directed mutagenesis are particularly sensitive to a given state of the reaction, such as the productive complex, and provide no atomic resolution. Hence, a full understanding of the reaction mechanism demands the consideration of the diverse data available, the timescales they involve and their strengths and limitations, followed by their integration under a single, unifying model.

ACKNOWLEDGEMENTS

We would like to thank all former and present members of the Biointeractomics group at the Institute for Plant Biochemistry and Photosynthesis (cicCartuja, University of Seville—CSIC). Financial support was provided by the Spanish Ministry of Economy and Competitiveness for several years (Grant No. BFU2003-00458/BMC, BFU2006-01361/BMC, BFU2009-07190/BMC and BFU2012-31670/BMC) and by the Andalusian Government (Grant PAI, BIO198, P07-CVI-02896 and P11-CVI-7216).

REFERENCES

Alberty, R. A., and G. G. Hammes. Application of the theory of diffusion-controlled reactions to enzyme kinetics. *The Journal of Physical Chemistry* 62 (1958): 154–59.

Ashe, D., T. Alleyne, and V. Sampson. Substrate binding-dissociation and intermolecular electron transfer in cytochrome *c* oxidase are driven by energy-dependent conformational changes in the enzyme and substrate. *Biotechnology and Applied Biochemistry* 59 (2012): 213–22.

Axelrod, H. L., and M. Y. Okamura. The structure and function of the cytochrome c_2: Reaction center electron transfer complex from *Rhodobacter sphaeroides*. *Photosynthesis Research* 85 (2005): 101–14.

Bashir, Q., S. Scanu, and M. Ubbink. Dynamics in electron transfer protein complexes. *FEBS Journal* 278 (2011): 1391–400.

Berg, O. G., and P. H. von Hippel. Diffusion-controlled macromolecular interactions. *Annual Review of Biophysics and Biophysical Chemistry* 14 (1985): 131–58.

Berg, O. G., R. B. Winter, and P. H. Von Hippel. Diffusion-driven mechanisms of protein translocation on nucleic acids. 1. Models and theory. *Biochemistry* 20 (1981): 6929–48.

Bloomfield, V. A., and S. Prager. Diffusion-controlled reactions on spherical surfaces. Application to bacteriophage tail fiber attachment. *Biophysical Journal* 27 (1979): 447–53.

Bottin, H., and P. Mathis. Interaction of plastocyanin with the photosystem I reaction centre: A kinetic study by flash absorption spectroscopy. *Biochemistry* 24 (1985): 6453–60.

Brown, G. C., H. V. Westerhoff, and B. N. Kholodenko. Molecular control analysis: Control within proteins and molecular processes. *Journal of Theoretical Biology* 182 (1996): 389–96.

Casaus, J. L., J. A. Navarro, M. Hervás, A. Lostao, M. A. De la Rosa, C. Gómez-Moreno, J. Sancho, and M. Medina. *Anabaena sp.* PCC 7119 flavodoxin as electron carrier from photosystem I to Ferredoxin-NADP(+) reductase—Role of Trp57 and Tyr94. *Journal of Biological Chemistry* 277 (2002): 22338–44.

Crowley, P. B., A. Díaz-Quintana, F. P. Molina-Heredia, P. M. Nieto, M. Sutter, W. Haehnel, M. A. De la Rosa, and M. Ubbink. The interactions of cyanobacterial cytochrome c_6 and cytochrome *f*, characterized by NMR. *Journal of Biological Chemistry* 277 (2002): 48685–9.

Crowley, P. B., G. Otting, B. G. Schlarb-Ridley, G. W. Canters, and M. Ubbink. Hydrophobic interactions in a cyanobacterial plastocyanin-cytochrome *f* complex. *Journal of the American Chemical Society* 123 (2001): 10444–53.

Cruz-Gallardo, I., I. Díaz-Moreno, A. Díaz-Quintana, and M. A. De la Rosa. The cytochrome *f*-plastocyanin complex as a model to study transient interactions between redox proteins. *FEBS Lett* 586 (2012): 646–52.

De la Cerda, B., A. Díaz-Quintana, J. A. Navarro, M. Hervás, and M. A. De la Rosa. Site-directed mutagenesis of cytochrome c_6 from *Synechocystis sp.* PCC 6803: The heme protein possesses a negatively charged area that may be isofunctional with the acidic patch of plastocyanin. *The Journal of Biological Chemistry* 274 (1999): 13292–97.

De la Cerda, B., J. A. Navarro, M. Hervás, and M. A. De la Rosa. Changes in the reaction mechanism of electron transfer from plastocyanin to photosystem I in the cyanobacterium *Synechocystis sp.* PCC 6803 as induced by site-directed mutagenesis of the copper protein. *Biochemistry* 36 (1997): 10125–30.

De la Rosa, M. A., J. A. Navarro, A. Díaz-Quintana et al. An evolutionary analysis of the reaction mechanisms of photosystem I reduction by cytochrome c_6 and plastocyanin. *Bioelectrochemistry and Bioenergetics* 55 (2002): 41–45.

De la Rosa, M. A., J. M. Vega, and W. G. Zumft. Composition and structure of assimilatory nitrate reductase from *Ankistrodesmus braunii. The Journal of Biological Chemistry* 256 (1981): 5814–19.

De March, M., N. Demitri, R. De Zorzi et al. Nitrate as a probe of cytochrome *c* surface: Crystallographic identification of crucial "hot spots" for protein-protein recognition. *Journal of Inorganic Biochemistry* 135 (2014): 58–67.

De Rienzo, F., R. R. Gabdoulline, M. C. Menziani, P. G. De Benedetti, and R. C. Wade. Electrostatic analysis and Brownian dynamics simulation of the association of plastocyanin and cytochrome *f*. *Biophysical Journal* 81 (2001): 3090–3104.

De Rienzo, F., R. R. Gabdoulline, M. C. Menziani, and R. C. Wade. Blue copper proteins: A comparative analysis of their molecular interaction properties. *Protein Science* 9 (2000): 1439–54.

De Rienzo, F., G. H. Grant, and M. C. Menziani. Theoretical descriptors for the quantitative rationalisation of plastocyanin mutant functional properties. *Journal of Computer-Aided Molecular Design* 16 (2002): 501–9.

Díaz, A., M. Hervás, J. A. Navarro, M. A. De La Rosa, and G. Tollin. A thermodynamic study by laser-flash photolysis of plastocyanin and cytochrome c_6 oxidation by photosystem I from the green alga *Monoraphidium braunii. European Journal of Biochemistry* 222 (1994a): 1001–7.

Díaz, A., F. Navarro, M. Hervás, J. A. Navarro, S. Chavez, F. J. Florencio, and M. A. De la Rosa. Cloning and correct expression in *Escherichia coli* of the petJ gene encoding cytochrome c_6 from *Synechocystis* 6803. *FEBS Letters* 347 (1994b): 173–7.

Díaz-Moreno, I., and M. A. De la Rosa. Transient interactions between biomolecules. *European Biophysics Journal* 40 (2011): 1273–74.

Díaz-Moreno, I., S. Díaz-Moreno, G. Subias, M. A. De la Rosa, and A. Díaz-Quintana. The atypical iron-coordination geometry of cytochrome *f* remains unchanged upon binding to plastocyanin, as inferred by XAS. *Photosynthesis Research* 90 (2006c): 23–28.

Díaz-Moreno, I., A. Díaz-Quintana, M. A. De la Rosa, P. B. Crowley, and M. Ubbink. Different modes of interaction in cyanobacterial complexes of plastocyanin and cytochrome *f*. *Biochemistry* 44 (2005a): 3176–83.

Díaz-Moreno, I., A. Díaz-Quintana, M. A. De la Rosa, and M. Ubbink. Structure of the complex between plastocyanin and cytochrome *f* from The cyanobacterium *Nostoc sp.* PCC 7119 as determined by paramagnetic NMR: The balance between electrostatic and hydrophobic interactions within the transient complex determines the relative orientation of the two proteins. *The Journal of Biological Chemistry* 280 (2005b): 18908–15.

Díaz-Moreno, I., A. Díaz-Quintana, S. Díaz-Moreno, G. Subias, and M. A. De la Rosa. Transient binding of plastocyanin to its physiological redox partners modifies the copper site geometry. *FEBS Letters* 580 (2006a): 6187–94.

Díaz-Moreno, I., A. Díaz-Quintana, F. P. Molina-Heredia et al. NMR analysis of the transient complex between membrane photosystem I and soluble cytochrome c_6. *The Journal of Biological Chemistry* 280 (2005c): 7925–31.

Díaz-Moreno, I., A. Díaz-Quintana, G. Subías, T. Mairs, M. A. De la Rosa, and S. Díaz-Moreno. Detecting transient protein-protein interactions by X-ray absorption spectroscopy: The cytochrome c_6-photosystem I complex. *FEBS Letters* 580 (2005d): 3023–8.

Díaz-Moreno, I., A. Díaz-Quintana, G. Subías, T. Mairs, M. A. De la Rosa, and S. Díaz-Moreno. Detecting transient protein-protein interactions by X-ray absorption spectroscopy: The cytochrome c_6-photosystem I complex. *FEBS Letters* 580 (2006b): 3023–28.

Díaz-Moreno, I., J. M. García-Heredia, A. Díaz-Quintana, and M. A. De la Rosa. Cytochrome *c* signalosome in mitochondria. *European Biophysics Journal* 40 (2011a): 1301–15.

Díaz-Moreno, I., J. M. García-Heredia, A. Diaz-Quintana, M. Teixeira, and M. A. De la Rosa. Nitration of tyrosines 46 and 48 induces the specific degradation of cytochrome *c* upon change of the heme iron state to high-spin. *Biochimica et Biophysica Acta— Bioenergetics* 1807 (2011b): 1616–23.

Díaz-Moreno, I., R. Hulsker, P. Skubak et al. The dynamic complex of cytochrome c_6 and cytochrome *f* studied with paramagnetic NMR spectroscopy. *Biochimica et Biophysica Acta—Bioenergetics* 1837 (2014): 1305–15.

Díaz-Moreno, I., F. J. Muñoz-López, E. Frutos-Beltrán, M. A. De la Rosa, and A. Díaz-Quintana. Electrostatic strain and concerted motions in the transient complex between plastocyanin and cytochrome *f* from the cyanobacterium *Phormidium laminosum*. *Bioelectrochemistry* 77 (2009): 43–52.

Díaz-Quintana, A., B. De la Cerda, M. Hervás, J. A. Navarro, and M. A. De la Rosa. Mutations in both leucine 12 and lysine 33 in plastocyanin from *Synechocystis sp.* PCC 6803 induce drastic changes in the hydrophobic interactions with photosystem I. *Photosynthesis Research* 72 (2002): 223–30.

Díaz-Quintana, A., J. A. Navarro, M. Hervás, F. P. Molina-Heredia, B. De la Cerda, and M. A. De la Rosa. A comparative structural and functional analysis of cyanobacterial plastocyanin and cytochrome c_6 as alternative electron donors to photosystem I. *Photosynthesis Research* 75 (2003): 97–110.

Dixon, D. W., X. Hong, and S. E. Woehler. Electrostatic and steric control of electron self-exchange in cytochromes *c*, c_{551}, and b_5. *Biophysical Journal* 56 (1989): 339–51.

Döpner, S., P. Hildebrandt, F. I. Rosell et al. The structural and functional role of lysine residues in the binding domain of cytochrome *c* in the electron transfer to cytochrome *c* oxidase. *European Journal of Biochemistry* 261 (1999): 379–91.

Drepper, F., P. Dorlet, and P. Mathis. Cross-linked electron transfer complex between cytochrome c_2 and the photosynthetic reaction centre of *Rhodobacter sphaeroides*. *Biochemistry* 36 (1997): 1418–27.

Drepper, F., M. Hippler, W. Nitschke, and W. Haehnel. Binding dynamics and electron transfer between plastocyanin and photosystem I. *Biochemistry* 35 (1996): 1282–95.

Eigen, M. Über die kinetik sehr schnell verlaufender lonenreaktionen in wässrige lösung. *Zeitschrift für Physikalische Chemie* 1 (1954): 176–200.

García-Heredia, J. M., I. Díaz-Moreno, A. Díaz-Quintana et al. Specific nitration of tyrosines 46 and 48 makes cytochrome *c* assemble a non-functional apoptosome. *FEBS Letters* 586 (2012): 154–8.

García-Heredia, J. M., I. Díaz-Moreno, P. M. Nieto et al. Nitration of tyrosine 74 prevents human cytochrome *c* to play a key role in apoptosis signaling by blocking caspase-9 activation. *Biochimica et Biophysica Acta—Bioenergetics* 1797 (2010): 981–3.

García-Heredia, J. M., A. Díaz-Quintana, M. Salzano et al. Tyrosine phosphorylation turns alkaline transition into a biologically relevant process and makes human cytochrome *c* behave as an anti-apoptotic switch. *Journal of Biological Inorganic Chemistry* 16 (2011): 1155–68.

García-Heredia, J. M., M. Hervás, M. A. De la Rosa, and J. A. Navarro. Acetylsalicylic acid induces programmed cell death in *Arabidopsis* cell cultures. *Planta* 228 (2008): 89–97.

Gross, E. L. A Brownian dynamics study of the interaction of *Phormidium laminosum* plastocyanin with *Phormidium laminosum* cytochrome *f*. *Biophysical Journal* 87 (2004): 2043–59.

Gross, E. L. A Brownian dynamics computational study of the interaction of spinach plastocyanin with turnip cytochrome *f*: The importance of plastocyanin conformational changes. *Photosynthesis Research* 94 (2007): 411–22.

Gross, E. L., and D. C. Pearson Jr. Brownian dynamics simulations of the interaction of *Chlamydomonas* cytochrome *f* with plastocyanin and cytochrome c_6. *Biophysical Journal* 85 (2003): 2055–68.

Gross, E. L., and I. Rosenberg. A Brownian Dynamics study of the Interaction of *Phormidium* cytochrome *f* with various cyanobacterial plastocyanins. *Biophysical Journal* 90 (2006): 366–80.

Haddadian, E. J., and E. L. Gross. Brownian dynamics study of cytochrome *f* interactions with cytochrome c_6 and plastocyanin in *Chlamydomonas reinhardtii* plastocyanin, and cytochrome c_6 mutants. *Biophysical Journal* 88 (2005): 2323–39.

Hannemann, F., A. Guyot, A. Zöllner, J. J. Müller, U. Heinemann, and R. Bernhardt. The dipole moment of the electron carrier adrenodoxin is not critical for redox partner interaction and electron transfer. *Journal of Inorganic Biochemistry* 103 (2009): 997–1004.

Hart, S. E., B. G. Schlarb-Ridley, C. Delon, D. S. Bendall, and C. J. Howe. Role of charges on cytochrome *f* from the cyanobacterium *Phormidium laminosum* in its interaction with plastocyanin. *Biochemistry* 42 (2003): 4829–36.

Hazzard, J. T., S. Y. Rong, and G. Tollin. Ionic strength dependence of the kinetics of electron transfer from bovine mitochondrial cytochrome *c* to bovine cytochrome *c* oxidase. *Biochemistry* 30 (1991): 213–22.

Hervás, M., M. A. De la Rosa, and G. Tollin. A comparative laser-flash absorption spectroscopy study of algal plastocyanin and cytochrome c_{552} photooxidation by photosystem I particles from spinach. *European Journal of Biochemistry* 203 (1992): 115–20.

Hervás, M., A. Díaz-Quintana, C. A. Kerfeld, D. W. Krogmann, M. A. De la Rosa, and J. A. Navarro. Cyanobacterial photosystem I lacks specificity in its interaction with cytochrome c_6 electron donors. *Photosynthesis Research* 83 (2005): 329–33.

Hervás, M., J. A. Navarro, A. Díaz, H. Bottin, and M. A. De la Rosa. Laser-flash kinetic analysis of the fast electron transfer from plastocyanin and cytochrome c_6 to photosystem I. Experimental evidence on the evolution of the reaction mechanism. *Biochemistry* 34 (1995): 11321–6.

Hervás, M., J. A. Navarro, A. Díaz, and M. A. De la Rosa. A comparative thermodynamic analysis by laser-flash absorption spectroscopy of photosystem I reduction by plastocyanin and cytochrome c_6 in *Anabaena* PCC 7119, *Synechocystis* PCC 6803, and spinach. *Biochemistry* 35 (1996): 2693–98.

Hervás, M., J. M. Ortega, J. A. Navarro, M. A. De la Rosa, and H. Bottin. Laser flash kinetic analysis of *Synechocystis* PCC 6803 cytochrome c_6 and plastocyanin oxidation by photosystem I. *Biochimica et Biophysica Acta—Bioenergetics* 1184 (1994): 235–41.

Hildebrandt, P., F. Vanhecke, G. Buse, T. Soulimane, and A. G. Mauk. Resonance Raman study of the interactions between cytochrome *c* variants and cytochrome *c* oxidase. *Biochemistry* 32 (1993): 10912–22.

Hom, K., Q.-F. Ma, G. Wolfe et al. NMR studies of the association of cytochrome b_5 with cytochrome *c*. *Biochemistry* 39 (2000): 14025–39.

Jackman, M. P., J. McGinnis, A. G. Sykes, C. A. Collyer, M. Murata, and H. C. Freeman. Kinetic studies on 1: 1 electron-transfer reactions involving blue copper proteins. Part 14. Reactions of poplar plastocyanin with inorganic complexes. *Journal of the Chemical Society, Dalton Transactions* 11 (1987): 2573–77.

Janin, J. Kinetics and thermodynamics of protein–protein interactions. In *Protein–Protein Recognition*, edited by C. Kleanthous, 1–32. Oxford: Oxford University Press, 2000.

Kastritis, P. L., and A. M. J. J. Bonvin. On the binding affinity of macromolecular interactions: Daring to ask why proteins interact. *Journal of the Royal Society Interface* 10 (2013): 20120835.

Kokhan, O., C. A. Wraight, and E. Tajkhorshid. The binding interface of cytochrome *c* and cytochrome c_1 in the bc_1 complex: Rationalizing the role of key residues. *Biophysical Journal* 99 (2010): 2647–56.

Kuhlgert, S., F. Drepper, C. Fufezan, F. Sommer, and M. Hippler. Residues PsaB Asp612 and PsaB Glu613 of photosystem I confer pH-dependent binding of plastocyanin and cytochrome c_6. *Biochemistry* 51 (2012): 7297–303.

Kyte, J., and R. F. Doolittle. A simple method for displaying the hydropathic character of a protein. *Journal of Molecular Biology* 157 (1982): 105–32.

Lange, C., T. Cornvik, I. Díaz-Moreno, and M. Ubbink. The transient complex of poplar plastocyanin with cytochrome *f*: Effects of ionic strength and pH. *Biochimica et Biophysica Acta—Bioenergetics* 1707 (2005): 179–88.

Lange, C., and C. Hunte. Crystal structure of the yeast cytochrome bc_1 complex with its bound substrate cytochrome *c*. *Proceedings of the National Academy of Sciences of the United States of America* 99 (2002): 2800–5.

Leys, D., J. Basran, F. Talfournier, M. J. Sutcliffe, and N. S. Scrutton. Extensive conformational sampling in a ternary electron transfer complex. *Nature Structural and Molecular Biology* 10 (2003): 219–25.

Lo Conte, L., C. Chothia, and J. Janin. The atomic structure of protein-protein recognition sites. *Journal of Molecular Biology* 285 (1999): 2177–98.

Ly, H. K., T. Utesch, I. Díaz-Moreno, J. M. García-Heredia, M. Á. De La Rosa, and P. Hildebrandt. Perturbation of the redox site structure of cytochrome *c* variants upon tyrosine nitration. *The Journal of Physical Chemistry B* 116 (2012): 5694–702.

Marcus, R. A., and N. Sutin. Electron transfers in chemistry and biology. *Biochimica et Biophysica Acta—Bioenergetics* 811 (1985): 265–322.

Martínez-Fábregas, J., I. Díaz-Moreno, K. González-Arzola et al. New *Arabidopsis thaliana* cytochrome *c* partners: a look into the elusive role of cytochrome *c* in programmed cell death in plants. *Molecular and Cellular Proteomics* 12 (2013): 3666–76.

Martínez-Fábregas, J., I. Díaz-Moreno, K. González-Arzola et al. Structural and functional analysis of novel human cytochrome *c* targets in apoptosis. *Molecular and Cellular Proteomics* 13 (2014a): 1439–56.

Martínez-Fábregas, J., I. Díaz-Moreno, K. González-Arzola, A. Díaz-Quintana, and M. A. De la Rosa. A common signalosome for programmed cell death in humans and plants. *Cell Death and Disease* 5 (2014b): e1314.

Mathis, P., and P. Setif. Near infra-red absorption spectra of the chlorophyll a cations and triplet state *in vitro* and *in vivo*. *Israel Journal of Chemistry* 21 (1981): 316–320.

Medina, M., A. Díaz, M. Hervás et al. A comparative laser-flash absorption-spectroscopy study of *Anabaena* PCC-7119 plastocyanin and cytochrome c_6 photooxidation by photosystem-I particles. *European Journal of Biochemistry* 213 (1993): 1133–38.

Medina, M., M. Hervás, J. A. Navarro, M. A. De la Rosa, C. Gómez-Moreno, and G. Tollin. A laser flash absorption spectroscopy study of *Anabaena sp.* PCC 7119 flavodoxin photoreduction by photosystem I particles from spinach. *FEBS Letters* 313 (1992): 239–42.

Meyer, T. E., C. T. Przysiecki, J. A. Watkins et al. Correlation between rate constant for reduction and redox potential as a basis for systematic investigation of reaction mechanisms of electron transfer proteins. *Proceedings of the National Academy of Sciences of the United States of America* 80 (1983): 6740–44.

Meyer, T. E., M. Rivera, F. A. Walker et al. Laser flash photolysis studies of electron transfer to the cytochrome b_5-cytochrome *c* complex. *Biochemistry* 32 (1993a): 622–27.

Meyer, T. E., Z.-G. Zhao, M. A. Cusanovich, and G. Tollin. Transient kinetics of electron transfer from a variety of *c*-type cytochromes to plastocyanin. *Biochemistry* 32 (1993b): 4552–9.

Mintseris, J., and Z. Weng. Atomic contact vectors in protein-protein recognition. *Proteins* 53 (2003): 629–39.

Molina-Heredia, F. P., A. Díaz-Quintana, M. Hervás, J. A. Navarro, and M. A. De la Rosa. Site-directed mutagenesis of cytochrome c_6 from *Anabaena* species PCC 7119: identification of surface residues of the hemeprotein involved in photosystem I reduction. *The Journal of Biological Chemistry* 274 (1999): 33565–70.

Molina-Heredia, F. P., M. Hervás, J. A. Navarro, and M. A. De la Rosa. Cloning and correct expression in Escherichia coli of the petE and petJ genes respectively encoding plastocyanin and cytochrome c_6 from the cyanobacterium *Anabaena* PCC 7119. *Biochemical and Biophysical Research Communications* 243 (1998): 302–6.

Molina-Heredia, F. P., M. Hervás, J. A. Navarro, and M. A. De la Rosa. A single arginyl residue in plastocyanin and in cytochrome c_6 from the cyanobacterium *Anabaena sp.* PCC 7119 is required for efficient reduction of photosystem I. *The Journal of Biological Chemistry* 276 (2001): 601–5.

Moreno-Beltrán, B., A. Díaz-Quintana, K. González-Arzola, A. Velázquez-Campoy, M. A. De la Rosa, I. Díaz-Moreno. Cytochrome c_1 exhibits two binding sites for cytochrome c in plants. *Biochimica et Biophysica Acta—Bioenergetics* (2014 in press). doi:10.1016/j.bbabio.2014.07.017.

Navarro, J. A., M. A. De la Rosa, and G. Tollin. Transient kinetics of flavin-photosensitized oxidation of reduced redox proteins. *European Journal of Biochemistry* 199 (1991): 239–43.

Navarro, J. A., M. Hervás, B. De la Cerda, and M. A. De la Rosa. Purification and physicochemical properties of the low-potential cytochrome c_{549} from the cyanobacterium *Synechocystis sp.* PCC 6803. *Archives of Biochemistry and Biophysics* 318 (1995): 46–52.

Nooren, I. M., and J. M. Thornton. Diversity of protein-protein interactions. *EMBO Journal* 22 (2003a): 3486–92.

Nooren, I. M., and J. M. Thornton. Structural characterisation and functional significance of transient protein-protein interactions. *Journal Molecular Biology* 325 (2003b): 991–1018.

Olesen, K., M. Ejdeback, M. M. Crnogorac, N. M. Kostic, and Ö. Hansson. Electron transfer to photosystem 1 from spinach plastocyanin mutated in the small acidic patch: Ionic strength dependence of kinetics and comparison of mechanistic models. *Biochemistry* 38 (1999): 16695–705.

Page, C. C., C. C. Moser, and P. L. Dutton. Mechanism for electron transfer within and between proteins. *Current Opinion in Chemical Biology* 7 (2003): 551–56.

Perkins, J. R., I. Diboun, B. H. Dessailly, J. G. Lees, and C. Orengo. Transient protein-protein interactions: Structural, functional, and network properties. *Structure* 18 (2010): 1233–43.

Piguet, C. Enthalpy-entropy correlations as chemical guides to unravel self-assembly processes. *Dalton Transactions* 40 (2011): 8059–71.

Prudêncio, M., and M. Ubbink. Transient complexes of redox proteins: Structural and dynamic details from NMR studies. *Journal of Molecular Recognition* 17 (2004): 524–39.

Qin, L., and N. M. Kostic. Electron-transfer reactions of cytochrome *f* with flavin semiquinones and with plastocyanin. Importance of protein-protein electrostatic interactions and of donor-acceptor coupling. *Biochemistry* 31 (1992): 5145–50.

Qin, L., and N. M. Kostic. Importance of protein rearrangement in the electron-transfer reaction between the physiological partners cytochrome *f* and plastocyanin. *Biochemistry* 32 (1993): 6073–80.

Rodríguez-Roldán, V., J. M. García-Heredia, J. A. Navarro, M. A. De la Rosa, and M. Hervás. Effect of nitration on the physicochemical and kinetic features of wild-type and monotyrosine mutants of human respiratory cytochrome *c*. *Biochemistry* 47 (2008): 12371–9.

Sampson, V., and T. Alleyne. Cytochrome *c*/cytochrome *c* oxidase interaction. *European Journal of Biochemistry* 268 (2001): 6534–44.

Scanu, S., J. M. Foerster, M. Timmer, G. M. Ullmann, and M. Ubbink. Loss of electrostatic interactions causes increase of dynamics within the plastocyanin–cytochrome *f* complex. *Biochemistry* 52 (2013a): 6615–26.

Scanu, S., J. M. Foerster, G. M. Ullmann, and M. Ubbink. Role of hydrophobic interactions in the encounter complex formation of the plastocyanin and cytochrome *f* complex revealed by paramagnetic NMR spectroscopy. *Journal of the American Chemical Society* 135 (2013b): 7681–92.

Schreiber, G., and A. E. Keating. Protein binding specificity versus promiscuity. *Current Opinion in Structural Biology* 21 (2011): 50–61.

Sharp, K. Entropy-enthalpy compensation: Fact or artifact?. *Protein Science* 10 (2001): 661–7.

Sommer, F., F. Drepper, W. Haehnel and M. Hippler. The hydrophobic recognition site formed by residues PsaA-Trp651 and PsaB-Trp627 of photosystem I in *Chlamydomonas reinhardtii* confers distinct selectivity for binding of plastocyanin and cytochrome c_6. *The Journal of Biological Chemistry* 279 (2004): 20009–17.

Sommer, F., F. Drepper, and M. Hippler. The luminal helix I of PsaB is essential for recognition of plastocyanin or cytochrome c_6 and fast electron transfer to photosystem I in *Chlamydomonas reinhardtii*. *The Journal of Biological Chemistry* 277 (2002): 6573–81.

Sun, J., W. Xu, M. Hervás, J. A. Navarro, M. A. De la Rosa, and P. R. Chitnis. Oxidizing side of the cyanobacterial photosystem I—Evidence for interaction between the electron donor proteins and a luminal surface helix of the PsaB subunit. *Journal of Biological Chemistry* 274 (1999): 19048–54.

Takabe, T., H. Ishikawa, S. Niwa, and S. Itoh. Electron transfer between plastocyanin and P700 in highly-purified photosystem I reaction centre complex. Effects of pH, cations, and subunit peptide composition. *The Journal of Biochemistry* 94 (1983): 1901–11.

Tang, C., J. Iwahara, and G. M. Clore. Visualization of transient encounter complexes in protein-protein association. *Nature* 444 (2006): 383–86.

Tollin, G., T. E. Meyer, and M. A. Cusanovich. Elucidation of the factors which determine reaction-rate constants and biological specificity for electron-transfer proteins. *Biochimica et Biophysica Acta—Bioenergetics* 853 (1986): 29–41.

Ubbink, M., M. Ejdebäck, B. G. Karlsson, and D. S. Bendall. The structure of the complex of plastocyanin and cytochrome *f*, determined by paramagnetic NMR and restrained rigid-body molecular dynamics. *Structure* 6 (1998): 323–35.

Van Amsterdam, I. M. C., M. Ubbink, O. Einsle et al. Dramatic modulation of electron transfer in protein complexes by crosslinking. *Nature Structural and Molecular Biology* 9 (2002): 48–52.

Van Leeuwen, J. W. The ionic strength dependence of the rate of a reaction between two large proteins with a dipole moment. *Biochimica et Biophysica Acta—Protein Structure and Molecular Enzymology* 743 (1983): 408–21.

Volkov, A. N., M. Ubbink, and N. A. J. van Nuland. Mapping the encounter state of a transient protein complex by PRE NMR spectroscopy. *Journal of Biomolecular NMR* 48 (2010): 225–36.

Volkov, A. N., J. A. R. Worrall, E. Holtzmann, and M. Ubbink. From the cover: Solution structure and dynamics of the complex between cytochrome *c* and cytochrome *c* peroxidase determined by paramagnetic NMR. *Proceedings of the National Academy of Sciences of the United States of America* 103 (2006): 18945–50.

Walker, M. C., J. J. Pueyo, J. A. Navarro, C. Gómez-Moreno, and G. Tollin. Laser flash photolysis studies of the kinetics of reduction of ferredoxins and ferredoxin-NADP+ reductases from *Anabaena* PCC 7119 and spinach: Electrostatic effects on intracomplex electron transfer. *Archives of Biochemistry and Biophysics* 287 (1991): 351–58.

Watkins, J. A., M. A. Cusanovich, T. E. Meyer, and G. Tollin. A "parallel plate" electrostatic model for bimolecular rate constants applied to electron transfer proteins. *Protein Science* 3 (1994): 2104–14.

Williams, P. A., V. Fulop, Y.-C. Leung et al. Pseudospecific docking surfaces on electron transfer proteins as illustrated by pseudoazurin, cytochrome c_{550} and cytochrome cd_1 nitrite reductase. *Nature Structural and Molecular Biology* 2 (1995): 975–82.

Worrall, J. A. R., U. Kolczak, G. W. Canters, and M. Ubbink. Interaction of yeast iso-1-cytochrome *c* with cytochrome *c* peroxidase investigated by [N^{15},H^1] heteronuclear NMR spectroscopy. *Biochemistry* 40 (2001): 7069–76.

CHAPTER 3

ATPases and Mitochondrial Supercomplexes

Sara Cogliati, Rebeca Acín-Pérez and José A. Enriquez

CONTENTS

3.1 INTRODUCTION

3.1.1 Mitochondria: Much More than Just the Powerhouse

The role of mitochondria in eukaryotic cells is so important and eclectic that defining them is not an easy task. In 1957, Philip Siekevitz coined the term *powerhouse of the cell* (Siekevitz, 1957) to describe mitochondria based on their ability to produce adenosine triphosphate (ATP) through the oxidative phosphorylation system (OXPHOS). Since then, countless studies have shown that mitochondria do much more than simply supply energy; they provide a myriad of services to the cell. Some of these include orchestrating calcium buffering (McBride and Scorrano, 2013), controlling cellular redox status by generation of reactive oxygen species (ROS), regulating critical pathways by releasing metabolites such as succinate and α-ketoglutarate (Stanley et al., 2013) and modulating cellular phenotypes through different signalling pathways to adapt to changes in substrate availability (Liesa and Shirihai, 2013). Moreover, mitochondria have a central role in regulating apoptosis by releasing key proteins such as cytochrome *c* and AIF1 (Scorrano and Korsmeyer, 2003) and serve as a launching platform for many pro- and anti-apoptotic proteins (e.g. BAX, BAK and VDAC) (Scorrano et al., 2003; Soriano and Scorrano, 2010). Additionally, recent findings have established that mitochondria can communicate with the nucleus, modulating the expression of many genes by a mechanism called *retrograde signalling* (Whelan and Zuckerbraun, 2013). Given this multitude of functions, it is reasonable to conclude that mitochondria are important epicentres for cellular homeostasis.

This versatility of mitochondria is mirrored by a very dynamic and complex structure. Mitochondria have two membranes that are highly specialized in structure and function. The outer membrane gate allows the passage of metabolites and nuclear-encoded proteins, and is also a platform for many proteins including mitochondrial-shaping proteins, import and assembly protein machinery and proteins of the apoptosis pathway. The outer membrane surrounds an inner membrane, which can be divided into an inner boundary membrane and numerous invaginations, or cristae. Cristae are connected to the inter-membrane space between the outer and inner membranes by narrow tubular structures.

The inner boundary membrane contains mitochondrial-shaping proteins and constitutes an important portal for the assembly and transport of proteins since it contains translocase and assembly machinery. Both outer and inner membranes are in structural and functional association through contact sites enriched in large protein complexes called MICOS/MINOS/MitOS (Alkhaja et al., 2012; Harner et al., 2011; Malsburg et al., 2011; Hoppins et al., 2011). Through these contact sites, exchange of metabolites, lipids and preproteins can occur. Cristae are very dynamic and heterogeneous structures that are separated from the intermembrane space by narrow junctions. Cristae morphology varies among different tissues, and during apoptosis cristae undergo complete remodelling, triggering the opening of cristae junctions to release pro-apoptotic factors such as cytochrome *c*.

The vast majority of the respiratory chain complexes are located in the cristae and this is also the case for cytochrome *c* since 85% of the total pool is stored in this compartment (Scorrano et al., 2002). Because of this, cristae are defined as the sites of oxidative phosphorylation (OXPHOS) (Gilkerson et al., 2003). Mitochondrial OXPHOS is a process carried out in animal mitochondria by three different complexes (complex I, III and IV; CI, CII and CII) and an additional group of enzymes that can collectively be called CII-type enzymes (see Section 3.2.1) that coordinate the transport of electrons by oxidation of reducing equivalents to generate a proton gradient that is used by ATP synthase (ATPase) to synthesize ATP. The intimate relationship between cristae and OXPHOS is beyond question since it has been extensively demonstrated that cristae morphology is dependent on energetic status (Packer, 1963). Indeed, the architecture of cristae determines the structural organization of the respiratory chain supercomplexes (SCs) (Cogliati et al., 2013). As an example, when ADP concentration is high and oxidative phosphorylation is triggered, cristae appear contracted and the matrix compartment is dense in a state termed *condensed*. Conversely, when ADP concentration is low, cristae shift to an orthodox state with expanded, less-dense matrix and more compact cristae compartments (Packer, 1963).

Apoptotic-cristae remodelling also causes the disorganization of respiratory chain SCs, resulting in an impairment of mitochondrial bioenergetics (Cogliati et al., 2013).

In order to satisfy the energetic requirement of cells, the OXPHOS is fine-tuned not only by modulation of cristae shape but also by a plethora of mechanisms such as post-translational modifications, selective degradation by proteases and mitophagy.

The aim of this chapter is to discuss in detail the structure and function of OXPHOS, summarizing the *state of the art* regarding the mechanisms of assembly and modulation of the complexes and SCs. Finally, an overview on physiology and disease related to OXPHOS is presented.

3.2 ELECTRON TRANSPORT CHAIN

3.2.1 Organization of the Respiratory Complexes

Mitochondrial respiratory complexes (CI to CIV) are responsible for electron transport generated by oxidation of reducing equivalents, in the form of reduced nicotinamide adenine dinucleotide (NADH) or reduced flavin adenine dinucleotide ($FADH_2$). These reducing equivalents originate in different metabolic pathways (glycolysis, fatty acid oxidation and the Krebs cycle). Oxidation of NADH and $FADH_2$ is coupled to the pumping of protons into the intermembrane space, and the resulting proton gradient is used by the ATPase (complex V; CV) to generate utilizable energy in the form of ATP. NADH-reducing equivalents enter the mitochondrial electron transport chain (mtETC) through CI, whereas $FADH_2$-reducing equivalents enter the mtETC through CII-type mitochondrial dehydrogenases, such as succinate dehydrogenase (SDH), electron-transferring-flavoprotein (ETF) dehydrogenase, glycerol 3-phosphate dehydrogenase (G3P), dihydroorotate dehydrogenase (DHODH) and others. The electrons are then passed to the redox-active carrier, coenzyme Q, and subsequently to CIII, cytochrome *c* and CIV, which passes them to oxygen as the final acceptor.

How these respiratory complexes are organized in the mitochondrial inner membrane has been an object of intense debate. Initially, a solid model of organization was proposed, where the respiratory complexes were closely packed to guarantee accessibility and thus high efficiency in electron transport (Keilin and Hartree, 1947; Slater, 2003). However, this original model was progressively abandoned and replaced by the fluid or random collision model (Hackenbrock et al., 1986). In the fluid model, the respiratory complexes are considered as independent entities embedded in the inner membrane, with CoQ and cytochrome *c* acting as mobile carriers that freely diffuse in the lipid membrane. In a pioneering study published in 1986, Hackenbrock and co-workers demonstrated how this model offered a better explanation for the structural organization of the mtETC (Hackenbrock et al., 1986). However, new experimental approaches based on applying native conditions to study protein interactions showed that it was possible to purify stable associations of respiratory complexes from yeast and mammalian mitochondria (Cruciat et al., 2000; Schägger and Pfeiffer, 2000). These observations led to the reformulation of the solid model where the respiratory complexes are organized in larger structures (respiratory SCs), allowing a more efficient transport of electrons. This new interpretation on how the mtETC was structured stimulated controversy and debate between the defenders of each of the models proposed. The outcome of this debate is that neither model can satisfactorily account for all of the experimental evidence (Lenaz et al., 2007), which further exacerbates the discrepancy between the two models (Strecker et al., 2010; Wittig and Schägger, 2009).

The principal lines of evidence supporting the existence of SCs are based on specific experimental procedures that allow the solubilization of the mitochondrial inner membrane with detergents such as digitonin, a mild detergent that binds cholesterol. The existence of SCs involves co-migration of respiratory complexes under native conditions and co-purification by sucrose gradient centrifugation (Eubel et al., 2004; Krause et al., 2004; Schägger and Pfeiffer, 2000). The fact that detergents were used to isolate both free complexes and SCs raised some concerns since not all detergents yield the same composition of SCs, and some detergents yield only free respiratory complexes. However, the same reasoning can be used for the opposite interpretation since the use of a stronger detergent, such as dodecyl-maltoside, only yielded isolated complexes and not SCs (Acín-Peréz et al., 2008). Regardless of the experimental issues with the use of detergents, biochemical studies based on the kinetics of electron transport between complexes failed to support the solid model (Lenaz and Genova, 2007). Considering this, and taken together the experimental evidence shown by others, none of the models, solid or fluid, provided a satisfactory prediction for how the respiratory complexes are distributed in the inner membrane. To prove the real existence of SCs as biological entities, these SCs, also called respirasomes, have to fulfil certain criteria: (1) The migration of a particular complex should be dependent on the presence of the other complexes with which it is proposed to interact; (2) there should be a temporal gap between the formation of the individual complexes and the SCs, accordingly, the formation of complexes and SCs should be asynchronous; (3) SCs should contain the electron carriers CoQ and cytochrome *c*; and (4) isolated SCs should transfer electrons from NADH to O_2, that is, they have to respire.

Through different and complementary experimental approaches, the existence for a functional SC or respirasome was recently demonstrated (Acín-Peréz et al., 2008). By using cell lines in which one complex was genetically ablated, we were able to determine that the migration of the other complexes in the putative SC assemblies was affected. This genetic analysis showed that most putative SCs do indeed display genuine interaction between complexes. There were, however, exceptions that confirmed the original concern that co-migration on gels or gradients is insufficient evidence of interaction. The temporal gap between the formation of complexes and SCs was addressed by metabolic labelling of mtDNA-encoded proteins, which revealed a time difference of several hours between the labelling of free complexes and the incorporation of labelled complexes into SCs (Acín-Peréz et al., 2008). Moreover, it was demonstrated that isolated respiratory SCs are able to transfer electrons among their components and from NADH to O_2, supporting a respiratory function for SCs (Acín-Peréz et al., 2008).

These observations led to the proposal of the so-called plasticity model, which accommodates the solid and the fluid models by regarding the organization of the respiratory complexes as a network of different associations as well as, in some cases, individual complexes.

Finally, some SCs have been purified and analysed by single particle electron microscopy and their structure has been defined (Althoff et al., 2011; Dudkina et al., 2011; Mileykovskaya et al., 2012).

3.2.2 Two Functional Pools of Coenzyme Q

Coenzyme Q (CoQ, ubiquinone) is required to transfer electrons from NADH or FAD-dependent enzymes to respiratory CIII within the inner mitochondrial membrane. The reduction/oxidation of CoQ is critical for energy production, redox balance, pyrimidine synthesis, amino acid and lipid metabolism and indirectly for apoptosis control and calcium handling. In a superb review published recently by Genova and Lenaz, the authors, in an impressive tour, gathered, chemical, biochemical, metabolic, structural and biophysical data from 1955 onwards to address the function and role of CoQ in mitochondria (Lenaz and Genova, 2012). With respect to the existence of a single CoQ pool in mitochondria, they stated that the concept of the CoQ pool has been universally accepted to the point that it "has gained place in all biochemistry textbooks", despite the existence of experimentally demonstrated deviations from this concept (Genova and Lenaz, 2011, p. 332). The issue of the CoQ pool was only recently linked to respiratory SCs because purified SCs containing CI and CIII also carried CoQ in quantities that are sufficient to efficiently transfer electrons between CI and CIII (Acín-Peréz et al., 2008). Thus, the fluid model used to explain the organization of the mtETC favours the idea of a single CoQ pool, while the solid model proposes different CoQ pools (Enriquez and Lenaz, 2014). However, this association although possible is not strictly necessary and the segmentation of pools or the existence of a single pool could theoretically be compatible with both structural solutions. Some reports studying the reduction of CoQ in isolated mitochondria determined that succinate and NADH could each reduce a limited and specific fraction of the total CoQ pool (Jørgensen et al., 1985; Lass and Sohal, 1998). In addition, it was estimated that a portion of CoQ (15% to 30% of the total pool) appeared to be bound to proteins (although the nature of the proteins was not determined), while the remaining fraction seemed to be free in the membrane (Lass and Sohal, 1999). Rossignol and co-workers clearly demonstrated that the CoQ content in the mitochondrial inner membrane is not homogenous (Benard et al., 2008). Thus, when they measured levels of reduced CoQ and cytochrome *c* in state 3 isolated mitochondria, they found that one fraction of CoQ was utilized during steady-state respiration, another fraction was mobilizable, that is, a reserve that is used in case of a perturbation to maintain the energy fluxes at normal values (e.g. as a consequence of inhibition of the respiratory complexes or in the case of mitochondrial disease), and a third fraction is not utilizable and is unable to participate in succinate-dependent respiration. They estimated that the succinate non-utilizable pool was 79% of total CoQ in muscle mitochondria and 21% in liver mitochondria. Unfortunately, they did not perform similar assays analysing NADH-dependent respiration. Nevertheless, these results are incompatible with the concept of a single CoQ pool (Benard et al., 2008; Bianchi et al., 2004).

More recently, Hirst's group has investigated the partitioning of the CoQ pool using spectroscopic and kinetic techniques (Blaza et al., 2014). Their results clearly show that the reduction of the CIII-containing cytochromes is higher when succinate and NADH are provided simultaneously rather than when only one substrate is used. Interestingly, the amount of reduced CIII-containing cytochromes with both substrates is lower than the sum obtained by each substrate used separately. These results are incompatible both with the existence of a single CoQ pool and with two fully independent pools, one for NADH and the other for succinate. Surprisingly, the authors mistakenly concluded that CoQ behaved as a single pool, communicating both substrates and reducing equivalents for NADH and succinate oxidation. Moreover, they claimed, again mistakenly, that these results can help to discriminate between the alternative models of organization of the mETC and concluded that the results provide experimental evidence in favour of the fluid model over the solid model. Explicitly, if the pure fluid and the pure solid models were identified with the single versus the double fully independent CoQ pool alternatives, neither would explain Hirst's observations (Figure 3.1).

Investigation of the potential functional role of the super assembly between CI and CIII, and hence the plasticity model, demonstrates that the decrease in the amount CIII below that of CI impairs the transfer of electrons from succinate and glycerol-3-phosphate, while

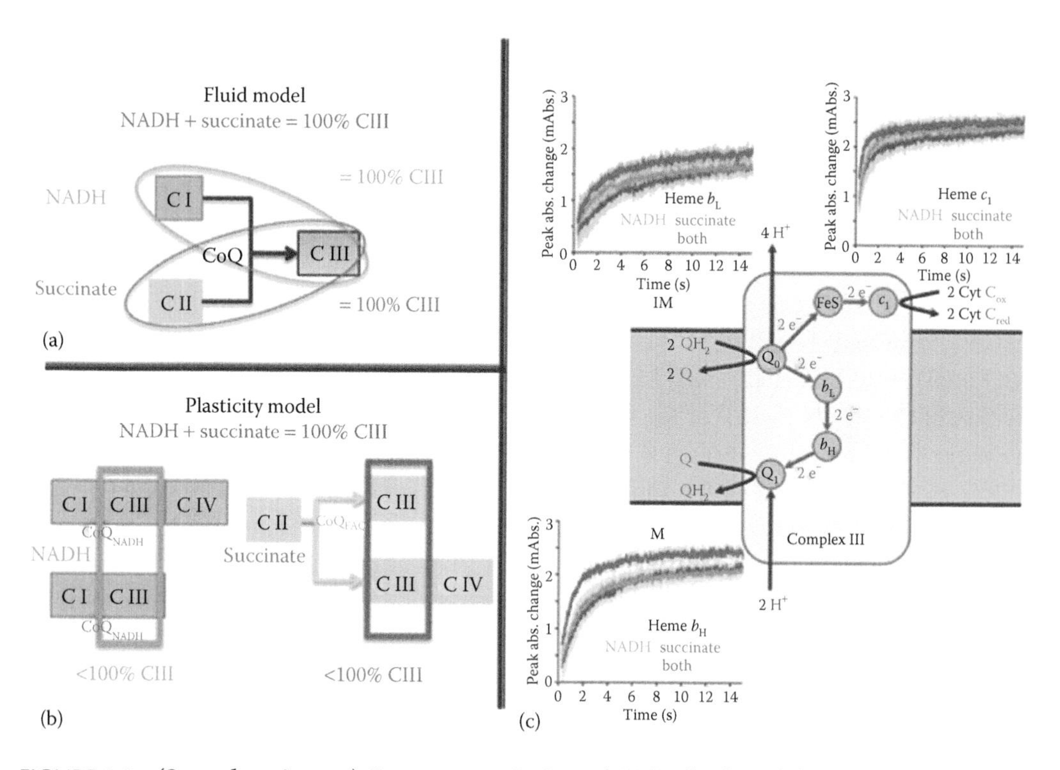

FIGURE 3.1 **(See colour insert.)** One or more CoQ pools? The fluid model requires a single, universally accessible Q pool that is not partitioned and the plasticity model proposes that different forms of superassembly define functionally dedicated CoQ pools (*cf.* Lapuente-Brun et al., 2013). Expected versus observed maximum reduction of CIII containing cytochromes as predicted from: (a) the fluid model: NADH or Succinate alone should be able to reduce the bulk of cytochromes in CIII; or (b) the plasticity model: since the CoQ pool is functionally segmented, to be able to reduce the bulk of CIII containing cytochromes both substrates (NADH and succinate) are required simultaneously. (c) Scheme representing CIII cytochrome reduction (bH, bL and c_1) showing the results obtained by Blaza et al. monitoring the reduction of CIII cytochromes and clearly showing that none of them can be fully reduced by NADH or succinate alone and that the simultaneous presence of both is required to reach complete reduction. (Reproduced from Figure 3A, 3C and 3D in Blaza, J.N. et al., *Proc. Nat. Acad. Sci. U.S.A.*, 111, 15735–15740, 2014. With permission.)

the transfer of electrons from NADH is unaltered. This observation, incompatible with the existence of a single CoQ pool, was explained by the fact that CI assembles with the majority of the available CIII and this structural arrangement significantly favours the oxidation of NADH over that of the FAD-dependent enzymes, SDH and G3PDH. These results led to the proposal of two functionally independent pools of CoQ: the CoQNADH and $CoQFADH_2$. Thus, the CoQNADH pool would be defined by the physical assembly between CI and CIII that would favour transfer between NADH and cyt *c*, and the $CoQFADH_2$ would use the remaining and majority of freely diffusible CoQ.

Like this, NADH oxidation and FAD oxidation would not crosstalk significantly, at least until the electrons reach cytochrome *c*. This observation, derived from the plasticity model, was predicted by Genova and Lenaz: "The non-homogeneity of the ubiquinone pool with respect to succinate and NADH oxidation may be interpreted today in terms of compartmentalization of CoQ in the I–III supercomplex in contrast with the free pool used for connecting complexes II and III" (Genova and Lenaz, 2011, p. 332). The existence of a single or two independent pools of CoQ has profound implications for the understanding of central physiological responses such as the adaptation to fasting by liver mitochondria. Among other molecular consequences, adaptation to different mitochondrial respiratory substrates that generate different proportions of NADH and FAD requires an optimization of the NADH route versus the FAD route. Thus, in fasting conditions when mitochondria have to rely on fatty acids rather than glucose, a modification in the proportion of respiratory SCs is required. How this is produced and what the consequences would be if this adaptation were impeded remains to be investigated and no obvious known mechanisms can be invoked. Nevertheless, an understanding of this behaviour would be of paramount importance since similar adaptations are expected to be necessary when there is an increase in fatty acid supply, either by high fat diet or during fasting. Impairment of this adaptation may then be relevant in pathological processes associated with obesity.

3.2.3 Modulation of the Respiratory Chain Complexes and Supercomplexes

3.2.3.1 Assembly Factors

The subunits of the complexes of the mtETC have a dual genetic origin since they are encoded by either nuclear or mitochondrial DNA. When complexes are formed, they can be further assembled into quaternary superstructures termed supercomplexes or SCs. The sequence of subunit assembly formation of complexes and SCs is finely tuned (Acín-Peréz et al., 2008). Many assembly factors and chaperones participate in the assembly of complexes, and they either play a role as assembly factors or are part of the complete complex. Many of these assembly factors have been identified because mutations in these proteins are associated with genetic diseases with a heterogeneous spectrum of symptoms including neurological manifestations, cardiomyopathy, muscle weakness, aciduria, liver failure and premature death—for an recent review, see Diaz et al. (2011). Mutations in the three functional domains of BCS1L, an assembly factor that assists the incorporation of Rieske protein and Qcr10q into CIII, can cause a variety of syndromes from the most severe GRACILE syndrome to the less severe Bjornstad syndrome. Surf1 is an assembly factor for CIV, and mutations in this gene have been associated with different manifestations of

Leigh syndrome, a neurometabolic disorder. Moreover, French-Canadian Leigh syndrome is associated with mutations in the *LRPPRC* gene, which is required for stability and translation of COX1 subunit of CIV (Mootha et al., 2003; Sasarman et al., 2010) but its mechanism is still unknown. Additionally, mutations in the *LRPPRC* gene have been associated with ATPase deficiency (Mourier et al., 2014). Many other assembly factors are involved in SC assembly and mutations in these factors are responsible for several genetic syndromes, many of which remain to be classified. Interestingly, it has been recently described that proteins interacting with the structural proteins of the MINOS complex could be important for the assembly of respiratory chain complexes (An et al., 2012; Darshi et al., 2011; Ott et al., 2012; Weber et al., 2013). Among them, C1orf163, which interacts with Sam50 and Mitofilin, was renamed as RESpiratory chain Assembly 1 (RESA1) (Kozjak-Pavlovic et al., 2014) since its depletion causes severe impairment of CIV assembly, with less effects observed in other complexes. This short summary serves to illustrate how the correct assembly and stabilization of complexes are regulated.

Since respiratory complexes are organized into different and dynamic SCs, additional assembly factors must be required to facilitate and regulate this process. Indeed, very recent publications from three independent laboratories described two related *Saccharomyces cerevisiae* proteins (rcf1 and rcf2) that are required for the assembly of the COX3 subunit into mature CIV, and may also be relevant for the assembly between CIII and CI (Chen et al., 2012; Strogolova et al., 2012; Vukotic et al., 2012). Two mammalian orthologs have been identified for rcf1 (HIG1A and HIG2A) (Chen et al., 2012). Results from gene-silencing experiments show that HIG1A does not influence the formation of respiratory complexes or SCs, but instead participates in the regulation of mitochondrial γ-secretase function (Hayashi et al., 2012). Further, HIG2A silencing not only impairs complex IV assembly, as occurs in yeast, but also results in a very moderate reduction in respirasome levels. Thus, in our view, these proteins should be regarded more as CIV assembly factors rather than SC assembly factors. A newly characterized protein, C11orf83, which binds cardiolipin and phosphatidic acid, interacts with the bc1 core complex and is required for its assembly/stabilization (Desmurs et al., 2015). Depletion of C11orf83 in HeLa cells causes a decrease in the III + IV supercomplex but not the I + III + IV supercomplex. This effect on supercomplex III + IV could be indirect due to its impact on CIII formation.

The first bona fide assembly factor identified for SC assembly in mammalian mitochondria is Cox7a2l. This protein is required for the stable interaction between CIII and CIV, but not for the assembly or function of either CIII or CIV. Consequently, Cox7a2l has been renamed SCAF1 for SC assembly factor 1 (Lapuente-Brun et al., 2013). A screening study of mouse strains showed that C57Bl/6J and Balb/cJ mice harbour a mutation in SCAF1 that renders it inactive, and as a result, mitochondria from these strains are unable to build respiratory SCs that require direct interaction between CIII and CIV. SCAF1 thus defines three populations of CIV: the fraction assembled with complexes CI and CIII in the respirasome, the fraction assembled with CIII alone and a non-interacting fraction. This segmentation leads to different SCs that allow each population of CIV to receive electrons preferentially from NADH, only from FAD-containing enzymes or from any substrate, depending of the type of SC. Interestingly, the absence of functional SCAF1 does not cause major bioenergetics problems and animals lacking SCAF1 are fertile and healthy, although they have lost regulatory options for the fine-tuning of their bioenergetic performance. Further work is needed to understand more fully the role of SCAF1 and its impact on organismal metabolism. Aside from SCAF1, new specific assembly factors for SCs remained unidentified.

The regulation of SC dynamics may require specific regulators, both positive and negative. Thus, MCJ/DnaJC15, a member of the DNAJC family of co-chaperones, is a negative modulator of CI activity and impairs the formation of respirasomes (Hatle et al., 2013). Indeed, mitochondria from MCJ/DnaJC15 knockout mice present an accumulation of SCs and an increase in respiration dependent on NADH. In normal conditions, this increase is inconsequential; however, in fasting conditions, it facilitates lipid metabolism by increasing lipid degradation and gluconeogenesis. Moreover, in a high-cholesterol diet, MCJ/DnaJC15 knockout mice display low levels of cholesterol compared with wild-type mice.

3.2.3.2 Mitochondrial Shape

As mentioned earlier, mitochondria have a well-structured architecture that continuously changes to maintain mitochondrial function according to the needs of cells. Indeed, mitochondria within the cell are subjected to a balanced dynamism of fusion and fission events, referred to as the *mitochondrial life cycle* (Wikstrom et al., 2014). The regulation of these processes allows for a change in mitochondrial architecture from a very elongated and interconnected network to a very fragmented state. Thus, mitochondrial function is adapted to the requirements of the cell. Recently, many laboratories have demonstrated that nutrient availability is an important factor that can modulate both mitochondrial shape and the behaviour of the mitochondrial network. Generally, nutrient depletion leads to mitochondria elongation in an attempt to increase the efficiency of energy production. Conversely, in conditions of nutrient excess, mitochondria fragment to increase uncoupled respiration and reduce bioenergetic efficiency (for a comprehensive review, see Liesa and Shirihai, 2013).

How mitochondrial fusion and fission events impact on the assembly and function of the respiratory chain is an interesting issue and many recent papers have confirmed that modification of mitochondrial morphology is an underlying mechanism regulating the assembly and structure of respiratory chain SCs. As mentioned, mitochondria elongate during starvation. This event protects mitochondria from autophagy and promotes the formation of more cristae. Increased numbers of cristae induce the dimerization of ATPase and so increase the production of ATP to support the lack of nutrients (Gomes et al., 2011). Cristae are very dynamic structures that undergo remodelling upon different stimuli and different metabolic states. Cristae house the OXPHOS complexes, creating specialized compartments for the electron transport chain (Althoff et al., 2011; Davies et al., 2011; Dudkina et al., 2005; Khalifat et al., 2008; Minauro-Sanmiguel et al., 2005; Rieger et al., 2014; Wilkens et al., 2013). Recently, a connection between mitochondrial-shaping proteins and the structure of OXPHOS complexes has been demonstrated, supporting the idea that the structure of the cristae is a key factor in the modulation of OXPHOS function. Several proteins are implicated in shaping the cristae architecture, and mutations in their respective genes, or gene silencing, affect both the shape of mitochondria and the structure of the complexes and SCs of the respiratory chain.

Opa1, a dynamin-related GTPase, is an inner membrane protein that is involved in the maintenance of tight cristae junctions, forming high molecular weight oligomers that are disrupted during apoptosis (Costa et al., 2010; Frezza et al., 2006). Apoptotic cristae-remodelling and genetic ablation of Opa1 lead to a disorganization of the cristae architecture that impairs the assembly of SCs and affects mitochondrial metabolism and cell growth (Cogliati et al., 2013). This constitutes the first experimental evidence for a direct link between mitochondrial inner membrane organization and SC arrangement. Furthermore, Opa1 responds to different energy demands by forming high molecular weight oligomers. During starvation, OPA1 oligomerization is increased, leading to a narrowing of cristae (Patten et al., 2014) in a response required to promote ATPase assembly and to sustain ATP production and cell survival (Gomes et al., 2011). Numerous studies have demonstrated that interactors and proteases of Opa1, which participate in cristae shaping, are needed for the correct formation and function of the respiratory chain SCs. Furthermore, a large protein complex named mitochondrial inner membrane organizing system (MINOS) (Alkhaja et al., 2012; Malsburg et al., 2011), also called mitochondrial contact site (MICOS) (Harner et al., 2011), or a mitochondrial organizing structure (MitOS) (Hoppins et al., 2011) has been shown to be a major player in the organization of the inner mitochondrial membrane architecture. The core proteins of the complex are Mitofilin/Fcj1 and MINOS1/Mio10, but additional components may also interact with the complex. Mutations in the core subunits of the MICOS complex lead to a severely altered inner membrane architecture and impairment of mitochondrial bioenergetics.

In conclusion, we can state that cristae shape is a key factor for the correct assembly and function of the mtETC, and consequently mitochondrial bioenergetics. Nevertheless, the molecular mechanisms that link mitochondrial shape and the assembly of SCs of the respiratory chain remain to be established and require further study.

3.2.3.3 Lipids

SC formation, stabilization and function are critically influenced by the lipid composition of the inner mitochondrial membrane. In particular, the mitochondrial inner membrane is enriched in cardiolipin, a non-bilayer lipid that is fundamental for the organization of protein-protein and protein-lipid microdomains, which are functional hot spots of mitochondria (Whitelegge, 2011).

In Barth syndrome patients, where cardiolipin remodelling is altered due to the mutation of the gene Tafazzin, SCs are unstable, leading to the mitochondrial functional impairment that underlies the disease (McKenzie et al., 2006). These changes have been demonstrated not only in human patient cells but also in yeast (Böttinger et al., 2012) and Drosophila (Acehan et al., 2011). Moreover, studies in Arabidopsis thaliana suggested that cardiolipin is able to strategically control the localization of functional protein complexes important for the coordination between assembly and degradation of respiratory chain SCs (Gehl and Sweetlove, 2014; Gehl et al., 2014). Those complexes are formed by prohibitins (PHB1,2), a family of proteins organized in the inner membrane with m-AAA proteases in long-ring complexes together with SLP2, a mitochondrial protein belonging to the stomatin like family. The importance of lipid composition in SC stabilization is supported by studies with reconstituted proteoliposomes (Bazán et al., 2013; Maranzana et al., 2013; Pfeiffer et al., 2003).

3.3 F_1-F_O-ATP SYNTHASE

F_1-F_0-ATP Synthase (also called F_1-F_0 ATPase, complex V or CV) is the molecular machine that manufactures ATP from ADP and phosphate by rotary catalysis using the driving force of the proton gradient generated by the electron transport chain. ATPase is formed by up to 18 subunits that comprise the two functional sectors: a soluble catalytic F_1 sector on the matrix side and a membranous F_O sector, which anchors the F_1 sector in the inner mitochondrial membrane and forms a specific proton-conducting channel.

Blue-native polyacrylamide gel electrophoresis (BNPAGE) and electron microscopy demonstrated that ATPase from yeast to mammalian mitochondria forms dimers and oligomers. In particular, cryo-electron microscopy has been useful to visualize dimers of ATPase, with the characteristic lollipop shape found in long rows sloping at an angle >70°, along the tightly curved membrane (Davies et al., 2011; Dudkina et al., 2005; Strauss et al., 2008). The localization of the ATPase at the cristae apex with the other complexes located along both sides drives the proton flow from the source (complexes) to the sink (ATPase).

3.3.1 Regulation of the Dimerization of ATPase

ATPase is fully active as a monomer, but studies of yeast to mammals have shown that it is more efficient in dimeric and oligomeric forms. Although dimerization is conserved among species, the actual mechanism(s) is still debated, doubtless because of species-specific differences in the processes. In yeast, ATPase subunits e and g maintain the dimeric form. Also, yeast ATPase subunit β is subject to a series of post-translational modifications that can modulate oligomerization. In particular, phosphorylation of the T58 residue, found on the surface of ATPase subunit β, leads to a decrease in the formation and stability of CV dimers, and thus ATPase activity (Kane et al., 2010).

In mammals, ATPase dimerization seems to be associated with the number and shape of the cristae. In starvation conditions, which increases demand for ATP, cells increase the cristae surface (Gomes et al., 2011) and narrow the junctions between cristae (Patten et al., 2014), resulting in an increase in ATPase oligomerization. Oligomerization of ATPase is regulated by IF1 (ATPase Inhibitory Factor 1). IF1 inhibits the hydrolytic activity of ATPase during hypoxia or ischemia, when the pH of the mitochondrial matrix drops below neutrality. Although the mechanism of action of IF1 is still under study, it has been suggested that IF1, by increasing the number and density of cristae, promotes oligomerization of ATPase (Cabezón et al., 2003). Conversely, other studies have suggested that IF1 stabilizes the formation of ATPase oligomers, which in turn increases membrane curvature and hence cristae density (Faccenda et al., 2013).

The question of what comes first, cristae or ATPase dimerization, is still open to debate and many results can be put forward in favour of one or the other hypothesis. In yeast and human, the deletion of subunits e and g, which abolishes dimerization and oligomerization of ATPase, results in alterations in cristae morphology (Habersetzer et al., 2013; Paumard et al., 2002). Additionally, ρ° cells, which are devoid of F_O, have abnormal cristae (Kao et al., 2012). Experiments using vesicles reconstituted with ATPase dimers reveal that oligomeric forms impose a highly positive curvature of the membrane, favouring the formation of cristae (Rabl et al., 2009). Collectively, these results support the hypothesis that ATPase dimers induce morphological changes in cristae. The opposite hypothesis that cristae shape modulates the dimerization of the ATPase is endorsed by the observation that cristae density increases during starvation (previously described). A second argument in favour of this hypothesis stems from findings in the ageing model *Podospora anserina*. In this organism, senescence leads to profound changes in membrane structure and macromolecular organization, with massive loss of cristae and disassembly of ATPase dimers (Daum et al., 2013).

Undoubtedly, a correlation exists between cristae structure and ATPase, and the molecular mechanisms that link them are very complex. Recently an interesting experiment in yeast mutants of fcj1 and e, g subunits proposed that these proteins work together to shape the cristae. Fcj1, an inner membrane protein, interferes with the formation of higher oligomers and induces a negative curvature of the membrane; on the other hand, the action of e and g subunits on the oligomerization of ATPase induces a positive curvature, leading to the formation of strong curvatures at the tip of the cristae (Rabl et al., 2009). These findings suggest that mitochondria contain a finely regulated community of proteins that work in concert to maintain the integrity and function of mitochondria.

3.4 SCs IN PHYSIOLOGY AND DISEASES

Given the fundamental and diverse roles of mitochondria, any malfunction is likely to trigger disease. Indeed, as previously described (Section 3.2.3), mutations in different mitochondrial assembly factors cause a broad spectrum of genetic diseases with heterogeneous clinical symptoms. Not surprisingly, these mutations primarily affect organs with high energy demands, leading in some cases to premature death—for a detailed review, see Diaz et al. (2011).

As described earlier, cristae morphology is important for the correct assembly and function of SCs. Opa1 is a mitochondrial-shaping protein important for cristae remodelling. Mutations in Opa1 cause autosomal dominant optic atrophy characterized by retinal ganglion cell degeneration, leading to optic neuropathy and neuromuscular problems such as ataxia and myopathy. Studies performed in primary cells from patients show that mitochondria present rearrangement of the cristae and an impairment of ATP production, correlating with the severity of symptoms (Lodi et al., 2004; Zanna et al., 2008). Moreover, expression of pathological mutations of Opa1 in different cell lines leads to defects in cristae morphology and OXPHOS (Agier et al., 2012), thus corroborating the hypothesis that impairment of respiratory chain function is responsible for the clinical symptoms. Indeed, SC disassembly is also a hallmark of Barth syndrome (McKenzie et al., 2006).

SC structure is extremely dynamic and this dynamism is apparent even under physiological conditions. SC formation is a characteristic phenotype of mesenchymal cells undergoing adipogenic differentiation (Hofmann et al., 2012). Our unpublished results demonstrate that the patterns of SC formation in neonates and adults are different; further, we have observed that SC structure in heart differs from that in liver, together suggesting that SCs can adapt and change to fit different metabolic requirements.

ROS are reactive chemical molecules containing oxygen that have an important role in maintaining mitochondrial activity. The mtETC is the direct source of ROS that, in the physiological range, participate in many homeostatic processes as modulators of growth factor signalling (Bratic and Larsson, 2013), activators of uncoupling proteins (Nakamura et al., 2012) and regulators of mitochondrial biogenesis (Moslehi et al., 2012). In contrast, when mitochondria are malfunctioning, ROS production exceeds the physiological threshold and causes deleterious pathological effects including lipid peroxidation, protein oxidation and mitochondrial DNA damage.

ROS have a strong influence on the maintenance and performance of SCs, and an ever-growing number of studies show that increased ROS levels decrease SC assembly and can cause disease. The vicious cycle of ROS and mitochondrial dysfunction has been extensively studied in models of ageing (Chan, 2012; Dai et al., 2012). Ageing-dependent decay of mitochondrial function is directly linked to increased ROS production, but little is known about the molecular mechanisms involved. Recent hypotheses propose that the decline in mitochondrial function with age is related to the decrease in the levels of SCs in heart (Salminen et al., 2012) and rat brain cortex (Marchi et al., 2012). Ageing *P. anserina* presents a massive loss of cristae and dissociation of ATP. In mammals, ageing causes remodelling of the cristae and disassembly of SCs (Daum et al., 2013), leading to the hypothesis that ageing impairs the structure and function of respiratory chain.

Contrasting with this explanation, in rat skeletal muscle the highest molecular weight SCs accumulate in older animals, perhaps as a consequence of a molecular mechanism that enhances the catalytic activity of the respirasome by better channelling of fuels and by preventing ROS generation (Sundaresan et al., 1995).

The mammalian heart is a high-energy demand tissue and is thus greatly dependent on mitochondria function. Mitochondria are involved in the progression of hypertrophy and heart failure (HF) (Echtay et al., 2002; Moreno-Loshuertos et al., 2006), but although it is clear that mitochondrial function is diminished in failing hearts, so far no detailed studies have investigated the relationship between HF and SC destabilization. A study using a canine model of HF reported a decrease in state 3 mitochondrial respiration in both subsarcolemmal and interfibrillar mitochondria, accompanied by a reduction in CI-containing SCs (Gómez and Hagen, 2012) that was due to changes in the phosphorylation of specific CIV subunits that altered protein-protein interactions or the stability of SCs containing CIV (Frenzel et al., 2010). Mitochondrial dysfunction has been correlated also with hypertension (Lopez-Campistrous et al., 2008). Mitochondria from whole brain of spontaneously hypertensive rats presented down-regulation of enzymes involved in cellular bioenergetics and also assembly defects of CI, III, IV and V.

In cancer, cell metabolism switches towards glycolysis (the Warburg effect) accompanied by a depression in OXPHOS. The mechanisms underlying this metabolic switch are not well understood. It is known that mitochondrial membrane potential and oxygen consumption decrease and ROS production is enhanced (Rosca et al., 2011), leading to decay of CI content and activity and impairment of ROS defences (de Cavanagh et al., 2009; Rosca et al., 2011).

This mitochondrial response translates to a switch towards glycolysis, lactate production and apoptosis, which leads to tumour formation and proliferation (Rosca et al., 2011), but how SC organization is altered in tumorigenesis is still unknown. The only model offered thus far proposes that ROS produced during tumorigenesis would alter SC formation, leading primarily to disruption of CI assembly and activity and thus promoting a second peak of ROS generation, which would amplify the mitochondrial defect (Gasparre et al., 2013).

3.5 CONCLUDING REMARKS

Considering the enormous amount of data published in the last 10 years, it is clear that mitochondria have reached centre stage as a key organelle for living cells. Mitochondrial physiology and biogenesis are deeply enmeshed in so many pathways that countless studies arrive finally at mitochondria in an effort to understand the molecular mechanisms of pathological situations, and also ageing and metabolism. Mitochondria modulate and fine-tune metabolism to optimize the performance of the cell. Additionally, they determine cell fate through internal crosstalk among signalling pathways that determines which fuel source to use and even whether to live or die. By reorganizing the mitochondrial network and inner membrane shape, they can respond to different stimuli, carbon sources or stress conditions. All of these tasks impact the SCs, which can be modified and re-organized to satisfy cellular requirements. Although much remains to be discovered about the function of SCs, it is clear, nonetheless, that they are more than simply protein structures. Modulation of SCs levels might be a new approach to control disease progression.

REFERENCES

Acehan, D., Malhotra, A., Xu, Y., Ren, M., Stokes, D.L., and Schlame, M. (2011). Cardiolipin affects the supramolecular organization of ATP synthase in mitochondria. *Biophys. J. 100*, 2184–2192.

Acín-Peréz, R., Fernández-Silva, P., Peleato, M.L., Pérez-Martos, A., and Enriquez, J.A. (2008). Respiratory active mitochondrial supercomplexes. *Mol. Cell 32*, 529–539.

Agier, V., Oliviero, P., Lainé, J., L'Hermitte-Stead, C., Girard, S., Fillaut, S., Jardel, C., Bouillaud, F., Bulteau, A.-L., and Lombes, A. (2012). Defective mitochondrial fusion, altered respiratory function, and distorted cristae structure in skin fibroblasts with heterozygous OPA1 mutations. *Biochim. Biophys. Acta 1822*, 1570–1580.

Alkhaja, A.K., Jans, D.C., Nikolov, M., Vukotic, M., Lytovchenko, O., Ludewig, F., Schliebs, W., Riedel, D., Urlaub, H., Jakobs, S. et al. (2012). MINOS1 is a conserved component of mitofilin complexes and required for mitochondrial function and cristae organization. *Mol. Biol. Cell 23*, 247–257.

Althoff, T., Mills, D.J., Popot, J.-L., and Kühlbrandt, W. (2011). Arrangement of electron transport chain components in bovine mitochondrial supercomplex I1III2IV1. *EMBO J. 30*, 4652–4664.

An, J., Shi, J., He, Q., Lui, K., Liu, Y., Huang, Y., and Sheikh, M.S. (2012). CHCM1/CHCHD6, novel mitochondrial protein linked to regulation of mitofilin and mitochondrial cristae morphology. *J. Biol. Chem. 287*, 7411–7426.

Bazán, S., Mileykovskaya, E., Mallampalli, V.K.P.S., Heacock, P., Sparagna, G.C., and Dowhan, W. (2013). Cardiolipin-dependent reconstitution of respiratory supercomplexes from purified *Saccharomyces cerevisiae* complexes III and IV. *J. Biol. Chem. 288*, 401–411.

Benard, G., Faustin, B., Galinier, A., Rocher, C., Bellance, N., Smolkova, K., Casteilla, L., Rossignol, R., and Letellier, T. (2008). Functional dynamic compartmentalization of respiratory chain intermediate substrates: Implications for the control of energy production and mitochondrial diseases. *Int. J. Biochem. Cell. Biol. 40*, 1543–1554.

Bianchi, C., Genova, M.L., Parenti Castelli, G., and Lenaz, G. (2004). The mitochondrial respiratory chain is partially organized in a supercomplex assembly: Kinetic evidence using flux control analysis. *J. Biol. Chem. 279*, 36562–36569.

Blaza, J.N., Serreli, R., Jones, A.J.Y., Mohammed, K., and Hirst, J. (2014). Kinetic evidence against partitioning of the ubiquinone pool and the catalytic relevance of respiratory-chain supercomplexes. *Proc. Nat. Acad. Sci. U.S.A. 111*, 15735–15740.

Böttinger, L., Horvath, S.E., Kleinschroth, T., Hunte, C., Daum, G., Pfanner, N., and Becker, T. (2012). Phosphatidylethanolamine and cardiolipin differentially affect the stability of mitochondrial respiratory chain supercomplexes. *J. Mol. Biol. 423*, 677–686.

Bratic, A., and Larsson, N.-G. (2013). The role of mitochondria in aging. *J. Clin. Invest. 123*, 951–957.

Cabezón, E., Montgomery, M.G., Leslie, A.G.W., and Walker, J.E. (2003). The structure of bovine F_1-ATPase in complex with its regulatory protein IF1. *Nat. Struct. Biol. 10*, 744–750.

Chan, D.C. (2012). Fusion and fission: Interlinked processes critical for mitochondrial health. *Annu. Rev. Genet. 46*, 265–287.

Chen, Y.-C., Taylor, E.B., Dephoure, N., Heo, J.-M., Tonhato, A., Papandreou, I., Nath, N., Denko, N.C., Gygi, S.P., and Rutter, J. (2012). Identification of a protein mediating respiratory supercomplex stability. *Cell Metab. 15*, 348–360.

Cogliati, S., Frezza, C., Soriano, M.E., Varanita, T., Quintana-Cabrera, R., Corrado, M., Cipolat, S., Costa, V., Casarin, A., Gomes, L.C. et al. (2013). Mitochondrial cristae shape determines respiratory chain supercomplexes assembly and respiratory efficiency. *Cell 155*, 160–171.

Costa, V., Giacomello, M., Hudec, R., Lopreiato, R., Ermak, G., Lim, D., Malorni, W., Davies, K.J.A., Carafoli, E., and Scorrano, L. (2010). Mitochondrial fission and cristae disruption increase the response of cell models of Huntington's disease to apoptotic stimuli. *EMBO Mol. Med. 2*, 490–503.

Cruciat, C.M., Brunner, S., Baumann, F., Neupert, W., and Stuart, R.A. (2000). The cytochrome bc1 and cytochrome c oxidase complexes associate to form a single supracomplex in yeast mitochondria. *J. Biol. Chem. 275*, 18093–18098.

Dai, D.-F., Rabinovitch, P.S., and Ungvari, Z. (2012). Mitochondria and cardiovascular aging. *Circ. Res. 110*, 1109–1124.

Darshi, M., Mendiola, V.L., Mackey, M.R., Murphy, A.N., Koller, A., Perkins, G.A., Ellisman, M.H., and Taylor, S.S. (2011). ChChd3, an inner mitochondrial membrane protein, is essential for maintaining crista integrity and mitochondrial function. *J. Biol. Chem. 286*, 2918–2932.

Daum, B., Walter, A., Horst, A., Osiewacz, H.D., and Kühlbrandt, W. (2013). Age-dependent dissociation of ATP synthase dimers and loss of inner-membrane cristae in mitochondria. *Proc. Nat. Acad. Sci. U.S.A. 110*, 15301–15306.

Davies, K.M., Strauss, M., Daum, B., Kief, J.H., Osiewacz, H.D., Rycovska, A., Zickermann, V., and Kühlbrandt, W. (2011). Macromolecular organization of ATP synthase and complex I in whole mitochondria. *Proc Nat Acad Sci. U.S.A. 108*, 14121–14126.

de Cavanagh, E.M., Ferder, M., Inserra, F., and Ferder, L. (2009). Angiotensin II, mitochondria, cytoskeletal, and extracellular matrix connections: An integrating viewpoint. *Am. J. Physiol. Heart Circ. Physiol. 296*, H550–H558.

Desmurs, M., Foti, M., Raemy, E., Vaz, F.M., Martinou, J.-C., Bairoch, A., and Lane, L. (2015). C11orf83, a mitochondrial cardiolipin-binding protein involved in bc1 complex assembly and supercomplex stabilization. *Mol. Cell. Biol. in press.*

Diaz, F., Kotarsky, H., Fellman, V., and Moraes, C.T. (2011). Mitochondrial disorders caused by mutations in respiratory chain assembly factors. *Semin. Fetal Neonatal Med. 16*, 197–204.

Dudkina, N.V., Heinemeyer, J., Keegstra, W., Boekema, E.J., and Braun, H.-P. (2005). Structure of dimeric ATP synthase from mitochondria: An angular association of monomers induces the strong curvature of the inner membrane. *FEBS Lett. 579*, 5769–5772.

Dudkina, N.V., Kudryashev, M., Stahlberg, H., and Boekema, E.J. (2011). Interaction of complexes I, III, and IV within the bovine respirasome by single particle cryoelectron tomography. *Proc. Nat. Acad. Sci. U.S.A. 108*, 15196–15200.

Echtay, K.S., Roussel, D., St-Pierre, J., Jekabsons, M.B., Cadenas, S., Stuart, J.A., Harper, J.A., Roebuck, S.J., Morrison, A., Pickering, S. et al. (2002). Superoxide activates mitochondrial uncoupling proteins. *Nature 415*, 96–99.

Enriquez, J.A., and Lenaz, G. (2014). Coenzyme q and the respiratory chain: Coenzyme q pool and mitochondrial supercomplexes. *Mol. Syndromol. 5*, 119–140.

Eubel, H., Heinemeyer, J., and Braun, H.-P. (2004). Identification and characterization of respirasomes in potato mitochondria. *Plant Physiol. 134*, 1450–1459.

Faccenda, D., Tan, C.H., Seraphim, A., Duchen, M.R., and Campanella, M. (2013). IF1 limits the apoptotic-signalling cascade by preventing mitochondrial remodelling. *Cell Death Differ. 20*, 686–697.

Frenzel, M., Rommelspacher, H., Sugawa, M.D., and Dencher, N.A. (2010). Ageing alters the supramolecular architecture of OxPhos complexes in rat brain cortex. *Exp. Gerontol. 45*, 563–572.

Frezza, C., Cipolat, S., Martins de Brito, O., Micaroni, M., Beznoussenko, G.V., Rudka, T., Bartoli, D., Polishuck, R.S., Danial, N.N., de Strooper, B. et al. (2006). OPA1 controls apoptotic cristae remodeling independently from mitochondrial fusion. *Cell 126*, 177–189.

Gasparre, G., Porcelli, A.M., Lenaz, G., and Romeo, G. (2013). Relevance of mitochondrial genetics and metabolism in cancer development. *Cold Spring Harb. Perspect. Biol. 5*, a011411–a011411.

Gehl, B., Lee, C.P., Bota, P., Blatt, M.R., and Sweetlove, L.J. (2014). An Arabidopsis stomatin-like protein affects mitochondrial respiratory supercomplex organization. *Plant Physiol. 164*, 1389–1400.

Gehl, B., and Sweetlove, L.J. (2014). Mitochondrial Band-7 family proteins: Scaffolds for respiratory chain assembly? *Front. Plant Sci. 5*, 141.

Genova, M.L., and Lenaz, G. (2011). New developments on the functions of coenzyme Q in mitochondria. *Biofactors 37*, 330–354.

Gilkerson, R.W., Selker, J.M.L., and Capaldi, R.A. (2003). The cristal membrane of mitochondria is the principal site of oxidative phosphorylation. *FEBS Lett. 546*, 355–358.

Gomes, L.C., Di Benedetto, G., and Scorrano, L. (2011). During autophagy mitochondria elongate, are spared from degradation and sustain cell viability. *Nat. Cell Biol. 13*, 589–598.

Gómez, L.A., and Hagen, T.M. (2012). Age-related decline in mitochondrial bioenergetics: Does supercomplex destabilization determine lower oxidative capacity and higher superoxide production? *Semin. Cell Dev. Biol. 23*, 758–767.

Habersetzer, J., Larrieu, I., Priault, M., Salin, B., Rossignol, R., Brèthes, D., and Paumard, P. (2013). Human F_1F_O ATP synthase, mitochondrial ultrastructure and OXPHOS impairment: A (super-) complex matter? *PLoS ONE 8*, e75429.

Hackenbrock, C.R., Chazotte, B., and Gupte, S.S. (1986). The random collision model and a critical assessment of diffusion and collision in mitochondrial electron transport. *J. Bioenerg. Biomembr. 18*, 331–368.

Harner, M., Körner, C., Walther, D., Mokranjac, D., Kaesmacher, J., Welsch, U., Griffith, J., Mann, M., Reggiori, F., and Neupert, W. (2011). The mitochondrial contact site complex, a determinant of mitochondrial architecture. *EMBO J. 30*, 4356–4370.

Hatle, K., Gummadidala, P., Navasa, N., Bernardo, E., Dodge, J., Silverstrim, B., Fortner, K., Burg, E., Suratt, B.T., Hammer, J. et al. (2013). MCJ/DnaJC15, an endogenous mitochondrial repressor of the respiratory chain that controls metabolic alterations. *Mol. Cell. Biol. 33*, 2302–2314.

Hayashi, H., Nakagami, H., Takeichi, M., Shimamura, M., Koibuchi, N., Oiki, E., Sato, N., Koriyama, H., Mori, M., Gerardo Araujo, R. et al. (2012). HIG1, a novel regulator of mitochondrial γ-secretase, maintains normal mitochondrial function. *FASEB J. 26*, 2306–2317.

Hofmann, A.D., Beyer, M., Krause-Buchholz, U., Wobus, M., Bornhäuser, M., and Rödel, G. (2012). OXPHOS supercomplexes as a hallmark of the mitochondrial phenotype of adipogenic differentiated human MSCs. *Plos ONE 7*, e35160.

Hoppins, S., Collins, S.R., Cassidy-Stone, A., Hummel, E., Devay, R.M., Lackner, L.L., Westermann, B., Schuldiner, M., Weissman, J.S., and Nunnari, J. (2011). A mitochondrial-focused genetic interaction map reveals a scaffold-like complex required for inner membrane organization in mitochondria. *J. Cell. Biol. 195*, 323–340.

Jørgensen, B.M., Rasmussen, H.N., and Rasmussen, U.F. (1985). Ubiquinone reduction pattern in pigeon heart mitochondria. Identification of three distinct ubiquinone pools. *Biochem. J 229*, 621–629.

Kane, L.A., Youngman, M.J., Jensen, R.E., and van Eyk, J.E. (2010). Phosphorylation of the F(1) F(o) ATP synthase beta subunit: Functional and structural consequences assessed in a model system. *Circ. Res. 106*, 504–513.

Kao, L.-P., Ovchinnikov, D., and Wolvetang, E. (2012). The effect of ethidium bromide and chloramphenicol on mitochondrial biogenesis in primary human fibroblasts. *Toxicol. Appl. Pharmacol. 261*, 42–49.

Keilin, D., and Hartree, E.F. (1947). Activity of the cytochrome system in heart muscle preparations. *Biochem. J. 41*, 500–502.

Khalifat, N., Puff, N., Bonneau, S., Fournier, J.-B., and Angelova, M.I. (2008). Membrane deformation under local pH gradient: Mimicking mitochondrial cristae dynamics. *Biophys. J. 95*, 4924–4933.

Kozjak-Pavlovic, V., Prell, F., Thiede, B., Götz, M., Wosiek, D., Ott, C., and Rudel, T. (2014). C1orf163/RESA1 is a novel mitochondrial intermembrane space protein connected to respiratory chain assembly. *J. Mol. Biol. 426*, 908–920.

Krause, F., Reifschneider, N.H., Vocke, D., Seelert, H., Rexroth, S., and Dencher, N.A. (2004). "Respirasome-" like supercomplexes in green leaf mitochondria of spinach. *J. Biol. Chem. 279*, 48369–48375.

Lapuente-Brun, E., Moreno-Loshuertos, R., Acín-Peréz, R., Latorre-Pellicer, A., Colás, C., Balsa, E., Perales-Clemente, E., Quirós, P.M., Calvo, E., Rodríguez-Hernández, M.A. et al. (2013). Supercomplex assembly determines electron flux in the mitochondrial electron transport chain. *Science 340*, 1567–1570.

Lass, A., and Sohal, R.S. (1998). Electron transport-linked ubiquinone-dependent recycling of alpha-tocopherol inhibits autooxidation of mitochondrial membranes. *Arch. Biochem. Biophys. 352*, 229–236.

Lass, A., and Sohal, R.S. (1999). Comparisons of coenzyme Q bound to mitochondrial membrane proteins among different mammalian species. *Free Radic. Biol. Med. 27*, 220–226.

Lenaz, G., Fato, R., Formiggini, G., and Genova, M.L. (2007). The role of Coenzyme Q in mitochondrial electron transport. *Mitochondrion 7(Suppl)*, S8–S33.

Lenaz, G., and Genova, M.L. (2007). Kinetics of integrated electron transfer in the mitochondrial respiratory chain: Random collisions vs. solid state electron channeling. *Am. J. Physiol. Cell. Physiol. 292*, C1221–C1239.

Lenaz, G., and Genova, M.L. (2012). Supramolecular organisation of the mitochondrial respiratory chain: A new challenge for the mechanism and control of oxidative phosphorylation. *Adv. Exp. Med. Biol. 748*, 107–144.

Liesa, M., and Shirihai, O.S. (2013). Mitochondrial dynamics in the regulation of nutrient utilization and energy expenditure. *Cell Metab. 17*, 491–506.

Lodi, R., Tonon, C., Valentino, M.L., Iotti, S., Clementi, V., Malucelli, E., Barboni, P., Longanesi, L., Schimpf, S., Wissinger, B. et al. (2004). Deficit of in vivo mitochondrial ATP production in OPA1-related dominant optic atrophy. *Ann. Neurol. 56*, 719–723.

Lopez-Campistrous, A., Hao, L., Xiang, W., Ton, D., Semchuk, P., Sander, J., Ellison, M.J., and Fernandez-Patron, C. (2008). Mitochondrial dysfunction in the hypertensive rat brain: Respiratory complexes exhibit assembly defects in hypertension. *Hypertension 51*, 412–419.

Malsburg, von der, K., Müller, J.M., Bohnert, M., Oeljeklaus, S., Kwiatkowska, P., Becker, T., Loniewska-Lwowska, A., Wiese, S., Rao, S., Milenkovic, D. et al. (2011). Dual role of mitofilin in mitochondrial membrane organization and protein biogenesis. *Dev. Cell 21*, 694–707.

Maranzana, E., Barbero, G., Falasca, A.I., Lenaz, G., and Genova, M.L. (2013). Mitochondrial respiratory supercomplex association limits production of reactive oxygen species from complex I. *Antioxid. Redox Signal. 19*, 1469–1480.

Marchi, S., Giorgi, C., Suski, J.M., Agnoletto, C., Bononi, A., Bonora, M., De Marchi, E., Missiroli, S., Patergnani, S., Poletti, F. et al. (2012). Mitochondria-ros crosstalk in the control of cell death and aging. *J. Signal Transduct. 2012*, 329635.

McBride, H., and Scorrano, L. (2013). Mitochondrial dynamics and physiology. *Biochim. Biophys. Acta 1833*, 148–149.

McKenzie, M., Lazarou, M., Thorburn, D.R., and Ryan, M.T. (2006). Mitochondrial respiratory chain supercomplexes are destabilized in Barth syndrome patients. *J. Mol. Biol. 361*, 462–469.

Mileykovskaya, E., Penczek, P.A., Fang, J., Mallampalli, V.K.P.S., Sparagna, G.C., and Dowhan, W. (2012). Arrangement of the respiratory chain complexes in *Saccharomyces cerevisiae* supercomplex III2IV2 revealed by single particle cryo-electron microscopy (EM). *J. Biol. Chem. 287*, 23095–23103.

Minauro-Sanmiguel, F., Wilkens, S., and García, J.J. (2005). Structure of dimeric mitochondrial ATP synthase: Novel F_O bridging features and the structural basis of mitochondrial cristae biogenesis. *Proc. Nat. Acad. Sci. U.S.A. 102*, 12356–12358.

Mootha, V.K., Lepage, P., Miller, K., Bunkenborg, J., Reich, M., Hjerrild, M., Delmonte, T., Villeneuve, A., Sladek, R., Xu, F. et al. (2003). Identification of a gene causing human cytochrome c oxidase deficiency by integrative genomics. *Proc. Nat. Acad. Sci. U.S.A. 100*, 605–610.

Moreno-Loshuertos, R., Acín-Peréz, R., Fernández-Silva, P., Movilla, N., Pérez-Martos, A., Rodríguez de Córdoba, S., Gallardo, M.E., and Enriquez, J.A. (2006). Differences in reactive oxygen species production explain the phenotypes associated with common mouse mitochondrial DNA variants. *Nat. Genet. 38*, 1261–1268.

Moslehi, J., Depinho, R.A., and Sahin, E. (2012). Telomeres and mitochondria in the aging heart. *Circ. Res. 110*, 1226–1237.

Mourier, A., Ruzzenente, B., Brandt, T., Kühlbrandt, W., and Larsson, N.-G. (2014). Loss of LRPPRC causes ATP synthase deficiency. *Hum. Mol. Genet. 23*, 2580–2592.

Nakamura, T., Cho, D.-H., and Lipton, S.A. (2012). Redox regulation of protein misfolding, mitochondrial dysfunction, synaptic damage, and cell death in neurodegenerative diseases. *Exp. Neurol. 238*, 12–21.

Ott, C., Ross, K., Straub, S., Thiede, B., Götz, M., Goosmann, C., Krischke, M., Mueller, M.J., Krohne, G., Rudel, T. et al. (2012). Sam50 functions in mitochondrial intermembrane space bridging and biogenesis of respiratory complexes. *Mol. Cell. Biol. 32*, 1173–1188.

Packer, L. (1963). Size and shape transforamation correlated with oxidative phosphorilatyon in mitochondria. I. Swelling-shrinkage mechanisms in intact mitochondria. *J. Cell. Biol. 18*, 487–494.

Patten, D.A., Wong, J., Khacho, M., Soubannier, V., Mailloux, R.J., Pilon-Larose, K., MacLaurin, J.G., Park, D.S., McBride, H.M., Trinkle-Mulcahy, L. et al. (2014). OPA1-dependent cristae modulation is essential for cellular adaptation to metabolic demand. *EMBO J. 33*, 2676–2691.

Paumard, P., Vaillier, J., Coulary, B., Schaeffer, J., Soubannier, V., Mueller, D.M., Brèthes, D., Di Rago, J.-P., and Velours, J. (2002). The ATP synthase is involved in generating mitochondrial cristae morphology. *EMBO J. 21*, 221–230.

Pfeiffer, K., Gohil, V., Stuart, R.A., Hunte, C., Brandt, U., Greenberg, M.L., and Schägger, H. (2003). Cardiolipin stabilizes respiratory chain supercomplexes. *J. Biol. Chem. 278*, 52873–52880.

Rabl, R., Soubannier, V., Scholz, R., Vogel, F., Mendl, N., Vasiljev-Neumeyer, A., Körner, C., Jagasia, R., Keil, T., Baumeister, W. et al. (2009). Formation of cristae and crista junctions in mitochondria depends on antagonism between Fcj1 and Su e/g. *J. Cell. Biol. 185*, 1047–1063.

Rieger, B., Junge, W., and Busch, K.B. (2014). Lateral pH gradient between OXPHOS complex IV and F(0)F(1) ATP-synthase in folded mitochondrial membranes. *Nature Comm. 5*, 3103.

Rosca, M., Minkler, P., and Hoppel, C.L. (2011). Cardiac mitochondria in heart failure: Normal cardiolipin profile and increased threonine phosphorylation of complex IV. *Biochim. Biophys. Acta 1807*, 1373–1382.

Salminen, A., Ojala, J., Kaarniranta, K., and Kauppinen, A. (2012). Mitochondrial dysfunction and oxidative stress activate inflammasomes: Impact on the aging process and age-related diseases. *Cell. Mol. Life Sci. 69*, 2999–3013.

Sasarman, F., Brunel-Guitton, C., Antonicka, H., Wai, T., Shoubridge, E.A., and LSFC Consortium. (2010). LRPPRC and SLIRP interact in a ribonucleoprotein complex that regulates posttranscriptional gene expression in mitochondria. *Mol. Biol. Cell 21*, 1315–1323.

Schägger, H., and Pfeiffer, K. (2000). Supercomplexes in the respiratory chains of yeast and mammalian mitochondria. *EMBO J. 19*, 1777–1783.

Scorrano, L., Ashiya, M., Buttle, K., Weiler, S., Oakes, S.A., Mannella, C.A., and Korsmeyer, S.J. (2002). A distinct pathway remodels mitochondrial cristae and mobilizes cytochrome c during apoptosis. *Dev. Cell. 2*, 55–67.

Scorrano, L., and Korsmeyer, S.J. (2003). Mechanisms of cytochrome c release by proapoptotic BCL-2 family members. *Biochem. Biophys. Res. Commun. 304*, 437–444.

Scorrano, L., Oakes, S.A., Opferman, J.T., Cheng, E.H., Sorcinelli, M.D., Pozzan, T., and Korsmeyer, S.J. (2003). BAX and BAK regulation of endoplasmic reticulum Ca2+: A control point for apoptosis. *Science 300*, 135–139.

Siekevitz, P. (1957). Powerhouse of the cell. *Sci. Am. 197*, 131–144.

Slater, E.C. (2003). Keilin, cytochrome, and the respiratory chain. *J. Biol. Chem. 278*, 16455–16461.

Soriano, M.E., and Scorrano, L. (2010). The interplay between BCL-2 family proteins and mitochondrial morphology in the regulation of apoptosis. *Adv. Exp. Med. Biol. 687*, 97–114.

Stanley, I.A., Ribeiro, S.M., Giménez-Cassina, A., Norberg, E., and Danial, N.N. (2013). Changing appetites: The adaptive advantages of fuel choice. *Trends Cell Biol. 24*, 118–127.

Strauss, M., Hofhaus, G., Schröder, R.R., and Kühlbrandt, W. (2008). Dimer ribbons of ATP synthase shape the inner mitochondrial membrane. *EMBO J. 27*, 1154–1160.

Strecker, V., Wumaier, Z., Wittig, I., and Schägger, H. (2010). Large pore gels to separate mega protein complexes larger than 10 MDa by blue native electrophoresis: Isolation of putative respiratory strings or patches. *Proteomics 10*, 3379–3387.

Strogolova, V., Furness, A., Robb-McGrath, M., Garlich, J., and Stuart, R.A. (2012). Rcf1 and Rcf2, members of the hypoxia-induced gene 1 protein family, are critical components of the mitochondrial cytochrome bc1-cytochrome c oxidase supercomplex. *Mol. Cell. Biol. 32*, 1363–1373.

Sundaresan, M., Yu, Z.X., Ferrans, V.J., Irani, K., and Finkel, T. (1995). Requirement for generation of H2O2 for platelet-derived growth factor signal transduction. *Science 270*, 296–299.

Vukotic, M., Oeljeklaus, S., Wiese, S., Vögtle, F.-N., Meisinger, C., Meyer, H.E., Zieseniss, A., Katschinski, D.M., Jans, D.C., Jakobs, S. et al. (2012). Rcf1 mediates cytochrome oxidase assembly and respirasome formation, revealing heterogeneity of the enzyme complex. *Cell Metab. 15*, 336–347.

Weber, T.A., Koob, S., Heide, H., Wittig, I., Head, B., van der Bliek, A., Brandt, U., Mittelbronn, M., and Reichert, A.S. (2013). APOOL is a cardiolipin-binding constituent of the Mitofilin/MINOS protein complex determining cristae morphology in mammalian mitochondria. *PLoS ONE 8*, e63683.

Whelan, S.P., and Zuckerbraun, B.S. (2013). Mitochondrial signaling: Forwards, backwards, and in between. *Oxid. Med. Cell. Longev. 2013*, 1–10.

Whitelegge, J. (2011). Structural biology. Up close with membrane lipid-protein complexes. *Science 334*, 320–321.

Wikstrom, J.D., Mahdaviani, K., Liesa, M., Sereda, S.B., Si, Y., Las, G., Twig, G., Petrovic, N., Zingaretti, C., Graham, A. et al. (2014). Hormone-induced mitochondrial fission is utilized by brown adipocytes as an amplification pathway for energy expenditure. *EMBO J. 33*, 418–436.

Wilkens, V., Kohl, W., and Busch, K. (2013). Restricted diffusion of OXPHOS complexes in dynamic mitochondria delays their exchange between cristae and engenders a transitory mosaic distribution. *J. Cell Sci. 126*, 103–116.

Wittig, I., and Schägger, H. (2009). Supramolecular organization of ATP synthase and respiratory chain in mitochondrial membranes. *Biochim. Biophys. Acta 1787*, 672–680.

Zanna, C., Ghelli, A., Porcelli, A.M., Karbowski, M., Youle, R.J., Schimpf, S., Wissinger, B., Pinti, M., Cossarizza, A., Vidoni, S. et al. (2008). OPA1 mutations associated with dominant optic atrophy impair oxidative phosphorylation and mitochondrial fusion. *Brain 131*, 352–367.

CHAPTER 4

Role of Cardiolipin in Mitochondrial Supercomplex Assembly

Eugenia Mileykovskaya and William Dowhan

CONTENTS

4.1 INTRODUCTION

The organization of the mitochondrial electron transport chain (ETC, or respiratory chain) individual complexes into large supermolecular complexes (SCs) or respirasomes was originally postulated by Chance (Chance and Williams 1955). As individual respiratory complexes were purified and shown to be functional, the respirasome model fell out of favour. In recent years SCs have been isolated and shown by biochemical and structural analysis to be composed of stoichiometric amounts of individual respiratory complexes thus reviving the respirasome. The respirasomes reside in the inner mitochondrial membrane (IMM) as integral membrane components requiring mild detergent disruption of the membrane for release as soluble components with associated phospholipid components. The phospholipid cardiolipin (CL), found almost exclusively in energy-transducing membranes, was found to be both an integral component within each of the individual respiratory complexes and necessary for formation and stability of higher order organization of these complexes into SCs. Following are the details of organization of the mitochondrial ETC and the role of CL in its organization and function.

4.2 SUPRAMOLECULAR ORGANIZATION OF THE MITOCHONDRIAL RESPIRATORY CHAIN

In most species the mitochondrial ETC is comprised of four multi-subunit electron transfer protein complexes: Complex I (CI; NADH:ubiquinone oxidoreductase), Complex II (CII; succinate:ubiquinone oxidoreductase), Complex III or cytochrome bc_1 complex (CIII; ubiqunol:cytochrome *c* oxidoreductase, which operates through the Q cycle) and Complex IV (CIV; cytochrome *c* oxidase, terminal oxidase). Lipid-soluble ubiquinone (UQ or coenzyme Q), which transfers electrons from CI or CII to CIII, and water-soluble cytochrome *c*, which shuttles electrons from CIII to CIV, are two small electron carriers that complete the ETC. Thus, there are two pathways of electron transfer: the first one starting from NADH through CI and the second one, coupled with the Krebs cycle in the matrix, starting from the oxidation of succinate by CII. CI, CIII and CIV electron transfer is coupled with proton pumping from the mitochondrial matrix to the mitochondrial inter-membrane space (IMS), which is between the IMM and outer mitochondrial membrane (OMM), resulting in the production of the electrochemical proton gradient across the IMM used by F_1F_0-ATPase (Complex V) for ATP synthesis.

The following is a brief description of the structure and function of CI, CII, CIII and CIV (see Sun et al. [2013] and accompanying references for more detailed information). CI is the first enzyme of the mitochondrial respiratory chain. It couples the transfer of two electrons from NADH to UQ with the translocation of four protons from the matrix to the IMS (Figure 4.1a). Similar enzymes are found in the membranes of many bacteria. CI is the largest protein complex (about 1 MDa and containing 44 subunits) in the mitochondrial ETC. It has an L-shape being formed by the peripheral arm extended into mitochondrial matrix (or cytosol in bacteria) and a hydrophobic membrane arm. The peripheral arm contains the NADH-binding centre at its distal end and a number of redox centres including one flavin mononucleotide and up to 10 iron-sulphur clusters. The membrane arm represents the transmembrane proton-translocating machinery, which works through conformational changes induced upon the reduction of UQ at its binding site located at the interface between the two arms (Brandt 2011; Efremov and Sazanov 2011).

CII consists of a hydrophilic catalytic heterodimer containing flavin adenine dinucleotide and three iron-sulphur clusters. This domain is connected to a transmembrane domain composed of two hydrophobic subunits containing one heme *b* and the UQ-binding site located proximal to the matrix side of the membrane. The enzyme oxidizes succinate and transfers electrons through iron-sulphur clusters to UQ without generating a proton gradient across the membrane (Sun et al. 2005).

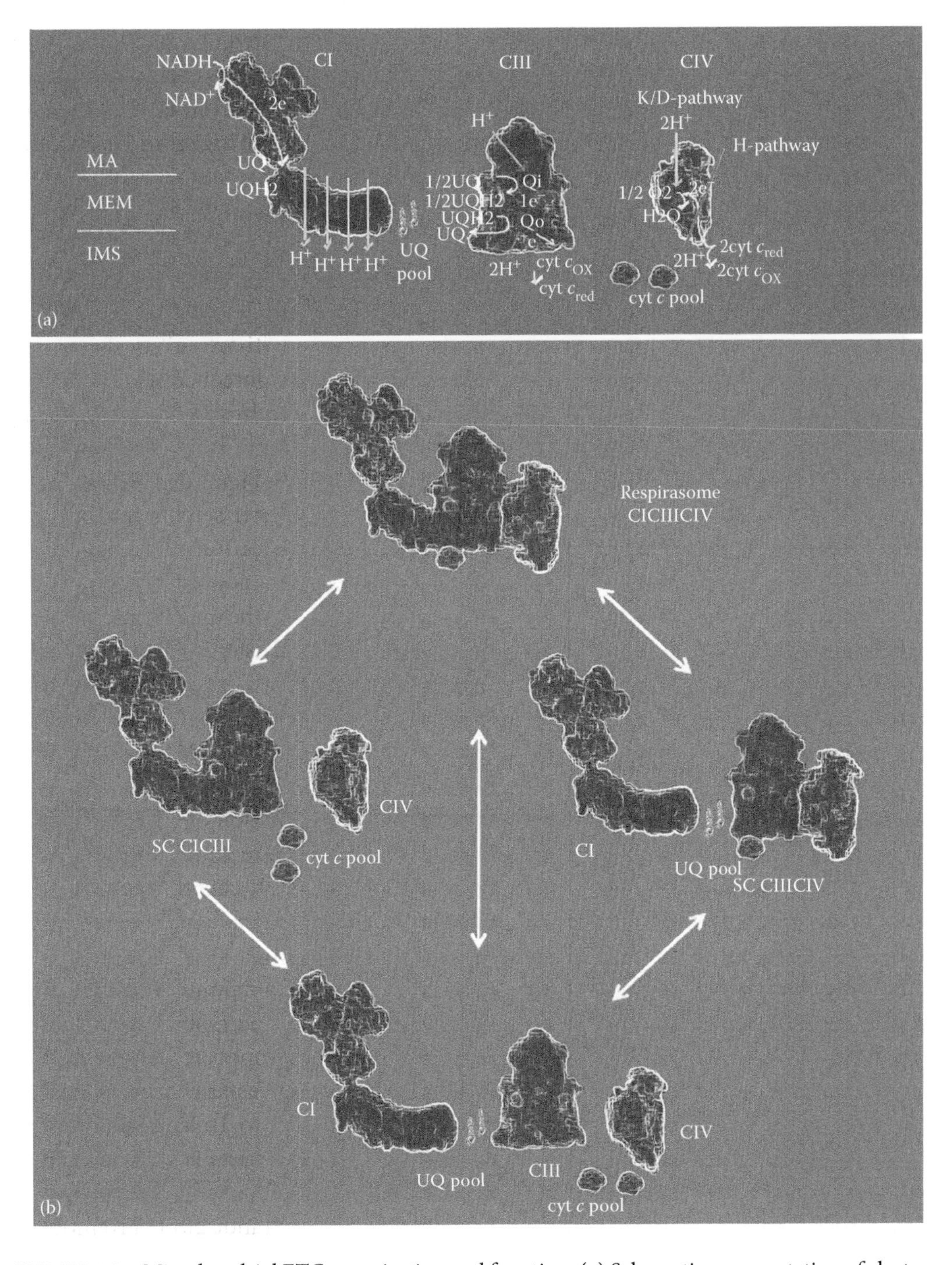

FIGURE 4.1 Mitochondrial ETC organization and function. (a) Schematic representation of electron transfer and proton pumping in CI, CIII and CIV. (b) Dynamic supramolecular organization of ETC complexes. Individual CI, CIII and CIV would co-exist with individual complexes organized into respirasomes in the mitochondrial membrane. The equilibrium among these different species would depend on the metabolic state of the cell. In the former case individual complexes would be connected by randomly diffusing pools of UQ and cytochrome *c*. In the latter case substrate channelling of UQ and cytochrome *c* between individual complexes within the respirasome would transfer electrons.

CIII catalyses the reduction of cytochrome *c* by oxidation of UQH_2 (reduced UQ) produced by CI and CII, and couples the transfer of two electrons with translocation of four protons across the IMM. Crystal structures of CIII from mitochondria of different species including bovine, chicken, rabbit and yeast have been determined (for review and references, see Xia et al. 2013). CIII is a structural and functional physiological dimer,* in which each monomer (about 250 kDa) is composed of 10 (yeast) or 11 (bovine) subunits including 3 catalytic subunits: cytochrome *b* (contains high-potential b_H and low-potential b_L hemes and is encoded by mitochondria DNA), cytochrome c_1 (containing heme c_1) and one iron-sulphur cluster binding protein (ISP, or Rieske protein). There are two UQ-binding sites in each monomer of CIII: one site is proximal to the matrix side (Q_i site) and another one is proximal to IMS (Q_o site). The UQH_2 is oxidized at the Q_o site with one electron transferring to cytochrome *c* via movement of ISP towards cytochrome c_1 heme and the second electron transferring to the Q_i site via hemes b_L and b_H. At the Q_i site the UQ is reduced to the UQH_2, and the newly reduced UQH_2 enters the UQ pool to be recycled to the Q_o site for oxidation (Figure 4.1a). In each cycle, there are two electrons transferred from UQH_2 to two molecules of cytochrome *c*. As a result, the proton translocation and electron transfer within CIII are coupled with a redox-driven Q-cycle mechanism (Hunte et al. 2003; Mitchell 1975; Osyczka et al. 2005).

CIV is the terminal component of the mitochondrial ETC that transfers electrons from cytochrome *c* to an oxygen molecule, resulting in the production of two water molecules (Figure 4.1a). This process is coupled with pumping of four protons from the matrix to the IMS. Bovine CIV (about 200 kDa) consists of 13 subunits, three of which (conserved subunits I, II and III that are encoded by mitochondrial DNA) form the core of this enzyme. All additional subunits of bovine CIV surround the core and stabilize the whole complex. Subunit II holds a Cu centre (CuA) with two Cu atoms that accept electrons from cytochrome *c*. Heme *a*, heme a_3 and a Cu centre termed CuB are located in subunit I. Heme a_3 and CuB form a binuclear centre where the oxygen molecule is reduced to form water. At the same time the protons from the matrix are pumped to IMS. Thus, two types of proton pathways lie within CIV: one type for delivery of matrix protons to the binuclear centre where they are consumed in water formation and the second type for proton pumping from the matrix to IMS. According to structure/function studies of mitochondrial wild-type and mutant CIV, three proton pathways, called K-pathway, D-pathway and H-pathway, were identified. It was proposed that both the K-pathway and D-pathway are responsible for delivering protons to the binuclear centre for water production and the H-pathway catalyses proton pumping from matrix to IMS in the process directly coupled to the redox conversion of heme *a* (see Sun et al. 2013). Alternatively, it was proposed that the protons are pumped through the D-pathway (see Rich and Marechal 2013).

CIV from *Saccharomyces cerevisiae* is composed of 11 subunits. All 11 subunits including 3 conserved core subunits (Cox1p, Cox2p and Cox3p) encoded by mitochondrial DNA

---

* When referring to CIII (a structural and functional dimer of two cytochrome bc_1 complexes) in an SC it is denoted as III_2.

and 8 nuclear-encoded subunits have their respective counterparts in the structure of bovine CIV. Two bovine subunits not present in the yeast structure are peripheral subunits VIIb and VIII. The crystal structure of yeast CIV was not reported yet. However, high similarity between subunits of yeast and bovine CIVs made it possible to build a 3D structural homology model for the yeast complex using the Modeller program (Marechal et al. 2012). The model demonstrates high structural and functional similarity of yeast and bovine CIVs including a possible H channel presence in the Cox1p subunit, which is an analogue of subunit I of the bovine enzyme.

The mitochondrial respiratory chain was originally considered to be a structural and functional unit composed of all respiratory complexes (Chance and Williams 1955). Subsequently, purified individual respiratory complexes were reconstituted together into liposomes and found to be fully capable of electron transfer. As a result, the initial model was replaced by the *random collision* model (Hackenbrock et al. 1986). In this model the individual respiratory complexes are envisioned independently imbedded in the lipid bilayer and connected by randomly diffusing UQ and cytochrome *c*. However, development of blue-native and colourless-native polyacrylamide gel electrophoretic techniques (BN-PAGE and CN-PAGE) of mitochondrial membrane extracts obtained by solubilization with the mild detergent digitonin revealed active stoichiometric assemblies of individual complexes into respiratory SCs. In case of mammalian mitochondria, BN-PAGE demonstrated association of CI with CIII (I_1III_2) or CI with CIII and one to four copies of CIV ($I_1III_2IV_{n=1-4}$) resulting in multiple SCs (Schagger 2001; Schagger and Pfeiffer 2000; Stuart 2008). *S. cerevisiae* lacks CI and reduces UQ utilizing peripheral membrane NADH dehydrogenases (not proton pumps) Nde1 and Nde2, which face the IMS and oxidize cytosolic NADH, and Ndi, which faces the matrix and oxidizes mitochondrial NADH (Yamashita et al. 2007). Electrons from UQH_2 are passed to two SCs, one composed of CIII and CIV (III_2IV) and another of CIII and two CIVs (III_2IV_2), which can be visualized by BN-PAGE and CN-PAGE (Schagger and Pfeiffer 2000; Stuart 2008). Metabolic flux control analysis of electron transfer through the mitochondrial ETC showed substrate channelling of UQ in bovine heart mitochondria and both UQ and cytochrome *c* in mitochondria of potato tuber, which demonstrated existence of structural and functional respiratory SCs (see Genova and Lenaz [2014]; Lenaz and Genova [2007] for reviews). A similar kinetic approach was applied to *S. cerevisiae* mitochondria isolated from cells grown on non-fermentable carbon sources. Both UQ and cytochrome *c* were found not to follow pool behaviour, thus supporting functional SC formation and channelling of small electron carriers. Treatment of mitochondria with chaotropic agents to disrupt SC formation resulted in pool behaviour kinetics (Boumans et al. 1998). The difference in kinetic behaviour of cytochrome *c* in mammalian and yeast mitochondria may stem from the difference in the levels of CIV not incorporated into SCs (Genova and Lenaz 2014). Indeed, in the case of mammalian mitochondria almost all CI and CIII are included into the SCs, but there is a population of free CIV, which is not included into the SCs (Wittig et al. 2006). In the case of yeast mitochondria grown on non-fermentable carbon source practically all CIV is included into SCs (Cruciat et al. 2000; Zhang et al. 2002).

On the other hand more recent studies (Trouillard et al. 2011) using time-resolved oxidation of cytochrome *c* by CIV in yeast have led the authors to the conclusion that transfer of electrons occurs by a random process not involving any kind of cytochrome *c* compartmentalization. The technique relied on reversible competitive inhibition of reduced CIV by carbon monoxide (CO), in which flash-induced dissociation of the CO-a_3 heme complex allows fast oxygen binding. This approach was used to study a number of isolated terminal oxidases to obtain detailed information on the electron pathway and proton pumping during the enzyme turnover (Einarsdottir and Szundi 2004; Kaila et al. 2010). However the approach, previously used in these isolated systems, did not take into account the oxygen-sensing function of CIV in whole cell, which might contribute to signalling pathways activated in the cells exposed to CO or anoxia (Kwast et al. 1999) and the ability of the ETC to dynamically reorganize with rapid dissociation of cytochrome *c* from respiratory SCs (Morin et al. 2003). Therefore, additional controls are required for the final interpretation of results obtained by this method in the whole cells.

Finally, it was demonstrated that SCs extracted from mitochondria of mouse cells and purified by BN-PAGE represented the entire structural and functional respirasome containing UQ and some amount of bound cytochrome *c* capable of electron transfer from NADH to oxygen, which was sensitive to specific inhibitors of the ETC (Acín-Pérez et al. 2008). In accordance with these experiments digitonin-extracted and sucrose-gradient-purified yeast SC composed of CIII and two CIVs (III_2IV_2) contained detectible amounts of cytochrome *c* and catalysed cyanide-sensitive (inhibits CIV) electron transfer from UQH_2 to oxygen (Mileykovskaya et al. 2012). This type or organization is important for the optimization of turnover of ETC enzymes, control of reactive oxygen species (ROS) production in ETC and stability of individual complexes (specifically CI) (Acín-Pérez et al. 2004, 2008; Maranzana et al. 2013).

A new *plasticity model*, which combines both extremes in the organization of the respiratory chain, was proposed based on experimental results. Alteration of the ratios of SCs and individual complexes by genetic manipulation (Barrientos and Ugalde 2013; Lapuente-Brun et al. 2013) revealed existence of UQ and cytochrome *c* pools dedicated to specific electron transfer pathways in mitochondria from these cells. Thus, electron transfer in the respiratory chain might occur either by direct substrate channelling of UQ and cytochrome *c* between respective individual complexes within a SC or via diffusion and random collision of these two electron carriers with the non-associated individual complexes (Figure 4.1b). SC assembly is dynamic and organizes electron flux to optimize the use of available substrates. This plasticity in the organization of the respiratory chain appears to be necessary for the cell to rapidly respond to changes in the physiological state and energy demands of the cell (Lapuente-Brun et al. 2013).

Three groups reported simultaneously the first identification of the protein factors necessary for CIII-CIV SC assembly in yeast (Chen et al. 2012; Strogolova et al. 2012; Vukotic et al. 2012), termed Rcf (respiratory supercomplex factor) 1 and 2. Either Rcf1p or Rcf2p is required for late-stage assembly of the Cox12p (mammalian CIV subunit VIb) and Cox13p subunits and for cytochrome *c* oxidase activity. These factors are members of the hypoxia-induced gene family and important for growth under hypoxic conditions. *Caenorhabditis elegans* and human orthologs of Rcf1p expressed in yeast are functional. It is not clear whether the Rcf proteins only support assembly of the SC or also directly participate in the stabilization of the SC structure. Another protein factor that participates in the formation and stabilization of respiratory SCs was recently identified in mammalian cells. Cox7a2l (CIV subunit VIIa polypeptide 2-like) was renamed SuperComplex Assembly Factor I (SCAFI) since its presence was mandatory for SC formation specifically for the interaction of CIV with CIII in the mouse mitochondria. Importantly, in this case SCAFI was not found in the individual CIV or CIII but was a component of the SCs (Lapuente-Brun et al. 2013). There is no analogue of this protein in *S. cerevisiae*. Another group demonstrated that the Cox7a-related protein in mice promotes stabilization of respiratory SCs. Homozygous mice with a knockout of the corresponding gene exhibited muscle weakness. Interestingly, this nuclear gene encodes a protein similar to polypeptides 1 and 2 of subunit VIIa (Cox7a-related proteins) in the C-terminal region, which is highly similar to the mouse Sig81 (estrogen receptor) protein sequence. It is possible that Cox7rp represents a regulatory subunit of CIV, which mediates the higher level of energy production in target cells in response to estrogen (Ikeda et al. 2013).

The majority of evidence supports the functional existence in mitochondria of higher order SCs composed of individual respiratory complexes. Although additional proteins appear to be involved in SC formation, lipids provide a more malleable and less stable interface between proteins, which would facilitate dynamic responses to changing physiological conditions. As will be outlined below (4.4–4.6), the phospholipid CL is specifically required to form respiratory SCs even in the presence of the earlier mentioned auxiliary proteins.

4.3 CARDIOLIPIN AS A UNIQUE PHOSPHOLIPID OF ENERGY-TRANSDUCING MEMBRANES

4.3.1 Biosynthesis of CL

CL (1′,3′-Bis-(1,2-diacyl-*sn*-glycero-3-phospho)-*sn*-glycerol) is a unique anionic phospholipid in that it contains four fatty acid chains, three glycerol residues and two phosphate moieties (see Figure 4.2 for structure), thus it is also referred to as diphosphatidylglycerol. The pathways for synthesis (see Figure 4.2) of CL in eukaryotic cells including yeast and mammalian sources are very similar with the final steps of biosynthesis occurring exclusively in the mitochondrial inner membrane by nuclear-encoded gene products.

Phosphatidic acid and CDP-diacylglycerol (-DAG) are common precursors to CL in all species. However, the location of enzymes responsible for the synthesis of these two precursors differs between mammalian cells and yeast. Eukaryotic cells contain multiple genes expressing the first and second acyltransferases involved in the formation of phosphatidic acid from glycerol. In yeast (Henry et al. 2012), the first and second acyltransferases are localized to the endoplasmic reticulum (ER) membrane. In mammalian cells (Wendel et al. 2009), the first acyltransferases are found in the OMM and ER membranes. The second acyltransferases reside only in the ER membrane. Mammalian cells express CDP-DAG synthases from two nuclear genes (*CDS1* and *CDS2*) whose products are localized to the ER membrane (Volta et al. 1999; Weeks et al. 1997). Since there is no

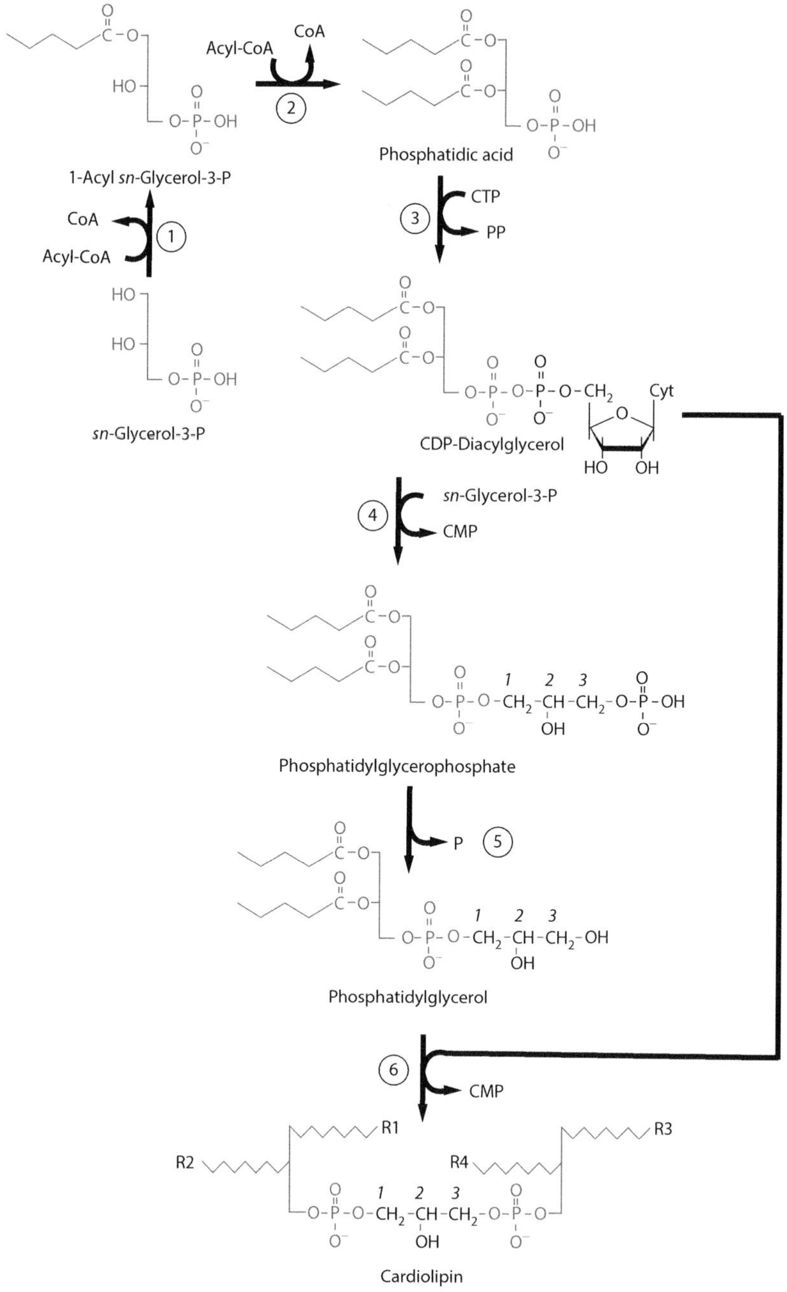

FIGURE 4.2 Biosynthetic pathway for the synthesis of CL. Numbers next to arrows indicate steps catalysed by the following enzymes: (1) *sn*-glycerol-3-phosphate:acyl-CoA acyltransferase, (2) 1-acyl-*sn*-glycerol-3-phosphate:acyl-CoA acyltranferase, (3) CDP-DAG synthase, (4) PGP synthase, (5) PGP phosphatase and (6) CL synthase. Numbers 1–3 over glycerol in PG and CL indicate the *sn* numbering. R1-R4 in the CL structure indicate distinct fatty acid species at each position.

evidence for a mitochondrial synthase, it appears that CDP-DAG must be transported to the IMM to complete CL synthesis. The yeast and mammalian *CDS1* gene products are highly homologous (Weeks et al. 1997) while the *CDS2* gene product appears to be a unique mammalian enzyme. The yeast ER CDP-DAG synthase is primarily responsible for supplying CDP-DAG for phosphatidylinositol synthesis in the ER, since the unique nuclear *TAM41* gene encodes a CDP-DAG localized to the IMM (Tamura et al. 2013). Null mutants in the yeast *CDS1* gene are lethal because of its necessity for phosphatidylinositol synthesis (Shen et al. 1996). However, null mutants in the *TAM41* gene are not lethal and only display a reduced ability to synthesize the downstream products (phosphatidylglycerol [PG] and CL) of the mitochondrial pathway (Tamura et al. 2013). This would suggest that there are mechanisms for transporting CDP-DAG from the ER to the mitochondria but not in the reverse or that additional mitochondrial-localized CDP-DAG synthases exist.

The synthesis of PG beginning with phosphatidic acid proceeds by identical pathways in all eukaryotic cells. The *PGS1* gene product (phosphatidylglycerophosphate [PGP] synthase) of yeast (Chang et al. 1998a) and mammalian cells (Kawasaki et al. 1999) shows a high degree of homology. PGP is rapidly dephosphorylated to PG in all systems leaving only trace amounts. The mammalian (*PTPMT1* gene) (Zhang et al. 2011) and yeast (*GEP4* gene) (Osman et al. 2010) gene products, which convert PGP to PG, display no significant homology. Eukaryotic cells appear to contain additional enzymes capable of dephosphorylating PGP since null mutants in what appear to be the primary genes still synthesize some PG.

The final step in the pathway is carried out by a single CL synthase. The human *hCLS1* (Chen et al. 2006; Lu et al. 2006) and yeast *CRD1* (Chang et al. 1998b; Jiang et al. 1997; Tuller et al. 1998) gene products are homologous and carry out the transfer of a phosphatidic acid moiety from CDP-DAG to the *sn*-1 hydroxyl of glycerol of PG. The equilibrium for the eukaryotic CL synthase catalysed step lies far to the formation of CL due to the conversion of a high-energy pyrophosphate bond to a lower-energy phosphate ester. The result is near complete conversion to CL with only trace amounts of PG in the mitochondria. Although PG and CL are partially interchangeable in supporting mitochondrial function, optimal function of mitochondria is dependent on CL.

In eukaryotic cells the fatty acid composition of CL is modified from that of its lipid precursors to a cell-dependent composition (Schlame 2013; Schlame et al. 2005). The most dramatic modification occurs in heart mitochondria where 80% of the fatty acid chains are 18 carbons long with double bonds between carbons 9/10 and carbons 12/13 (linoleic acid, $18{:}2^{\Delta 9,12}$). Although there is considerable variation between cell types and across species, the modifications always result in almost exclusively unsaturated fatty acids and only a few unique CL species. For instance the yeast CL pool of fatty acids is almost exclusively palmitoleic acid ($16{:}1^{\Delta 9}$) and oleic acid ($18{:}1^{\Delta 9}$). The exact mechanism by which this modification occurs has not been completely resolved, but the process involves continuous action of a calcium-independent phospholipase A_2, resulting in the formation of monolyso-CL (MLCL) followed by a direct transfer of a fatty acid from the *sn*-2 position of phosphatidylcholine (PC) to MLCL by the enzyme tafazzin (encoded by the *TAZ1* gene) (Malhotra et al. 2009).

The phospholipase is highly specific for saturated fatty acids of nascent CL while the tafazzin PC substrate is highly enriched in unsaturated fatty acids at the *sn*-2 position resulting in the increase in unsaturated fatty acid content of CL. A model based on thermodynamic stability of various CL species suggests that high enrichment of unsaturated fatty acids in CL would occur without fatty acid specificity for tafazzin (Schlame 2009). However, there are also additional fatty acid transferases in mitochondria with high specificity for unsaturated fatty acids. Interestingly, tafazzin is localized to the outer leaflet of the IMM, which would require transverse flipping of CL from its site of synthesis within the inner leaflet (Baile et al. 2014a).

4.3.2 Chemical and Physical Properties of CL

Given its structure, CL displays unique physical and chemical properties (Schlame et al. 2005). The phosphatidyl moieties flanking the central glycerol are both derived from the natural pool of phosphatidic acid. Although each contains a chiral centre at the *sn*-2 position of glycerol, these two groups have the same stereochemistry. Thus, the central free hydroxyl is a prochiral centre (i.e. it is only potentially a chiral centre) when the fatty acids R1 = R3 and R2 = R4. However, if either of these fatty acid pairs are not the same, the central hydroxyl becomes a chiral centre. Additionally, even the prochiral form displays six unique surface areas depending on the direction an enzyme binding site approaches CL (i.e. right, left, top, bottom, front or back). Therefore, one function of CL remodelling may be to reduce the number of stereoisomers to maximize the pool of effective CL species required for specific processes. The presence of two phosphatidic acid moieties provides the potential of CL to interaction with at least two lipid-binding sites to stabilize multisubunit complexes. Symmetrical CL molecules are also at a lower energy state due to increased mobility within the molecule (Schlame 2009). This is a factor considered in the mathematical model by which symmetry is generated without assigning fatty acid specificity to tafazzin. CL has been postulated to act as a proton buffer due to an initial determination that the two phosphates have very different pKas of around 2 and 7 due to hydrogen bonding of the phosphate groups with the central hydroxyl. This would suggest a singly charged CL near physiological pH (Haines 2009). Only symmetrical CL species would support the singly protonated form of CL because CL species with an asymmetric distribution of fatty acids would distort the angles necessary to maximize hydrogen bonding with the central hydroxyl. However, recent reports indicate that there may be little difference in pKas (Olofsson and Sparr 2013).

CL is considered to be a potential non-bilayer forming lipid, which can disrupt the regular bilayer arrangements of phospholipids or result in phase separation into CL-enriched domains that induce high curvature in a normally planar lipid bilayer. Lipids whose hydrophobic fatty acid domain and hydrophilic head group domain carve out a cylindrical space (such as PC or PG) readily form lipid bilayers in solution. However, lipids with a large disparity in lateral space occupied by these two domains form alternative organizations in solution. CL at low pH or chelated with divalent cations displays a small head group relative to the fatty acid domain resulting in non-bilayer properties—for review, see Lewis and McElhaney (2009). Mitochondria maintain significant pH and calcium gradients, which would support the non-bilayer properties, suggesting that CL may be enriched in lipid domains or found in high concentrations at the points of high membrane curvature such as at the IMM-OMM contact sites or within the highly curved regions of mitochondrial cristae—for review, see Mileykovskaya and Dowhan (2009) and references.

4.3.3 Consequences of Mutations in the CL Biosynthetic Pathway

Mutants in the eukaryotic pathway for CL biosynthesis display increasing degrees of severity the earlier the defect lies in the pathway. Mutations in higher eukaryotes display more severe phenotypes than those observed in lower eukaryotes. All mutations except for null mutants in the yeast CL deacylase (Δ*cld1* that encodes a CL-specific phospholipase A_2) (Baile et al. 2014b) show reduced mitochondrial function and disruption of mitochondrial morphology. Lack of the yeast mitochondrial CDP-DAG synthase (Δ*tam41*) and PGP phosphatase (Δ*gep4*) results in only partial defects in mitochondrial function apparently due to some transport of CDP-DAG from the ER (Tamura et al. 2013) and the presence of other phosphatases (Osman et al. 2010), respectively. Lack of the yeast PGP synthase (Δ*pgs1*) has the most severe consequences in yeast (Chang et al. 1998a) with complete loss of mitochondrial function due to loss of mitochondrial DNA (Zhong and Greenberg 2005) and lack of synthesis of mitochondria-encoded subunits of CIII and CIV and the nuclear-encoded subunit Cox4p of CIV (Ostrander et al. 2001). Cells are temperature sensitive and grow only by fermentation due to lack of a functional ETC. Yeast mutants lacking CL synthase (Δ*crd1*) are temperature sensitive for growth and display compromised growth on non-fermentable carbon sources (i.e. slower growth in exponential phase and a reduction in final cell density) (Zhang et al. 2002), impaired iron homeostasis (Patil et al. 2013; Raja and Greenberg 2014), and reduced membrane potential (Jiang et al. 2000). CL makes up about 20%–25% of the IMM phospholipid while PG comprises 1% or less (Horvath and Daum 2013). In Δ*crd1* mutant CL is undetectable and PG levels rise to about 20%–25% suggesting that the increase in PG partially substitutes in functions that are optimal when CL is present. This partial suppression of CL-dependent phenotypes by PG complicates assigning specific roles for CL in mitochondrial function. Most examples of mitochondrial dysfunction due to reduced CL levels in higher eukaryotes do not show increased PG levels making such cells behave more like partial mutations in PGP synthase. Therefore, use of Δ*crd1* yeast cells to study CL function must consider the role PG plays in substituting for CL.

4.4 DIRECT INVOLVEMENT OF CL IN THE MITOCHONDRIAL SC FORMATION

4.4.1 Yeast as Model System

Yeast mutants in CL biosynthesis display many of the mitochondrial phenotypes observed in higher eukaryotes. Therefore, coupled with the ease of genetic manipulation, growth and biochemical analysis, *S. cerevisiae* is an excellent model organism to study the role of CL in mitochondrial function. To study the role of CL in the SC organization of the ETC, strains of *S. cerevisiae*, in which the CL content can be regulated *in vivo*, were constructed. For this purpose, a plasmid-borne copy of the *CRD1* gene under transcriptional down-regulation by doxycycline in the growth medium was introduced into a Δ*crd1* yeast strain.

Addition of the repressor doxycycline to the growth medium resulted in a dose-dependent decrease in CL levels accompanied by a parallel increase in the levels of its immediate precursor, PG (Zhang et al. 2002).

BN-PAGE of digitonin extracts of mitochondria revealed that in Δ*crd1* yeast mutants completely lacking CL, but containing elevated amounts of its precursor PG, did not form the stable SC III_2IV_2 as observed in the wild type parental strain (Pfeiffer et al. 2003; Zhang et al. 2002). Importantly, use of the mutant strain, with controllable levels of CL, revealed a direct correlation between the levels of CL in the cell, the ratio of SC to individual respiratory complexes and the growth-rate on non-fermentable carbon sources thus showing direct dependence of the efficiency of the mitochondrial energetic system on the level of CL. The total amount of individual respiratory complexes was not affected by the lack of CL (Zhang et al. 2002).

Based on these results it was concluded that CL is required for association of CIII and CIV to form a SC and that PG cannot effectively substitute for CL (Zhang et al. 2002). The more gentle approach utilizing CN-PAGE, which does not use an anionic dye during electrophoresis, revealed existence of loosely associated SCs in the mutant strain lacking CL with highly elevated abnormal levels of PG, which probably partially substitutes for CL in the SC formation (Pfeiffer et al. 2003). However, it should be noted that in mammals under physiological conditions where CL levels are reduced, there is no compensatory increase in PG levels to partially substitute for CL. Importantly, direct correlation between CL levels and energy efficiency of *S. cerevisiae* mutants demonstrates that the weakly associated CIII and CIV are not fully functionally competent (Zhang et al. 2002).

Kinetic studies of NADH oxidation in wild-type mitochondria and mitochondria isolated from the Δ*crd1* mutant further supports this conclusion (Zhang et al. 2005). Titration with the CIII-specific inhibitor antimycin A, which affects the rate-limiting reduction rate of cytochrome *c*, was used to distinguish between the organization in the lipid bilayer of the CIII and CIV into a single functional unit performing substrate-channelling of cytochrome *c* and the non-associated individual CIII and CIV connected by randomly diffusing cytochrome *c*. This classical kinetic approach takes into account that when the diffusion rate of a mobile electron carrier is faster than the rates of its reduction and oxidation during steady-state respiration, the mobile carrier behaves kinetically as a homogeneous pool (Boumans et al. 1998). In this case the overall respiration rate should show a hyperbolic relation to the rate of either the reduction or oxidation of the carrier. However, this relation approaches linearity (non-pool behaviour) when the substrate channelling takes place between the mobile carrier and the individual respiratory complexes organized into a SC. Titration of whole ETC activity with antimycin A resulted in the linear decrease of overall respiration rate (i.e. titrated as a single functional unit), which is in good agreement with organization of CIII and CIV into a SC in wild-type mitochondria (Boumans et al. 1998; Zhang et al. 2005). Using the same kinetic analysis, it was found that cytochrome *c* in intact mitochondria lacking CL displayed pool behaviour identical to that observed for wild-type mitochondria treated with a chaotropic agent to dissociate the SC (Zhang et al. 2005). The presence of PG may be sufficient to support activity and a weak association of the individual CIII and CIV into the SC tetramer, but it is not efficient in formation of the stable functional SC. All these finding together revealed a central role of CL in yeast mitochondria in association CIII and CIV into a structural and functional SC (Mileykovskaya et al. 2005).

Experiments comparing *crd1Δ* mutants with their parental wild-type strains directly demonstrated that CL is required for the maintenance of a mitochondrial membrane potential and for tight coupling of ATP production by mitochondrial oxidative phosphorylation (see Joshi et al. [2009] and accompanying references). In yeast mitochondria CL also supports an association of the respiratory SCs with the ADP/ATP carrier into the supramolecular structure termed the interactome. The interactome is composed of an ADP/ATP carrier dimer, one CIII and two CIVs (Claypool et al. 2008). The authors suggested that such organization not only should enhance the efficiency of electron transport, but also, CL as a proton trap, could locally contribute to creating a high membrane potential to drive ATP export from mitochondria in exchange for ADP import through the ADP/ATP carrier across the IMM. Thus CL increases the efficiency of the whole oxidative phosphorylation system.

4.4.2 Disruption of the Mitochondrial Redox SCs under Pathological Conditions

A similar requirement for CL in respirasome organization in mammalian cells is supported by studies of mitochondria from patients with Barth syndrome, an X-linked genetic disorder characterized by cardiomyopathy, skeletal myopathy, neutropenia and growth retardation (see Hauff and Hatch [2006]; Schlame and Ren [2006] for review). On the molecular level mutations in *TAZ* gene, result in an overall decrease in mitochondrial CL levels accompanied by the absence of tetralinoleoyl $(18{:}2^{\Delta 9,12})_4$CL, accumulation of MLCL, and a decrease in unsaturated fatty acyl species of CL. Barth syndrome is classified as a disorder of CL metabolism (Schlame and Ren 2006) and the MLCL/CL ratio is used in the diagnosis of this disorder (Houtkooper et al. 2009). Analysis of the $I_1III_2IV_1$ SC in the extracts from lymphoblasts from Barth syndrome patients by BN-PAGE revealed an increase in the level of free CIV monomer and a decrease in the level of the I_1III_2 and I_1III_2 IV SCs (McKenzie et al. 2006). Dramatic changes in SC organization due to Barth syndrome were demonstrated by use of a pluripotent stem cell model system of this disorder. Alterations in mitochondrial morphology, respiratory chain malfunction with decreased membrane potential and increased oxidative stress were shown in the stem cell model system (Dudek et al. 2013). Interestingly, immortalized lymphoblasts from Barth syndrome patients displayed an increase in the content of mitochondria per cell to compensate for the decrease in the energy production due to a decrease in the content of the respiratory complexes and SCs (Gonzalvez et al. 2013). Recent results from experiments with *TAZ* gene knockdown mice indicate that mechanisms of dynamic regulation of whole-body energy metabolism are affected by CL deficiency in mitochondria. Specifically, the mice were not able to endure high intensity exercise. Application of open circuit calorimetry during exercise revealed that in contrast to wild-type mice, which dynamically shifted their reliance on metabolic substrates from predominantly carbohydrates to mixed substrates, *TAZ* mutant mice did not possess this metabolic switch (Powers et al. 2013). As was discussed earlier, the respiratory chain is a dynamic supramolecular assembly, and its organization depends on substrate availability and energy demands (Acín-Pérez and Enriquez 2014; Lapuente-Brun et al. 2013). Therefore, it is tempting to speculate that a primary factor, which may affect energy metabolism in Barth syndrome, is abnormalities in dynamic reorganization of the CL-supported supramolecular structure of the respiratory chain in response to changes in energy demand.

Recent studies from yeast suggest that the molecular basis for the defect in Barth Syndrome may not be the lack of remodelling but may be a reduction in the total amount of CL. Null mutants in *TAZ1* of yeast display many of the mitochondrial defects seen in mammalian cells (Brandner et al. 2005; Li et al. 2007), but a null mutant in the phospholipase A_2 (*CLD1* gene product) responsible for initiating the CL remodelling cascade displayed no obvious phenotypes even though there was no remodelling of CL (Baile et al. 2014b). The wide variation in fatty acid compositions of CL between cell types within the same mammalian species also suggests that extensive remodelling may not be necessary in itself to support mitochondrial function. Given the evolutionary retention of the remodelling pathway, it must serve an important function possibly in ridding cells of oxidized CL species.

Reduced CL levels associated with reduced formation of individual respiratory complexes and SCs results in reduced efficiency of mitochondrial energy metabolism in ageing (Frenzel et al. 2010; Gomez and Hagen 2012; Paradies et al. 2010) and in such pathological states as neurodegenerative diseases (Paradies et al. 2011), and cancer (Dumas et al. 2013). It is well documented that CI and CIII are involved in superoxide production during electron transfer (Bleier and Drose 2013; Grivennikova and Vinogradov 2013; Koopman et al. 2010; Lanciano et al. 2013). Under normal physiological conditions the ROS generated at low levels play an important signalling role in cell physiology. Under pathological conditions the high levels of ROS cause cell damage. Oxidative stress accompanied by lipid peroxidation, particularly for CL with its polyunsaturated acyl chains, occurs in the above-mentioned diseases and results in reduced levels of SCs. Therefore, reduction in coupling of complexes within the respiratory chain amplifies oxidative stress, which further adversely affects energy production with the release of cytochrome *c* to the cytoplasm accompanied by induction of apoptosis (for comprehensive detailed review, see Lenaz 2012; Paradies et al. 2014). Therefore, evidence from yeast to humans supports a role for CL in supramolecular organization of respiratory chain.

4.5 STRUCTURAL ANALYSIS IN SUPPORT OF CL-DEPENDENT SC FORMATION

Analysis of the structural organization of the respiratory SCs, and specifically mutual orientation of the individual complexes inside the SC structures, is required for understanding the mechanism by which CL performs its task in *gluing* the individual complexes together. Indeed, structural data provide information on: (1) Relative orientation of individual complexes inside a SC, (2) Points of protein contact, (3) Spaces between individual complexes that may contain loosely bound CLs important for SC assembly, (4) Positions in the SC of the tightly bound CL in the individual complexes and (5) Possible sites of loosely bound CLs important for SC assembly.

Structural organization of the respirasome from bovine heart mitochondria composed of one CI, one CIII and one CIV ($I_1III_2IV_1$) was investigated by two different electron

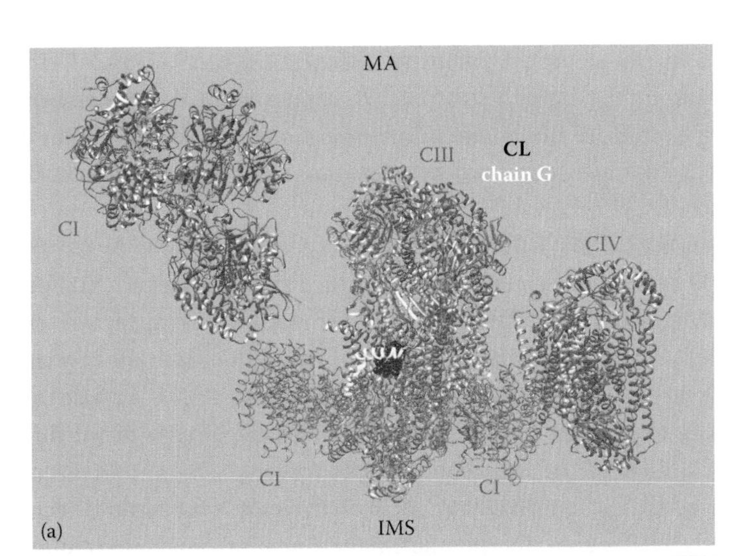

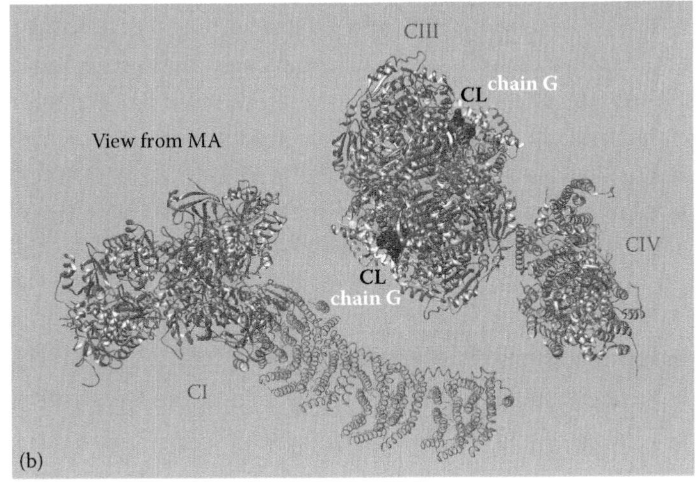

FIGURE 4.3 Arrangement of CI, CIII and CIV in the bovine respirasome. (Derived from the structure reported in Althoff, T. et al., *EMBO J.*, 30(22), 4652–4664, 2011); PDB ID: 2YBB. (a) Side view and (b) view from matrix (MA). Two CL molecules (in *black*, only one visible in [a]) in the cavity of each monomer of CIII formed by cytochromes c_1 and b and closed by the belt formed by QP-C protein (chain G, *white*).

microscopy (EM) approaches. Cryo-electron tomography combined with sub-volume averaging was used to study the digitonin-solubilized respirasome (Dudkina et al. 2011). Cryo-EM and single particle analysis of the amphipol-solubilized respirasome was also performed (Althoff et al. 2011). Docking of available X-ray crystal structures of individual CI, CIII and CIV into 3D electron density maps produced pseudo-atomic models of the bovine respirasome (Figure 4.3). The structures obtained by two independent approaches showed a good correlation in the arrangement of the individual complexes in the SC. The resulting structure shows the specific arrangement of the CI, CIII and CIV and suggests the pathways for UQ and cytochrome c substrate channelling within the respirasome (Althoff et al. 2011).

Interestingly, in the bovine SC mutual orientation of CIV and CIII differs significantly from their orientation in the yeast *S. cerevisiae* SC composed of CIII and two CIVs (III_2IV_2) (yeast does not make CI, as mentioned earlier). Pseudo-atomic models for the yeast SC were also obtained by two independent approaches. One model was suggested based on 2D projection averages obtained with negative stain single particle EM (Heinemeyer et al. 2007). The second structure was obtained by using single particle cryo-EM and 3D density map reconstruction (Mileykovskaya et al. 2012). Docking of the X-ray crystal structures of the yeast CIII and the bovine CIV (no crystal structure of yeast CIV is available) into the 3D density map confirmed the difference in the arrangement of CIII and CIV in yeast and bovine SCs (Figures 4.3 and 4.4). At first this difference in orientation may be surprising, but this may be a result of significant differences in the composition of the yeast and mammalian respirasome. The mammalian CIII interacts with both CIV and CI, the latter being absent in yeast. One should also keep in mind that based on BN-PAGE analysis the mammalian respirasomes might contain up to four CIVs (Schagger 2001). The arrangement of CIVs in such a structure is not known. Both SCs have gaps at the interfaces of individual complexes. These gaps lie within the membrane imbedded domains of the SC and are therefore most likely filled with lipids (see Figure 4.4). The mammalian SC has larger gaps than the yeast SC (Althoff et al. 2011; Dudkina et al. 2011; Mileykovskaya et al. 2012).

In the yeast symmetrical structure (C2 symmetry) the dimeric CIII is flanked by two monomers of CIV. Each CIV is oriented towards CIII with its convex sides ([Heinemeyer

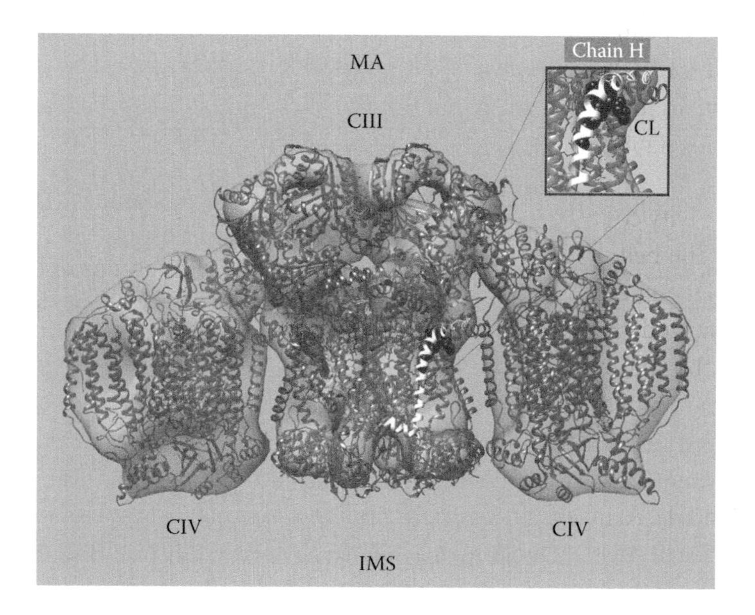

FIGURE 4.4 Arrangement of CIII and CIV in the yeast tetrameric SC III_2IV_2. (Derived from the structure reported in Mileykovskaya, E. et al., *J. Biol. Chem.*, 287, 23095–23103, 2012). One CL molecule (in *black*) in the cavity of each monomer of CIII formed by cytochromes c_1 and b and closed by the belt formed by Qcr8p subunit (chain H, *white*), which is homologue of chain G in bovine CIII.

et al. 2007; Mileykovskaya et al. 2012] and Figure 4.4), which is opposite to the side involved in the formation of CIV dimers in its crystal structure (Tsukihara et al. 1996). CIII faces CIV with the side containing a cavity formed by transmembrane helices of cytochrome c_1 and cytochrome *b*. This cavity contains a *tightly bound CL* resolved in the crystal structures of CIII (designated as CDL in PDB ID: 1KB9 (Lange et al. 2001); or CN3 in PDB ID: 3CX5 ([Solmaz and Hunte 2008] and Figure 4.4). Several tightly bound CLs remain associated with CIII and CIV after solubilization with dodecyl maltoside, which dissociates the SC and removes all lipid (*loosely bound*) that is not integral to the individual respiratory complexes. The cavity was designated as the CL_i site (Wenz et al. 2009) due to its close proximity to the Q_i site (UQ reduction site), which suggests direct involvement of the bound CL in proton uptake by CIII (Hunte et al. 2003). The role of CL in the CL_i site in SC stability was investigated by constructing single, double and triple replacement variants of the lysine residues (K288L, K289L and K296L) of cytochrome c_1 subunit of CIII, which are important for CL binding (Wenz et al. 2009). BN-PAGE analysis revealed CIII/CIV SCs in all mutants. When these variants of CIII were introduced into the Δ*crd1* mutant lacking CL, SC formation was found largely restored in two double replacement variants (K288L/K289L and K289L/K296L) and in the triple replacement variant. Since replacement of positively charged residues with hydrophobic residues replaced the CL requirement of SC formation, the authors concluded that the role of CL is in neutralization of the positive charges in the cytochrome c_1 subunit of CIII, thus allowing stable interaction of this domain with its counterpart in the CIV. However, the authors noted that the final interpretation of their results was complicated by the presence of elevated levels of PG in the Δ*crd1* mutant, which may be more effective in stabilizing SC formation with the variant CIIIs (Wenz et al. 2009).

In the bovine SC the interface of CIII with CIV is at the periphery of the interface between monomers of the CIII (PDB ID: 2YBB [Althoff et al. 2011; Dudkina et al. 2011]). Thus in contrast to CL_i site in the yeast structure, the homologous site of bovine CIII is not oriented towards CIV (and Figure 4.3). Consequently CL bound in this site cannot participate in association of CIII and CIV in the bovine structure (Mileykovskaya and Dowhan 2014). In bovine CIII this cavity contains two CLs, which appear to be stabilized by a helical belt that caps this cavity. The belt formed by chain G (subunit VII or UQ-binding protein QP-C, PDB ID: 1PP9 [Huang et al. 2005] corresponds to chain H [subunit Qcr8] in yeast, PDB ID: 3CX5 [Solmaz and Hunte 2008]). This belt contributes additional positive charges and closes the cavity thus more firmly stabilizing CL association. In yeast CIII, the belt contains fewer positive charges, which might allow exchange of bound CL with CL from the membrane. This suggestion (Dibrova et al. 2013) was supported by coarse-grained molecular dynamic simulation (CGMDS) of CL binding to the yeast and bovine CIII (Arnarez et al. 2013b; see Figure 4.4). An interesting hypothesis was formulated that the role of two trapped CL molecules in addition to its role in the catalytic events in Q_i centre is to serve as a sensor of the level of ROS production in this site of bovine CIII (Dibrova et al. 2013). Indeed, CIII functioning through the Q cycle with formation of stabilized semiquinone intermediates is considered as a source of ROS (Bleier and Drose 2013; Lanciano et al. 2013).

A second site with conserved tightly bound CL was identified in the hydrophobic cleft close to the CIII homodimer interface (Palsdottir and Hunte 2004). It was suggested that

CL bound in this site might promote diffusion of UQ to the active sites and/or exchange between sites of UQ/UQH_2 catalysis within the CIII (Palsdottir and Hunte 2004).

Two tightly bound CLs were resolved in the crystal structure of bovine CIV (Shinzawa-Itoh et al. 2007; PDB ID: 2DYR). One of them together with two phosphatidylethanolamines and one PG stabilizes the CIV dimer making strong contact between monomers in the crystal structure. The four acyl chains of CL interact through van der Waals contacts with hydrophobic amino acid residues belonging to both monomers, and the two phosphate groups interact with both monomers via hydrogen bonds. A third CL, found by photolabeling with arylazido-containing CL analogues (Sedlak et al. 2006), is located near the entrance to the putative proton-pumping channel (Branden et al. 2006) and thus might facilitate proton entry into the channel (Musatov and Robinson 2014).

An important feature in the arrangement of the individual complexes in the bovine and yeast SCs revealed by cryo-EM structural studies is the presence of 2–5 nm spaces between transmembrane domains of individual complexes at their interface (Figures 4.3b and 4.4). These gaps are most likely filled with lipids since they lie within the lipid bilayer. Protein contact sites lie outside the lipid bilayer (Althoff et al. 2011; Dudkina et al. 2011; Mileykovskaya et al. 2012). In the case of bovine respirasome it was suggested that the lipid filled space between CI and CIII could serve as a diffusion microdomain, which restricts movement of UQ facilitating its channelling inside the SC (Chaban et al. 2014).

Application of quantitative mass spectral lipid analysis revealed 50 CL molecules in the purified yeast III_2IV_2 supercomplex, which is in large excess of the tightly bound CLs (Mileykovskaya et al. 2012). The CL associated with the digitonin solubilized and purified SC has exactly the same proportion among the five major CL species with acyl chains $(16{:}1^{\Delta9})_4$, $(16{:}1^{\Delta9})_3(18{:}1^{\Delta9})_1$, $(16{:}1^{\Delta9})_2(18{:}1^{\Delta9})_2$ and $(18{:}1^{\Delta9})_4$ as CL in the IMM (Mileykovskaya et al. 2012). About 200 CL molecules were found in the purified bovine respirasome (Althoff et al. 2011). The higher level of CL in the bovine respirasome is consistent with its much larger size due to the presence of CI and larger spaces between the three individual complexes. The most important finding is that in both bovine and yeast SCs the level of CL is in great excess over the amount of CL molecules tightly associated and integrated into the structure of the individual purified complexes. This finding raises the possible role of bilayer CL in organization of respiratory SCs.

4.6 *IN VITRO* AND *IN SILICO* APPROACHES FOR MODELLING OF CL-DEPENDENT FORMATION OF THE SCs

To further study the direct role of CL in connecting individual respiratory complexes into SCs, specifically by filling the spaces between transmembrane domains of CIII and CIV, a minimal system for *in vitro* SC formation was developed (Bazán et al. 2013). The purpose of this system was to completely exclude the possibility of indirect influence of CL levels in the membrane on SC formation, which might take place *in vivo*. Purified CIII and CIV from *S. cerevisiae* mitochondria, which contained mostly tightly bound phospholipids including CL (as determined by quantitative mass spectral lipid analysis) were incorporated into liposomes of different phospholipid composition. For the first time, the reconstitution of the trimeric III_2IV_1 and tetrameric III_2IV_2 SCs from individual

CIII and CIV in proteoliposomes was achieved. Both reconstituted structures retain CIII and CIV enzymatic activity. Formation of the SC tetramer (III_2IV_2) but not the SC trimer (III_2IV) showed direct dependence on the presence of CL in liposomes, which could not be substituted by any other anionic phospholipids including PG. Moreover, the structural arrangement of CIII and CIV in the reconstituted SC tetramer as IV_1-III_2-IV_1 displayed the same orientation of CIII and CIV as in native SC, which was confirmed by negative-stain EM and single particle analysis. Finally, the reconstituted SC tetramer catalysed electron transfer from UQH_2 to oxygen in the presence of low concentrations of cytochrome *c*, that is, under conditions reported (Schagger and Pfeiffer 2000) to involve its channelling within the SC. These experiments clearly demonstrated that CL tightly bound and integral to individual complexes is not sufficient for the organization of CIII and CIV into the structural and functional SC and that additional CL from the bilayer is required (Bazán et al. 2013).

The above conclusion is supported by molecular modelling of CL binding to CIII and CIV (Arnarez et al. 2013a,b). The authors tested the hypothesis that CLs present in the bulk membrane play an active role in the function and structural organization of the respiratory chain in coordination with the tightly bound CLs found in the purified complexes. To characterize the interaction between CLs of the bulk membrane with the respiratory complexes and the role of these CLs in the SC formation, the authors performed a series of CGMD simulations of CL binding to wild-type bovine and yeast CIII; several CIII mutants including the mutants in K288, K289, K296 of the cytochrome c_1 subunit, and bovine CIV embedded in mixed PC/CL bilayers. Starting structures included the integral CL molecules found in the crystal structures. Importantly, the simulations reproduced the known CL-binding sites, and in addition, revealed the existence of well-defined non-integral CL-binding sites on the membrane-exposed protein surface of CIII and CIV. All six CL binding sites established per monomer of dimeric CIII were conserved in CIII from bovine and yeast mitochondria. These binding sites are enriched in positively charged residues. All are located on the matrix side of the CIII although CLs are present in both leaflets of the membrane and uniformly visit both matrix and IMS sides of CIII. Interestingly, in the yeast crystal structure in contrast to bovine only one tightly bound CL fills the site close to Q_i site. During an extended simulation timescale an additional CL found its way into the cavity of this site in one of the monomers of the yeast CIII (Arnarez et al. 2013b). In the case of CIV the authors identified the precise position of a number of high-affinity and low-affinity sites. Two of the binding sites are located at the matrix entrance of the known proton uptake pathways for the cytochrome *c* oxidase consistent with the role of CL in facilitating proton translocation activity of CIV (Arnarez et al. 2013a).

The author made exploratory simulations, which demonstrated how CL bound only in the membrane-exposed protein surface of bovine CIII and CIV would stabilize interactions between CIII and CIV in the bovine respirasome (Arnarez et al. 2013b). Thus *in silico* modelling of CL dependent SC formation supports *in vitro* experiments with proteoliposomes, which demonstrated the role of non-integral bound CL on the membrane-exposed protein surface in the assembly of respiratory SCs.

4.7 CONCLUSIONS

The mitochondrial phospholipid CL plays a specific and essential role in the supramolecular organization of the ETC into the SCs. The respiratory SCs contain higher numbers of CL molecules over CL integrated into the individual respiratory complexes. 3D density maps of bovine and yeast SCs obtained by single particle cryo-EM analysis revealed spaces between the transmembrane domains of individual complexes that face each other, which are most likely filled with lipid molecules. Kinetic and biochemical studies comparing yeast wild-type cells and mutants lacking CL in combination with *in vitro* reconstitution of the yeast SCs dependent on addition of CL demonstrate a direct role for CL in the higher-order organization of the respiratory chain. The correlation between reduced CL levels and disruption of respiratory SCs in many diseases strongly indicates a similar role for CL in mammalian mitochondria. It also suggests that changes in CL levels might be a metabolic regulatory signal acting through the low-affinity CL-binding sites that participate in SC formation. Molecular dynamics simulations of CL-binding sites in the respiratory complexes and of CL-dependent formation of the SCs suggest a molecular mechanism for this process, which involves interactions through CL bound at the low-affinity binding sites on the surfaces of individual complexes. To test this mechanism more precise information should be obtained by a combination of higher resolution structural studies of the SCs, biochemical determination of CL-binding sites using *in vitro* reconstitution system, genetic perturbation of CL-binding sites determined in *in vitro* studies and predicted by molecular dynamics simulations.

REFERENCES

Acín-Pérez, R., M. P. Bayona-Bafaluy, P. Fernandez-Silva et al. Respiratory complex III is required to maintain complex I in mammalian mitochondria. *Mol Cell* 13, no. 6 (2004): 805–15.

Acín-Pérez, R., and J. A. Enriquez. The function of the respiratory supercomplexes: The plasticity model. *Biochim Biophys Acta* 1837, no. 4 (2014): 444–50.

Acín-Pérez, R., P. Fernandez-Silva, M. L. Peleato, A. Perez-Martos, and J. A. Enriquez. Respiratory active mitochondrial supercomplexes. *Mol Cell* 32, no. 4 (2008): 529–39.

Althoff, T., D. J. Mills, J. L. Popot, and W. Kuhlbrandt. Arrangement of electron transport chain components in bovine mitochondrial supercomplex $I_1III_2IV_1$. *EMBO J* 30, no. 22 (2011): 4652–64.

Arnarez, C., S. J. Marrink, and X. Periole. Identification of cardiolipin binding sites on cytochrome c oxidase at the entrance of proton channels. *Sci Rep* 3 (2013a): 1263.

Arnarez, C., J. P. Mazat, J. Elezgaray, S. J. Marrink, and X. Periole. Evidence for cardiolipin binding sites on the membrane-exposed surface of the cytochrome bc_1. *J Am Chem Soc* 135, no. 8 (2013b): 3112–20.

Baile, M. G., Y. W. Lu, and S. M. Claypool. The topology and regulation of cardiolipin biosynthesis and remodeling in yeast. *Chem Phys Lipids* 179 (2014a): 25–31.

Baile, M. G., M. Sathappa, Y. W. Lu et al. Unremodeled and remodeled cardiolipin are functionally indistinguishable in yeast. *J Biol Chem* 289, no. 3 (2014b): 1768–78.

Barrientos, A., and C. Ugalde. I function, therefore I am: Overcoming skepticism about mitochondrial supercomplexes. *Cell Metab* 18, no. 2 (2013): 147–9.

Bazán, S., E. Mileykovskaya, V. K. Mallampalli et al. Cardiolipin-dependent reconstitution of respiratory supercomplexes from purified *Saccharomyces cerevisiae* complexes III and IV. *J Biol Chem* 288, no. 1 (2013): 401–11.

Bleier, L., and S. Drose. Superoxide generation by complex III: From mechanistic rationales to functional consequences. *Biochim Biophys Acta* 1827, no. 11–12 (2013): 1320–31.

Boumans, H., L. A. Grivell, and J. A. Berden. The respiratory chain in yeast behaves as a single functional unit. *J Biol Chem* 273, no. 9 (1998): 4872–7.

Branden, G., R. B. Gennis, and P. Brzezinski. Transmembrane proton translocation by cytochrome c oxidase. *Biochim Biophys Acta* 1757, no. 8 (2006): 1052–63.

Brandner, K., D. U. Mick, A. E. Frazier et al. Taz1, an outer mitochondrial membrane protein, affects stability and assembly of inner membrane protein complexes: Implications for Barth Syndrome. *Mol Biol Cell* 16, no. 11 (2005): 5202–14.

Brandt, U. A two-state stabilization-change mechanism for proton-pumping complex I. *Biochim Biophys Acta* 1807, no. 10 (2011): 1364–9.

Chaban, Y., E. J. Boekema, and N. V. Dudkina. Structures of mitochondrial oxidative phosphorylation supercomplexes and mechanisms for their stabilisation. *Biochim Biophys Acta* 1837, no. 4 (2014): 418–26.

Chance, B., and G. R. Williams. Respiratory enzymes in oxidative phosphorylation. IV. The respiratory chain. *J Biol Chem* 217, no. 1 (1955): 429–38.

Chang, S. C., P. N. Heacock, C. J. Clancey, and W. Dowhan. The PEL1 gene (renamed PGS1) encodes the phosphatidylglycero-phosphate synthase of *Saccharomyces cerevisiae*. *J Biol Chem* 273, no. 16 (1998a): 9829–36.

Chang, S. C., P. N. Heacock, E. Mileykovskaya, D. R. Voelker, and W. Dowhan. Isolation and characterization of the gene (CLS1) encoding cardiolipin synthase in *Saccharomyces cerevisiae*. *J Biol Chem* 273, no. 24 (1998b): 14933–41.

Chen, D., X. Y. Zhang, and Y. Shi. Identification and functional characterization of hCLS1, a human cardiolipin synthase localized in mitochondria. *Biochem J* 398, no. 2 (2006): 169–76.

Chen, Y. C., E. B. Taylor, N. Dephoure et al. Identification of a protein mediating respiratory supercomplex stability. *Cell Metab* 15, no. 3 (2012): 348–60.

Claypool, S. M., Y. Oktay, P. Boontheung, J. A. Loo, and C. M. Koehler. Cardiolipin defines the interactome of the major ADP/ATP carrier protein of the mitochondrial inner membrane. *J Cell Biol* 182, no. 5 (2008): 937–50.

Cruciat, C. M., S. Brunner, F. Baumann, W. Neupert, and R. A. Stuart. The cytochrome bc1 and cytochrome c oxidase complexes associate to form a single supracomplex in yeast mitochondria. *J Biol Chem* 275, no. 24 (2000): 18093–8.

Dibrova, D. V., D. A. Cherepanov, M. Y. Galperin, V. P. Skulachev, and A. Y. Mulkidjanian. Evolution of cytochrome bc complexes: From membrane-anchored dehydrogenases of ancient bacteria to triggers of apoptosis in vertebrates. *Biochim Biophys Acta* 1827, no. 11–12 (2013): 1407–27.

Dudek, J., I. F. Cheng, M. Balleininger et al. Cardiolipin deficiency affects respiratory chain function and organization in an induced pluripotent stem cell model of Barth syndrome. *Stem Cell Res* 11, no. 2 (2013): 806–19.

Dudkina, N. V., M. Kudryashev, H. Stahlberg, and E. J. Boekema. Interaction of complexes I, III, and IV within the bovine respirasome by single particle cryoelectron tomography. *Proc Natl Acad Sci U S A* 108, no. 37 (2011): 15196–200.

Dumas, J. F., L. Peyta, C. Couet, and S. Servais. Implication of liver cardiolipins in mitochondrial energy metabolism disorder in cancer cachexia. *Biochimie* 95, no. 1 (2013): 27–32.

Efremov, R. G., and L. A. Sazanov. Respiratory complex I: "Steam engine" of the cell? *Curr Opin Struct Biol* 21, no. 4 (2011): 532–40.

Einarsdottir, O., and I. Szundi. Time-resolved optical absorption studies of cytochrome oxidase dynamics. *Biochim Biophys Acta* 1655, no. 1–3 (2004): 263–73.

Frenzel, M., H. Rommelspacher, M. D. Sugawa, and N. A. Dencher. Ageing alters the supramolecular architecture of OxPhos complexes in rat brain cortex. *Exp Gerontol* 45, no. 7–8 (2010): 563–72.

Genova, M. L., and G. Lenaz. Functional role of mitochondrial respiratory supercomplexes. *Biochim Biophys Acta* 1837, no. 4 (2014): 427–43.

Gomez, L. A., and T. M. Hagen. Age-related decline in mitochondrial bioenergetics: Does supercomplex destabilization determine lower oxidative capacity and higher superoxide production? *Semin Cell Dev Biol* 23, no. 7 (2012): 758–67.

Gonzalvez, F., M. D'Aurelio, M. Boutant et al. Barth syndrome: Cellular compensation of mitochondrial dysfunction and apoptosis inhibition due to changes in cardiolipin remodeling linked to tafazzin (TAZ) gene mutation. *Biochim Biophys Acta* 1832, no. 8 (2013): 1194–206.

Grivennikova, V. G., and A. D. Vinogradov. Partitioning of superoxide and hydrogen peroxide production by mitochondrial respiratory complex I. *Biochim Biophys Acta* 1827, no. 3 (2013): 446–54.

Hackenbrock, C. R., B. Chazotte, and S. S. Gupte. The random collision model and a critical assessment of diffusion and collision in mitochondrial electron transport. *J Bioenerg Biomembr* 18, no. 5 (1986): 331–68.

Haines, T. H. A new look at Cardiolipin. *Biochim Biophys Acta* 1788, no. 10 (2009): 1997–2002.

Hauff, K. D., and G. M. Hatch. Cardiolipin metabolism and Barth Syndrome. *Prog Lipid Res* 45, no. 2 (2006): 91–101.

Heinemeyer, J., H. P. Braun, E. J. Boekema, and R. Kouril. A structural model of the cytochrome C reductase/oxidase supercomplex from yeast mitochondria. *J Biol Chem* 282, no. 16 (2007): 12240–8.

Henry, S. A., S. D. Kohlwein, and G. M. Carman. Metabolism and regulation of glycerolipids in the yeast *Saccharomyces cerevisiae*. *Genetics* 190, no. 2 (2012): 317–49.

Horvath, S. E., and G. Daum. Lipids of mitochondria. *Prog Lipid Res* 52, no. 4 (2013): 590–614.

Houtkooper, R. H., R. J. Rodenburg, C. Thiels et al. Cardiolipin and monolysocardiolipin analysis in fibroblasts, lymphocytes, and tissues using high-performance liquid chromatography-mass spectrometry as a diagnostic test for Barth syndrome. *Anal Biochem* 387, no. 2 (2009): 230–7.

Huang, L. S., D. Cobessi, E. Y. Tung, and E. A. Berry. Binding of the respiratory chain inhibitor antimycin to the mitochondrial bc1 complex: A new crystal structure reveals an altered intramolecular hydrogen-bonding pattern. *J Mol Biol* 351, no. 3 (2005): 573–97.

Hunte, C., H. Palsdottir, and B. L. Trumpower. Protonmotive pathways and mechanisms in the cytochrome bc1 complex. *FEBS Lett* 545, no. 1 (2003): 39–46.

Ikeda, K., S. Shiba, K. Horie-Inoue, K. Shimokata, and S. Inoue. A stabilizing factor for mitochondrial respiratory supercomplex assembly regulates energy metabolism in muscle. *Nat Commun* 4 (2013): 2147.

Jiang, F., H. S. Rizavi, and M. L. Greenberg. Cardiolipin is not essential for the growth of *Saccharomyces cerevisiae* on fermentable or non-fermentable carbon sources. *Mol Microbiol* 26, no. 3 (1997): 481–91.

Jiang, F., M. T. Ryan, M. Schlame et al. Absence of cardiolipin in the crd1 null mutant results in decreased mitochondrial membrane potential and reduced mitochondrial function. *J Biol Chem* 275, no. 29 (2000): 22387–94.

Joshi, A. S., J. Zhou, V. M. Gohil, S. Chen, and M. L. Greenberg. Cellular functions of cardiolipin in yeast. *Biochim Biophys Acta* 1793, no. 1 (2009): 212–8.

Kaila, V. R., M. I. Verkhovsky, and M. Wikstrom. Proton-coupled electron transfer in cytochrome oxidase. *Chem Rev* 110, no. 12 (2010): 7062–81.

Kawasaki, K., O. Kuge, S. C. Chang et al. Isolation of a chinese hamster ovary (CHO) cDNA encoding phosphatidylglycerophosphate (PGP) synthase, expression of which corrects the mitochondrial abnormalities of a PGP synthase-defective mutant of CHO-K1 cells. *J Biol Chem* 274, no. 3 (1999): 1828–34.

Koopman, W. J., L. G. Nijtmans, C. E. Dieteren et al. Mammalian mitochondrial complex I: Biogenesis, regulation, and reactive oxygen species generation. *Antioxid Redox Signal* 12, no. 12 (2010): 1431–70.

Kwast, K. E., P. V. Burke, B. T. Staahl, and R. O. Poyton. Oxygen sensing in yeast: Evidence for the involvement of the respiratory chain in regulating the transcription of a subset of hypoxic genes. *Proc Natl Acad Sci U S A* 96, no. 10 (1999): 5446–51.

Lanciano, P., B. Khalfaoui-Hassani, N. Selamoglu et al. Molecular mechanisms of superoxide production by complex III: A bacterial versus human mitochondrial comparative case study. *Biochim Biophys Acta* 1827, no. 11–12 (2013): 1332–9.

Lange, C., J. H. Nett, B. L. Trumpower, and C. Hunte. Specific roles of protein-phospholipid interactions in the yeast cytochrome bc1 complex structure. *EMBO J* 20, no. 23 (2001): 6591–600.

Lapuente-Brun, E., R. Moreno-Loshuertos, R. Acín-Pérez et al. Supercomplex assembly determines electron flux in the mitochondrial electron transport chain. *Science* 340, no. 6140 (2013): 1567–70.

Lenaz, G. Mitochondria and reactive oxygen species. Which role in physiology and pathology? *Adv Exp Med Biol* 942 (2012): 93–136.

Lenaz, G., and M. L. Genova. Kinetics of integrated electron transfer in the mitochondrial respiratory chain: Random collisions vs. solid state electron channeling. *Am J Physiol Cell Physiol* 292, no. 4 (2007): C1221–39.

Lewis, R. N., and R. N. McElhaney. The physicochemical properties of cardiolipin bilayers and cardiolipin-containing lipid membranes. *Biochim Biophys Acta* 1788, no. 10 (2009): 2069–79.

Li, G., S. Chen, M. N. Thompson, and M. L. Greenberg. New insights into the regulation of cardiolipin biosynthesis in yeast: Implications for Barth syndrome. *Biochim Biophys Acta* 1771, no. 3 (2007): 432–41.

Lu, B., F. Y. Xu, Y. J. Jiang et al. Cloning and characterization of a cDNA encoding human cardiolipin synthase (hCLS1). *J Lipid Res* 47, no. 6 (2006): 1140–5.

Malhotra, A., Y. Xu, M. Ren, and M. Schlame. Formation of molecular species of mitochondrial cardiolipin. 1. A novel transacylation mechanism to shuttle fatty acids between sn-1 and sn-2 positions of multiple phospholipid species. *Biochim Biophys Acta* 1791, no. 4 (2009): 314–20.

Maranzana, E., G. Barbero, A. I. Falasca, G. Lenaz, and M. L. Genova. Mitochondrial Respiratory Supercomplex Association Limits Production of Reactive Oxygen Species from Complex I. *Antioxid Redox Signal* 19, no. 13 (2013): 1469–80.

Marechal, A., B. Meunier, D. Lee, C. Orengo, and P. R. Rich. Yeast cytochrome c oxidase: A model system to study mitochondrial forms of the haem-copper oxidase superfamily. *Biochim Biophys Acta* 1817, no. 4 (2012): 620–8.

McKenzie, M., M. Lazarou, D. R. Thorburn, and M. T. Ryan. Mitochondrial respiratory chain supercomplexes are destabilized in Barth Syndrome patients. *J Mol Biol* 361, no. 3 (2006): 462–9.

Mileykovskaya, E., and W. Dowhan. Cardiolipin membrane domains in prokaryotes and eukaryotes. *Biochim Biophys Acta* 1788, no. 10 (2009): 2084–91.

Mileykovskaya, E., and W. Dowhan. Cardiolipin-dependent formation of mitochondrial respiratory supercomplexes. *Chem Phys Lipids* 179, (2014): 42–8.

Mileykovskaya, E., P. A. Penczek, J. Fang et al. Arrangement of the respiratory chain complexes in *Saccharomyces cerevisiae* supercomplex III2IV2 revealed by single particle cryo-electron microscopy. *J Biol Chem* 287, no. 27 (2012): 23095–103.

Mileykovskaya, E., M. Zhang, and W. Dowhan. Cardiolipin in energy transducing membranes. *Biochemistry (Mosc)* 70, no. 2 (2005): 154–8.

Mitchell, P. Protonmotive redox mechanism of the cytochrome b-c1 complex in the respiratory chain: protonmotive ubiquinone cycle. *FEBS Lett* 56, no. 1 (1975): 1–6.

Morin, C., R. Zini, and J. P. Tillement. Anoxia-reoxygenation-induced cytochrome c and cardiolipin release from rat brain mitochondria. *Biochem Biophys Res Commun* 307, no. 3 (2003): 477–82.

Musatov, A., and N. C. Robinson. Bound cardiolipin is essential for cytochrome c oxidase proton translocation. *Biochimie* 105, (2014): 159–64.

Olofsson, G., and E. Sparr. Ionization constants pKa of cardiolipin. *PloS One* 8, no. 9 (2013): e73040.

Osman, C., M. Haag, F. T. Wieland, B. Brugger, and T. Langer. A mitochondrial phosphatase required for cardiolipin biosynthesis: The PGP phosphatase Gep4. *EMBO J* 29, no. 12 (2010): 1976–87.

Ostrander, D. B., M. Zhang, E. Mileykovskaya, M. Rho, and W. Dowhan. Lack of mitochondrial anionic phospholipids causes an inhibition of translation of protein components of the electron transport chain. A yeast genetic model system for the study of anionic phospholipid function in mitochondria. *J Biol Chem* 276, no. 27 (2001): 25262–72.

Osyczka, A., C. C. Moser, and P. L. Dutton. Fixing the Q cycle. *Trends Biochem Sci* 30, no. 4 (2005): 176–82.

Palsdottir, H., and C. Hunte. Lipids in membrane protein structures. *Biochim Biophys Acta* 1666, no. 1–2 (2004): 2–18.

Paradies, G., V. Paradies, V. De Benedictis, F. M. Ruggiero, and G. Petrosillo. Functional role of cardiolipin in mitochondrial bioenergetics. *Biochim Biophys Acta* 1837, no. 4 (2014): 408–17.

Paradies, G., G. Petrosillo, V. Paradies, and F. M. Ruggiero. Oxidative stress, mitochondrial bioenergetics, and cardiolipin in aging. *Free Radic Biol Med* 48, no. 10 (2010): 1286–95.

Paradies, G., G. Petrosillo, V. Paradies, and F. M. Ruggiero. Mitochondrial dysfunction in brain aging: Role of oxidative stress and cardiolipin. *Neurochem Int* 58, no. 4 (2011): 447–57.

Patil, V. A., J. L. Fox, V. M. Gohil, D. R. Winge, and M. L. Greenberg. Loss of cardiolipin leads to perturbation of mitochondrial and cellular iron homeostasis. *J Biol Chem* 288, no. 3 (2013): 1696–705.

Pfeiffer, K., V. Gohil, R. A. Stuart et al. Cardiolipin stabilizes respiratory chain supercomplexes. *J Biol Chem* 278, no. 52 (2003): 52873–80.

Powers, C., Y. Huang, A. Strauss, and Z. Khuchua. Diminished Exercise Capacity and Mitochondrial bc1 Complex Deficiency in Tafazzin-Knockdown Mice. *Front Physiol* 4 (2013): 74.

Raja, V., and M. L. Greenberg. The functions of cardiolipin in cellular metabolism-potential modifiers of the Barth syndrome phenotype. *Chem Phys Lipids* 179 (2014): 49–56.

Rich, P. R., and A. Marechal. Functions of the hydrophilic channels in protonmotive cytochrome c oxidase. *J R Soc Interface* 10, no. 86 (2013): 20130183.

Schagger, H. Respiratory chain supercomplexes. *IUBMB Life* 52, no. 3–5 (2001): 119–28.

Schagger, H., and K. Pfeiffer. Supercomplexes in the respiratory chains of yeast and mammalian mitochondria. *EMBO J* 19, no. 8 (2000): 1777–83.

Schlame, M. Formation of molecular species of mitochondrial cardiolipin 2. A mathematical model of pattern formation by phospholipid transacylation. *Biochim Biophys Acta* 1791, no. 4 (2009): 321–5.

Schlame, M. Cardiolipin remodeling and the function of tafazzin. *Biochim Biophys Acta* 1831, no. 3 (2013): 582–8.

Schlame, M., and M. Ren. Barth syndrome, a human disorder of cardiolipin metabolism. *FEBS Lett* 580, no. 23 (2006): 5450–5.

Schlame, M., M. Ren, Y. Xu, M. L. Greenberg, and I. Haller. Molecular symmetry in mitochondrial cardiolipins. *Chem Phys Lipids* 138, no. 1–2 (2005): 38–49.

Sedlak, E., M. Panda, M. P. Dale, S. T. Weintraub, and N. C. Robinson. Photolabeling of cardiolipin binding subunits within bovine heart cytochrome c oxidase. *Biochemistry* 45, no. 3 (2006): 746–54.

Shen, H., P. N. Heacock, C. J. Clancey, and W. Dowhan. The CDS1 gene encoding CDP-diacylglycerol synthase in *Saccharomyces cerevisiae* is essential for cell growth. *J Biol Chem* 271, no. 2 (1996): 789–95.

Shinzawa-Itoh, K., H. Aoyama, K. Muramoto et al. Structures and physiological roles of 13 integral lipids of bovine heart cytochrome c oxidase. *EMBO J* 26, no. 6 (2007): 1713–25.

Solmaz, S. R., and C. Hunte. Structure of complex III with bound cytochrome c in reduced state and definition of a minimal core interface for electron transfer. *J Biol Chem* 283, no. 25 (2008): 17542–9.

Strogolova, V., A. Furness, M. Robb-McGrath, J. Garlich, and R. A. Stuart. Rcf1 and Rcf2, members of the hypoxia-induced gene 1 protein family, are critical components of the mitochondrial cytochrome bc1-cytochrome c oxidase supercomplex. *Mol Cell Biol* 32, no. 8 (2012): 1363–73.

Stuart, R. A. Supercomplex organization of the oxidative phosphorylation enzymes in yeast mitochondria. *J Bioenerg Biomembr* 40, no. 5 (2008): 411–7.

Sun, F., X. Huo, Y. Zhai et al. Crystal structure of mitochondrial respiratory membrane protein complex II. *Cell* 121, no. 7 (2005): 1043–57.

Sun, F., Q. Zhou, X. Pang, Y. Xu, and Z. Rao. Revealing various coupling of electron transfer and proton pumping in mitochondrial respiratory chain. *Curr Opin Struct Biol* 23, no. 4 (2013): 526–38.

Tamura, Y., Y. Harada, S. Nishikawa et al. Tam41 is a CDP-diacylglycerol synthase required for cardiolipin biosynthesis in mitochondria. *Cell Metab* 17, no. 5 (2013): 709–18.

Trouillard, M., B. Meunier, and F. Rappaport. Questioning the functional relevance of mitochondrial supercomplexes by time-resolved analysis of the respiratory chain. *Proc Natl Acad Sci U S A* 108, no. 45 (2011): E1027–34.

Tsukihara, T., H. Aoyama, E. Yamashita et al. The whole structure of the 13-subunit oxidized cytochrome c oxidase at 2.8 A. *Science* 272, no. 5265 (1996): 1136–44.

Tuller, G., C. Hrastnik, G. Achleitner et al. YDL142c encodes cardiolipin synthase (Cls1p) and is non-essential for aerobic growth of *Saccharomyces cerevisiae*. *FEBS Lett* 421, no. 1 (1998): 15–8.

Volta, M., A. Bulfone, C. Gattuso et al. Identification and characterization of CDS2, a mammalian homolog of the Drosophila CDP-diacylglycerol synthase gene. *Genomics* 55, no. 1 (1999): 68–77.

Vukotic, M., S. Oeljeklaus, S. Wiese et al. Rcf1 mediates cytochrome oxidase assembly and respirasome formation, revealing heterogeneity of the enzyme complex. *Cell Metab* 15, no. 3 (2012): 336–47.

Weeks, R., W. Dowhan, H. Shen et al. Isolation and expression of an isoform of human CDP-diacylglycerol synthase cDNA. *DNA Cell Biol* 16, no. 3 (1997): 281–9.

Wendel, A. A., T. M. Lewin, and R. A. Coleman. Glycerol-3-phosphate acyltransferases: Rate limiting enzymes of triacylglycerol biosynthesis. *Biochim Biophys Acta* 1791, no. 6 (2009): 501–6.

Wenz, T., R. Hielscher, P. Hellwig et al. Role of phospholipids in respiratory cytochrome bc(1) complex catalysis and supercomplex formation. *Biochim Biophys Acta* 1787, no. 6 (2009): 609–16.

Wittig, I., H. P. Braun, and H. Schagger. Blue native PAGE. *Nat Protoc* 1, no. 1 (2006): 418–28.

Xia, D., L. Esser, W. K. Tang et al. Structural analysis of cytochrome bc1 complexes: Implications to the mechanism of function. *Biochim Biophys Acta* 1827, no. 11–12 (2013): 1278–94.

Yamashita, T., E. Nakamaru-Ogiso, H. Miyoshi, A. Matsuno-Yagi, and T. Yagi. Roles of bound quinone in the single subunit NADH-quinone oxidoreductase (Ndi1) from *Saccharomyces cerevisiae*. *J Biol Chem* 282, no. 9 (2007): 6012–20.

Zhang, J., Z. Guan, A. N. Murphy et al. Mitochondrial phosphatase PTPMT1 is essential for cardiolipin biosynthesis. *Cell Metab* 13, no. 6 (2011): 690–700.

Zhang, M., E. Mileykovskaya, and W. Dowhan. Gluing the respiratory chain together. Cardiolipin is required for supercomplex formation in the inner mitochondrial membrane. *J Biol Chem* 277, no. 46 (2002): 43553–6.

Zhang, M., E. Mileykovskaya, and W. Dowhan. Cardiolipin is essential for organization of complexes III and IV into a supercomplex in intact yeast mitochondria. *J Biol Chem* 280, no. 33 (2005): 29403–8.

Zhong, Q., and M. L. Greenberg. Deficiency in mitochondrial anionic phospholipid synthesis impairs cell wall biogenesis. *Biochem Soc Trans* 33, no. Pt 5 (2005): 1158–61.

CHAPTER 5

Mitochondrial Supercomplexes and ROS Regulation

Implications for Ageing

Maria L. Genova and Giorgio Lenaz

CONTENTS

5.1 INTRODUCTION

The simplistic concept that mitochondria are merely the power factories of the cell by way of performing ATP synthesis through oxidative phosphorylation (OXPHOS), and as such that they are discrete semi-autonomous organelles, has given way to the concept of a dynamic network that fuses and divides and is strictly linked to the rest of the cell structures, thereby directing a variety of functions central to cellular life, death and differentiation.[1] It is therefore not surprising that mitochondrial dysfunction has emerged as a key factor in a myriad of diseases, including common degenerative and metabolic disorders.[2] Moreover, mitochondrial dysfunction is certainly related to the ageing process,[3] although the mechanism is still strongly debated.

A major common determinant of the involvement of mitochondria in the aetiology and pathogenesis of so many diseases has been considered their key role in the generation of reactive oxygen species (ROS); although ROS are generated by several other cellular systems, mitochondrial ROS arising from the respiratory chain appear to be strategic for the development of pathological states. Nevertheless, contrary to previous understanding that ROS are deleterious by-products whose production should be avoided, it is now clear that ROS are physiological messengers acting through redox modifications in signalling proteins. For this reason it is believed that ageing may be, at least in part, the result of alterations of signalling pathways such as those involved in mitochondrial biogenesis and apoptosis, induced by an increasing ROS production.[4,5]

Among the factors controlling ROS generation by mitochondria, it is becoming increasingly clear that a major role is played by the supramolecular structure of mitochondrial respiratory complexes.

The FMN- and CoQ-binding sites of NADH-Coenzyme Q reductase (Complex I) and the Q_o site (at the outer or positive side) of ubiquinol-cytochrome *c* reductase (Complex III) are often invoked as the most important mitochondrial superoxide producers. In the classic model of the electron transfer chain,[6] complexes I and III are randomly distributed in the inner mitochondrial membrane (IMM) together with the two other major multi-subunit complexes, designated as succinate-CoQ reductase (Complex II) and cytochrome *c* oxidase (Complex IV). The enzyme complexes are functionally connected by two redox-active molecules, that is, a lipophilic quinone (Coenzyme Q or ubiquinone, CoQ) embedded in the membrane lipid bilayer, and a hydrophilic heme protein (cytochrome *c*) localized on the external surface of the IMM.

Contrary to the view of a random organization of the respiratory chain complexes, prevailing in the last decades of the past century,[6] evidence has now accumulated that a large proportion of the respiratory complexes in a variety of organisms is arranged in supramolecular assemblies called supercomplexes or respirasomes.[7–10]

The natural assembly of the respiratory Complexes I, III and IV into supramolecular stoichiometric entities, such as $I_1III_2IV_{0-4}$ can have deep functional implications on the properties of the respiratory chain, possibly being enzymatic channelling the most striking consequence of supercomplex association.[10–12]

There is increasing evidence that supercomplexes are not static assemblies of the individual complexes, but are present in a dynamic state, described by the plasticity model.[13]

A newly discovered consequence of disruption of supercomplex association is an increase of ROS production.[14]

In this chapter, we would like to provide experimental evidence that the dynamic nature of supercomplexes is at the basis of a control mechanism of the ROS concentration in the cell. An alteration of this finely tuned mechanism would induce a catastrophic loss of control of ROS generation, culminating in the establishment of a vicious circle of mitochondrial damage and ROS generation that is at the basis of pathological changes. In particular we will provide evidence pertaining to the role that such a vicious circle may have in the ageing process.

5.2 RESPIRATORY SUPERCOMPLEXES: STRUCTURE AND FUNCTION

5.2.1 Molecular Composition

In bovine heart mitochondria, Schagger and Pfeiffer[15] first described supercomplexes (respirasomes) comprising Complex I, a Complex III dimer and different copy numbers of Complex IV (Figure 5.1; cf. Chapter 4 in this book).

The fundamental features of the respirasome are conserved in all higher eukaryotes; it contains all the redox enzymes required for the complete pathway of electron transfer from NADH to molecular oxygen. Analysis of a multitude of data in the literature obtained by blue-native polyacrylamide gel electrophoresis (BN-PAGE) reveals that supercomplexes containing only two partners are also present in detectable quantities, like the respiratory assemblies comprising only Complex III and Complex IV and the supercomplex I-III_2 where Complex IV is not present.[10]

The other respiratory enzymes not comprising the *core* of the proton translocation machinery (e.g. Complex II) appear not to be associated in supercomplexes,[10] although some quinone reductases have not been investigated in detail.

Partial elucidation of the interaction of the individual respiratory complexes within the supercomplex was achieved by single-particle electron microscopy.[16–20] Subsequently, new 3D maps at nanoscale resolution allowed interpretation of the architecture of mammalian respirasomes at the level of secondary structure. In a new model by Dudkina et al.[17] (Figure 5.1) the single complexes I, III_2 and IV appear to be at some distance, suggesting that there is a little close contact. In particular, the section through the model on the level of the membrane demonstrates gaps between complexes III_2 and IV within the membrane, whereas the same two complexes appear to contact each other in their matrix portions close to the membrane.

Althoff et al.[20] also proposed that only few points of direct contact are allowed between the three complexes in the mammalian supercomplex $I_1III_2IV_1$ because average distances exceed 2 nm. Moreover, at 19 Å resolution the membrane-embedded part of the supercomplex shows intermediate values of density between that of soluble protein and the hydrophobic membrane interior, suggesting that the supercomplex is held together at least partly by lipid-protein interaction. Likely, a gap filled with membrane lipid would also facilitate the diffusion of ubiquinol/ubiquinone between Complex I and Complex III and vice versa. Three-dimensional density maps of yeast and bovine supercomplexes by electron

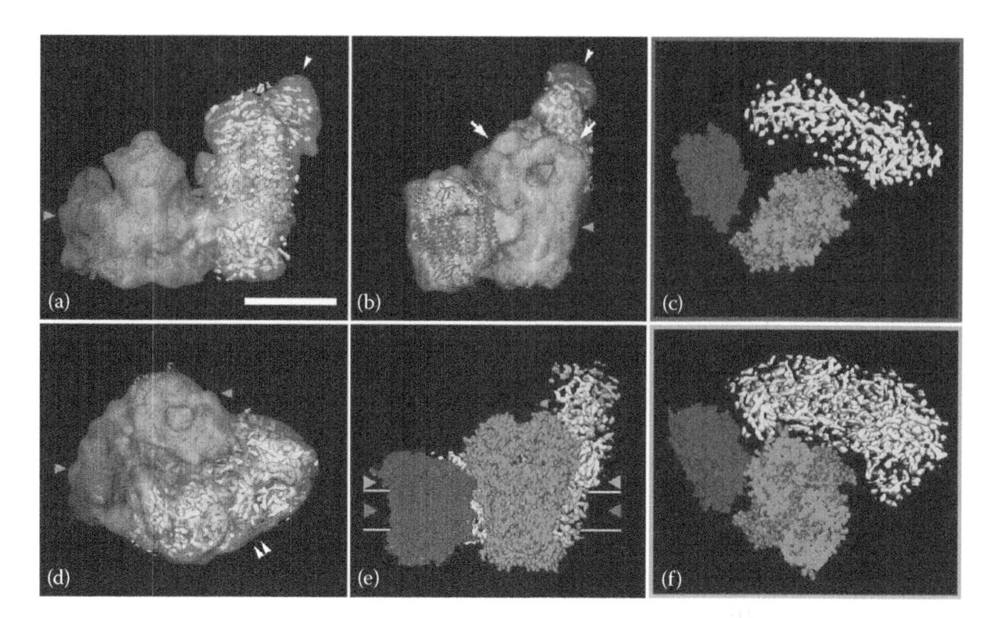

FIGURE 5.1 **(See colour insert.)** Fitting of the high- and medium-resolution structures of complexes I, III_2, and IV to the 3D cryo-EM map of $I_1III_2IV_1$ supercomplex: (a) side view, arrowhead points to flavoproteins; (b) side view from the membrane, arrows point to core I and II subunits of complex III_2, arrowhead to flavoproteins; (c) section through the space-filling model of respirasome on the level of membrane, demonstrating gaps between complexes within the supercomplex; (d) top view from the intermembrane space, double arrowhead points to the bend of complex I in membrane; (e) space-filling model of respirasome seen from the membrane, red and light-blue arrowheads show the level of sections in *C* and *F* and (f) section through the space-filling model of respirasome on the level of matrix. In green, X-ray structure of the bovine dimeric complex III; in purple, X-ray structure of bovine monomeric complex IV; in yellow, the density map of complex I from *Yarrowia lipolytica*. Horizontal lines on (e) indicate the position of the membrane. Orange arrowheads on (a), (b) and (d) point to the position of detergent micelles (scale bar: 10 nm). (Reprinted from Dudkina, N.V. et al., Interaction of complexes I, III, and IV within the bovine respirasome by single particle cryoelectron tomography, *Proc. Natl. Acad. Sci. U.S.A*, 108, 15196–15200. Copyright 2011 National Academy of Sciences. With permission.)

cryo-microscopy show gaps between the transmembrane-localized interfaces of individual complexes (Figure 5.1) consistent with the large excess of cardiolipin (CL) in isolated supercomplexes over that integrated into the structure of the individual respiratory complexes.[21,22]

Lipids are essential determinants of supercomplex association[9]; in particular the studies by Ragan[23,24] and from our laboratory[25] have shown that reconstitution of Complexes I/III at a high lipid-to-protein ratio prevents formation of the supercomplex I_1III_2, as shown by BN-PAGE and kinetic studies. In other words, it appears that dilution of the proteins with an excess of phospholipids may weaken the forces holding together the respiratory complexes.

Among lipids, CL and phosphatidyl ethanolamine are crucial for mitochondrial functions; they are both non-bilayer-forming phospholipids, due to their small polar heads compared with the bulky non-polar tails.[26,27] The phospholipids in closest vicinity to the protein surface, as well as those in the free bilayer, are highly mobile and free to exchange, but CL is tightly bound being more likely buried within the protein complexes, as in Complex IV and Complex III.[28,29] There are now extensive indications that CL stabilizes respiratory supercomplexes as well as the individual complexes.[30]

Further evidence concerning this subject has been collected for the III_2-IV_2 supercomplex in yeast mutants lacking CL.[31–33] Site-directed mutagenesis investigations[34] indicated that CL stabilizes supercomplex formation by neutralizing the charges of lysine residues in the interaction domain of Complex III with cytochrome oxidase.

In addition the stability and assembly of Complex IV were found to be reduced in yeast cells lacking Taz1,[35,36] the orthologous of human Tafazzin, the acyl transferase involved in the maturation of CL.

Mutations of Tafazzin in humans result in Barth syndrome, a cardio-skeletal myopathy with neutropenia, characterized by respiratory chain dysfunction. Patient mitochondria display lower CL content and a polydispersity in acyl chain composition of CL.[37] BN-PAGE revealed an increase in free Complex IV monomer and a decrease in the supercomplex $I_1III_2IV_1$ in patient lymphocytes.[38,39] Gonzalvez et al.[40] confirmed that in immortalized lymphoblasts from Barth syndrome patients the amount of supercomplexes is decreased, as well as the amount of individual complexes I and IV; these changes were compensated by increasing mitochondrial mass.

The dependence on CL for supercomplex formation was clearly demonstrated in a recent study by Bazán et al.[41] by reconstituting purified complexes III and IV from *Saccharomyces cerevisiae* and liposomes of different phospholipid composition.

The shape of the cristae also determines the assembly and stability of respiratory supercomplexes and hence mitochondrial respiratory efficiency. Genetic ablation or overexpression of OPA1 with consequent disorganization of cristae structure affects the assembly and activity of supercomplexes.[42]

5.2.2 Dynamic Nature and Kinetic Advantage of Supercomplexes

The first demonstration that respiratory supercomplexes are indeed functional and capable of full activity came from the chromatographic isolation of a functional respirasome from *Paracoccus denitrificans*.[43] The respiratory activity of a respirasome isolated from mammalian mitochondria was reported by Acín-Pérez et al.[8] who performed an extensive study on isolated mouse liver mitochondria and mitochondria from mouse fibroblast cell lines. They demonstrated that BN-PAGE high-molecular weight bands of mitochondrial protein extracts containing both CoQ_9 (the major CoQ homologue in rodents) and cytochrome *c* showed complete NADH oxidase activity.

Substrate channelling is the direct transfer of an intermediate between the active sites of two enzymes catalysing consecutive reactions[44]; in the case of electron transfer, this means direct transfer of electrons between two consecutive enzymes by successive reduction and reoxidation of the intermediate without its diffusion in the bulk medium. Some evidence for possible channelling comes from the 3D structure of the mitochondrial supercomplex $I_1III_2IV_1$[20]; a unique arrangement of the three component complexes indicates the pathways along which ubiquinone and cytochrome *c* can travel to shuttle electrons between their respective protein partners.

A direct comparison of the effect of channelling with respect to CoQ-pool behaviour was performed in a simple experimental condition in our laboratory.

A system, obtained by fusing a crude mitochondrial fraction enriched in Complex I and Complex III with different amounts of phospholipids and CoQ_{10}[45] was used in our laboratory to discriminate whether the reconstituted protein fraction behaves as individual enzymes (CoQ-pool behaviour) or as assembled supercomplexes depending on the lipid content. The comparison of the experimentally determined NADH-cytochrome *c* reductase activity with the values expected by theoretical calculation applying the pool equation[46] showed overlapping results at phospholipids-protein ratios (w/w) higher than 10:1, that is, for theoretical distances >50 nm. On the contrary, the observed rates of NADH-cytochrome *c* reductase activity were higher than the theoretical values[25,45,47] at low lipid-protein ratio (1:1 w/w).[48]

Moreover, when the same proteoliposomes at 1:1 lipid-protein ratios were treated with dodecyl maltoside to destroy the supercomplex organization, the NADH-cytochrome *c* reductase activity fell dramatically, whereas both C_I and C_{III} individual activities were unchanged[14]; an analogous behaviour was detected by treating bovine heart mitochondria with the same detergent.

Evidence for channelling between Complex I and Complex III also derives from the demonstration that selective decrease of Complex III content under a given threshold induces a preferential deficiency in the transfer of electrons between complexes II and III while the transfer of electrons between complexes I and III remains unaltered.[49]

These studies clearly demonstrate that electron transfer between Complex I and Complex III can take place both by CoQ channelling within the I_1III_2 supercomplex and by the less-efficient collision-based pool behaviour, depending on the experimental conditions.

In addition, the characteristics of the respiratory supercomplexes were investigated on a functional basis by exploiting flux control analysis in bovine heart mitochondrial membranes.[47,50] The flux control coefficients[51] of the complexes involved in aerobic NADH oxidation (I, III, IV) and in succinate oxidation (II, III, IV) were calculated after enzyme inhibition with specific inhibitors: both Complex I and Complex III were found to be highly rate-controlling over NADH oxidation.[50] According to the principles described by Kholodenko and Westerhoff,[51] this is a strong kinetic evidence suggesting the existence of functionally relevant association of Complex I with Complex III, so that electron transfer through Coenzyme Q is accomplished by channelling between the two complexes.

On the contrary, Complex II showed as fully rate-limiting for succinate oxidation, clearly indicating the absence of substrate channelling towards complexes III and IV.[50]

A few other studies that addressed to the functional aspects of supercomplexes using metabolic control analysis[52–54] confirmed that the respiratory chain, at least under certain conditions, is organized in functionally relevant supramolecular structures.

However, the mechanism of electron channelling in supercomplexes is still uncertain.[55] In the interaction between Complex I and Complex III within a supercomplex, we can speculate that CoQ microdiffusion takes place within a lipid milieu (cf. Section 5.2.1), although we cannot exclude that the two active sites are put together by movement of CoQ on the protein or by movement of the protein itself.

Nevertheless, Blaza et al.[56] have very recently discussed their spectroscopic and kinetic data in support of the hypothesis that the metabolic pathways for NADH and succinate oxidation in mammalian mitochondria communicate and catalyse via a single, universally accessible ubiquinone/ubiquinol pool that is not partitioned or channelled. This also points out that the functional and catalytic advantages of the respiratory supercomplexes are still disputed in the scientific community.

It is worth noting that, from our flux control analysis using cyanide inhibition in bovine heart mitochondrial membranes,[50] it appears that a large excess of active Complex IV exists in free form therefore we concluded that channelling of cytochrome *c* is unlikely to occur in the pathway from either NADH or succinate to oxygen.

Using a different approach, Enriquez and coworkers recently demonstrated that at least part of Complex IV forms a functional respirasome with channelling of cytochrome *c*.[49] In particular, they demonstrated in isolated liver mitochondria that the assembling of Complex IV in supercomplexes defines three types of Complex IV, one dedicated exclusively to receive electrons from NADH oxidation (forming supercomplex I+III+IV), another dedicated to receive electrons from FAD-dependent enzymes (forming supercomplex III+IV) and a third major one that is in free form and that is able to receive electrons from both NADH and $FADH_2$ oxidization.

The group of Enriquez has also proposed an integrated model, *the plasticity model*, for the organization of the mitochondrial electron transport chain (ETC)[8,13] where the stoichiometry of the complexes and the variable stability of free versus associated structures under different physiological conditions would determine a variety of different options for the supramolecular organization of the respiratory chain. This view reconciles the previous opposed models, solid versus fluid; in fact they would represent two extreme situations of a dynamic range of molecular associations between respiratory complexes.

A fundamental prediction of the plasticity model is that, in vivo, the mitochondrial respiratory chain should be able to work both when supercomplexes are present and when the formation of supercomplexes is prevented. Indeed, some observations in the literature indirectly support the view that the supercomplex organization may not be fixed but in equilibrium with randomly dispersed complexes in living cells under physiological conditions. Besides phospholipid composition that may change by genetic or dietary reasons in a relatively long timescale, some biochemical parameters have been suggested to affect the supramolecular structure of the respiratory complexes at a shorter timescale. These parameters are the mitochondrial membrane potential and the phosphorylation state of the protein subunits of the complexes.

A study of the influence of the mitochondrial transmembrane potential ($\Delta\mu_{H+}$) on cytochrome *c* oxidase and respiration in intact cells and isolated mitochondria indicated that the control strength of the oxidase (that indicates the extent of control exerted by the enzyme) is decreased under conditions mimicking state 4 respiration in respect to endogenous state 3 respiration.[57] More recently[52] the extension of these studies to the other proton-translocating respiratory complexes (I and III) revealed that at high potential the flux control coefficients were much lower than under conditions of low potential (i.e., state 3) when the sum of the flux control coefficients exceeded 1 (i.e. enzymes form supramolecular units, according to Kholodenko and Westerhoff[51]). Although the interpretation of the results in such a complex system is very difficult, the authors suggest that such a change in control strength might be featured in terms of supercomplex plasticity; since the respiratory rate is high in state 3 conditions, the supercomplex organization would produce an extra advantage by raising the rate by channelling.

Concerning the role of the post-translational modifications of the protein subunits of the complexes, it is tempting to speculate that endocrine alterations may affect the assembly state of Complex IV, by hyper- or hypo-phosphorylation of some subunits in the complex.[58–60] Indeed, cAMP- and PKA-dependent phosphorylation of Complex IV in heart mitochondria[61] was found to be higher in free Complex IV, not associated in form of supercomplex, than in the bound enzyme, thus suggesting that phosphorylation prevents supercomplex association.

Analysis of the state of supercomplexes in human patients with an isolated deficiency of Complex III[62,63] and in cultured cell models harbouring either Complex III[62] or Complex IV[64] or cytochrome *c*[65] depletion leads to propose that the formation of respirasomes may be essential for the assembly/stability of Complex I. Genetic alterations leading to a loss of Complex III led to secondary loss of Complex I, therefore primary Complex III assembly deficiencies are presented as Complex III/I defects.[62,66,67]

In fibroblasts lacking the Rieske iron-sulphur protein,[68] the mechanism for Complex I destabilization in the absence of critical subunits of complexes III or IV was found to be an enhanced ROS generation, supporting the current knowledge that Complex I is particularly sensitive to ROS damage. It will be shown in Section 5.4 that supercomplex plasticity is involved in the regulation of ROS generation by Complex I.

5.3 ROS GENERATION BY MITOCHONDRIA

ROS is a collective term including oxygen derivatives, either radical or non-radical, that are oxidizing agents and/or are easily converted into radicals.[69]

If a single electron is supplied to O_2, it enters one of the π^* orbitals to form an electron pair there, thus leaving only one unpaired electron in the superoxide radical anion O_{2-}; addition of another electron gives the peroxide ion, which is a weaker acid and is protonated to hydrogen peroxide H_2O_2; addition of two more electrons breaks the molecule producing water H_2O. If one single electron is added to H_2O_2 by a reduced metal ion (e.g. Fe^{2+}), the hydroxyl radical OH• is produced by the *Fenton reaction*. The hydroxyl radical is extremely reactive with a half-life of less than 1 ns; thus it reacts close to its site of formation.

ROS arise in cells from exogenous and endogenous sources. Exogenous sources of ROS include UV and visible light, ionizing radiation, drugs and environmental toxins. Among endogenous sources there are xanthine oxidase, cytochrome P-450 enzymes in the endoplasmic reticulum, peroxisomal flavin oxidases and plasma membrane NADPH oxidases.[70]

Nevertheless, the mitochondrial respiratory chain in the IMM is usually considered one of the major sources of ROS, although other enzyme systems in mitochondria can be important contributors to ROS generation.[71–73] Among these, we mention here dihydrolipoamide dehydrogenase (a subunit of the α-ketoglutarate and pyruvate dehydrogenase complexes), monoamine oxidase, and mitochondrial nitric oxide synthase. A special mention has to be given to the adaptor protein $p66^{Shc}$, a 66 kDa Src collagen homologue (Shc) protein that is one of three main isoforms encoded by the *SHC1* gene ($p46^{Shc}$, $p52^{Shc}$, $p66^{Shc}$).[74] Expression of $p66^{Shc}$ is required for mitochondrial depolarization and release of cytochrome *c* after a variety of pro-apoptotic signals.[75] This effect is due to the intrinsic property of $p66^{Shc}$ to act as a redox protein accepting electrons from cytochrome *c* and directly producing hydrogen peroxide.[76]

It is worth noting that mitochondria from different tissues may vary conspicuously in their capacity to produce ROS using different substrates,[77] and this capacity is also related to animal species and age.

Murphy[78] carefully analysed the thermodynamics of mitochondrial superoxide production concluding that, *in vivo*, the one-electron reduction of O_2 to O_2^- is thermodynamically favoured by the existing steady-state concentrations of O_2 and O_2^-, allowing O_2 to be theoretically reducible by a wide range of electron donors within mitochondria. However, only a small proportion of mitochondrial electron carriers with the thermodynamic potential to reduce O_2 to O_2^- do so. The reason is that electron transfer to oxygen requires the protein-linked potential donors to be at a distance compatible with electron tunnelling, according to the Markus theory.[79]

5.3.1 Respiratory Chain as a Source of ROS

The FMN and CoQ-binding sites of Complex I and the Q_o site (at the outer or positive side of the IMM) of Complex III are often invoked as the most important mitochondrial superoxide producers, but other sites have also been defined, including glycerol phosphate dehydrogenase, ETF-CoQ reductase, and pyruvate and α-ketoglutarate dehydrogenases (Figure 5.2). The absolute and relative contribution of each site differs greatly with different substrates (reviewed in Lenaz[80]).

The topology of the sites is important because it establishes whether the ROS are produced on the matrix side of the IMM, where they may damage mitochondrial DNA (mtDNA), or are released out of the mitochondria, where they can damage other systems or act as signalling molecules.

Most of superoxide is generated at the matrix side of the IMM, as appears from the observation that superoxide is detected in submitochondrial particles (SMP), which are inside-out with respect to mitochondria. However, Han et al.[81] demonstrated the formation of superoxide radical also in mitoplasts, indicating that a significant aliquot of this species is released at the outer face of the IMM. It is likely that Complex I releases ROS in the matrix while Complex III mostly in the intermembrane space (IMS). The simple passive diffusion of hydrophilic ROS across lipid membranes is limited due to permeability restrictions, but an anion channel related to the voltage-dependent anion channel (VDAC) may export the superoxide anion from the IMS directly to the cytoplasm.[82] Interestingly, a new function of uncoupling proteins UCP2 and UCP3 was suggested by Wojtczak et al.[83] who inferred that these proteins also act as anion channels to export superoxide from the

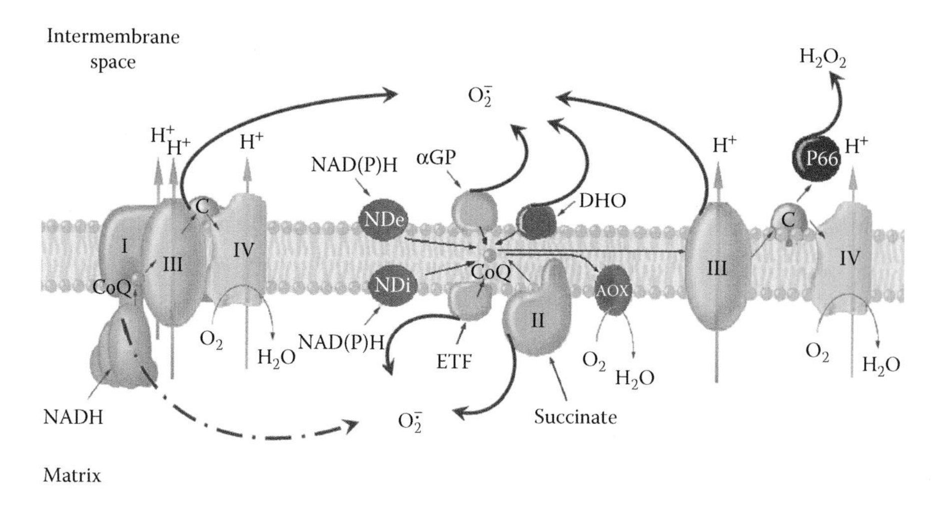

FIGURE 5.2 A schematic drawing of the respiratory chain depicting the protein complexes and their substrates. Complex I is depicted as a component of the $I_1III_2IV_1$ supercomplex; whereas Complex III and Complex IV are also shown in their free form. Thick arrows indicate the sources of superoxide at different sites in relation to the inner mitochondrial membrane. I, NADH-ubiquinone oxidoreductase; II, succinate-ubiquinone oxidoreductase; III, ubiquinol-cytochrome *c* oxidoreductase; IV, cytochrome oxidase; NDi and NDe, internal and external alternative NAD(P)H dehydrogenases; AOX, alternative oxidase; αGP, glycerol-3-phosphate; ETF, electron transfer flavoprotein; DHO, dihydroorotate; CoQ, Coenzyme Q (ubiquinone); C, cytochrome *c*; P66, adaptor protein p66 (see Section 5.3 for details).

mitochondrial matrix. These observations, however, suggest that ROS export is largely a protective device, potentially allowing ROS detoxification elsewhere in the cell without consuming mitochondrial reduction equivalents.[84]

5.3.1.1 Complex I

Complex I is a major source of superoxide in several types of mitochondria. Several prosthetic groups in the enzyme have been suggested to be the direct reductants of oxygen. These include FMN,[85–87] ubisemiquinone,[88,89] and iron-sulphur cluster N2.[90,91]

Grivennikova and Vinogradov[92] investigated ROS generation by Complex I either isolated or membrane-bound at different NAD^+/NADH levels. At the optimal NADH concentration of 50 μM, Complex I produced both superoxide and hydrogen peroxide at a 0.7 ratio O_2^-/H_2O_2. The production of superoxide was attributed presumably to FeS cluster N2, whereas hydrogen peroxide was interpreted to derive from 2-electron oxidation of fully reduced FMN.

Superoxide production by Complex I is enhanced during the energy-dependent reverse electron transfer from succinate to NAD^+, which occurs at high membrane potential, and is inhibited by processes dissipating the membrane potential.[93–96] An interesting characteristic of Complex I from some organisms is the ability to adopt two distinct states: the so-called catalytically active (A) form can undergo deactivation (dormant state, D) in response to oxygen deprivation.[97] The A/D transition could represent an intrinsic mechanism that provides a fast response of the mitochondrial respiratory chain to anoxia, probably as a means to protect mitochondria from ROS generation due to the rapid burst of respiration following reoxygenation.[98] Studies on the molecular mechanism and driving force behind the A/D transition of the enzyme are still ongoing, but several subunits (e.g. NDUFA9, ND3 and ND1 located in the region of the quinone-binding site) are most likely involved in the conformational rearrangements.[99] Notably, in the D-form of Complex I, a single residue of cysteine (cys-39) of the mitochondrially encoded ND3 subunit is exposed by conformational changes of the protein complex[100] and can undergo reversible redox modification (cf. also Section 5.5.2) exerting the A/D transition of the enzyme.

5.3.1.2 Complex III

The formation of superoxide in Complex III depends on the peculiar mechanism of the so-called Q-cycle[101] that involves the biphasic oxidation of ubiquinol: one electron is given to the Rieske iron-sulphur cluster and then to cytochromes c_1 and c, while the other electron reduces low potential cytochrome b (b_L) of Complex III. In the controlled state 4, when $\Delta\mu_{H+}$ is high, the subsequent electron transfer from cytochrome b_L to b_H is strongly retarded, since it occurs against the electrical gradient (from the positive to the negative side). This retardation, prolonging the lifetime of the semiquinone (Q) at the outer Q_o site, has been interpreted to allow reaction of the semiquinone with O_2, thus forming superoxide.[96]

Antimycin A (AA), an inhibitor acting at the inner or negative side of the membrane (Q_i site), fully abolishes the reoxidation of cytochrome b_H by ubiquinone and stimulates the production of ROS. The latter phenomenon is inhibited by the inhibitors acting at Q_o site. Thus, we may locate the site of one-electron reduction of oxygen in presence of antimycin at a component located at Q_o, presumably ubisemiquinone.[102]

Recent observations suggest that the source of the electron to reduce oxygen is the semiquinone formed in the so-called semi-reverse reaction in which the electron moves back from cytochrome b_L and reduces the fully oxidized quinone. In fact superoxide formation is stimulated by the presence of oxidized quinone[103] and by mutations that do not allow interaction of the newly formed ubisemiquinone with the Rieske cluster,[104] thus favouring the reaction with oxygen.

5.3.1.3 Other Respiratory Enzymes

The highest rate of ROS production in isolated mitochondria occurs with succinate as substrate.[105] Contrary to common opinion, Moreno-Sanchez et al.[106] produced experimental results that seem to exclude a significant ROS production catalysed by reverse electron flow through Complex I, thus allowing ROS production by Complex II as the only plausible explanation. Actually, several other observations suggest that Complex II may be a source of ROS.[72,107–111] However, it is worth noting that mitochondria in intact cells oxidize predominantly NAD-linked substrates so that neither Complex II nor reverse electron transfer can be very important contributors to superoxide formation in vivo.[106]

Other enzymes that feed electrons to the respiratory chain have been shown to be sources of ROS, such as glycerol phosphate dehydrogenase,[112–115] dihydroorotate dehydrogenase,[116,117] electron transfer flavoprotein (ETF) and ETF dehydrogenase, involved in fatty acid oxidation.[118,119] An extensive review of the sites in the respiratory chain that are responsible for ROS generation can be found in Lenaz.[73]

5.4 MITOCHONDRIAL ROS AS SIGNALS: TARGETS AND MECHANISMS

ROS that can be formed during aerobic respiration have traditionally been viewed as detrimental because they can cause oxidative damage to cellular components. However, it is clear nowadays that there is a threshold below which ROS are essential signalling molecules.[4,5,120–127] At moderate concentrations, ROS play an important role as regulatory mediators in a complex array of control mechanisms in mitochondrial metabolism and many others processes. Indeed, they are required for satiety signalling, adipocyte differentiation, insulin release and signalling, hypoxic signalling.[128–131] Many of the ROS-mediated responses actually protect the cell against oxidative stress and re-establish redox homeostasis. The cost/benefit dichotomy depends on the concentration and types of ROS. For example ROS may be oncogenic and promote proliferation, invasiveness, angiogenesis and metastasis,[132] but they may also be anti-oncogenic and promote cell cycle stasis, senescence and apoptosis.[133]

Because most ROS are oxidants, they can influence the redox state of signalling proteins through reactions with specific sulfhydryl groups (Figure 5.3), but enzymatic catalysis is usually required to allow the modification to occur under physiological conditions.[134] For this reason, not all ROS are equally suitable for signal transduction: O_2^{-}, H_2O_2 and lipid peroxidation derivatives like 4-hydroxy-2-nonenal (HNE) are extensively employed as signalling molecules whereas the OH• radical is too unspecific to undergo catalysed reactions. Mitochondrial ROS levels reflect the balance between their rate of generation and of removal. Among the arsenal of antioxidants and detoxifying systems existing in mitochondria (cf. Lenaz[73]), thioredoxin (Trx; cf. Chapter 11 in this book) and the non-protein thiol antioxidant glutathione (GSH) emerge as the main line of defence for the maintenance of the appropriate mitochondrial redox environment.[135,136] GSH is transported into mitochondria from the cytosol via dicarboxylate and α-ketoglutarate carriers.[137] GSH importance is based not only on its abundance, but also on its versatility to counteract hydrogen peroxide, lipid hydroperoxides or xenobiotics, mainly as a cofactor of enzymes such as glutathione peroxidase or glutathione-S-transferase.[138] The involvement of GSH in redox pathways converts GSH in its oxidized form (GSSG).

The concentration of ROS within cells is largely dependent on the redox state of glutathione (GSH/GSSG); the mitochondrial GSH/GSSG ratio is generally greater than 100:1 and is widely used as an indicator of the redox status of the cell.[139]

Thiols exposed on the surface of mitochondrial proteins[5] are very reactive since the relatively high matrix pH favours the dissociation to thiolate anions; moreover, the high proportion of vicinal thiols facilitates the formation of disulphide bonds.[5,78,140]

The ROS-mediated oxidative modification of proteins is usually followed by other post-translational changes (e.g. phosphorylation, acetylation, ubiquitination, and SUMOylation among others[126]) in the same protein and in other proteins of the signalling cascade.

Redox regulation of phosphatases and kinases is used to control the activity of select eukaryotic signalling pathways, making ROS important second messengers that regulate

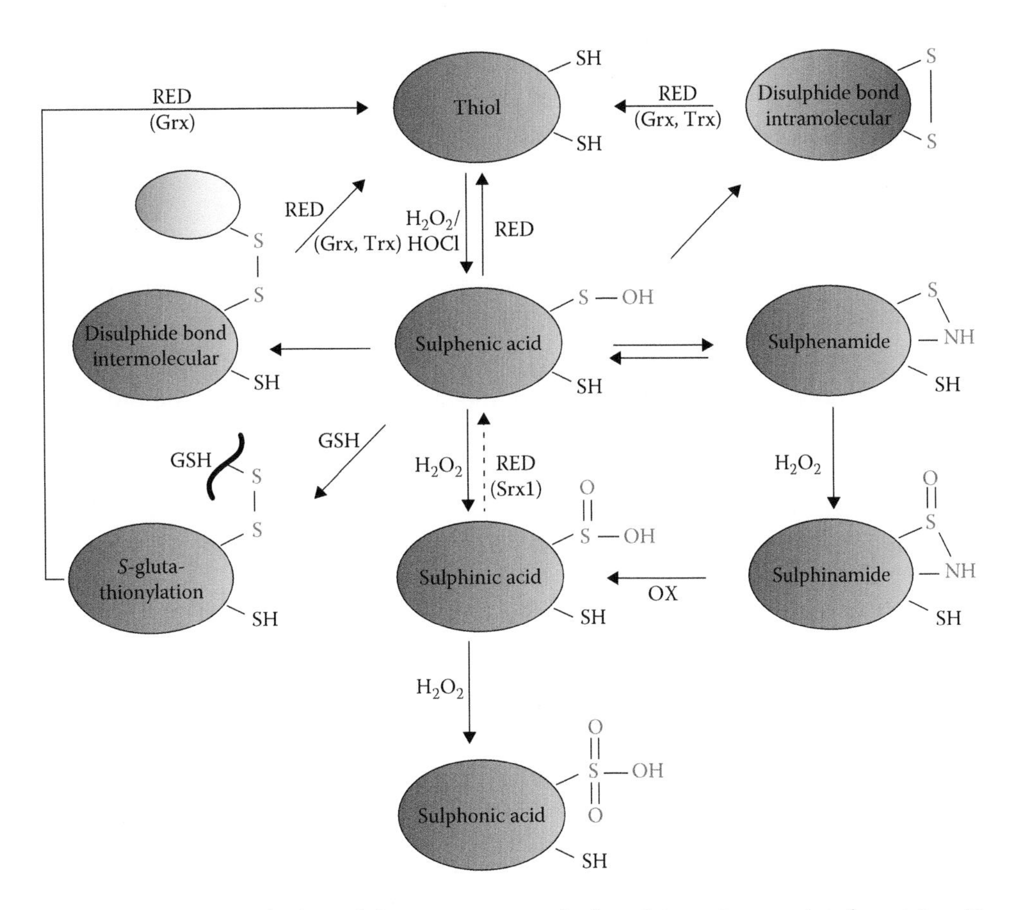

FIGURE 5.3 Oxidative thiol modifications commonly found in redox-regulated proteins. Upon reaction with peroxide (H_2O_2) or hypochlorous acid (HOCl), redox-sensitive thiol groups (RSH) rapidly form sulphenic acids (RSOH). These sulfenates are highly reactive and tend to quickly react with nearby cysteine thiols to form inter- or intramolecular disulphide bonds (RSSR). Alternatively, they form mixed disulphides with the small tripeptide glutathione (GSH) (RSSG), or undergo cyclic sulphenamide formation (RSNHR). These oxidative thiol modifications are fully reversible, and reduction (RED) is catalysed by members of the glutaredoxin (Grx) or thioredoxin (Trx) system. Further oxidation of sulphenic acid to sulphinic acid (RSO_2H), sulphinamide (RSONHR) or sulphonic acid (RSO_3H) is irreversible. One exception is the active site sulphinic acid in peroxiredoxin, whose reduction is mediated by the highly specialized ATP-dependent sulfiredoxin Srx1. (Reprinted from *Biochimica et Biophysica Acta*, 1844, Groitl, B., and Jakob, U., Thiol-based redox switches, 1335–1343, Copyright 2014, with permission from Elsevier.)

growth, development and differentiation.[141] A number of transcription factors contain redox-sensitive cysteine residues at their DNA-binding sites.[142,143] The paradigm of mitochondrial signalling leading to changes in nuclear gene expression is relatively novel and is considered *mitochondrial retrograde signalling*.[144] The retrograde signalling pathway interacts with several other signalling pathways, such as target of rapamycin (TOR) and ceramide signalling. All of these pathways respond to stress, including metabolic stress.[145] ROS can be part of an important retrograde signal by stimulating the antioxidant response element (ARE) of cytoprotective genes. One notable example is Nrf2 that, in the presence of ROS, is translocated from the cytoplasm to the nucleus. There, it binds the ARE of genes involved in the antioxidant response (e.g. heme oxygenase, NRF-1 and other inducers of mitochondrial biogenesis).[146]

Signalling proteins modified by ROS include phosphoprotein phosphatases (PTPs), Ras, large G-proteins, serine/threonine kinases of the MAPK families, transcription factors as AP-1, NFκB, p53 and others. The effect of oxidants is different with different proteins: PTPs are inhibited, nuclear transcription factors are activated.[147] In addition, ROS appear to activate the hypoxia-inducible factor HIF-1α[148,149] by inhibiting prolyl-4-hydroxylase (that hydroxylates a critical proline directing the factor to proteolytic digestion) similarly to what happens in anoxia for lack of O_2 required for HIF hydroxylation.

Mitochondrial ROS can also act as important signals to regulate the inflammatory response by activating the inflammasome[150,151] and to regulate autophagic processes including mitophagy.[152]

5.5 REGULATION OF MITOCHONDRIAL ROS GENERATION

The generation of ROS by isolated mitochondria accounts for 0.1%–0.2% of oxygen consumed and may reach up to 2%–3% under particular conditions that may not be found physiologically[153]; the rate and extent of ROS generation greatly vary in different tissues (see also Panov et al.[154]) and specifically depends upon the substrate employed. For example, succinate is important for ROS production in brain, heart, kidney and skeletal muscle, while fatty acids are major generators of ROS in kidney and liver.[155]

The factors controlling mitochondrial ROS levels are linked to their rate of generation and of removal. In the former case, the forces directly associated with respiratory activity, which are the redox potential of the NAD^+/NADH couple and the proton-motive force, are powerful regulators of the steady-state concentration of the redox species responsible for electron leaking.[124] These forces, on the other hand, are regulated by the redox supply to the chain, by the degree of coupling, by physiological or pathological constraints to electron transfer, such as enzyme phosphorylation, cytochrome *c* removal, Complex IV inhibition, oxygen concentration and so on.

The *Redox-Optimized ROS Balance hypothesis* (R-ORB) described by Aon et al.[156] attempts to explain at a mechanistic level the link between mitochondrial respiration and ROS emission. The hypothesis is based on the observation that ROS levels (as the net result of production and scavenging) attain a minimum when mitochondria maximize their energetic output (i.e. maximal state 3 respiration) under conditions of intermediate values of the redox environment (i.e. summation of the redox potential and concentration of the redox couples present in the system, such as NADH/NAD^+ and GSH/GSSG), values that are more oxidized than the corresponding values in state 4 respiration. On the other hand, ROS overflow will occur at both highly reduced or highly oxidized redox environment, albeit governed by the increased probability of proton leak and by the compromised scavenging capacity, respectively. Cortassa et al.[157] tested this tenet by assessing, in parallel, the rates of mitochondrial respiration, ROS emission, and the redox environment in isolated guinea pig heart mitochondria under forward electron transport; the authors confirmed that mitochondria are able to keep ROS emission to a minimum, likely compatible with signalling, while maximizing their energetic output.

In addition, mitochondrial ROS release is modulated by a series of nuclear-encoded proteins[123] and in response to external stimuli, such as TNFα,[158] hypoxia,[148] serum deprivation,[159] oxidative stress (ROS-induced ROS release[160]) and others. The ROS generation and release from mitochondria are controlled by p53, p66Shc, the Bcl-2 family and Romo-1.[123]

5.5.1 Role of Mitochondrial Membrane Potential

Mitochondrial ROS production is enhanced in state 4 and when the rate of electron transfer is lowered.[161] The rationale is in a more reduced state of the respiratory carriers capable of donating electrons to oxygen. To this purpose, uncoupling and release of excessive membrane proton potential may protect mitochondria from damage induced by excessive free radical production. In rat hepatocytes the futile cycle of proton pumping and proton leak may be responsible for 20%–25% of respiration[162]; in perfused rat muscle the value is even greater. Uncoupling may be obtained by activating proton leak through uncoupling proteins.[102] In such way, a tissue may dissipate a conspicuous part of the energy conserved by its mitochondria and keep the mitochondrial respiratory chain under more oxidized conditions preventing the formation of free radicals.

However, the notion that mild uncoupling may be a protective mechanism that lowers mitochondrial $\Delta\Psi$ and thus alleviates oxidative stress[163] has been challenged on the theoretical basis that the mitochondrial membrane potential within living cells is much lower (100–130 mV) than that measured in isolated mitochondria.[164] Accordingly, Johnson-Cadwell et al.[165] found that no significant change in matrix superoxide occurred after treatment of cerebellar neurons with the uncoupler FCCP.

It is worth mentioning that Selivanov et al.[166] found that matrix pH per se is an essential factor defining ROS production by the respiratory chain even in the absence of pH gradient, and that pH increase in the matrix induces the increase in ROS generation. This observation is in agreement with a previous report by Turrens and Boveris[167] showing that the rate of superoxide generation increased in conjunction with an increase of medium pH from 7 to 9.2 in the presence of respiratory substrates and appropriate inhibitors (e.g. NADH and rotenone or succinate and antimycin A). Moreover, Selivanov et al.[166] reported that the decrease of matrix pH induced by the addition of P_i and nigericin is accompanied by an increase of $\Delta\Psi$, which was expected to stimulate ROS production. Conversely, the ROS generation rate decreased in Selivanov's experimental conditions, thus indicating that the effect of pH dominates over the opposite effect of $\Delta\Psi$.

5.5.2 Role of Post-Translational Modifications

Events leading to decrease of the rate of electron flow in Complex I lead to overproduction of ROS. Physiological states, such as subunit phosphorylation that reversibly inhibits Complex I activity, may modify its ROS generating capacity.[168–170] A different mechanism, centred on cytochrome *c* oxidase, has been proposed by Kadenbach et al.[171,172]: the cAMP-dependent phosphorylation of subunit I strongly activates the allosteric inhibition of the enzyme by ATP. The result of this inhibition is decrease of the H^+/e^- stoichiometry of the enzyme from

1 to 0.5, with resulting decrease of $\Delta\Psi m$ and of ROS generation. It was also proposed that stress conditions would induce dephosphorylation of the enzyme with transient increase of membrane potential and a burst of ROS generation.

It is therefore tempting to speculate that endocrine alterations may affect the capacity of ROS formation by hyper- or hypo-phosphorylation of the respiratory complexes. The signals inducing phosphorylation and dephosphorylation of the respiratory chain are still however poorly understood.

Other protein modifications in mitochondrial complexes can modulate ROS production,[4] such as thiol oxidation or S-nitrosation or S-glutathionylation of Complex I, or Complex I and Complex II acetylation that is regulated by sirtuins (NAD^+-dependent deacetylases). (cf. also Mailloux et al.[173] for a comprehensive review on how cysteine oxidation reactions regulate key mitochondrial functions including oxidative phosphorylation, and Dröse et al.[99] for particular emphasis on Complex I).

In addition, ROS production by mitochondria is under signalling control through the pathways leading to mitochondrial uptake and activation of $p66^{Shc}$ under conditions of oxidative stress.[174]

5.5.3 Hypoxia and ROS Production

Mitochondria have long been suspected to be the site of oxygen sensing, since they bind O_2 at cytochrome oxidase and they represent the primary site of oxygen consumption in the cell. However, a conceptual hurdle has been that cytochrome oxidase activity, as indicated by the rate of oxygen consumption by isolated mitochondria, does not become limited by O_2 availability until the oxygen concentration falls to about 1 μM (0.1% O_2). This behaviour would appear to make the ETC suitable as a detector of anoxia, but unsuitable as a sensor of moderate hypoxia. Guzy and Schumacker[175] point out that studies from their laboratory suggested that the ETC acts as an O_2 sensor by releasing ROS in response to hypoxia.

It is seemingly paradoxical that a decrease in a required substrate, O_2, would result in an increase in ROS production. However, inhibition of cytochrome *c* oxidase by hypoxia[176] has been found to enhance ROS formation by the respiratory chain; this result may be due to the increased membrane potential[177] or by determining a more reduced state of cytochrome c, facilitating its interaction with $p66^{Shc}$ and further amplification of ROS production (cf. Section 5.3), and preventing its superoxide scavenging activity.

More recently, studies using the genetically encoded H_2O_2 biosensor pHyPer[178] have found that hypoxia affects mitochondrial ROS generation in isolated pulmonary artery smooth muscle cells by directly causing an increase of ROS production from Complex III.[179] Consistent with previous reports using the traditional ROS detection approach based on measurements of the DCF-derived fluorescence signal,[148,180,181] the result by Korde et al.[179] demonstrates that the mitochondrial Complex III has an inherent hypoxia-sensing capability. The Rieske iron-sulphur protein is possibly imperative for this hypoxic response that does not necessarily depend on the availability of oxygen at cytochrome *c* site and reduced electron flux through the ETC.[179]

The role of Complex III was also confirmed in a transgenic mouse model where deletion of the Rieske protein of Complex III[182] abolishes the hypoxia-induced increase in ROS signalling.

The content of oxidized cytochrome *c* seems to be an essential factor in controlling ROS release from mitochondria[183,184]; in fact, cytochrome *c* in the IMS oxidizes superoxide produced by Complex III to O_2, thus preventing generation of harmful H_2O_2. The loss of cytochrome *c* due to outer membrane permeabilization during hypoxia and ischemia[185] was considered a major factor of ROS generation under pathological conditions.[184] The ROS released during hypoxia act as signalling agents that trigger diverse functional responses, including activation of gene expression through the stabilization of the transcription factor hypoxia-inducible-α (HIF-α).[148,186]

Nevertheless, one important challenge will be to identify the signalling pathway linking the mitochondrial oxidant generation to the regulation of HIF-α; additional studies are needed to fill this gap in understanding (see Movafagh et al.[187] for an updated review).

In addition, it has to be considered that the enhanced ROS production during hypoxia has been ascribed not only to mitochondria, but also to the plasma membrane NADPH oxidase (e.g. Mittal et al.[188]).

5.5.4 Role of Supercomplexes

An implication of supercomplex organization as the missing link between oxidative stress and energy was first suggested by Lenaz and Genova.[9] They speculated that dissociation of supercomplex I_1III_2 occurs under conditions of oxidative stress, with loss of facilitated electron channelling and resumption of a less efficient random diffusional behaviour, causing electron transfer to depend upon the collisional encounters of the free ubiquinone molecules with the partner complexes.

As predicted by Lenaz and Genova,[9] dissociation of supercomplexes might have further deleterious consequences, such as disassembly of Complex I and Complex III subunits and loss of their catalytic activity. As a consequence, the alteration of electron transfer may elicit further induction of ROS generation. Following this line of thought, the different susceptibility of different types of cells and tissues to ROS damage may be interpreted in terms of extent and tightness of supercomplex organization in their respiratory chains, which also depend on the phospholipid composition of their mitochondrial membranes. Supercomplex disorganization eventually leading to destabilization of Complex I and Complex III would both decrease NAD-linked respiration and ATP synthesis and increase the capacity of these enzymes of producing superoxide. Here we briefly summarize the experimental evidence pertaining to this hypothesis.

The possibility that loss of supercomplex organization may enhance ROS generation by the respiratory chain has been advanced on theoretical grounds,[9,10,189] in the sense that a tighter organization of the respiratory enzymes may hide auto-oxidizable prosthetic groups hindering their reaction with oxygen, or alternatively that slowing electron flow in the chain may keep the prosthetic groups in a more reduced form allowing them to interact with oxygen.

Although the molecular structure of the individual complexes does not allow to envisage a close apposition of the matrix arm of Complex I, where the prosthetic groups are localized, with either Complex III or IV,[16,19,190] the actual shape of the $I_1III_2IV_1$ supercomplex from bovine heart[19] suggests a slightly different conformation of Complex I in the supercomplex with a smaller angle of the matrix arm with the membrane arm showing a higher bending towards the membrane (and presumably Complex III), in line with the notion that Complex I may undergo important conformational changes.[191] Moreover, the observed destabilization of Complex I in the absence of supercomplex may render the 51 kDa subunit containing the FMN more *loose*, allowing it to interact with oxygen.

On the basis of their studies on rat brain mitochondria oxidizing different substrates, Panov et al.[154] suggested that supercomplex organization of Complex I within the chain prevents excessive superoxide production on oxidation of NAD-linked substrates because the efficient channelling helps maintaining the chain in the oxidized state (see also below in this section), whereas on succinate oxidation the backward electron flow keeps the centres in Complex I more reduced, favouring production of superoxide; to this purpose it is interesting to note that Complex II is not forming a respirasome.

Circumstantial evidence on the role of supercomplex organization comes from the observation that high mitochondrial membrane potential elicits ROS generation, while uncoupling strongly reduces ROS production[96,192]; although other explanations may be given to these observations, they are compatible with the suggestion by Capitanio's group[52,57] that high membrane potential may dissociate the supercomplexes into the individual units. In a recent study, Maranzana et al.[14] obtained the first direct demonstration that loss of supercomplex organization causes an enhancement of ROS generation by Complex I. In a model system of reconstituted binary Complex I/Complex III at high lipid to protein ratio (30:1), where formation of the supercomplex I_1III_2 is prevented, the generation of superoxide is several fold higher than in the same system reconstituted at a 1:1 ratio, which is rich in supercomplexes (Figure 5.4). In agreement with this finding, dissociation with the detergent dodecyl maltoside of the supercomplex present in the latter system, as well as in mitochondrial membranes, induces a strong enhancement of ROS generation.

It is worth noting that, in the experiments reported by Maranzana et al.,[14] ROS production is investigated in the presence of inhibitors (mucidin and rotenone) that prevent electron transfer to any possible acceptor; therefore the redox centres in Complex I are maximally reduced both in the situations where supercomplexes are maintained and in the situations where Complex I is free. Consequently, the above-mentioned reasoning by Panov et al.[154] cannot be taken as the only explanation of the role of supercomplexes in regulating ROS formation by Complex I.

Several additional observations in cellular and animal models link together supercomplex dissociation and enhanced ROS production.

A strong decrease of high molecular weight supercomplexes correlating with higher ROS generation was observed in mouse fibroblasts expressing the activated form of the k-ras oncogene, in comparison with wild-type fibroblasts.[193,194]

Diaz et al.[68] showed that diminished stability of supercomplexes is associated with increased levels of ROS in mouse lung fibroblasts lacking the Rieske iron-sulphur protein of Complex III and hence devoid of the supercomplexes containing Complex I.

The CL defect in Barth syndrome, a cardio-skeletal myopathy with neutropenia which is characterized by respiratory chain dysfunction, results in the destabilization of the supercomplexes by weakening the interactions between respiratory complexes. Remarkably,

Sample ID	Relative amount of bound Complex I (% of total)	ROS production (%)
1 proteoliposomes	92	224 ± 83
2 proteoliposomes DIL	24	851 ± 243
3 proteoliposomes + DDM	21	672 ± 257
4 BHM	87	121 ± 20
5 BHM + DDM	2	223 ± 86

FIGURE 5.4 Production of ROS by mitochondrial Complex I in different situations where supercomplexes are maintained or disassembled. The percentage value of ROS production measured in all the samples listed in the table (*lower panel*) is plotted in the graph (*upper panel*). The ratio of bound Complex I versus total Complex I was determined by densitometric analysis of immunoblots obtained after 2D BN/SDS-PAGE. The NADH-stimulated production of ROS was measured as the relative fluorescence intensity of dichlorofluorescein in the presence of 1.8 μM mucidin and 4 μM rotenone, and expressed as percentage value of the corresponding reference samples. Notes—proteoliposomes (1:1 w:w, cf. text for details) and BHM (bovine heart mitochondria) respectively assayed in the presence of 1.8 μM mucidin only. In the case of BHM, the existence of endogenous systems operating to reduce ROS levels in the mitochondrial sample might have counteracted the dramatic effects of the complete dissociation of Complex I, thus leading to a twofold only increase of the measured ROS production. DIL, dilution at high lipid to protein ratio (30:1 w:w); DDM, dodecyl maltoside. (Data from Maranzana, E. et al., *Antioxid. Redox Signal.*, 19, 1469–1480, 2013.)

higher basal levels of superoxide anion production, detected by hydroethidine staining, have been observed in lymphoblasts from patients, compared to control cells.[38,40]

The availability of CL-lacking yeast mutants provided the opportunity to demonstrate alterations in stabilization of supercomplexes similar to those found in Barth syndrome. CL-deficient strains of *S. cerevisiae* harbour significantly less stable supercomplex III_2IV_2 than the parental strain.[31,32] Moreover, Chen et al.[195] observed that *taz1*Δ as well as *crd1*Δ yeast mutants, which cannot synthesize CL, exhibit increased protein carbonylation, an indicator of ROS. The increase in ROS is most likely not due to defective oxidant defence systems, as the CL mutants do not display sensitivity to paraquat, menadione or hydrogen peroxide (H_2O_2).

Similarly, enhanced ROS generation and oxidative stress were found in yeast mutants lacking the supercomplex assembly factor Rcf1 and thus devoid of supercomplexes III-IV,[196–198]; since the yeast *S. cerevisiae* lacks Complex I, in this case we may consider the origin of the extra ROS being presumably Complex III.

The large evidence that supercomplexes physiologically exist in dynamic equilibrium with isolated complexes (cf. Section 5.2.2) raises the puzzling question that also ROS generation may be subjected to physiological changes. It is tempting to suggest that these changes may be aimed to regulation of ROS levels in the cell, in view of the well-documented role of ROS in cellular redox signalling (cf. Section 5.4).

5.6 ROS AND MITOCHONDRIAL QUALITY CONTROL

In order to maintain a functional mitochondrial network and avoid cell damage, cells are endowed with elaborate mechanisms of mitochondrial quality control, ranging from removal of damaged mitochondrial proteins within the organelle to elimination of proteins exposed in the outer membrane via the cytosolic ubiquitin-proteasome system, and to removal of the entire mitochondria through mitophagy, in the case of extensive organelle damage and dysfunction.[199]

Since the mitochondria are a major source of ROS, mitochondrial proteins are especially exposed to oxidative modification, so that elimination of oxidized proteins is crucial for maintaining the integrity of this organelle. The mitochondrial protein quality control system is very efficient and keeps mitochondrial proteins functional as long as damage does not reach a certain threshold and the components of this system themselves are not excessively damaged.[200] It consists of chaperones, which counteract protein aggregation through binding and refolding the misfolded polypeptides, and of both membrane-bound and soluble ATP-dependent proteases that are involved in degradation of damaged proteins.

Several lines of evidence suggest the presence of a mitochondria-associated degradation pathway that regulates mitochondrial protein quality control.[201] Internal mitochondrial proteins may be retrotranslocated to the outer mitochondrial membrane where multiple E3 ubiquitin ligases are present; Cdc48/p97 is recruited to stressed mitochondria, extracts ubiquitinated proteins from the outer membrane and presents them to the proteasome for degradation.

Protein degradation and failure of enzymatic reversal of protein oxidation have been implicated in the age-related accumulation of oxidized proteins. Within the mitochondrial matrix, the ATP-stimulated mitochondrial Lon-protease (Pim-1 in yeast) is believed to play an important role in the degradation of oxidized protein[202,203]; age-associated impairment of Lon-like protease activity has been suggested to contribute to oxidized protein build-up in the mitochondria.[204,205] Lon-protease is involved in mitochondrial quality control in removing damaged or oxidized proteins, but is also devoted to regulatory mechanisms by selective protein degradation; for example during hypoxia Lon participates in remodelling cytochrome oxidase by selectively degrading the Cox4-1 subunit.[206]

Cellular structures and organelles undergo turnover and after a suitable life time are directed to autophagy by lysosomal digestion.[207–209]

Both macroautophagy and chaperone-mediated autophagy, differing in the way that substrate proteins are delivered to the lysosome,[210] are inducible and are involved in the removal of oxidatively damaged organelles and proteins.[211,212]

The macroautophagy systems recognize damaged organelles, as mitochondria, and preferentially address them to digestion (mitophagy).[211]

Mitophagy is required for the turnover of mitochondria as well as elimination of damaged or dysfunctional mitochondria.[199,213,214] Mitophagy is also implicated in developmental processes as it allows the removal of undamaged mitochondria during maturation of reticulocytes and adapts the amount of mitochondria to the cell energy requirements.[215]

In yeast and mammalian cells mitophagy is preceded by mitochondrial fission, which divides elongated mitochondria into pieces of manageable size for encapsulation and segregation of damaged mitochondrial material prior to selective removal by mitophagy. The *PARK2* gene product, Parkin, is a ubiquitin ligase that contains a ubiquitin-like domain and two special zinc finger domains.[216] It promotes the ubiquitinylation and proteasomal degradation of mitofusins, large GTPases anchored in the membrane and responsible for mitochondrial fusion.[217] Damage-induced loss of mitofusins prevents fusion of dysfunctional mitochondria, and targets them to the autophagy pathway.

Non-selective autophagy and mitophagy have been shown to be triggered by ROS in response to nerve growth factor deprivation, rapamycin or starvation.[152,218] Frank et al.[219] showed that moderate levels of ROS specifically trigger mitophagy but are insufficient to trigger non-selective autophagy. Expression of a dominant-negative variant of the fission factor DRP1 blocked mitophagy induction by mild oxidative stress as well as by starvation. This study showed that selective mitophagy can occur in a very specific way, suggesting the presence of a very specific and selective signalling cascade initiated by ROS. On the other hand, mitochondria hyperfuse are protected from mitophagy when non-selective autophagy is induced by starvation or by other means.[220,221]

5.7 IMPLICATIONS FOR AGEING

The supercomplex organization of the mitochondrial ETC imposes functional features to mitochondrial physiology that are a consequence of the new properties that supercomplexes assume with respect to individual complexes. Rate changes that depend on the compartmentalization of the respiratory assemblies may address metabolic pathways to preferential routes. Moreover, the possibility to modulate ROS production and ATP levels by changing the supramolecular organization may contribute to the control of the cellular activity by interfering with energy-sensitive and redox-sensitive signal transduction pathways. These considerations allow to predict that supercomplex organization may be involved in pathological changes where energy loss and oxidative stress are involved, in particular in the ageing process.

The mitochondrial theory of ageing[222–224] states that the original damage to mtDNA is induced by the continuous generation of ROS and other toxic species; thus it is not necessary that an increase of ROS generation occurs in ageing, since it is the damage that would accumulate even at a steady ROS generation with time.[3] Nevertheless, a vicious circle[225,226] can be established only if the accumulated damage (to the respiratory chain) would enhance ROS generation. This event is theoretically expected when electron transfer within a respiratory complex competent in ROS generation is hindered; it is well known that inhibition of Complex I by rotenone[90,227] and of Complex III by antimycin A[228] strongly enhance ROS production.

Many reports demonstrated that the rate of production of ROS from mitochondria increases with age in mammalian tissues.[229–234] Hepatocytes from old rats were found, by dichlorofluorescin diacetate labeling and flow cytometric analysis, to have a higher peroxide level than hepatocytes from young animals.[235,236] In addition, the peroxide production after an oxidative stress induced by adriamycin was much higher in old animals. ROS production was also found to increase with ageing in resting skeletal muscle and to be potentiated by strenuous exercise to exhaustion.[237]

ROS production increases in fibroblasts and other cultured cells during replicative cell senescence, considered to represent a plausible model of in vivo ageing,[238–241] as well as in cultured cells from old individuals.[242–247]

To our knowledge, only a few studies report no increase of ROS production with ageing.[248–250] It has to be noted, however, that the occasional observed lack of increase with ageing of the rate of ROS generation may be a consequence of the decrease of content and activity of respiratory complexes, so that there is a decreased number of respiratory units producing ROS even if each unit has an enhanced ROS generation; for this reason the rate of ROS generation should be related to the content of the enzymes that are responsible for it.

A different aspect that unambiguously relates ageing to ROS generation is the strong negative correlation of animal longevity with the rate of mitochondrial ROS generation[251–255] (reviewed by Barja[3]) and with the degree of fatty acid unsaturation of cellular membranes[3,256–259]; the latter correlation is explained by the higher oxidative damage (peroxidizability) that may be induced in unsaturated membrane fatty acids. According to Barja[3] only the mitochondrial-free radical theory of ageing can explain these correlations.

The major source of ROS involved in ageing has been considered to be Complex I,[3] since ROS generated by Complex I are released in the matrix and may damage mitochondrial proteins and mtDNA; however, ROS released by Complex III must also be taken in serious consideration, since they are released in the intramembrane space and may be exported in the cytosol, where they can activate, or interfere with, proteins involved in signal transduction.

All biomolecules in the cell are targets of ROS and undergo chemical modifications that accumulate with age[73,260]: protein carbonylation and methionine oxidation,[261,262] advanced glycation end-products (AGE),[263–266] lipid peroxidation,[267,268] and nucleotide modifications.[269,270] However, it is not completely understood which species are responsible for the damage in vivo and how the increased availability of ROS translates into the accumulation of specific oxidative damage.[271] For instance, some proteins are better targets than others for oxidative damage.[262,272–274]

In a recent review Kazachkova et al.[275] summarized the evidence on mtDNA copy number, deletions and point mutations during ageing in humans and mice; both humans and mice demonstrated a clear pattern of age-dependent and tissue-specific accumulation of mtDNA deletions. Deletions increase with age, and the highest amount of deletions has been observed in brain tissues both in humans and in mice. On the other hand, mtDNA point mutations accumulation has been clearly associated with age in humans, but not in mice.

A minimal threshold level of 50%–95% mutated mtDNA is usually necessary to impair respiratory chain function, depending on the type of mutation and the tissue affected.[276,277] Since the proportions of mutated mtDNA within ageing human tissues (except the D-loop mutations) rarely exceed 1%, it has been questioned how these levels may cause significant bioenergetic effects. Hayakawa et al.[278] found that mtDNA in elderly subjects is extensively fragmented in mini circles with different sizes. As a result, the amount of mtDNA mutations may reach such a high level that it could cause significant impairment of oxidative phosphorylation.[279]

In agreement with the ROS origin of mtDNA defects, it was found that hydrogen peroxide induces large-scale deletions of mtDNA through formation of double-strand DNA breaks.[280] Moreover, the frequency of mtDNA deletions was significantly decreased in mice that express the human catalase gene targeted at mitochondria.[281]

Contrary to most evidence, in a recent study Itsara et al.[282] propose that oxidative stress is not a major cause of somatic mtDNA mutations, since only a few G-C to T-A transversions, a signature of oxidative damage, were observed in *Drosophila*, which is considered a representative animal model of ageing in vertebrates. The authors therefore suggest that somatic mtDNA mutations arise primarily from errors that occur during mtDNA replication.

Indeed, in a basic study Trifunovic et al.[283] showed that expression of a proof-reading deficient mtDNA polymerase in a homozygous knockin mouse strain (the mutator mouse) leads to increased levels of somatic mtDNA mutations, causing progressive respiratory chain deficiency; the mice develop symptoms strikingly reminiscent of ageing. This is the most striking demonstration that mtDNA mutations can cause ageing.

Even if the mutator mice were not shown to have an enhanced ROS production[284,285] and their premature ageing results from the severe extent of mutation, it is likely that the *natural* way to induce mutations is ROS attack. In a recent study, Logan et al.[286] found that the level of hydrogen peroxide was the same in the young mutator and control mice, but as the mutator mouse aged, hydrogen peroxide increased; the authors suggest that the consequences of mtDNA mutations increase ROS generation that further contributes to the accelerated ageing phenotype.

There is wide evidence that the bioenergetic function of mitochondria declines with ageing, especially in post-mitotic tissues.[3] This decline may be also induced by ROS, as demonstrated by the decrease of respiratory activities in MnSOD-deficient mice.[287]

Edgar et al.[288] performed molecular analyses to determine the mechanism whereby mtDNA mutations impair respiratory chain function in the mutator mice, concluding that the amino acid substitutions in mtDNA-encoded respiratory chain subunits lead to decreased stability of the respiratory chain complexes and respiratory chain deficiency.

Hauser et al.[289] have recently found that the mitochondrial respiratory chain proteins are specifically decreased in abundance in the brains of the mutator mice harbouring extensive mtDNA mutations; the changes were not attributed to decreased transcription but to post-translational effects associated with the Polg mutation.

A mosaic pattern of cytochrome oxidase alterations in muscular tissues from elderly individuals has been found by histochemical investigations.[290–292]

The absent or scant decrease of activity of individual respiratory complexes observed in many studies[248,293,294] is compatible with the observed mosaic pattern of the respiratory lesions that cannot be appreciated in studies on mitochondria isolated from bulk tissues; it may also be a reflection of the existence of threshold effects.[277]

The effects of ROS are not limited to structural damage to cell components, but are now recognized to involve redox changes in molecules involved in signalling pathways, and ROS themselves are seen today as physiological signals. We may therefore ask whether the structural damage induced by ROS is sufficient to explain the ageing phenotype, or whether more complex factors involving the derangement of signalling pathways are the major driving force for the metabolic failure characterizing ageing.

Changes in the redox status of cellular components by oxidative stress during ageing is considered the cause of the observed increased contents or DNA-binding activities of such transcription factors as NF-kB, AP-1 and HIF-1,[295–297] of heat-shock proteins[298] and of heme oxygenase.[299] Their increased activity is considered a compensatory mechanism for cellular protection and may depend either upon direct alteration of the factor or indirectly through activation of related transduction pathways[297] by means of mechanisms still scarcely understood.

Gomes et al.[300] suggested that ageing is accompanied by a pseudo-hypoxic state with induction of the HIF pathway; they showed that there is a specific loss of mitochondrial-encoded, but not nuclear-encoded, OXPHOS subunits with age. Deleting SIRT1 accelerates this process, whereas raising NAD(+) levels in old mice restores mitochondrial function to that of a young mouse in a SIRT1-dependent manner. It must be considered to this purpose that also ROS appear to activate HIF1α (see Section 5.3).

In accordance with the importance of oxidative stress in activation of redox-sensitive transcription factors, caloric restriction, the main known factor recognized to delay ageing,[254,301] was found to prevent their activation.[297] Similarly, hepatocytes from old mice[302] and rats[303] showed reduced activation of ERK by H_2O_2, and the effect was suppressed by caloric restriction.[303] Available data on redox-responsive transcription factors suggest that their uncontrolled activation in ageing could lead to serious chronic pathogenic conditions characterized by what has been called *molecular inflammation*[304] or *inflammageing*.[305]

5.7.1 Unifying Hypothesis Involving Supercomplex Destabilization in Ageing

The notion that the respiratory chain is mainly controlled at the level of C_I suggests that the main alterations due to ageing must be found at the level of this enzyme.[306,307] Flux control analysis in aerobic respiration in coupled liver mitochondria[308] showed that Complex I has little control in young rats but very high control in the old animals, meaning that ageing induces a profound alteration of Complex I that is reflected on the entire OXPHOS.

Analysis of the occurrence of respiratory supercomplexes comprised of various stoichiometries of complexes I, III and IV reveals age-related variations, suggesting that destabilization of their supramolecular organization may be crucial for the development of the ageing-phenotype.[189,309,310]

In mitochondria of rat cortex, Frenzel et al.[311] quantified profound age-associated changes in the proportion of supramolecular assemblies of the respiratory complexes as well as of the FoF1 ATP synthase. Notably, the overall decline with age (40%) in the Complex I-containing supercomplexes is caused to large extent by the pronounced decline (58%) of abundance of the supercomplex I_1III_2.

The progeroid profile of the mutator mice (see Section 5.7) is accompanied by decreased respiration and impeded assembly of respiratory complexes.[288] Despite that only Complex IV subunits were significantly reduced, Complex I activity was also strongly decreased, in line with the idea that a decreased assembly of Complex IV may affect the stability of Complex I through disruption of the supercomplex organization.[67]

The model that we propose poses supercomplex dissociation amongst a double influence of ROS: on one hand ROS contribute to dissociate supercomplexes, and on the other hand supercomplex dissociation enhances ROS generation (Figure 5.5). This means that if these events are not tightly controlled they may lead to the establishment of a vicious circle of ROS generation. Is this the series of events that conduce to ageing?

Taken together, the observations collected in this review locate supercomplex dissociation in a physiological signalling network that can be easily altered and lead to a catastrophic event when the generation of ROS loses control.

Under physiological conditions we may envisage supercomplex association/dissociation to occur, according to the plasticity model,[8] under such stimuli as mitochondrial membrane potential and protein phosphorylation/dephosphorylation of the respiratory complexes; the ensuing changes in ROS generation modulate the signalling pathways that are initiated by ROS. These changes are reversible and are kept under strict control from changes in the starting conditions.

We propose that the primary event responsible for ageing is the structural damage induced by ROS in mitochondria, as predicted by the original mitochondrial theory of ageing.

A possible series of events might be the following. Progressive damage is induced by ROS to the mitochondrial membrane lipids and proteins. MtDNA mutations, although present, may not necessarily be an early phenomenon in the ageing process.

The level of ROS may be affected by such factors as the nutrition state and the activity of the mTOR and insulin/IGF pathways. Direct protein damage and increased CL peroxidation

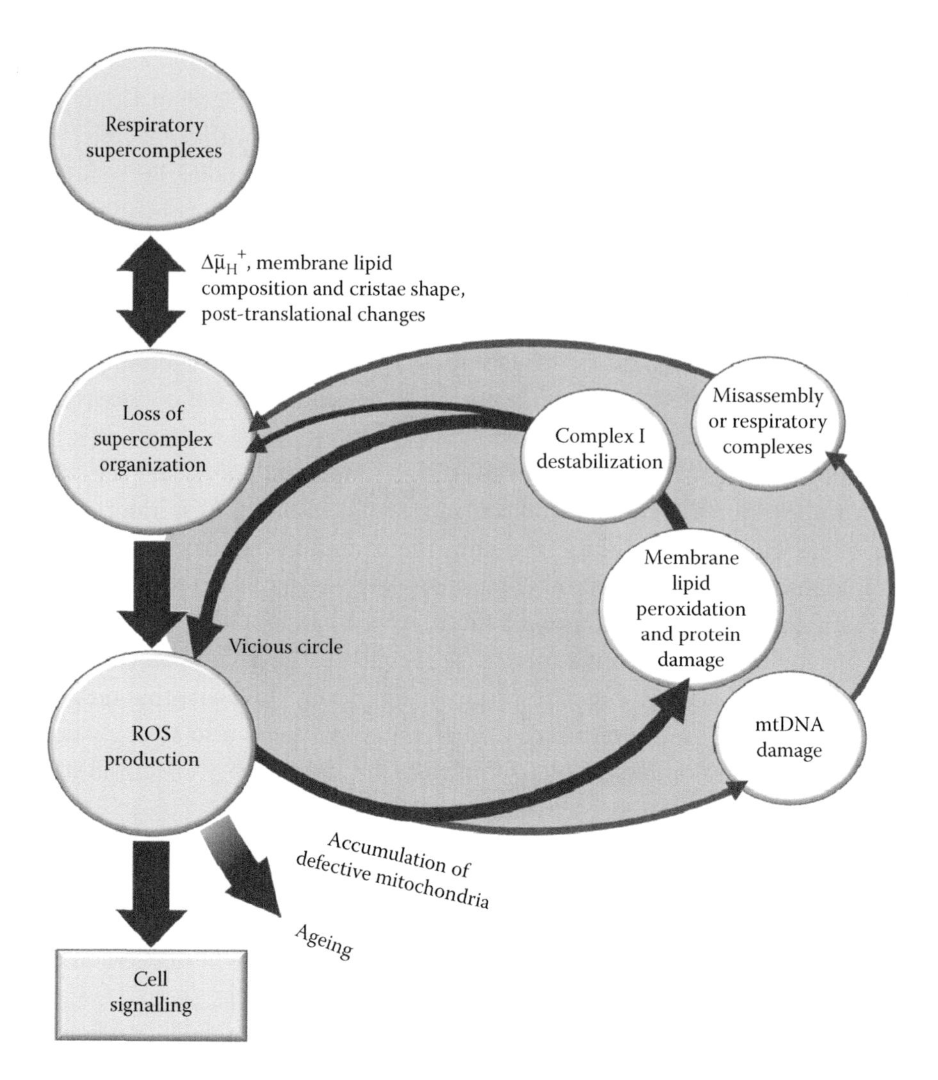

FIGURE 5.5 Scheme showing how the loss of supercomplex organization may be involved in a vicious circle of oxidative stress and energy failure. ROS production by Complex I is enhanced as a consequence of supercomplex disassembling. Membrane phospholipid peroxidation, mtDNA damage and subsequent misassembly of the respiratory complexes with further loss of supercomplex organization may occur due to enhanced mitochondrial oxidative stress, thus perpetuating the vicious circle. Depending on the amount produced, ROS can operate as signalling molecules from mitochondria to the nucleus. See text for explanations.

hamper supercomplex association,[25] leading to further increase of ROS generation. ROS at low concentration may induce retrograde signals evoking compensatory mechanisms that attempt to counteract the ROS generation and consequent risk of damage (see Section 5.4), but at higher concentration they may induce further damage with loss of coordination of the signalling pathways. Mutations in mtDNA at a later stage would make the overall process irreversible and lead to the final ageing phenotype.

REFERENCES

1. McBride HM, Neuspiel M, and Wasiak S. Mitochondria: More than just a powerhouse. *Curr Biol* 16 (2006): R551–60.
2. Wallace DC. A mitochondrial paradigm of metabolic and degenerative diseases, aging, and cancer: A dawn for evolutionary medicine. *Annu Rev Genet* 39 (2005): 359–407.
3. Barja G. Updating the mitochondrial free radical theory of aging: An integrated view, key aspects, and confounding concepts. *Antioxid Redox Signal* 19 (2013): 1420–45.
4. Handy DE, and Loscalzo J. Redox regulation of mitochondrial function. *Antioxid Redox Signal* 16 (2012): 1323–67.
5. Murphy MP. Mitochondrial thiols in antioxidant protection and redox signaling: Distinct roles for glutathionylation and other thiol modifications. *Antioxid Redox Signal* 16 (2012): 476–95.
6. Hackenbrock CR, Chazotte B, and Gupte SS. The random collision model and a critical assessment of diffusion and collision in mitochondrial electron transport. *J Bioenerg Biomembr* 18 (1986): 331–68.
7. Schägger H. Respiratory chain supercomplexes of mitochondria and bacteria. *Biochim Biophys Acta* 1555 (2002): 154–59.
8. Acín-Pérez R, Fernández-Silva P, Peleato ML, Pérez-Martos A, and Enriquez JA. Respiratory active mitochondrial supercomplexes. *Mol Cell* 32 (2008): 529–39.
9. Lenaz G, and Genova ML. Kinetics of integrated electron transfer in the mitochondrial respiratory chain: Random collisions vs. solid state electron channeling. *Am J Physiol Cell Physiol* 292 (2007): C1221–39.
10. Lenaz G, and Genova ML. Structure and organization of mitochondrial respiratory complexes: A new understanding of an old subject. *Antioxid Redox Signal* 12 (2010): 961–1008.
11. Genova ML, and Lenaz G. A critical appraisal of the role of respiratory supercomplexes in mitochondria. *Biol Chem* 394 (2013): 631–39.
12. Genova ML, and Lenaz G. Functional role of mitochondrial respiratory supercomplexes. *Biochim Biophys Acta* 1837 (2014): 427–43.
13. Acín-Pérez R, and Enriquez JA. The function of the respiratory supercomplexes: The plasticity model. *Biochim Biophys Acta* 1837 (2014): 444–50.
14. Maranzana E, Barbero G, Falasca AI, Lenaz G, and Genova ML. Mitochondrial respiratory supercomplex association limits production of reactive oxygen species from complex I. *Antioxid Redox Signal* 19 (2013): 1469–80.
15. Schägger H, and Pfeiffer K. The ratio of oxidative phosphorylation complexes I-V in bovine heart mitochondria and the composition of respiratory chain supercomplexes. *J Biol Chem* 276 (2001): 37861–67.
16. Dudkina NV, Eubel H, Keegstra W, Boekema EJ, and Braun HP. Structure of a mitochondrial supercomplex formed by respiratory-chain complexes I and III. *Proc Natl Acad Sci USA* 102 (2005): 3225–29.
17. Dudkina NV, Kudryashev M, Stahlberg H, and Boekema EJ. Interaction of complexes I, III, and IV within the bovine respirasome by single particle cryoelectron tomography. *Proc Natl Acad Sci USA* 108 (2011): 15196–200.
18. Schäfer E, Seelert H, Reifschneider NH, Krause F, Dencher NA, and Vonck J. Architecture of active mammalian respiratory chain supercomplexes. *J Biol Chem* 281 (2006): 15370–75.
19. Schäfer E, Dencher NA, Vonck J, and Parcej DN. Three-dimensional structure of the respiratory chain supercomplex I1III2IV1 from bovine heart mitochondria. *Biochemistry* 46 (2007): 12579–85.
20. Althoff T, Mills DJ, Popot JL, and Kühlbrandt W. Arrangement of electron transport chain components in bovine mitochondrial supercomplex I(1)III(2)IV(1). *EMBO J* 30 (2011): 4652–64.

21. Mileykovskaya E, Penczek PA, Fang J, Mallampalli VK, Sparagna GC, and Dowhan W. Arrangement of the respiratory chain complexes in Saccharomyces cerevisiae supercomplex III2IV2 revealed by single particle cryo-electron microscopy. *J Biol Chem* 287 (2012): 23095–103.
22. Mileykovskaya E, and Dowhan W. Cardiolipin-dependent formation of mitochondrial respiratory supercomplexes. *Chem Phys Lipids* 179 (2014): 42–48.
23. Ragan CI, and Heron C. The interaction between Mitochondrial NADH-Ubiquinone Oxidoreductase and Ubiquinol-Cytochrome c Oxidoreductase—evidence for stoicheiometric association. *Biochem J* 174 (1978): 783–90.
24. Heron C, Ragan CI, and Trumpower BL. The interaction between Mitochondrial NADH-Ubiquinone Oxidoreductase and Ubiquinol-Cytochrome c Oxidoreductase—Restoration of Ubiquinone-pool behaviour. *Biochem J* 174 (1978): 791–800.
25. Genova ML, Baracca A, Biondi A et al. Is supercomplex organization of the respiratory chain required for optimal electron transfer activity? *Biochim Biophys Acta* 1777 (2008): 740–46.
26. Van den Brink-van der Laan E, Killian JA, and de Kruijff B. Nonbilayer lipids affect peripheral and integral membrane proteins via changes in the lateral pressure profile. *Biochim Biophys Acta* 1666 (2004): 275–88.
27. Osman C, Voelker DR, and Langer T. Making heads or tails of phospholipids in mitochondria. *J Cell Biol* 192 (2011): 7–16.
28. Kang SY, Gutowsky HS, Hsung JC et al. Nuclear magnetic resonance investigation of the cytochrome oxidase-phospholipid interaction: A new model for boundary lipid. *Biochemistry* 18 (1979): 3257–67.
29. Lange C, Nett JH, Trumpower BL, and Hunte C. Specific roles of protein-phospholipid interactions in the yeast cytochrome bc1 complex structure. *EMBO J* 20 (2001): 6591–600.
30. Paradies G, Paradies V, De Benedictis V, Ruggiero FM, and Petrosillo G. Functional role of cardiolipin in mitochondrial bioenergetics. *Biochim Biophys Acta* 1837 (2014): 408–17.
31. Pfeiffer K, Gohil V, Stuart RA et al. Cardiolipin stabilizes respiratory chain supercomplexes. *J Biol Chem* 278 (2003): 52873–80.
32. Zhang M, Mileykovskaya E, and Dowhan W. Gluing the respiratory chain together. Cardiolipin is required for supercomplex formation in the inner mitochondrial membrane. *J Biol Chem* 277 (2002): 43553–56.
33. Mileykovskaya E, Zhang M, and Dowhan W. Cardiolipin in energy transducing membranes. *Biochemistry (Mosc)* 70 (2005): 154–58.
34. Wenz T, Hielscher R, Hellwig P, Schägger H, Richers S, and Hunte C. Role of phospholipids in respiratory cytochrome bc(1) complex catalysis and supercomplex formation. *Biochim Biophys Acta* 1787 (2009): 609–16.
35. Brandner K, Mick DU, Frazier AE, Taylor RD, Meisinger C, and Rehling P. Taz1, an outer mitochondrial membrane protein, affects stability and assembly of inner membrane protein complexes: implications for Barth Syndrome. *Mol Biol Cell* 16 (2005): 5202–14.
36. Li G, Chen S, Thompson MN, and Greenberg ML. New insights into the regulation of cardiolipin biosynthesis in yeast: Implications for Barth syndrome. *Biochim Biophys Acta* 1771 (2007): 432–41.
37. Schlame M, and Ren M. Barth syndrome, a human disorder of cardiolipin metabolism. *FEBS Lett* 580 (2006): 5450–55.
38. McKenzie M, Lazarou M, Thorburn DR, and Ryan MT. Mitochondrial respiratory chain supercomplexes are destabilized in Barth syndrome patients. *J Mol Biol* 361 (2006): 462–69.
39. Dudek J, Cheng IF, Balleininger M et al. Cardiolipin deficiency affects respiratory chain function and organization in an induced pluripotent stem cell model of Barth syndrome. *Stem Cell Res* 11 (2013): 806–19.
40. Gonzalvez F, D'Aurelio M, Boutant M et al. Barth syndrome: Cellular compensation of mitochondrial dysfunction and apoptosis inhibition due to changes in cardiolipin remodeling linked to tafazzin (TAZ) gene mutation. *Biochim Biophys Acta* 1832 (2013): 1194–206.

41. Bazán S, Mileykovskaya E, Mallampalli VK, Heacock P, Sparagna GC, and Dowhan W. Cardiolipin-dependent reconstitution of respiratory supercomplexes from purified Saccharomyces cerevisiae complexes III and IV. *J Biol Chem* 288 (2013): 401–11.
42. Cogliati S, Frezza C, Soriano ME et al. Mitochondrial cristae shape determines respiratory chain supercomplexes assembly and respiratory efficiency. *Cell* 155 (2013): 160–71.
43. Stroh A, Anderka O, Pfeiffer K et al. Assembly of respiratory complexes I, III, and IV into NADH oxidase supercomplex stabilizes complex I in Paracoccus denitrificans. *J Biol Chem* 279 (2004): 5000–07.
44. Ovàdi J. Physiological significance of metabolic channelling. *J Theor Biol* 152 (1991): 135–41.
45. Lenaz G, Fato R, Di Bernardo S et al. Localization and mobility of coenzyme Q in lipid bilayers and membranes. *Biofactors* 9 (1999): 87–93.
46. Kröger A, and Klingenberg M. The kinetics of the redox reactions of ubiquinone related to the electron-transport activity in the respiratory chain. *Eur J Biochem* 34 (1973): 358–68.
47. Bianchi C, Fato R, Genova ML, Parenti Castelli G, and Lenaz G. Structural and functional organization of complex I in the mitochondrial respiratory chain. *Biofactors* 18 (2003): 3–9.
48. Lenaz G. Role of mobility of redox components in the inner mitochondrial membrane. *J Membr Biol* 104 (1988): 193–209.
49. Lapuente-Brun E, Moreno-Loshuertos R, Acín-Peréz R et al. Supercomplex assembly determines electron flux in the mitochondrial electron transport chain. *Science* 340 (2013): 1567–70.
50. Bianchi C, Genova ML, Parenti Castelli G, and Lenaz G. The mitochondrial respiratory chain is partially organized in a supercomplex assembly: Kinetic evidence using flux control analysis. *J Biol Chem* 279 (2004): 36562–69.
51. Kholodenko NB, and Westerhoff HV. Metabolic channelling and control of the flux. *FEBS Lett* 320 (1993): 71–74.
52. Quarato G, Piccoli C, Scrima R, and Capitanio N. Variation of flux control coefficient of cytochrome c oxidase and of the other respiratory chain complexes at different values of protonmotive force occurs by a threshold mechanism. *Biochim Biophys Acta* 1807 (2011): 1114–24.
53. Kaambre T, Chekulayev V, Shevchuk I et al. Metabolic control analysis of cellular respiration in situ in intraoperational samples of human breast cancer. *J Bioenerg Biomembr* 44 (2012): 539–58.
54. Kaambre T, Chekulayev V, Shevchuk I et al. Metabolic control analysis of respiration in human cancer tissue. *Front Physiol* 4 (2013): 151.
55. Genova ML, and Lenaz G. New developments on the functions of coenzyme Q in mitochondria. *Biofactors* 37 (2011): 330–54.
56. Blaza JN, Serreli R, Jones AJ, Mohammed K, and Hirst J. Kinetic evidence against partitioning of the ubiquinone pool and the catalytic relevance of respiratory-chain supercomplexes. *Proc Natl Acad Sci USA* 111 (2014): 15735–40.
57. Piccoli C, Scrima R, Boffoli D, and Capitanio N. Control by cytochrome c oxidase of the cellular oxidative phosphorylation system depends on the mitochondrial energy state. *Biochem J* 396 (2006): 573–83.
58. Acín-Peréz R, Salazar E, Brosel S, Yang H, Schon EA, and Manfredi G. Modulation of mitochondrial protein phosphorylation by soluble adenylyl cyclase ameliorates cytochrome oxidase defects. *EMBO Mol Med* 1 (2009): 392–406.
59. Acín-Peréz R, Salazar E, Kamenetsky M, Buck J, Levin LR, and Manfredi G. Cyclic AMP produced inside mitochondria regulates oxidative phosphorylation. *Cell Metab* 9 (2009): 265–76.
60. Ramzan R, Weber P, Kadenbach B, and Vogt S. Individual biochemical behaviour versus biological robustness: spotlight on the regulation of cytochrome c oxidase. *Adv Exp Med Biol* 748 (2012): 265–81.
61. Rosca M, Minkler P, and Hoppel CL. Cardiac mitochondria in heart failure: Normal cardiolipin profile and increased threonine phosphorylation of complex IV. *Biochim Biophys Acta* 1807 (2012): 1373–82.

62. Acín-Peréz R, Bayona-Bafaluy MP, Fernández-Silva P et al. Respiratory complex III is required to maintain complex I in mammalian mitochondria. *Mol Cell* 13 (2004): 805–15.
63. Schägger H, de Coo R, Bauer MF, Hofmann S, Godinot C, and Brandt U. Significance of respirasomes for the assembly/stability of human respiratory chain complex I. *J Biol Chem* 279 (2004): 36349–53.
64. Diaz F, Fukui H, Garcia S, and Moraes CT. Cytochrome c oxidase is required for the assembly/stability of respiratory complex I in mouse fibroblasts. *Mol Cell Biol* 26 (2006): 4872–81.
65. Vempati U, Han X, and Moraes CT. Lack of cytochrome c in mouse fibroblasts disrupts assembly/stability of respiratory complexes I and IV. *J Biol Chem* 284 (2009): 4383–91.
66. Lamantea E, Carrara F, Mariotti C, Morandi L, Tiranti V, and Zeviani M. A novel nonsense mutation (Q352X) in the mitochondrial cytochrome b gene associated with a combined deficiency of complexes I and III. *Neuromuscul Disord* 12 (2002): 49–52.
67. D'Aurelio M, Gajewski CD, Lenaz G, and Manfredi G. Respiratory chain supercomplexes set the threshold for respiration defects in human mtDNA mutant cybrids. *Hum Mol Genet* 15 (2006): 2157–69.
68. Diaz F, Enriquez JA, and Moraes CT. Cells lacking Rieske iron-sulfur protein have a reactive oxygen species-associated decrease in respiratory complexes I and IV. *Mol Cell Biol* 32 (2012): 415–29.
69. Halliwell B. Oxidative stress and neurodegeneration: Where are we now? *J Neurochem* 97 (2006): 1634–58.
70. Lenaz G, Strocchi P. Reactive oxygen species in the induction of toxicity. In B. Ballantyne, T. Marrs, and T. Syversen (Eds.), *General and Applied Toxicology* (Wiley Chichester, 2009), vol. 1, Chapter 15.
71. Brand MD. The sites and topology of mitochondrial superoxide production. *Exp Gerontol* 45 (2010): 466–72.
72. Quinlan CL, Perevoshchikova IV, Hey-Mogensen M, Orr AL, and Brand MD. Sites of reactive oxygen species generation by mitochondria oxidizing different substrates. *Redox Biol* 1 (2013): 304–12.
73. Lenaz G. Role of mitochondria in the generation of reactive oxygen species. In M. Suzuk and S. Yamamoto (Eds.), *Handbook of Reactive Oxygen Species (ROS)* (New York: Nova Biomedical, 2014), 1–108.
74. Migliaccio E, Mele S, Salcini AE et al. Opposite effects of the p52shc/p46shc and p66shc splicing isoforms on the EGF receptor-MAP kinase-fos signalling pathway. *EMBO J* 16 (1997): 706–16.
75. Trinei M, Giorgio M, Cicalese A et al. A p53-p66Shc signalling pathway controls intracellular redox status, levels of oxidation-damaged DNA and oxidative stress-induced apoptosis. *Oncogene* 21 (2002): 3872–78.
76. Giorgio M, Migliaccio E, Orsini F et al. Electron transfer between cytochrome c and p66Shc generates reactive oxygen species that trigger mitochondrial apoptosis. *Cell* 122 (2005): 221–33.
77. Kwong LK, and Sohal RS. Substrate and site specificity of hydrogen peroxide generation in mouse mitochondria. *Arch Biochem Biophys* 350 (1998): 118–26.
78. Murphy MP. How mitochondria produce reactive oxygen species. *Biochem J* 417 (2009): 1–13.
79. Moser CC, Page CC, and Dutton PL. Tunneling in PSII. *Photochem Photobiol Sci* 4 (2005): 933–39.
80. Lenaz G. Mitochondria and reactive oxygen species. Which role in physiology and pathology? *Adv Exp Med Biol* 942 (2012): 93–136.
81. Han D, Williams E, and Cadenas E. Mitochondrial respiratory chain-dependent generation of superoxide anion and its release into the intermembrane space. *Biochem J* 353 (2001): 411–16.
82. Han D, Antunes F, Canali R, Rettori D, and Cadenas E. Voltage-dependent anion channels control the release of the superoxide anion from mitochondria to cytosol. *J Biol Chem* 278 (2003): 5557–63.

83. Wojtczak L, Lebiedzińska M, Suski JM, Więckowski MR, and Schönfeld P. Inhibition by purine nucleotides of the release of reactive oxygen species from muscle mitochondria: Indication for a function of uncoupling proteins as superoxide anion transporters. *Biochem Biophys Res Commun* 407 (2011): 772–76.
84. Bienert GP, Schjoerring JK, and Jahn TP. Membrane transport of hydrogen peroxide. *Biochim Biophys Acta* 1758 (2006): 994–1003.
85. Galkin A, and Brandt U. Superoxide radical formation by pure complex I (NADH:ubiquinone oxidoreductase) from Yarrowia lipolytica. *J Biol Chem* 280 (2005): 30129–35.
86. Kussmaul L, and Hirst J. The mechanism of superoxide production by NADH:ubiquinone oxidoreductase (complex I) from bovine heart mitochondria. *J Proc Natl Acad Sci USA* 103 (2006): 7607–12.
87. Esterházy D, King MS, Yakovlev G, and Hirst J. Production of reactive oxygen species by complex I (NADH:ubiquinone oxidoreductase) from Escherichia coli and comparison to the enzyme from mitochondria. *Biochemistry* 47 (2008): 3964–71.
88. Lambert AJ, and Brand MD. Inhibitors of the quinone-binding site allow rapid superoxide production from mitochondrial NADH:ubiquinone oxidoreductase (complex I). *J Biol Chem* 279 (2004): 39414–20.
89. Lambert AJ, Buckingham JA, Boysen HM, and Brand MD. Diphenyleneiodonium acutely inhibits reactive oxygen species production by mitochondrial complex I during reverse, but not forward electron transport. *Biochim Biophys Acta* 1777 (2008): 397–403.
90. Genova ML, Ventura B, Giuliano G et al. The site of production of superoxide radical in mitochondrial Complex I is not a bound ubisemiquinone but presumably iron-sulfur cluster N2. *FEBS Lett* 505 (2001): 364–68.
91. Fato R, Bergamini C, Bortolus M et al. Differential effects of Complex I inhibitors on production of reactive oxygen species. *Biochim Biophys Acta* 1787 (2009): 384–92.
92. Grivennikova VG, and Vinogradov AD. Partitioning of superoxide and hydrogen peroxide production by mitochondrial respiratory complex I. *Biochim Biophys Acta* 1827 (2013): 446–54.
93. Korshunov SS, Skulachev VP, and Starkov AA. High protonic potential actuates a mechanism of production of reactive oxygen species in mitochondria. *FEBS Lett* 416 (1997): 15–18.
94. Kushnareva Y, Murphy AN, and Andreyev A. Complex I-mediated reactive oxygen species generation: modulation by cytochrome c and NAD(P)+ oxidation-reduction state. *Biochem J* 368 (2002): 545–53.
95. Starkov AA, and Fiskum G. Regulation of brain mitochondrial H_2O_2 production by membrane potential and NAD(P)H redox state. *J Neurochem* 86 (2003): 1101–07.
96. Jezek P, and Hlavata L. Mitochondria in homeostasis of reactive oxygen species in cell, tissues, and organism. *Int J Biochem Cell Biol* 37 (2005): 2478–503.
97. Maklashina E, Kotlyar AB, and Cecchini G. Active/de-active transition of respiratory complex I in bacteria, fungi, and animals. *Biochim Biophys Acta* 1606 (2003): 95–103.
98. Babot M, Birch A, Labarbuta P, and Galkin A. Characterisation of the active/de-active transition of mitochondrial complex I. *Biochim Biophys Acta* 1837 (2014): 1083–92.
99. Dröse S, Brandt U, and Wittig I. Mitochondrial respiratory chain complexes as sources and targets of thiol-based redox-regulation. *Biochim Biophys Acta* 1844 (2014): 1344–54.
100. Galkin A, Meyer B, Wittig I et al. Identification of the mitochondrial ND3 subunit as a structural component involved in the active/deactive enzyme transition of respiratory complex I. *J Biol Chem* 283 (2008): 20907–13.
101. Crofts AR. The cytochrome bc1 complex: Function in the context of structure. *Annu Rev Physiol* 66 (2004): 689–733.
102. Casteilla L, Rigoulet M, and Pénicaud L. Mitochondrial ROS metabolism: Modulation by uncoupling proteins. *IUBMB Life* 52 (2001): 181–88.
103. Dröse S, and Brandt U. The mechanism of mitochondrial superoxide production by the cytochrome bc1 complex. *J Biol Chem* 283 (2008): 21649–54.

104. Sarewicz M, Borek A, Cieluch E, Swierczek M, and Osyczka A. Discrimination between two possible reaction sequences that create potential risk of generation of deleterious radicals by cytochrome bc_1. Implications for the mechanism of superoxide production. *Biochim Biophys Act.* 1797 (2010): 1820–27.
105. Ralph SJ, Moreno-Sánchez R, Neuzil J, and Rodríguez-Enríquez S. Inhibitors of succinate: Quinone reductase/Complex II regulate production of mitochondrial reactive oxygen species and protect normal cells from ischemic damage but induce specific cancer cell death. *Pharm Res* 28 (2011): 2695–730.
106. Moreno-Sánchez R, Hernández-Esquivel L, Rivero-Segura NA et al. Reactive oxygen species are generated by the respiratory complex II—evidence for lack of contribution of the reverse electron flow in complex I. *FEBS J* 280 (2013): 927–38.
107. Ishii N, Fujii M, Hartman PS et al. A mutation in succinate dehydrogenase cytochrome b causes oxidative stress and ageing in nematodes. *Nature* 394 (1998): 694–97.
108. Ishii T, Miyazawa M, Hartman PS, and Ishii N. Mitochondrial superoxide anion (O(2)(-)) inducible "mev-1" animal models for aging research. *BMB Rep* 44 (2011): 298–305.
109. Ishii T, Miyazawa M, Onouchi H, Yasuda K, Hartman PS, and Ishii N. Model animals for the study of oxidative stress from complex II. *Biochim Biophys Acta* 1827 (2013): 588–97.
110. Zhang L, Yu L, and Yu CA. Generation of superoxide anion by succinate cytochrome c reductase from bovine heart mitochondria. *J Biol Chem* 273 (1998): 33972–76.
111. Lemarie A, Huc L, Pazarentzos E, Mahul-Mellier AL, and Grimm S. Specific disintegration of complex II succinate:ubiquinone oxidoreductase links pH changes to oxidative stress for apoptosis induction. *Cell Death Differ* 18 (2011): 338–49.
112. Drahota Z, Chowdhury SK, Floryk D et al. Glycerophosphate-dependent hydrogen peroxide production by brown adipose tissue mitochondria and its activation by ferricyanide. *J Bioenerg Biomembr* 34 (2002): 105–13.
113. Miwa S, St-Pierre J, Partridge L, and Brand MD. Superoxide and hydrogen peroxide production in *Drosophila* mitochondria. *Free Radic Biol Med* 35 (2003): 938–48.
114. Tretter L, Takacs K, Hegedus V, and Adam-Vizi V. Characteristics of alpha-glycerophosphate-evoked H_2O_2 generation in brain mitochondria. *J Neurochem* 100 (2007): 650–63.
115. Mráček T, Holzerová E, Drahota Z et al. ROS generation and multiple forms of mammalian mitochondrial glycerol-3-phosphate dehydrogenase. *Biochim Biophys Acta* 1837 (2014): 98–111.
116. Forman HJ, and Kennedy J. Superoxide production and electron transport in mitochondrial oxidation of dihydroorotic acid. *J Biol Chem* 250 (1975): 4322–26.
117. Hail N Jr, Chen P, Kepa JJ, Bushman LR, and Shearn C. Dihydroorotate dehydrogenase is required for N-(4-hydroxyphenyl)retinamide-induced reactive oxygen species production and apoptosis. *Free Radic Biol Med* 49 (2010): 109–16.
118. St-Pierre J, Buckingham JA, Roebuck SJ, and Brand MD. Topology of superoxide production from different sites in the mitochondrial electron transfer chain. *J Biol Chem* 277 (2002): 44784–90.
119. Seifert EL, Estey C, Xuan JY, and Harper ME. Electron transport chain-dependent and -independent mechanisms of mitochondrial H_2O_2 emission during long-chain fatty acid oxidation. *J Biol Chem* 285 (2010): 5748–58.
120. Dröge W. Free radicals in the physiological control of cell function. *Physiol Rev* 82 (2002): 47–95.
121. Forman HJ, Fukuto JM, Miller T, Zhang H, Rinna A, and Levy S. The chemistry of cell signalling by reactive oxygen and nitrogen species and 4-hydroxynonenal. *Arch Biochem Biophys* 477 (2008): 183–95.
122. Janssen-Heininger YM, Mossman BT, Heintz NH et al. Redox-based regulation of signal transduction: Principles, pitfalls, and promises. *Free Radic Biol Med* 45 (2008): 1–17.
123. Bae YS, Oh H, Rhee SG, and Yoo YD. Regulation of reactive oxygen species generation in cell signaling. *Mol Cells* 32 (2011): 491–509.

124. Rigoulet M, Yoboue ED, and Devin A. Mitochondrial ROS generation and its regulation: Mechanisms involved in H(2)O(2) signaling. *Antioxid Redox Signal* 14 (2011): 459–68.
125. Gough DR, and Cotter TG. Hydrogen peroxide: a Jekyll and Hyde signalling molecule. *Cell Death Dis* 2 (2011): e213.
126. Wang Y, Yang J, and Yi J. Redox sensing by proteins: Oxidative modifications on cysteines and the consequent events. *Antioxid Redox Signal* 16 (2012): 649–57.
127. Corcoran A, and Cotter TG. Redox regulation of protein kinases. *FEBS J* 280 (2013): 1944–65.
128. Tormos KV, Anso E, Hamanaka RB et al. Mitochondrial complex III ROS regulate adipocyte differentiation. *Cell Metab* 14 (2011): 537–44.
129. Loh K, Deng H, Fukushima A et al. Reactive oxygen species enhance insulin sensitivity. *Cell Metab* 10 (2009): 260–72.
130. Mailloux RJ, Fu A, Robson-Doucette C et al. Glutathionylation state of uncoupling protein-2 and the control of glucose-stimulated insulin secretion. *J Biol Chem* 287 (2012): 39673–85.
131. Al-Mehdi AB, Pastukh VM, Swiger BM et al. Perinuclear mitochondrial clustering creates an oxidant-rich nuclear domain required for hypoxia-induced transcription. *Sci Signal* 5 (2012): 47.
132. Wu WS. The signaling mechanism of ROS in tumor progression. *Cancer Metastasis Rev* 25 (2006): 695–705.
133. Halliwell B. Oxidative stress and cancer: Have we moved forward? *Biochem J* 401 (2007): 1–11.
134. D'Autréaux B, and Toledano MB. ROS as signalling molecules: Mechanisms that generate specificity in ROS homeostasis. *Nat Rev Mol Cell Biol* 8 (2007): 813–24.
135. Mailloux RJ, McBride SL, and Harper ME. Unearthing the secrets of mitochondrial ROS and glutathione in bioenergetics. *Trends in Biochemical Science* 38 (2013): 592–602.
136. Cox AG, Winterbourn CC, and Hampton MB. Mitochondrial peroxiredoxin involvement in antioxidant defence and redox signalling. *Biochem J* 425 (2010): 313–25.
137. Chen Z, Putt DA, and Lash LH. Enrichment and functional reconstitution of glutathione transport activity from rabbit kidney mitochondria: Further evidence for the role of the dicarboxylate and 2-oxoglutarate carriers in mitochondrial glutathione transport. *Arch Biochem Biophys* 373 (2000): 193–202.
138. Marí M, Morales A, Colell A, García-Ruiz C, and Fernández-Checa JC. Mitochondrial glutathione, a key survival antioxidant *Antioxid Redox Signal* 11 (2009): 2685–700.
139. Jones DP. Redox potential of GSH/GSSG couple: Assay and biological significance. *Methods Enzymol* 348 (2002): 93–112.
140. Marino SM, and Gladyshev VN. Cysteine function governs its conservation and degeneration and restricts its utilization on protein surfaces. *J Mol Biol* 404 (2010): 902–16.
141. Groitl B, and Jakob U. Thiol-based redox switches. *Biochim Biophys Acta* 1844 (2014): 1335–43.
142. Haddad JJ. Antioxidant and prooxidant mechanisms in the regulation of redox(y)-sensitive transcription factors. *Cell Signal* 14 (2002): 879–97.
143. Trachootham D, Lu W, Ogasawara MA, Nilsa RD, and Huang P. Redox regulation of cell survival. *Antioxid Redox Signal* 10 (2008): 1343–74.
144. Whelan SP, and Zuckerbraun BS Mitochondrial signaling: forwards, backwards, and in between. *Oxid Med Cell Longev* 2013 (2013): 351613.
145. Jazwinski SM. The retrograde response: When mitochondrial quality control is not enough. *Biochim Biophys Acta* 1833 (2013): 400–09.
146. Itoh K, Wakabayashi N, Katoh Y et al. Keap1 represses nuclear activation of antioxidant responsive elements by Nrf2 through binding to the amino-terminal Neh2 domain. *Genes Dev* 13 (1999): 76–86.
147. Valko M, Rhodes CJ, Moncol J, Izakovic M, and Mazur M. Free radicals, metals and antioxidants in oxidative stress-induced cancer. *Chemico-Biological Interactions* 160 (2006): 1–40.
148. Chandel NS, McClintock DS, Feliciano CE et al. Reactive oxygen species generated at mitochondrial complex III stabilize hypoxia-inducible factor-1alpha during hypoxia: A mechanism of O_2 sensing *J Biol Chem* 275 (2000): 25130–38.

149. Kietzmann T, and Gorlach A. Reactive oxygen species in the control of hypoxia-inducible factor-mediated gene expression. *Seminars Cell Dev Biol* 16 (2005): 474–86.
150. Finkel T. Signal transduction by mitochondrial oxidants. *J Biol Chem* 287 (2012): 4434–40.
151. Pelletier M, Lepow TS, Billingham LK, Murphy MP, and Siegel RM. New tricks from an old dog: Mitochondrial redox signaling in cellular inflammation. *Semin Immunol* 24 (2012): 384–92.
152. Scherz-Shouval R, and Elazar Z. Regulation of autophagy by ROS: Physiology and pathology. *Trends Biochem Sci* 36 (2011): 30–8.
153. Tahara EB, Navarete FD, and Kowaltowski AJ. Tissue-, substrate-, and site-specific characteristics of mitochondrial reactive oxygen species generation. *Free Radic Biol Med* 46 (2009): 1283–97.
154. Panov A, Dikalov S, Shalbuyeva N, Hemendinger R, Greenamyre JT, and Rosenfeld J. Species- and tissue-specific relationships between mitochondrial permeability transition and generation of ROS in brain and liver mitochondria of rats and mice. *Am J Physiol Cell Physiol* 292 (2007): C708–18.
155. Tahara EB, Barros MH, Oliveira GA, Netto LES, and Kowaltowski AJ. Dihydrolipoyl dehydrogenase as a source of reactive oxygen species inhibited by caloric restriction and involved in *Saccharomyces cerevisiae* aging. *FASEB J* 21 (2007): 274–83.
156. Aon MA, Cortassa S, and O'Rourke B. Redox-optimized ROS balance: A unifying hypothesis. *Biochim Biophys Acta* 1797 (2010): 865–77.
157. Cortassa S, O'Rourke B, and Aon MA. Redox-Optimized ROS Balance and the relationship between mitochondrial respiration and ROS. *Biochim Biophys Acta* 1837 (2014): 287–95.
158. Kim JJ, Lee SB, Park JK, and Yoo YD. TNF-alpha-induced ROS production triggering apoptosis is directly linked to Romo1 and Bcl-X(L). *Cell Death Differ* 17 (2010): 1420–34.
159. Lee SB, Kim JJ, Kim TW, Kim BS, Lee MS, and Yoo YD. Serum deprivation-induced reactive oxygen species production is mediated by Romo1. *Apoptosis* 15 (2010): 204–18.
160. Zorov DB, Juhaszova M, and Sollott SJ. Mitochondrial ROS-induced ROS release: An update and review. *Biochim Biophys Acta* 1757 (2006): 509–17.
161. Skulachev VP. Role of uncoupled and non-coupled oxidations in maintenance of safely low levels of oxygen and its one-electron reductants. *Q Rev Biophys* 29 (1996): 169–202.
162. Brand MD. Uncoupling to survive? The role of mitochondrial inefficiency in ageing. *Exp Gerontol* 35 (2000): 811–20.
163. Brand MD, Affourtit C, Esteves TC et al. Mitochondrial superoxide: Production, biological effects, and activation of uncoupling proteins. *Free Radic Biol Med* 37 (2004): 755–67.
164. Zhang H, Huang HM, Carson RC, Mahmood J, Thomas HM, and Gibson GE. Assessment of membrane potentials of mitochondrial populations in living cells. *Anal Biochem* 298 (2001): 170–80.
165. Johnson-Cadwell LI, Jekabsons MB, Wang A, Polster BM, and Nicholls DG. "Mild Uncoupling" does not decrease mitochondrial superoxide levels in cultured cerebellar granule neurons but decreases spare respiratory capacity and increases toxicity to glutamate and oxidative stress. *J Neurochem* 101 (2007): 1619–31.
166. Selivanov VA, Zeak JA, Roca J, Cascante M, Trucco M, and Votyakova TV. The role of external and matrix pH in mitochondrial reactive oxygen species generation. *J Biol Chem* 283 (2008): 29292–300.
167. Turrens JF, and Boveris A. Generation of superoxide anion by the NADH dehydrogenase of bovine heart mitochondria. *Biochem J* 191 (1980): 421–27.
168. Raha S, Myint AT, Johnstone L, and Robinson BH. Control of oxygen free radical formation from mitochondrial complex I: Roles for protein kinase A and pyruvate dehydrogenase kinase. *Free Radic Biol Med* 32 (2002): 421–30.
169. Maj MC, Raha S, Myint AT, and Robinson BH. Regulation of NADH/CoQ oxidoreductase: Do phosphorylation events affect activity? *Protein J* 23 (2004): 25–32.

170. Scacco S, Petruzzella V, Bestini E et al. Mutations in structural genes of complex I associated with neurological diseases. *Ital J Biochem* 55 (2006): 254–62.
171. Kadenbach B, Ramzan R, and Vogt S. Degenerative diseases, oxidative stress and cytochrome c oxidase function. *Trends Mol Med* 15 (2009): 139–47.
172. Kadenbach B, Ramzan R, and Vogt S. High efficiency versus maximal performance—the cause of oxidative stress in eukaryotes: A hypothesis. *Mitochondrion* 13 (2013): 1–6.
173. Mailloux RJ, Jin X, and Willmore WG. Redox regulation of mitochondrial function with emphasis on cysteine oxidation reactions. *Redox Biology* 2 (2014): 123–39.
174. Diogo CV, Suski JM, Lebiedzinska M et al. Cardiac mitochondrial dysfunction during hyperglycemia—the role of oxidative stress and p66Shc signaling. *Int J Biochem Cell Biol* 45 (2013): 114–22.
175. Guzy RD, and Schumacker PT. Oxygen sensing by mitochondria at complex III: The paradox of increased reactive oxygen species during hypoxia. *Exp Physiol* 91 (2006): 807–19.
176. Chandel N, Budinger GR, Kemp RA, and Schumacker PT. Inhibition of cytochrome-c oxidase activity during prolonged hypoxia. *Am J Physiol* 268 (1995): L918–25.
177. Zuckerbraun BS, Chin BY, Bilban M et al. Carbon monoxide signals via inhibition of cytochrome c oxidase and generation of mitochondrial reactive oxygen species. *FASEB J* 21 (2007): 1099–106.
178. Belousov VV, Fradkov AF, Lukyanov KA et al. Genetically encoded fluorescent indicator for intracellular hydrogen peroxide. *Nat Methods* 3 (2006): 281–86.
179. Korde AS, Yadav VR, Zheng YM, and Wang YX. Primary role of mitochondrial Rieske iron-sulfur protein in hypoxic ROS production in pulmonary artery myocytes. *Free Radic Biol Med* 50 (2011): 945–52.
180. Guzy RD, Hoyos B, Robin E et al. Mitochondrial complex III is required for hypoxia-induced ROS production and cellular oxygen sensing. *Cell Metab* 1 (2005): 401–8.
181. Mansfield D, Guzy RD, Pan Y et al. Mitochondrial dysfunction resulting from loss of cytochrome *c* impairs cellular oxygen sensing and hypoxic HIF-alpha activation *Cell Metab* 1 (2005): 393–99.
182. Waypa GB, Marks JD, Guzy RD et al. Superoxide generated at mitochondrial complex III triggers acute responses to hypoxia in the pulmonary circulation. *Am J Respir Crit Care Med* 187 (2013): 424–32.
183. Cai J, and Jones DP. Superoxide in apoptosis. Mitochondrial generation triggered by cytochrome c loss. *J Biol Chem* 273 (1998): 11401–4.
184. Pasdois P, Parker JE, Griffiths EJ, and Halestrap AP. The role of oxidized cytochrome c in regulating mitochondrial reactive oxygen species production and its perturbation in ischaemia. *Biochem J* 436 (2011): 493–505.
185. Chen Q, Moghaddas S, Hoppel CL, and Lesnefsky EJ. Ischemic defects in the electron transport chain increase the production of reactive oxygen species from isolated rat heart mitochondria. *Am J Physiol Cell Physiol* 294 (2008): C460–66.
186. Chandel NS, Maltepe E, Goldwasser E, Mathieu CE, Simon MC, and Schumacker PT Mitochondrial reactive oxygen species trigger hypoxia-induced transcription *Proc Natl Acad Sci USA* 95 (1998): 11715–20.
187. Movafagh S, Crook S, and Vo K. Regulation of hypoxia-inducible factor-1a by reactive oxygen species: New developments in an old debate. *J Cell Biochem* 2014 Dec 25. doi:10.1002/jcb.25074. [Epub ahead of print].
188. Mittal M, Gu XQ, Pak O et al. Hypoxia induces Kv channel current inhibition by increased NADPH oxidase-derived reactive oxygen species. *Free Radic Biol Med* 52 (2012): 1033–42.
189. Dencher NA, Frenzel M, Reifschneider NH, Sugawa M, and Krause F. Proteome alterations in rat mitochondria caused by aging. *Ann N Y Acad Sci* 1100 (2007): 291–98.
190. Peters K, Dudkina NV, Jänsch L, Braun HP, and Boekema EJ. A structural investigation of complex I and I+III2 supercomplex from Zea mays at 11–13 Å resolution: Assignment of the carbonic anhydrase domain and evidence for structural heterogeneity within complex I. *Biochim Biophys Acta* 1777 (2008): 84–93.

191. Radermacher M, Ruiz T, Clason T, Benjamin S, Brandt U, and Zickermann V. The three-dimensional structure of complex I from Yarrowia lipolytica: A highly dynamic enzyme. *J Struct Biol* 154 (2006): 269–79.
192. Lenaz G. The mitochondrial production of reactive oxygen species: Mechanisms and implications in human pathology. *IUBMB Life* 52 (2001): 159–64.
193. Baracca A, Chiaradonna F, Sgarbi G, Solaini G, Alberghina L, and Lenaz G. Mitochondrial Complex I decrease is responsible for bioenergetic dysfunction in K-ras transformed cells. *Biochim Biophys Acta* 1797 (2010): 314–23.
194. Lenaz G, Baracca A, Barbero G et al. Mitochondrial respiratory chain super-complex I-III in physiology and pathology. *Biochim Biophys Acta* 1797 (2010): 633–40.
195. Chen S, He Q, and Greenberg ML. Loss of tafazzin in yeast leads to increased oxidative stress during respiratory growth. *Mol Microbiol* 68 (2008): 1061–72.
196. Chen Y-C, Taylor EB, Dephoure N et al. Identification of a Protein Mediating Respiratory Supercomplex Stability. *Cell Metab* 15 (2012): 348–60.
197. Strogolova V, Furness A, Robb-McGrath M, Garlich J, and Stuart RA. Rcf1 and Rcf2, members of the hypoxia induced gene 1 protein family, are critical components of the mitochondrial cytochrome bc1-cytochrome c oxidase supercomplex. *Mol Cell Biol* 32 (2012): 1363–73.
198. Vukotic M, Oeljeklaus S, Wiese S et al. Rcf1 mediates cytochrome oxidase assembly and respirasome formation, revealing heterogeneity of the enzyme complex. *Cell Metab* 15 (2012): 336–47.
199. Campello S, Strappazzon F, and Cecconi F. Mitochondrial dismissal in mammals, from protein degradation to mitophagy. *Biochim Biophys Acta* 1837 (2014): 451–60.
200. Luce K, Weil AC, and Osiewacz HD. Mitochondrial protein quality control systems in aging and disease. *Adv Exp Med Biol* 694 (2010): 108–25.
201. Taylor EB, and Rutter J. Mitochondrial quality control by the ubiquitin-proteasome system. *Biochem Soc Trans* 39 (2011): 1509–13.
202. Bayot A, Gareil M, Rogowska-Wrzesinska A, Roepstorff P, Friguet B, and Bulteau AL. Identification of novel oxidized protein substrates and physiological partners of the mitochondrial ATP-dependent lon-like protease Pim1. *J Biol Chem* 285 (2010): 11445–57.
203. Venkatesh S, Lee J, Singh K, Lee I, and Suzuki CK. Multitasking in the mitochondrion by the ATP-dependent Lon protease. *Biochim Biophys Acta* 1823 (2012): 56–66.
204. Bulteau AL, Szweda LI, and Friguet B. Mitochondrial protein oxidation and degradation in response to oxidative stress. *Exp Gerontol* 41 (2006): 653–57.
205. Ugarte N, Petropoulos I, and Friguet B. Oxidized mitochondrial protein degradation and repair in aging and oxidative stress *Antioxid Redox Signal* 13 (2009): 539–49.
206. Fukuda R, Zhang H, Kim JW, Shimoda L, Dang CV, and Semenza GL. HIF-1 regulates cytochrome oxidase subunits to optimize efficiency of respiration in hypoxic cells. *Cell* 129 (2007): 111–22.
207. Levine B, and Klionsky DJ. Development by self-digestion: Molecular mechanisms and biological functions of autophagy. *Dev Cell* 6 (2004): 463–77.
208. Bergamini E. Autophagy: A cell repair mechanism that retards ageing and age-associated diseases and can be intensified pharmacologically. *Mol Aspects Med* 27 (2006): 403–10.
209. Chen Y, and Gibson SB. Is generation of reactive oxygen species a trigger for autophagy? *Autophagy* 4 (2008): 246–48.
210. Kiffin R, Bandyopadhyay U, and Cuervo AM. Oxidative stress and autophagy. *Antioxid Redox Signal* 8 (2006): 152–62.
211. Lemasters JJ. Selective mitochondrial autophagy, or mitophagy, as a targeted defense against oxidative stress, mitochondrial dysfunction, and aging. *Rejuvenation Res* 8 (2005): 3–5.
212. Kiffin R, Christian C, Knecht E, and Cuervo AM. Activation of chaperone-mediated autophagy during oxidative stress. *Mol Biol Cell* 15 (2004): 4829–40.
213. Kim I, Rodriguez-Enriquez S, and Lemasters JJ. Selective degradation of mitochondria by mitophagy. *Arch Biochem Biophys* 462 (2007): 245–53.

214. Youle RJ, and Narendra DP. Mechanisms of mitophagy. *Nat Rev Mol Cell Biol* 12 (2011): 9–14.
215. Mortensen M, Ferguson DJ, Edelmann M et al. Loss of autophagy in erythroid cells leads to defective removal of mitochondria and severe anemia in vivo. *Proc Natl Acad Sci USA* 107 (2010): 832–37.
216. Cookson MR, Lockhart PJ, McLendon C, O'Farrell C, Schlossmacher M, and Farrer MJ. RING finger 1 mutations in Parkin produce altered localization of the protein. *Hum Mol Genet* 12 (2003): 2957–65.
217. Zorzano A, Liesa M, Sebastian D, Segales J, and Palacin M. Mitochondrial fusion proteins: dual regulators of morphology and metabolism. *Semin Cell Dev Biol* 21 (2010): 566–74.
218. Lee J, Giordano S, and Zhang J. Autophagy, mitochondria and oxidative stress: Cross-talk and redox signalling. *Biochem J* 441 (2012): 523–40.
219. Frank M, Duvezin-Caubet S, Koob S et al. Mitophagy is triggered by mild oxidative stress in a mitochondrial fission dependent manner. *Biochim Biophys Acta* 1823 (2012): 2297–310.
220. Gomes LC, Di Benedetto G, and Scorrano L. During autophagy mitochondria elongate, are spared from degradation and sustain cell viability. *Nat Cell Biol* 13 (2011): 589–98.
221. Rambold AS, Kostelecky B, Elia N, and Lippincott-Schwartz J. Tubular network formation protects mitochondria from autophagosomal degradation during nutrient starvation. *Proc Natl Acad Sci USA* 108 (2011): 10190–5.
222. Harman DJ. Aging: a theory based on free radical and radiation chemistry. *Gerontol* 11 (1956): 298–300.
223. Miquel J, Economos AC, Fleming J, and Johnson JE Jr. Mitochondrial role in cell aging. *Exp Gerontol* 15 (1980): 575–91.
224. Linnane AW, Marzuki S, Ozawa T, and Tanaka M. Mitochondrial DNA mutations as an important contributor to ageing and degenerative diseases. *Lancet* i (1989): 642–45.
225. Ozawa T. Genetic and functional changes in mitochondria associated with aging. *Physiol Rev* 77 (1997): 425–64.
226. Dlasková A, Hlavatá L, and Jezek P. Oxidative stress caused by blocking of mitochondrial complex I H(+) pumping as a link in aging/disease vicious cycle. *Int J Biochem Cell Biol* 40 (2008): 1792–805.
227. Herrero A, and Barja G. Localization of the site of oxygen radical generation inside the Complex I of heart and nonsynaptic brain mammalian mitochondria. *J Bioenerg Biomembr* 32 (2000): 609–16.
228. Boveris A, and Cadenas E. Mitochondrial production of superoxide anions and its relationship to the antimycin insensitive respiration. *FEBS Lett* 54 (1975): 311–14.
229. Sohal RS, and Sohal BH. Hydrogen peroxide release by mitochondria increases during aging. *Mech Ageing Dev* 57 (1991): 187–202.
230. Sohal RS, Ku HH, Agarwal S, Forster MJ, and Lal H. Oxidative damage, mitochondrial oxidant generation and antioxidant defenses during aging and in response to food restriction in the mouse. *Mech Ageing Dev* 74 (1994): 121–33.
231. Richter C. Oxidative damage to mitochondrial DNA and its relationship to ageing. *Int J Biochem Cell Biol* 27 (1995): 647–53.
232. Capel F, Buffiere C, Patureau Mirand P, and Mosoni L. Differential variation of mitochondrial H_2O_2 release during aging in oxidative and glycolytic muscles in rats. *Mech Ageing Dev* 125 (2004): 367–73.
233. Judge S, Jang YM, Smith A, Hagen T, and Leeuwenburgh C. Age-associated increases in oxidative stress and antioxidant enzyme activities in cardiac interfibrillar mitochondria: Implications for the mitochondrial theory of aging. *FASEB J* 19 (2005): 419–21.
234. Muller FL, Lustgarten MS, Jang Y, Richardson A, and Remmen HV. Trends in oxidative aging theories. *Free Radic Biol Med* 43 (2007): 477–503.
235. Barogi S, Baracca A, Cavazzoni M, Parenti Castelli G, and Lenaz G. Effect of the oxidative stress induced by adriamycin on rat hepatocyte bioenergetics during ageing. *Mech Ageing Dev* 113 (2000): 1–21.

236. Cavazzoni M, Barogi S, Baracca A, Parenti Castelli G, and Lenaz G. The effect of aging and an oxidative stress on peroxide levels and the mitochondrial membrane potential in isolated rat hepatocytes. *FEBS Lett* 440 (1999): 53–56.
237. Bejma J, and Ji LL. Aging and acute exercise enhance free radical generation in rat skeletal muscle. *J Appl Physiol* 87 (1999): 465–70.
238. Hayflick L. Living forever and dying in the attempt. *Exp Gerontol* 38 (2003): 1231–41.
239. Hayflick L. Aging, longevity, and immortality in vitro. *Exp Gerontol* 27 (1992): 363–68.
240. Beckman KB, and Ames BN. The free radical theory of aging matures. *Physiol Rev* 78 (1998): 547–81.
241. Ford JH. Saturated fatty acid metabolism is key link between cell division, cancer, and senescence in cellular and whole organism aging. *Age (Dordr)* 32 (2010): 231–37.
242. Hutter E, Unterluggauer H, Uberall F, Schramek H, and Jansen-Durr P. Replicative senescence of human fibroblasts: The role of Ras-dependent signaling and oxidative stress. *Exp Gerontol* 37 (2002): 1165–74.
243. Lee HC, Yin PH, Chi CW, and Wei YH. Increase in mitochondrial mass in human fibroblasts under oxidative stress and during replicative cell senescence. *J Biomed Sci* 9 (2002): 517–26.
244. Lee YH, Lee JC, Moon HJ et al. Differential effect of oxidative stress on the apoptosis of early and late passage human diploid fibroblasts: Implication of heat shock protein 60. *Cell Biochem Funct* 26 (2008): 502–8.
245. Wei YH, Wu SB, Ma YS, and Lee HC. Respiratory function decline and DNA mutation in mitochondria, oxidative stress and altered gene expression during aging. *Chang Gung Med J* 32 (2009): 113–32.
246. Lebiedzinska M, Duszynski J, Rizzuto R, Pinton P, and Wieckowski MR. Age-related changes in levels of p66Shc and serine 36-phosphorylated p66Shc in organs and mouse tissues. *Arch Biochem Biophys* 486 (2009): 73–80.
247. Suski JM, Karkucinska-Wieckowska A, Lebiedzinska M et al. p66Shc aging protein in control of fibroblasts cell fate. *Int J Mol Sci* 12 (2011): 5373–89.
248. Hansford RG, Hogue BA, and Mildaziene V. Dependence of H_2O_2 formation by rat heart mitochondria on substrate availability and donor age. *J Bioenerg Biomembr* 29 (1997): 89–95.
249. López-Torres M, Gredilla R, Sanz A, and Barja G. Influence of aging and long-term caloric restriction on oxygen radical generation and oxidative DNA damage in rat liver mitochondria. *Free Radic Biol Med* 32 (2002): 882–89.
250. Sanz A, Caro P, Ibañez J, Gómez J, Gredilla R, and Barja G. Dietary restriction at old age lowers mitochondrial oxygen radical production and leak at complex I and oxidative DNA damage in rat brain. *J Bioenerg Biomembr* 37 (2005): 83–90.
251. Cutler RG. Peroxide-producing potential of tissues: Inverse correlation with longevity of mammalian species. *Proc Natl Aca. Sci USA* 82 (1985): 4798–802.
252. Sohal RS, Sohal BH, and Orr WC. Mitochondrial superoxide and hydrogen peroxide generation, protein oxidative damage, and longevity in different species of flies. *Free Radic Biol Med* 19 (1995): 499–504.
253. Barja G, Cadenas S, Rojas C, López-Torres M, and Pérez-Campo R. A decrease of free radical production near critical sites as the main cause of maximum longevity in animals. *Comp Biochem Physiol* 108B (1994): 501–12.
254. Barja G. Aging in vertebrates, and the effect of caloric restriction: A mitochondrial free radical production-DNA damage mechanism? *Biol Rev* 79 (2004): 235–51.
255. Lambert A, Boysen H, Buckingham JA et al. Low rates of hydrogen peroxide production by isolated heart mitochondria associate with long maximum lifespan in vertebrate homeotherms. *Aging Cell* 6 (2007): 607–18.
256. Pamplona R, Portero-Otín M, Ruiz C, Gredilla R, Herrero A, and Barja G. Double bond content of phospholipids and lipid peroxidation negatively correlate with maximum longevity in the heart of mammals. *Mech Ageing Dev* 112 (1999): 169–83.

257. Pamplona R, Portero-Otín M, Ruiz C, Prat J, Bellmunt MJ, and Barja G. Mitochondrial membrane peroxidizability index is inversely related to maximum life span in mammals. *J Lipid Res* 39 (1998): 1989–94.
258. Pamplona R, Prat J, Cadenas S et al. Low fatty acid unsaturation protects against lipid peroxidation in liver mitochondria from longevous species: The pigeon and human case. *Mech Ageing Dev* 86 (1996): 53–66.
259. Naudí A, Jové M, Ayala V, Portero-Otín M, Barja G, and Pamplona R. Membrane lipid unsaturation as physiological adaptation to animal longevity. *Front Physiol* 4 (2013): 372.
260. Diplock AT. Antioxidants and disease prevention. *Molec Aspects Med* 15 (1994): 293–376.
261. Stadtman ER. Protein oxidation and aging. *Science* 257 (1992): 1220–24.
262. Schoneich C. Reactive oxygen species and biological aging: A mechanistic approach. *Exp Gerontol* 34 (1999): 19–34.
263. De Groot J. The AGE of the matrix: Chemistry, consequence and cure. *Curr Opinion Pharmacol* 4 (2004): 301–05.
264. Smit AJ, and Lutgers HL. The clinical relevance of advanced glycation endproducts (AGE) and recent developments in pharmaceutics to reduce AGE accumulation. *Curr Med Chem* 11 (2004): 2767–84.
265. Gkogkolou P, and Böhm M. Advanced glycation end products: Key players in skin aging? *Dermatoendocrinol* 4 (2012): 259–70.
266. Rojas A, Delgado-López F, González I, Pérez-Castro R, Romero J, and Rojas I. The receptor for advanced glycation end-products: A complex signaling scenario for a promiscuous receptor. *Cell Signal* 25 (2013): 609–14.
267. Balazy M, and Nigam S. Aging, lipid modifications and phospholipases—new concepts. *Ageing Res Rev* 2 (2003): 191–209.
268. Paradies G, Petrosillo G, Paradies V, and Ruggiero FM. Oxidative stress, mitochondrial bioenergetics, and cardiolipin in aging. *Free Radic Biol Med* 48 (2010).: 1286–95.
269. Dizdaroglu M, Jaruga P, Birincioglu M, and Rodriguez H. Free radical-induced damage to DNA: mechanisms and measurement. *Free Radic Biol Med* 32 (2002): 1102–15.
270. Karanjawala ZE, and Lieber MR. DNA damage and aging. *Mech Ageing Dev* 125 (2004): 405–16.
271. Schoneich C. Methionine oxidation by reactive oxygen species: Reaction mechanisms and relevance to Alzheimer's disease. *Biochim Biophys Acta* 1703 (2005): 111–19.
272. Agarwal S, and Sohal RS. Differential oxidative damage to mitochondrial proteins during aging. *Mech Ageing Dev* 85 (1995): 55–63.
273. Hunzinger C, Wozny W, Schwall GP et al. Comparative profiling of the mammalian mitochondrial proteome: Multiple aconitase-2 isoforms including N-formylkinurenine modifications as part of a protein biomarker signature for reactive oxygen species. *J. Proteome Res* 5 (2006): 625–33.
274. Rexroth S, Poetsch A, Rögner M et al. Reactive oxygen species target specific tryptophan site in the mitochondrial ATP synthase. *Biochim Biophys Acta* 1817 (2012): 381–87.
275. Kazachkova N, Ramos A, Santos C, and Lima M. Mitochondrial DNA damage patterns and aging: Revising the evidences for humans and mice. *Aging Dis* 4 (2013): 337–50.
276. Schon EA, Bonilla E, and DiMauro S. Mitochondrial DNA mutations and pathogenesis. *J Bioenerg Biomembr* 29 (1997): 131–49.
277. Rossignol R, Faustin B, Rocher C, Malgat M, Mazat JP, and Letellier T. Mitochondrial threshold effects. *Biochem J* 370 (2003): 751–62.
278. Hayakawa M, Katsumata K, Yoneda M, Tanaka M, Sugiyama S, and Ozawa T. Age-related extensive fragmentation of mitochondrial DNA into minicircles. *Biochem Biophys Res Commun* 226 (1996): 369–77.
279. Nagley P, and Wei YH. Ageing and mammalian mitochondrial genetics. *Trends Genet* 14 (1998): 513–17.

280. Srivastava S, and Moraes CT. Double-strand breaks of mouse muscle mtDNA promote large deletions similar to multiple mtDNA deletions in humans. *Hum Mol Genet* 14 (2005): 893–902.
281. Schriner SE, Linford NJ, Martin GM et al. Extension of murine life span by overexpression of catalase targeted to mitochondria. *Science* 308 (2005): 1909–11.
282. Itsara LS, Kennedy SR, Fox EJ et al. Oxidative stress is not a major contributor to somatic mitochondrial DNA mutations. *PLoS Genet* 10 (2014): e1003974.
283. Trifunovic A, Wredenberg A, and Falkenberg M. Premature ageing in mice expressing defective mitochondrial DNA polymerase. *Nature* 429 (2004): 417–23.
284. Trifunovic A. Mitochondrial DNA and ageing. *Biochim Biophys Acta* 1757 (2006): 611–17.
285. Trifunovic A, and Larsson NG. Mitochondrial dysfunction as a cause of ageing. *J Intern Med* 263 (2008): 167–78.
286. Logan A, Shabalina IG, Prime TA et al. In vivo levels of mitochondrial hydrogen peroxide increase with age in mtDNA mutator mice. *Aging Cell* 13 (2014): 765–68.
287. Melov S, Coskun P, Patel M et al. Mitochondrial disease in superoxide dismutase 2 mutant mice. *Proc Natl Acad Sci USA* 96 (1999): 846–51.
288. Edgar D, Shabalina I, Camara Y et al. Random point mutations with major effects on protein-coding genes are the driving force behind premature aging in mtDNA mutator mice. *Cell Metab* 10 (2009): 131–38.
289. Hauser DN, Dillman AA, Ding J, Li Y, and Cookson MR. Post-translational decrease in respiratory chain proteins in the polg mutator mouse brain. *PLoS One* 9 (2014): e94646.
290. Müller-Höcker J. Cytochrome-c-oxidase deficient cardiomyocytes in the human heart--an age-related phenomenon. A histochemical ultracytochemical study. *Am J Pathol* 134 (1989): 1167–73.
291. Müller-Höcker J, Schneiderbanger K, Stefani FH, and Kadenbach B. Progressive loss of cytochrome c oxidase in the human extraocular muscles in ageing—a cytochemical-immunohistochemical study. *Mutat Res* 275 (1992): 115–24.
292. Wanagat J, Cao Z, Pathare P, and Aiken JM. Mitochondrial DNA deletion mutations colocalize with segmental electron transport system abnormalities, muscle fiber atrophy, fiber splitting, and oxidative damage in sarcopenia. *FASEB J* 15 (2001): 322–32.
293. Allen RG, Keogh BP, Tresini M et al. Development and age-associated differences in electron transport potential and consequences for oxidant generation. *J Biol Chem* 272 (1997): 24805–12.
294. Mirò O, Casademont J, Casals E et al. Aging is associated with increased lipid peroxidation in human hearts, but not with mitochondrial respiratory chain enzyme defects. *Cardiovasc Res* 47 (2000): 624–31.
295. Lavrovsky Y, Chatterjee B Clark RA, and Roy AK. Role of redox-regulated transcription factors in inflammation, aging and age-related diseases. *Exp Gerontol* 35 (2000): 521–32.
296. Calabrese V, Scapagnini G, Colombrita C et al. Redox regulation of heat shock protein expression in aging and neurodegenerative disorders associated with oxidative stress: A nutritional approach. *Amino Acids* 25 (2003): 437–44.
297. Kim HJ, Jung KJ, Yu BP, Cho CG, Choi JS, and Chung HY. Modulation of redox-sensitive transcription factors by calorie restriction during aging. *Mech Ageing Dev* 123 (2002): 1589–95.
298. Calabrese V, Scapagnini G, Ravagna A et al. Increased expression of heat shock proteins in rat brain during aging: relationship with mitochondrial function and glutathione redox state. *Mech Ageing Dev* 125 (2004): 325–35.
299. Calabrese V, Stella AM, Butterfield DA, and Scapagnini G. Redox regulation in neurodegeneration and longevity: role of the heme oxygenase and HSP70 systems in brain stress tolerance. *Antioxid Redox Signal* 6 (2004): 895–913.
300. Gomes AP, Price NL, Ling AJ et al. Declining NAD(+) induces a pseudohypoxic state disrupting nuclear-mitochondrial communication during aging. *Cell* 155 (2013): 1624–38.

301. Fontana L1, Partridge L, and Longo VD. Extending healthy life span--from yeast to humans. *Science* 328 (2010): 321–26.
302. Li J, and Holbrook NJ. Common mechanisms for declines in oxidative stress tolerance and proliferation with aging. *Free Radic Biol Med* 35 (2003): 292–99.
303. Ikeyama N, Kokkonen G, Martindale JL, Wang XT, Gorospe M, and Holbrook NJ. Effects of aging and calorie restriction of Fischer 344 rats on hepatocellular response to proliferative signals. *Exp Gerontol* 38 (2003): 431–39.
304. Chung HY, Kim HJ, Kim KW, Choi JS, and Yu BP. Molecular inflammation hypothesis of aging based on the anti-aging mechanism of calorie restriction. *Micros Res Techniq* 59 (2002): 264–72.
305. Franceschi C, Capri M, Monti D et al. Inflammaging and anti-inflammaging: a systemic perspective on aging and longevity emerged from studies in humans. *Mech Ageing Dev* 128 (2007): 92–105.
306. Barrientos A, and Moraes CT. Titrating the effects of mitochondrial complex I impairment in the cell physiology. *J Biol Chem* 274 (1999): 16188–97.
307. Lenaz G, D'Aurelio M, Merlo Pich M et al. Mitochondrial bioenergetics in aging. *Biochim Biophys Acta* 1459 (2000): 397–404.
308. Ventura B, Genova ML, Bovina C, Formiggini G, and Lenaz G. Control of oxidative phosphorylation by Complex I in rat liver mitochondria: Implications for aging. *Biochim Biophys Acta* 1553 (2002): 249–60.
309. Gómez LA, Monette JS, Chavez JD, Maier CS, and Hagen TM. Supercomplexes of the mitochondrial electron transport chain decline in the aging rat heart. *Arch Biochem Biophys* 490 (2009): 30–35.
310. Gómez LA, and Hagen TM. Age-related decline in mitochondrial bioenergetics: Does supercomplex destabilization determine lower oxidative capacity and higher superoxide production? *Semin Cell Dev Biol* 23 (2012): 758–67.
311. Frenzel M, Rommelspacherr H, Sugawa MD, and Dencher NA. Ageing alters the supramolecular architecture of OxPhos complexes in rat brain cortex. *Exp Gerontol* 45 (2010): 563–72.

CHAPTER 6

Mitochondrial Respiratory Supercomplexes in Physiology and Diseases

Anna M. Ghelli, Valentina C. Tropeano
and Michela Rugolo

CONTENTS

In eukaryotic cells, mitochondria play the fundamental role of adenosine triphosphate (ATP) production during the process of oxidative phosphorylation (OXPHOS). However, these cytosolic organelles also have several other important physiological functions, including sugar and fatty acid catabolism, amino acid metabolism, buffering of the cytosolic calcium concentration (Rizzuto et al., 2012), regulation and execution of different types of cell death (Galluzzi et al., 2012) and arrangement of adaptive responses to perturbations of intracellular homeostasis (Liesa and Shirihai, 2013). Furthermore, mitochondria are able to discharge a range of intracellular signals including reactive oxygen species (ROS), mitochondrial DNA (mtDNA) and specific proteins, thus operating as fundamental hubs of a wide array of signalling pathways (Galluzzi et al., 2012).

6.1 OXPHOS COMPLEXES

Mitochondria are surrounded by two membranes, the outer and inner membranes, which enclose the matrix, a dense milieu including enzymes of intermediary metabolism and multiple copies of the mtDNA. This small circular genome encodes for 13 inner membrane components of the respiratory chain and several tRNAs and rRNAs required for mitochondrial protein translation. In contrast with the outer mitochondrial membrane (OMM), the inner mitochondrial membrane (IMM) is heavily folded forming the cristae, which are connected by narrow structures to the peripheral surface of the IMM. The IMM houses many copies of the respiratory chain components that together with the ATP synthase (complex V, CV) form the molecular machinery for ATP production. The respiratory complexes are large multi-subunit enzymes that catalyse the transport of electrons coupled to protons extrusion, with the exception of complex II (CII), to generate an electrochemical proton gradient across the IMM that is used by CV to produce ATP from ADP and inorganic phosphate.

Electrons are delivered from reduced nicotinamide adenine dinucleotide (NADH) and reduced flavin adenine dinucleotide ($FADH_2$), mostly generated during the oxidation of metabolites at the Krebs cycle, at the matrix faces of complex I (CI; NADH: ubiquinone oxidoreductase, EC 1.6.5.3) and CII (succinate:ubiquinone oxidoreductase, EC 1.3.5.1), respectively. Electrons are then delivered by CI and CII to coenzyme Q_{10} (CoQ), which transfers them to complex III (CIII; ubiquinol:cytochrome *c* oxidoreductase, EC 1.10.2.2). Other enzymes can also provide electrons to the CoQ pool, namely the electron-transferring flavoprotein ubiquinone oxidoreductase, operating during the catabolism of fatty acids, the glycerophosphate dehydrogenase and dihydroorotate dehydrogenase, present only in some types of mitochondria (Nicholls and Ferguson, 2002). From CIII, electrons flow to cytochrome *c* and then to complex IV (CIV; cytochrome *c* oxidase, EC1.9.3.1), where molecular oxygen is reduced to water.

6.2 ORGANIZATION OF RESPIRATORY SUPERCOMPLEXES

Since the pioneering studies on the catalytic mechanisms of the respiratory chain (Chance and Williams, 1955; Green and Tzagoloff, 1966), the question of the possible organization of these multienzymatic complexes in the IMM has been hotly debated. As illustrated in Figure 6.1a and b, two opposite models have been proposed. The fluid or *random collision model of electron transfer* states that each complex acts as an individual entity. Accordingly,

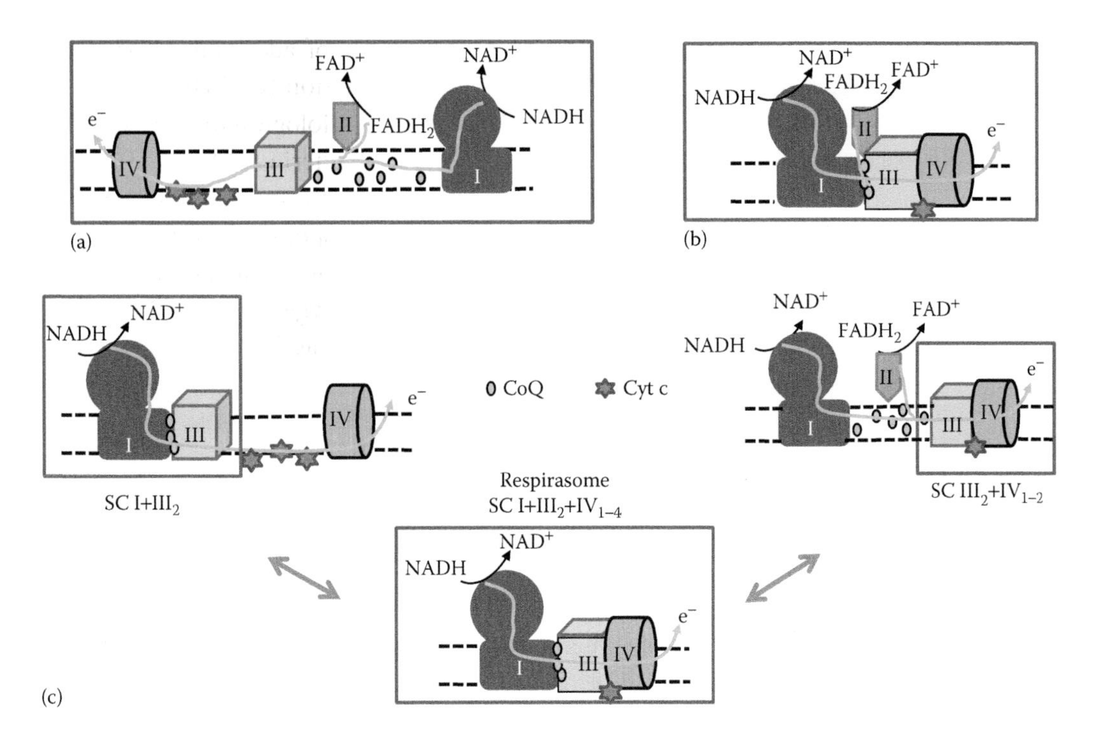

FIGURE 6.1 Different models proposed to explain the supramolecular organization of the respiratory complexes in the inner membrane: (a) the random-collision model or fluid model, (b) the solid model and (c) the plasticity model.

CoQ and cytochrome *c* freely diffuse as mobile carriers in the lipid bilayer, and electron transfer occurs during random and transient collision events (Hackenbrock et al., 1986). By contrast, the solid state model stems from the early observations reporting that CI and CIII preferentially associated in the native membranes (Hatefi et al., 1962). Only during the last two decades the solid-state model has received strong support from the specific co-migration of digitonin-solubilized assemblies of respiratory complexes, referred to as supercomplexes (SCs), on blue native gel electrophoresis (BN-PAGE) (Schägger and Pfeiffer, 2000; Krause et al., 2004a) and also their co-purification by sucrose-gradient centrifugation (Eubel et al., 2004). The hypothesis of the random diffusion model has been questioned by a number of experiments; among them flux-control measurements combined with inhibitor titration revealed that in yeast the electron transport chain behaves as a single functional unit (Boumans et al., 1998). Flux control analysis performed in bovine heart mitochondria and sub-mitochondrial particles could not discard any model and supported partially the solid model (Bianchi et al., 2004). (For more details on this issue, refer to Genova and Lenaz, 2014.)

Furthermore, respiratory SCs have been described as very stable entities in different eukaryotic organisms, algae (van Lis et al., 2003), fungi (Nubel et al., 2004; Krause et al., 2004b; Heinemeyer et al., 2007), plants (Eubel et al., 2003), mammals (Shägger and Pfeiffer, 2000) and also bacteria (Berry et al., 1985; Sone et al., 1987; Iwasaki et al., 1995), suggesting that their organization in supramolecular assemblies is a common and advantageous feature for OXPHOS systems.

Different types of SCs have been described, dependent on their components. CI, CIII and CIV were found to assemble into I + III_2, I + III_2 + IV_{1-4} and III_2 + IV_{1-2} SCs. CII was never found associated with other respiratory complexes (Genova and Lenaz, 2014), with the exception of mitochondria from the mice tissues (Acín-Pérez et al., 2008). It has to be noticed that CII is the only respiratory complex participating also to the Krebs cycle, and this could explain why it is not found in association with SCs in most tissues.

CV is not associated with respiratory SCs also, but it can self-organize in dimers that assemble into rows responsible for the formation of highly curved ridges, thus shaping the mitochondrial cristae (Paumard et al., 2003; Davies et al., 2012). (For a detailed description of CV dimeric organization, please refer to Chaban et al., 2014.)

The I + III_2 SC is mainly found in plant mitochondria (Eubel et al., 2003). In some organisms $CIII_2$ was found to associate with one or two copies of CIV (Krause et al., 2004b; Bultema et al., 2009). The $CIII_2$ + IV_{1-2} SC is the most stable in *Saccharomyces (S.) cerevisiae*, which lacks CI. No free monomeric or dimeric forms of CIV could be detected upon detergent solubilization, indicating that all CIV makes part of the SC (Heinemeyer et al., 2007).

The largest respiratory SCs comprise CI, $CIII_2$ and CIV, present up to four copies (Schägger and Pfeiffer, 2000). They are referred to as *respirasomes* because they represent the minimal unit able to perform complete respiration from NADH to oxygen. That SCs are competent in respiration has been demonstrated by Acín-Pérez et al. (2008), who were able to directly measure the NADH oxidase activity of bands excised from BN-PAGE by using a Clark electrode. This was the proof that the respirasomes, which are enriched in CoQ and cytochrome *c*, do indeed exhibit respiratory activity and therefore must provide some advantage in electron transfer.

Recently, by using both cryo-electron tomography of digitonin-solubilized respirasomes (Dudkina et al., 2011) and cryo-electron microscopy of amphipol-solubilized respirasomes (Althoff et al., 2011), the structure of the bovine respirasome CI + III_2 + IV has been determined, demonstrating that the $CIII_2$ is in the arc of the membrane arm of CI while CIV is attached to the tip of NADH dehydrogenase.

6.3 ADVANTAGES OF SCs ORGANIZATION

The organization in macromolecular assemblies is predicted to provide functional advantages to the mitochondrial respiratory function.

6.3.1 Substrate Channelling

An obvious consequence of SCs organization is that the active sites of two enzymes catalysing consecutive reactions are in close proximity and face each other; in the case of electron transport, this means that direct transfer of electrons occurs between two adjacent enzymes by successive reduction and re-oxidation of the intermediate, without its diffusion in the membrane. This is referred as to substrate channelling. In the case of the specific interaction between CI and CIII, direct channelling of CoQ should enhance electron flow increasing the kinetic efficiency of the reaction.

Direct evidence for possible channelling derives from the 3D structure of the mitochondrial SC I + III_2 + IV_1 obtained by electron cryo-microscopy (Althoff et al., 2011). The arrangement of the three component complexes indicates the pathways along which CoQ and cytochrome *c* can move to transfer electrons between their respective protein partners. The CoQ-binding sites in CI and in CIII face each other, being separated by a 13 nm gap within the membrane core of the SC, which is presumably filled with membrane lipids. More concern existed on the cytochrome *c* channelling because of the experimental difficulties to test its functional role in $CIII_2$ + IV SC, but recently it has been demonstrated that at least part of CIV forms a functional SC channelling cytochrome *c* (Lapuente-Brun et al., 2013).

6.3.2 Protection from ROS Damage

CI and CIII are the two potential sites of the respiratory chain where partial oxygen reduction can occur; therefore another advantage of SC organization is to limit the extent of superoxide generation during the electron transfer. In fact a tighter organization in SCs may protect auto-oxidizable prosthetic groups, hampering their reaction with oxygen. According to Panov et al. (2007), SC organization prevents excessive superoxide production by oxidation of NAD-linked substrates, because the efficient channelling helps to maintain the chain in the oxidized state. In agreement with this is the observation that high mitochondrial membrane potential elicits ROS generation, whereas uncoupling strongly reduces it (Lenaz, 2001).

Recently, this issue has been directly addressed by determining ROS generation by CI under experimental conditions impairing the supramolecular organization of SCs. Under these conditions, production of ROS by CI was strongly increased, supporting the view that CI and CIII association prevents superoxide generation (Maranzana et al., 2013). Accordingly, Suthammarak et al. (2013) recently reported that in *Caenorhabditis elegans* the mitochondrial superoxide dismutase 2 (SOD2) is associated with I + III_2 + IV_{1-4} SCs, confirming the importance of confining ROS scavenge close to the potential ROS production site.

6.3.3 Stability of CI

CI is believed to be regularly bound to CIII under physiologic conditions to preserve its structural integrity and activity (Schagger and Pfeiffer, 2001; Schagger et al., 2004). This view stems from several investigations on the state of SCs in cells derived from patients with isolated deficiencies of single complexes, providing strong evidence that the formation of SCs is essential for the activity and assembly of CI (see Section 6.7). Studies in ad hoc-generated mice and human cell models confirmed this assumption. In particular, CI was shown to be very unstable in absence of CIII in mice-derived cells (Acín-Peréz et al., 2004) and to be stabilized by CIII in NDUFS4 subunit knock-out mice (Calvaruso et al., 2012). Furthermore, the assembly of CI was defective in cells lacking CIII or CIV (D'Aurelio et al., 2006; Diaz et al., 2006, 2012), suggesting that critical amounts of CIII or IV are required for SCs formation and function. Despite evidence for strict dependence of CI stability on SCs organization, there is no consensus on the role of CI in the dynamics of SCs formation

(Acín-Pérez and Enriquez, 2014). According to Acín-Peréz et al. (2008), free complexes are assembled independently and then incorporated and stabilized into SCs, whereas the Ugalde's group favours the hypothesis that CI assembly is completed only after interactions with CIII and CIV, excluding the formation of a fully assembled CI as an independent unit (Moreno-Lastres et al., 2012). Additional work is required to define in detail this issue.

It has to be noticed that the CI stabilization into the respirasomes seems to be very conserved during evolution, since also in *Paracoccus denitrificans* mutant strains lacking CIII or IV, the complete dissociation of CI has been detected by BN-PAGE analysis (Stroh et al., 2004). However this is not true in fungi, because a *Podospora anserina* mutant lacking both CIII and IV exhibited normal CI, likely as a consequence of peculiar fungal features, such as the presence of a dimeric CI and of the alternative oxidase (Maas et al., 2009).

6.4 PLASTICITY MODEL

The finding that different types of associations are formed, that is, CI + III or CIII + IV, in addition to CI + III_2 + IV_{1-4}, leads Acín-Pérez et al. (2008) to suggest that these assemblies can likely co-exist with free complexes and that a variable combination of SCs versus free complexes can occur under diverse physiological conditions, determining a variety of different structural options. These considerations were rationalized in the plasticity model, which integrates the solid and the fluid models by regarding the organization of the respiratory complexes as a dynamic array of different associations as well as individual complexes (schematically represented in Figure 6.1c). According to this model, the mitochondrial respiratory chain is predicted to be able to work both when SCs are present and when the formation of SCs is prevented. In agreement, recently the presence of two distinct CoQ pools have been proposed: one dedicated to reducing equivalents coming from NADH through CI-containing SCs and a second independent pool coming from $FADH_2$ through CII and other FAD-dependent enzymes to free CIII (Lapuente-Brun et al., 2013). The notion that CoQ could be compartmentalized in mitochondria has been previously suggested and may have profound implications in cell metabolic adaptation; however, this issue is still a matter of intense and controversial debate (Blaza et al., 2014; Enriquez and Lenaz, 2014).

6.5 FACTORS INVOLVED IN MAINTAINING THE SCs ORGANIZATION

As previously noticed, respirasomes appear to be quite stable, being easily isolated as whole entities without significant degradation; therefore factors must exist capable of maintaining strictly and stably together the respirasome components.

6.5.1 Lipids

It has been widely recognized that, among the forces influencing SC association, the content and composition of phospholipids play a crucial role. Major phospholipids comprise cardiolipin, phosphatidyl choline, phosphatidyl ethanolamine and minor amounts of neutral lipids and phosphatidyl inositol. Phospholipids can have a general solubilization effect, but also fulfil specific catalytic roles, as in the case of the tightly bound cardiolipin. In addition, phospholipids can positively contribute to build the lipophilic environment for the interaction of the electron carrier CoQ in the CI + III containing SCs. The role of lipids has been evidenced when the respiratory proteins have been diluted with an excess of phospholipids, and consequently the forces holding together the respiratory complexes weakened (Genova et al., 2008). Furthermore, the cryo-EM analysis demonstrated that the space within the CI + III_2 + IV SC is mostly filled with lipids (Dudkina et al., 2011). In this regard, it has to be remembered that the IMM is the only cellular membrane containing cardiolipin, the unique anionic phospholipid formed by two phosphatidyl groups bound by a glycerol. It was previously shown that cardiolipin is required for the stability of single CIII and CIV and formation of the III_2 + IV_2 SC (Pfeiffer et al., 2003; Zang et al., 2002). A content of about 50 molecules of cardiolipins per *S. cerevisiae* SC was determined by mass spectrometry analysis (Mileykovskaya et al., 2012). Due to larger distances between complexes, the bovine respirasome can potentially accommodate several times more cardiolipin molecules (Althoff et al., 2011).

6.5.2 SCs Assembly Factors

Contrary to the well-established role of lipids, our understanding of the proteins involved in the assembly and stability of SCs is in its infancy. A few proteins both related and unrelated to the respiratory chain have been described in the last years.

6.5.2.1 Respiratory Supercomplex Factor 1 (Rcf1 and Rcf2)

Two related *S. cerevisiae* proteins, renamed as rcf1 and rcf2, were described to be relevant for CIII + IV SC assembly (Chen et al., 2012; Strogolova et al., 2012; Vukotic et al., 2012), although they are also required for complete assembly of mature CIV, as shown by the finding that their ablation impairs CIV assembly. The 20 kDa Rcf1 is highly conserved from eukaryotes to bacteria (e.g. CcoH in *Rhodobacter capsulatus*), and its mammalian homologues (e.g. HIG2A in mice and human) belong to the hypoxia-inducible gene family (Chen et al., 2012). However, HIG2A interference has been reported to cause a very moderate reduction in the respirasome levels, but similar to what occurs in yeast, CIV assembly was also impaired to the same extent (Chen et al., 2012). Given that a SC chaperone is expected to allow assembly of SCs without influencing the individual complexes, these proteins cannot be properly referred to as SC assembly factors but rather as CIV late-stage assembly factors.

6.5.2.2 Supercomplex Assembly Factor I

An important contribution to this topic has been recently provided by Lapuente-Brun et al. (2013), who identified the cytochrome *c* oxidase subunit VIIa polypeptide 2-like (Cox7-a2l, renamed by the authors as SC assembly factor I, SCAFI) in a screen for proteins present in SCs but not in free complexes of mice fibroblasts mitochondria. SCAFI is present as short and long isoforms, of 111 and 113 amino acids, respectively, and the absence of the long isoform has been correlated with the lack of CIII + IV interaction in several mice strains. In this way SCAFI would define distinct CIV populations, free or associated with SCs, minimizing the competitive inhibition of respiration between FAD- and NAD-dependent substrates, thus allowing metabolic adaptation to variation in carbon sources (Lapuente-Brun et al., 2013).

Interestingly, in a simultaneous report, Ikeda et al. (2013) described the generation of a SCAFI (here named COX7a-related protein, COX7RP) knock-out mice exhibiting compromised exercise performance and reduced CIV activity, when determined in embryo fibroblasts derived from the same mice. Overexpression of COX7RP in 293T cells increased CIV activity and assembly of the CI + III_2 + IV_n SCs, whereas the contribution to the assembly of CIII + IV SC was not apparent (Ikeda et al., 2013). The identification of SCAFI as SCs assembly factor is very exciting; however, its interactions with SCs components are controversial and require further investigations (Barrientos and Ugalde, 2013).

6.5.2.3 MCJ/DnaJC15

Another protein has been recently shown to be involved in the reversible regulation of SCs association and activity, MCJ/DnaJC15 (methylation-controlled J protein), identified as a distinct co-chaperone localized at the IMM, where it interacts preferentially with CI (Hatle et al., 2013). MCJ impairs the formation of SCs, acting as a negative regulator of the respiratory chain. In fact, the loss of MCJ leads to increased CI activity, mitochondrial hyperpolarization, and enhanced ATP production, in line with an enhanced association of respiratory CI into SCs. Conversely, in the presence of MCJ, CI activity is decreased by dissociating the SCs. Interestingly, *MCJ* gene expression has been shown to be negatively modulated by methylation in malignant tumours (ovarian cancer, brain tumour and melanoma), emphasizing a possible novel epigenetic regulation of mitochondrial energetic function (Hatle et al., 2013).

6.6 MITOCHONDRIAL ULTRASTRUCTURE AND SCs

Studies of EM tomography have clearly established that the IMM can be divided into sub-compartments: the inner boundary membrane, facing the OMM, and the cristae membrane, separated by narrow tubular junctions referred to as cristae junctions (Frey and Mannella, 2000; Perkins et al., 2010). The cristae membrane is believed to be the site of the OXPHOS complexes and SCs, although their exact distribution in the IMM compartments was not defined. This issue has been investigated for the first time in cells in vivo by using the super-resolution fluorescence microscopy, showing an overlapping distribution profile for CI, CIII and CIV in the cristae membranes, whereas CII and CV were significantly enriched also in the inner boundary membrane and cristae junctions (Wilkens et al., 2013). The presence of these complexes in the cristae junctions is new, but is in agreement with the notion that only CII and CV would fit into the negatively curved membrane of cristae junctions, whereas the intramembrane portions of CI, CIII and CIV are too large to be integrated, especially when assembled into SCs (Althoff et al., 2011). The composition of cristae membrane was shown to be preserved during fusion of mitochondrial network, due to the restricted diffusion of the proteins in the IMM sub-compartments (Wilkens et al., 2013), in agreement with the suggestion that the cristae junctions act as diffusion barriers (Sukhorukov and Bereiter-Han, 2009).

A crucial component of the cristae junction regions is the mitochondrial shaping protein optic atrophy 1 (OPA1), which forms oligomers of soluble and membrane-bound forms, both required to keep well tight the cristae junctions (Frezza et al., 2006). Dramatic remodelling of cristae junctions occurs during activation of apoptosis and severe cristae disorganization was associated with the loss OPA1 protein (Frezza et al., 2006). Noticeably, OPA1 is the only protein responsible for fusion at the IMM, whereas the highly homologous dynamin-related proteins Mitofusins 1 and 2 orchestrate fusion at the OMM and the cytoplasmic dynamin-related protein 1 (DRP1) mediates the fission process (for recent reviews, see Chan et al., 2012; Youle and Van der Blieck, 2012). Given that under physiological conditions the mitochondrial network undergoes continuous cycles of fusion and fission, it follows that relocation of complexes/SCs must also occur in order to maintain proper OXPHOS function. Recently, Cogliati et al. (2013) provided the first experimental evidence of a direct link between OPA1, IMM organization and SCs arrangement, evidencing that SCs assembly and function require an intact cristae structure. In agreement with this, disruption of cristae junctions during activation of apoptosis leads to SCs destabilization, which is also observed when OPA1 protein is lost.

In this regard, we have previously reported that OPA1 protein can directly interact with CI, CII and CIII, but not with CIV (Zanna et al., 2008). It is noteworthy, however, that pathogenic missense mutations in *OPA1* gene causing dominant optic atrophy induced abnormal cristae organization and reduced CI-driven ATP synthesis, but failed to alter the respiratory complexes (Zanna et al., 2008) and SCs pattern (Zanna, C., pers. comm.). These findings suggest caution in extrapolating the results obtained in the complete absence of OPA1 protein with those observed in the presence of reduced quantities or of genetically modified forms.

A large hetero-oligomeric protein complex of the IMM, highly conserved from yeast to humans, has been identified and recently renamed as the *mitochondrial contact site and cristae organizing system* (MICOS) (Pfanner et al., 2014). MICOS plays crucial roles in the maintenance of IMM ultrastructure, formation of contact sites between OMM and IMM and regulation of mitochondrial dynamics, biogenesis and inheritance (Harner et al., 2011). It is likely that MICOS components will also be deeply involved in maintenance of SCs organization.

6.7 MITOCHONDRIAL DISEASES

Mitochondrial disorders are a group of genetically heterogeneous diseases, caused by mutations in mitochondrial or nuclear DNA, which encompass almost all fields of medicine. These diseases exhibit clinical heterogeneity and have tissue-specific manifestations, so that mutations in the same mitochondrial protein lead to different pathological phenotypes (for a recent exhaustive review, see DiMauro et al., 2013). The unifying theme of mitochondrial diseases was mainly ascribed to decreased OXPHOS efficiency and damaging consequences of ROS over-production (Lenaz et al., 2006).

6.7.1 Loss of SC Assembly in Mitochondrial Diseases

In the following section, we will focus on some genetic alterations causing mitochondrial dysfunctions specifically associated with alterations in the supramolecular organization of respiratory complexes involved in the respirasomes.

6.7.1.1 CI

CI dysfunction is the most frequent respiratory chain defect found in neuro-muscular mitochondrial disorders, which can be caused by mutations in mtDNA or nDNA genes coding for CI structural subunits and assembly chaperones. These pathogenic CI mutations can induce either subtle or severe reduction of CI activity without or with alteration of its structural stability, depending on the type of mutation (Iommarini et al., 2013). Noticeably, most CI isolated deficiencies are caused by disassembling mutations in nDNA-encoded subunits and chaperones (Nouws et al., 2012). The only mtDNA pathogenic mutation inducing a strong effect on CI assembly is the m.14487T > C/MT-ND6 identified in neuro-muscular disorders (Ugalde et al., 2003). An extensive analysis of the state of respirasomes in patients with an isolated deficiency of single complexes was provided by Schägger et al. (2004), showing that mutations that induce primary genetic assembly defects of CI completely lack respirasomes, whereas CIII, CIV and CV were found in normal amounts and assembled. Accordingly, these authors proposed that in the respirasome CI acts as a scaffold for the interaction with CIII and IV, without interfering with their activities. However, the real role of CI on the supermolecular organization of respiratory complexes is still debated, considering that some CI disassembling mutations could also alter the stability of CIII (Budde et al., 2000, 2003; Ugalde et al., 2004).

The importance of CI assembly in mitochondrial physiology has been pointed out in oncocytic tumours, a distinctive class of rather benign cancers characterized by cells with a striking degree of mitochondrial hyperplasia (Tallini, 1998) and with high frequency of disruptive mutations mostly in CI subunits (Gasparre et al., 2007). One of these mutations has been deeply investigated showing that it causes CI disassembly (Genova et al., 2008), which is associated with destabilization of HIF1α, the major player of tumourigenic transformation (Porcelli et al., 2010). Interestingly, CI impairment was associated with altered α-ketoglutarate/succinate ratio, these metabolites being the main regulators of HIF1α stability through the activity of prolyl-hydroxylase (Porcelli et al., 2010). Thus CI disassembly, perturbing the content of these Krebs's cycle metabolites, is likely responsible for the benign nature of oncocytic neoplasms through lack of HIF1α stabilization, indicating that structural and functional integrity of CI and OXPHOS is required for hypoxic adaptation and tumour progression (Porcelli et al., 2010; Iommarini et al., 2014).

6.7.1.2 CIII

Among the pathogenic mutations in respiratory complexes, those affecting CIII are the rarest (Bénit et al., 2009). However, studying CIII dysfunction is of primary importance not only to understand the CIII associated diseases, but also to acquire a better knowledge of the relationships between respiratory complexes.

Most CIII pathogenic mutations affect the mtDNA-encoded gene *cytochrome b* (*MT-CYTB*). These mutations are mainly somatic and limited to the skeletal muscle inducing exercise intolerance and lactic acidosis, whereas only a few *MT-CYTB* mutations are associated with multisystem disorders that are rather difficult to classify in a defined pathology (Bénit et al., 2009; Meunier et al., 2013). Although several studies reported that the presence of CIII is required for the assembly/stability of CI (Acín-Peréz et al., 2004;

Shägger et al., 2004; Blakely et al., 2005), only two pathogenic mutations in *MT-CYTB* have been demonstrated to directly affect SCs formation in cellular models. In 2006, D'Aurelio and colleagues reported that an out-of-frame Δ4-*MT-CYTB* pathogenic mutation, inducing the loss of *cytochrome b* (Rana et al., 2000), completely abolished the formation of CI + $CIII_2$ + CIV_n and $CIII_2$ + CIV SCs, demonstrating the essentiality of assembled CIII. More recently, we have reported that the Y278C homoplasmic missense mutation in *MT-CYTB* causative of a multisystem disorder (Wibrand et al., 2001) alters the supramolecular organization of SCs (Ghelli et al., 2013). In particular, we detected decreased amounts of $CIII_2$ and $CIII_2$ + IV SCs, probably due to enhanced degradation induced by ROS overproduction. In parallel, we demonstrated that mutated CIII was protected by the oxygen-damaging effects when included within the I_1 + III_2 + IV_n SC, providing direct evidence that this supramolecular structure can also be effective in mitigating the detrimental effects of pathogenic mutations (Ghelli et al., 2013).

The nuclear mutations leading to CIII deficiency have been localized mostly in the *BCS1L* gene, which encodes an IMM protein required for CIII biogenesis (Cruciat et al., 1999). *BCS1L* mutations lead to different clinical phenotypes that range from the relatively mild Björnstad syndrome to the severe GRACILE syndrome (Morán et al., 2010). Accordingly, it has been reported that different mutations could induce a spectrum of biochemical effects ranging from unaffected CIII activity to severe impairment of both activity/assembly. Interestingly, Tamai et al. (2008) reported that the knock-down of *BCS1L* in human cells caused disassembly of CIII and consequent loss of SCs, indicating that only the complete lack of assembled CIII is deleterious for respirasome formation. All together these findings are in agreement with the recent proposal that CI has a very high affinity for CIII and this association is preferred when a partial loss of CIII occurs in order to maintain NADH-linked substrate oxidation (Lapuente-Brun et al., 2013).

6.7.1.3 CIV

Defects in CIV are the other most common deficiencies of the OXPHOS system together with CI-associated diseases (Diaz, 2010). CIV dysfunctions could arise from mutations in the 3 subunits encoded by the mtDNA, in the 10 structural subunits encoded by the nucleus or in any of the CIV-specific assembly factors. Phenotypically, the clinical symptoms are variable and could affect single or multiple organs. Many of these CIV defects are reported to cause disassembly, but, although the supramolecular organization of SCs was not investigated, it should not be affected since CI and CIII activities were normal (Bruno et al., 1999; Papadopoulou et al., 1999; Rahman et al., 1999; Valnot et al., 2000; Antonicka et al., 2003; Massa et al., 2008). Conversely, it has been reported that also CI + CIII SC and CI stability are affected in mice and cellular models devoid of CIV (D'Aurelio et al., 2006; Diaz et al., 2006, 2012), indicating that relatively small amounts of CIV can promote CI + $CIII_2$ + CIV_n SCs stability.

6.7.2 Barth Syndrome

The Barth syndrome is a fatal X-linked mitochondrial myopathy and cardiopathy caused by mutations in the nuclear gene–encoding tafazzin (Bione et al., 1996), a transacylase catalysing the terminal maturation of cardiolipin (Schlame and Ren, 2006). The four acyl chains of this unique phospholipid are usually mono- and di-unsaturated fatty acids in higher animals. Cardiolipin remodelling, by deacylation and successive reacylations, controls the final, specific acyl composition of the molecule (Xu et al., 2003). In lymphoblasts from Barth syndrome patients, the reduced cardiolipin content led to the destabilization of the I + III_2 + IV_n, I + III and III + IV SCs (McKenzie et al., 2006; Gonzalves et al., 2013). These results further support the notion that cardiolipin is an essential component of the IMM, acting as a sort of *glue* holding the mitochondrial respiratory chain together (Pfeiffer et al., 2003; Zang et al., 2002).

6.8 CONCLUDING REMARKS

During the last years several exciting reports have opened astonishing perspectives concerning the IMM ultrastructure and the supramolecular organization of OXPHOS.

It is becoming more and more apparent that reorganization of SCs in response to different stimuli, carbon sources or stress conditions is an important and subtle adaptive mechanism controlled by mitochondria. Here we have focused our attention on the major findings related to the functional role of respiratory SCs and described pathologic dysfunctions specifically associated with their derangement. Although we are aware that many questions are still far from being resolved, these findings disclose novel and challenging approaches for mitigating the detrimental effects of diseases, pointing to identify mechanisms capable to adjust SCs association to better address the cellular energetic requirements.

ACKNOWLEDGEMENTS

We thank Professor Fevzi Daldal, University of Philadelphia, Pennsylvania, for his continuous support. CVT was supported by NIH grant GM 38237 and from the MIUR-PRIN grant 20107Z8XBW.

REFERENCES

Acín-Peréz, R., Bayona-Bafaluy, M. P., Fernández-Silva, P., Moreno-Loshuertos, R., Pérez-Martos, A., Bruno, C., Respiratory complex III is required to maintain complex I in mammalian mitochondria, *Mol. Cell* 13 (2004): 805–815.

Acín-Pérez, R., and Enriquez, J. A., The function of the respiratory supercomplexes: The plasticity model, *Biochim. Biophys. Acta.* 1837 (2014): 444–450.

Acín-Pérez, R., Fernández-Silva, P., Peleato, M. L., Pérez-Martos, A., Enriquez, J. A., Respiratory active mitochondrial supercomplexes, *Mol. Cell* 32 (2008): 529–539.

Althoff, T., Mills, D. J., Popot, J. L., and Kühlbrandt, W., Arrangement of electron transport chain components in bovine mitochondrial supercomplex I1III2IV1, *EMBO J.* 30 (2011): 4652–4664.

Antonicka, H., Leary, S. C., Guercin, G. H. et al., Mutations in COX10 result in a defect in mitochondrial heme A biosynthesis and account for multiple, early-onset clinical phenotypes associated with isolated COX deficiency, *Hum. Mol. Genet.* 12 (2003): 2693–2702.

Barrientos, A., Ugalde, C., I function, therefore I am: Overcoming skepticism about mitochondrial supercomplexes, *Cell Metab.* 18 (2013): 147–149.

Bénit, P., Lebon, S., Rustin, P., Respiratory-chain diseases related to complex III deficiency, *Biochim. Biophys. Acta.* 1793 (2009): 181–185.

Berry, E. A., Trumpower, B. L., Isolation of ubiquinol oxidase from Paracoccus denitrificans and resolution into cytochrome bc1 and cytochrome c-aa3 complexes, *J. Biol. Chem.* 260 (1985): 2458–2467.

Bianchi, C., Genova, M. L., Parenti, C G., Lenaz, G., The mitochondrial respiratory chain is partially organized in a supercomplex assembly: Kinetic evidence using flux control analysis, *J. Biol. Chem.* 279 (2004): 36562–36569.

Bione, S., D'Adamo, P., Maestrini, E., Gedeon, A. K., Bolhuis, P. A., Toniolo, D. A., Novel X-linked gene, G4.5. is responsible for Barth syndrome, *Nat. Genet.* 12 (1996): 385–389.

Blakely, E. L., Mitchell, A. L., Fisher, N. et al., A mitochondrial cytochrome b mutation causing severe respiratory chain enzyme deficiency in humans and yeast, *FEBS J.* 272 (2005): 3583–3592.

Blaza, J. N., Serreli, R., Jones, A. J. Y., Mohammed, K., Hirst, J., Kinetic evidence against partitioning of the ubiquinone pool and the catalytic relevance of respiratory-chain supercomplexes, *Proc. Natl. Acad. Sci. U. S. A.* (2014): 15735–15740.

Boumans, H., Grivell, L. A., Berden, J. A., The respiratory chain in yeast behaves as a single functional unit, *J. Biol. Chem.* 273 (1998): 4872–4877.

Bruno, C., Martinuzzi, A., Tang, Y. et al., A stop-codon mutation in the human mtDNA cytochrome c oxidase I gene disrupts the functional structure of complex IV, *Am. J. Hum. Genet.* 65 (1999): 611–620.

Budde, S. M., van den Heuvel, L. P., Janssen, A. J. et al., Combined enzymatic complex I and III deficiency associated with mutations in the nuclear encoded NDUFS4 gene, *Biochem. Biophys. Res. Commun.* 275 (2000): 63–68.

Budde, S. M., van den Heuvel, L. P. W. J., Smeets, R. J. P. et al., Clinical heterogeneity in patients with mutations in the NDUFS4 gene of mitochondrial complex I, *J. Inherit. Metab. Dis.* 26 (2003): 813–815.

Bultema, J. B., Braun, H. P., Boekema, E. J., Kouril, R., Megacomplex organization of the oxidative phosphorylation system by structural analysis of respiratory supercomplexes from potato, *Biochim. Biophys. Acta* 1787 (2009): 60–67.

Calvaruso, M. A., Willems, P., van den Brand, M. et al., Mitochondrial complex III stabilizes complex I in the absence of NDUFS4 to provide partial activity, *Hum. Mol. Genet.* 21 (2012): 115–120.

Chaban, Y., Boekema, E. J., Dudkina, N. V., Structures of mitochondrial oxidative phosphorylation supercomplexes and mechanisms for their stabilization, *Biochim. Biophys. Acta* 1837 (2014): 418–426.

Chan, D. C., Fusion and fission: Interlinked processes critical for mitochondrial health, *Annu. Rev. Genet.* 46 (2012): 265–287.

Chance, B., Williams, G. R., A method for the localization of sites for oxidative phosphorylation. *Nature* 176 (1955): 250–254.

Chen, Y. C., Taylor, E. B., Dephoure, N. et al., Identification of a protein mediating respiratory supercomplex stability, *Cell Metab.* 15 (2012): 348–360.

Cogliati, S., Frezza, C., Soriano, M. E. et al., Mitochondrial cristae shape determines respiratory chain supercomplexes assembly and respiratory efficiency, *Cell* 155 (2013): 160–171.

Cruciat, C. M., Hell, K., Folsch, H., Neupert, W., Stuart, R. A., Bcs1p, an AAA-family member, is a chaperone for the assembly of the cytochrome bc(1) complex, *EMBO J.* 18 (1999): 5226–5233.

D'Aurelio, M., Gajewski, C. D., Lenaz, G., Manfredi, G., Respiratory chain supercomplexes set the threshold for respiration defects in human mtDNA mutant cybrids, *Hum. Mol. Genet.* 15 (2006): 2157–2169.

Davies, K. M., Anselmi, C., Wittig, I., Faraldo-Gómezb, J. D., Kühlbrandt, W., Structure of the yeast F1Fo-ATP synthase dimer and its role in shaping the mitochondrial cristae, *Proc. Natl. Acad. Sci. U. S. A.* 109 (2012): 13602–13607.

Diaz, F., Cytochrome c oxidase deficiency: Patients and animal models, *Biochim. Biophys. Acta* 1802 (2010): 100–110.

Diaz, F., Fukui, H., Garcia, S., Moraes, C. T., Cytochrome c oxidase is required for the assembly/stability of respiratory complex I in mouse fibroblasts, *Mol. Cell. Biol.* 26 (2006): 4872–4881.

Diaz, F., Garcia, S., Padgett, K. R., Moraes, C. T., A defect in the mitochondrial complex III, but not complex IV, triggers early ROS-dependent damage in defined brain regions, *Hum. Mol. Genet.* 21 (2012): 5066–5077.

DiMauro, S., Schon, E. A., Carelli, V., Hirano, M., The clinical maze of mitochondrial neurology, *Nat. Rev. Neurol.* 9 (2013): 429–444.

Dudkina, N. V., Kudryashev, M., Stahlberg, H., Boekema, E. J., Interaction of complexes I, III, and IV within the bovine respirasome by single particle cryo-electron tomography, *Proc. Natl. Acad. Sci. U. S. A.* 108 (2011): 15196–15200.

Enriquez, J. A., and Lenaz, G., Coenzyme Q and the respiratory chain: Coenzyme Q pool and mitochondrial supercomplexes, *Mol. Syndromol.* 5 (2014): 119–140.

Eubel, H., Heinemeyer, J., Braun, H. P., Identification and characterization of respirasomes in potato mitochondria, *Plant Physiol.* 134 (2004): 1450–1459.

Eubel, H., Jansch, L., Braun, H. P., New insights into the respiratory chain of plant mitochondria. Supercomplexes and a unique composition of complex II, *Plant Physiol.* 133 (2003): 274–286.

Frey, T. G., Mannella, C. A., The internal structure of mitochondria, *Trends Biochem. Sci.* 25 (2000): 319–324.

Frezza, C., Cipolat, S., Martins de Brito, O. et al., OPA1 controls apoptotic cristae remodeling independently from mitochondrial fusion, *Cell* 126 (2006): 177–189.

Galluzzi, L., Kepp, O., Trojel-Hansen, C., Kroemer, G., Mitochondrial control of cellular life, stress, and death, *Circ. Res.* 111 (2012): 1198–1207.

Gasparre, G., Porcelli, A. M., Bonora, E. et al., Disruptive mitochondrial DNA mutations in complex I subunits are markers of oncocytic phenotype in thyroid tumors, *Proc. Natl. Acad. Sci. U.S.A.* 104 (2007): 9001–9006.

Genova, M. L., Baracca, A., Biondi, A. et al., Is supercomplex organization of the respiratory chain required for optimal electron transfer activity? *Biochim. Biophys. Acta* 1777 (2008): 740–746.

Genova, M. L., Lenaz, G., Functional role of mitochondrial respiratory supercomplexes, *Biochim. Biophys. Acta.* 1837 (2014): 427–443.

Ghelli, A., Tropeano, C. V., Calvaruso, M. A. et al., The cytochrome b p.278Y > C mutation causative of a multisystem disorder enhances superoxide production and alters supramolecular interactions of respiratory chain complexes, *Hum. Mol. Genet.* 22 (2013): 2141–2151.

Gonzalvez, F., D'Aurelio, M., Boutant, M. et al., Barth syndrome: Cellular compensation of mitochondrial dysfunction and apoptosis inhibition due to changes in cardiolipin remodeling linked to tafazzin (TAZ) gene mutation, *Biochim. Biophys. Acta* 1832 (2013): 1194–1206.

Green, D. E., Tzagoloff, A., The mitochondrial electron transfer chain, *Arch. Biochem. Biophys.* 116 (1966): 293–304.

Hackenbrock, C. R., Chazotte, B., Gupte, S. S., The random collision model and a critical assessment of diffusion and collision in mitochondrial electron transport, *J. Bioenerg. Biomembr.* 18 (1986): 331–368.

Harner, M., Körner, C., Walther, D. et al., The mitochondrial contact site complex, a determinant of mitochondrial architecture, *EMBO J.* 30 (2011): 4356–4370.

Hatefi, Y., Haavik, A. G., Fowler, L. R., Griffiths, D. E., Studies on the electron transfer system. XLII. Reconstitution of the electron transfer system, *J. Biol. Chem.* 237 (1962): 2661–2669.

Hatle, K. M., Gummadidala, P., Navasa, N. et al., MCJ/DnaJC15, an endogenous mitochondrial repressor of the respiratory chain that controls metabolic alterations, *Mol. Cell. Biol.* 33 (2013): 2302–2314.

Heinemeyer, J., Braun, H. P., Boekema, E. J., Kouril, R., A structural model of the cytochrome c reductase/oxidase supercomplex from yeast mitochondria, *J. Biol. Chem.* 282 (2007): 12240–12248.

Ikeda, K., Shiba, S., Horie-Inoue, K., Shimokata, K., Inoue, S., A stabilizing factor for mitochondrial respiratory supercomplex assembly regulates energy metabolism in muscle, *Nat. Commun.* 4 (2013): 2147–2155.

Iommarini, L., Calvaruso, M. A., Kurelac, I., Gasparre, G., Porcelli, A. M. Complex I impairment in mitochondrial diseases and cancer: Parallel roads leading to different outcomes, *Int. J. Biochem. Cell. Biol.* 45 (2013): 47–63.

Iommarini, L., Kurelac, I., Capristo, M. et al., mtDNA mutations modify tumor progression in dependence of the degree of respiratory complex I impairment, *Hum. Mol. Genet.* 23 (2014): 1453–1466.

Iwasaki, T., Matsuura, K., Oshima, T., Resolution of the aerobic respiratory system of the thermoacidophilic archaeon, Sulfolobus sp. strain 7. I. The archaeal terminal oxidase supercomplex is a functional fusion of respiratory complexes III and IV with no c-type cytochromes, *J. Biol. Chem.* 270 (1995): 30881–30892.

Krause, F., Reifschneider, N. H., Vocke, D., Seelert, H., Rexroth, S., Dencher, N. A., "Respirasome-" like supercomplexes in green leaf mitochondria of spinach, *J. Biol. Chem.* 279 (2004a): 48369–48375.

Krause, F., Scheckhuber, C. Q., Werner, A. et al., Supramolecular organization of cytochrome c oxidase- and alternative oxidase-dependent respiratory chains in the filamentous fungus Podospora anserina, *J. Biol. Chem.* 279 (2004b): 26453–26461.

Lapuente-Brun, E., Moreno-Loshuertos, R., Acín-Peréz, R. et al., Supercomplex assembly determines electron flux in the mitochondrial electron transport chain, *Science* 340 (2013): 1567–1570.

Lenaz, G., The mitochondrial production of reactive oxygen species: Mechanisms and implications in human pathology, *IUBMB Life* 52 (2001): 159–164.

Lenaz, G., Baracca, A., Fato, R., Genova, M. L., Solaini, G., New insights into structure and function of mitochondria and their role in ageing and disease, *Antioxid. Redox. Signal.* 8 (2006): 417–437.

Liesa, M., Shirihai, O. S., Mitochondrial dynamics in the regulation of nutrient utilization and energy expenditure, *Cell Metab.* 17 (2013): 491–506.

Maas, M. F., Krause, F., Dencher, N. A., Sainsard-Chanet, A., Respiratory complexes III and IV are not essential for the assembly-stability of complex I in fungi, *J. Mol. Biol.* 387 (2009): 259–269.

Maranzana, E., Barbero, A., Falasca, A., Lenaz, G., Genova, M. L., Mitochondrial respiratory supercomplex association limits production of reactive oxygen species from complex I, *Antioxid. Redox Signal.* 19 (2013): 1469–1480.

Massa, V., Fernandez-Vizarra, E., Alshahwan, S. et al., Severe infantile encephalomyopathy caused by a mutation in COX6B1, a nucleus encoded subunit of cytochrome c oxidase, *Am. J. Hum. Genet.* 82 (2008): 1281–1289.

McKenzie, M., Lazarou, M., Thorburn, D. R., Ryan, M. T., Mitochondrial respiratory chain supercomplexes are destabilized in Barth syndrome patients, *J. Mol. Biol.* 361 (2006): 462–469.

Meunier, B., Fisher, N., Ransac, S., Mazat, J. P., Brasseur, G., Respiratory complex III dysfunction in humans and the use of yeast as a model organism to study mitochondrial myopathy and associated diseases, *Biochim. Biophys. Acta* 1827 (2013): 1346–1361.

Mileykovskaya, E., Penczek, P. A., Fang, J., Mallampalli, V. K., Sparagna, G. C., Dowhan, W., Arrangement of the respiratory chain complexes in Saccharomyces cerevisiae supercomplex III2IV2 revealed by single particle cryo-electron microscopy, *J. Biol. Chem.* 287 (2012): 23095–23103.

Morán, M., Marín-Buera, L., Gil-Borlado, M. C. et al., Cellular pathophysiological consequences of BCS1L mutations in mitochondrial complex III enzyme deficiency, *Hum. Mutat.* 31 (2010): 930–941.

Moreno-Lastres, D., Fontanesi, F., García-Consuegra, I. et al., Mitochondrial complex I plays an essential role in human respirasome assembly, *Cell Metab.* 15 (2012): 324–335.

Nicholls, D. G., Ferguson, S. J., *Bioenergetics 3*. 2nd ed. Amsterdam, the Netherlands: Elsevier, 2002.

Nouws, J., Nijtmans, L. G., Smeitink, J. A., Vogel, R. O., Assembly factors as a new class of disease genes for mitochondrial complex I deficiency: Cause, pathology and treatment options, *Brain* 135 (2012): 12–22.

Nubel, E., Wittig, I., Kerscher, S., Brandt, U., Schägger, H., Two-dimensional native electrophoretic analysis of respiratory supercomplexes from Yarrowia lipolytica, *Proteomics* 9 (2009): 2408–2418.

Panov, A., Dikalov, S., Shalbuyeva, N., Hemendinger, R., Greenamyre, J. T., Rosenfeld, J., Species- and tissue specific relationships between mitochondrial permeability transition and generation of ROS in brain and liver mitochondria of rats and mice, *Am. J. Physiol. Cell Physiol.* 292 (2007): C708–C718.

Papadopoulou, L. C., Sue, C. M., Davidson, M. M. et al., Fatal infantile cardioencephalomyopathy with COX deficiency and mutations in SCO2, a COX assembly gene, *Nat. Genet.* 23 (1999): 333–337.

Paumard, P., Vaillier, B., Coulary, J. et al., Cardiolipin stabilizes respiratory chain supercomplexes, *J. Biol. Chem.* 278 (2003): 52873–52880.

Perkins, G. A., Tjong, J., Brown, J. M. et al., The micro-architecture of mitochondria at active zones: Electron tomography reveals novel anchoring scaffolds and cristae structured for high-rate metabolism, *J. Neurosci.* 30 (2010): 1015–1026.

Pfanner, N., van der Laan, M., Amati, P. et al., Uniform nomenclature for the mitochondrial contact site and cristae organizing system, *J. Cell. Biol.* 204 (2014): 1083–1086.

Pfeiffer, K., Gohil, V., Stuart, R. A. et al., Cardiolipin stabilizes respiratory chain supercomplexes, *J. Biol. Chem.* 278 (2003): 52873–52880.

Porcelli, A. M., Ghelli, A., Ceccarelli, C. et al., The genetic and metabolic signature of oncocytic transformation implicates HIF1alpha destabilization, *Hum. Mol. Genet.* 19 (2010): 1019–1032.

Rahman, S., Taanman, J. W., Cooper, J. M. et al., A missense mutation of cytochrome oxidase subunit II causes defective assembly and myopathy, *Am. J. Hum. Genet.* 65 (1999): 1030–1039.

Rana, M., de Coo, I., Diaz, F., Smeets, H., Moraes, C. T., An out-of-frame cytochrome b gene deletion from a patient with parkinsonism is associated with impaired complex III assembly and an increase in free radical production, *Ann. Neurol.* 48 (2000): 774–781.

Rizzuto, R., De Stefani, D., Raffaello, A., Mammucari, C., Mitochondria as sensors and regulators of calcium signaling, *Nat. Rev. Mol. Cell. Biol.* 13 (2012): 566–578.

Schägger, H., de Coo, R., Bauer, M. F., Hofmann, S., Godinot, C., Brandt, U., Significance of respirasomes for the assembly-stability of human respiratory chain complex I, *J. Biol. Chem.* 279 (2004): 36349–36353.

Schägger, H., Pfeiffer, K., Supercomplexes in the respiratory chains of yeast and mammalian mitochondria, *EMBO J.* 19 (2000): 1777–1783.

Schagger, H., Pfeiffer, K., The ratio of oxidative phosphorylation complexes I-V in bovine heart mitochondria and the composition of respiratory chain supercomplexes, *J. Biol. Chem.* 276 (2001): 37861–37867.

Schlame, M., Ren, M., Barth syndrome, a human disorder of cardiolipin metabolism, *FEBS Lett.* 580 (2006): 5450–5455.

Sone, N., Sekimachi, M., Kutoh, E., Identification and properties of a quinol oxidase super-complex composed of a bc1 complex and cytochrome oxidase in the thermophilic bacterium PS3, *J. Biol. Chem.* 262 (1987): 15386–15391.

Strogolova, V., Furness, A., Robb-McGrath, M., Garlich, J., Stuart, R. A., Rcf1 and Rcf2, members of the hypoxia induced gene 1 protein family, are critical components of the mitochondrial cytochrome bc1–cytochrome c oxidase supercomplex, *Mol. Cell. Biol.* 32 (2012): 1363–1373.

Stroh, A., Anderka, O., Pfeiffer, K., Yagi, T., Finel, M., Ludwig, B., Schägger, H., Assembly of respiratory complexes I, III, and IV into NADH oxidase supercomplex stabilizes complex I in Paracoccus denitrificans, *J. Biol. Chem.* 279 (2004): 5000–5007.

Sukhorukov, V. M., Bereiter-Han, J., Anomalous diffusion of induced by cristae geometry in the inner mitochondrial membrane, *PLoS One* 4 (2009): e4604.

Suthammarak, W., Somerlot, B. H., Opheim, E., Sedensky, M., Morgan, P. G., Novel interactions between mitochondrial superoxide dismutases and the electron transport chain, *Aging Cell.* 12 (2013): 1132–1140.

Tallini, G., Oncocytic tumours, *Virchows Arch.* 433 (1998): 5–12.

Tamai, S., Iida, H., Yokota, S. et al., Characterization of the mitochondrial protein LETM1, which maintains the mitochondrial tubular shapes and interacts with the AAA-ATPase BCS1L, *J. Cell. Sci.* 121 (2008): 2588–2600.

Ugalde, C., Janssen, R. J., Van den Heuvel, L. P., Smeitink, J. A., Nijtmans, L. G., Differences in the assembly and stability of complex I and other OXPHOS complexes in inherited complex I deficiency, *Hum. Mol. Genet.* 13 (2004): 659–667.

Ugalde, C., Triepels, R. H., Coenen, M. et al., Impaired complex I assembly in a Leigh syndrome patient with a novel missense mutation in the ND6 gene, *Ann. Neurol.* 54 (2003): 665–669.

Valnot, I., Osmond, S., Gigarel, N. et al., Mutations of the SCO1 gene in mitochondrial cytochrome c oxidase deficiency with neonatal onset hepatic failure and encephalopathy, *Am. J. Hum. Genet.* 67 (2000): 1104–1109.

van Lis, R., Atteia, A., Mendoza-Hernandez, G., Gonzalez-Halphen, D., Identification of novel mitochondrial protein components of Chlamydomonas reinhardtii. A proteomic approach, *Plant Physiol.* 132 (2003): 318–330.

Vukotic, M., Oeljeklaus, S., Wiese, S. et al., Rcf1 mediates cytochrome oxidase assembly and respirasome formation, revealing heterogeneity of the enzyme complex, *Cell Metab.* 15 (2012): 336–347.

Wibrand, F., Ravn, K., Schwartz, M., Rosenberg, T., Horn, N., Vissing, J., Multisystem disorder associated with a missense mutation in the mitochondrial cytochrome b gene, *Ann. Neurol.* 50 (2001): 540–543.

Wilkens, V., Kohl, W., Busch, K., Restricted diffusion of OXPHOS complexes in dynamic mitochondria delays their exchange between cristae and engenders a transitory mosaic distribution, *J. Cell Sci.* 126 (2013): 103–116.

Xu, Y., Kelley, R. I., Blanck, T. J., Schlame, M., Remodeling of cardiolipin by phospholipid transacylation, *J. Biol. Chem.* 278 (2003): 51380–51385.

Youle, R. J., Van del Bliek, A. M., Mitochondrial fission, fusion, and stress, *Science* 337 (2012): 1062–1065.

Zanna, C., Ghelli, A., Porcelli, A.M. et al., OPA1 mutations associated with dominant optic atrophy impair oxidative phosphorylation and mitochondrial fusion, *Brain* 131 (2008): 352–367.

Zhang, M., Mileykovskaya, E., Dowhan, W. Gluing the respiratory chain together. Cardiolipin is required for supercomplex formation in the inner mitochondrial membrane. *J Biol Chem* 277 (2002): 43553–43556.

CHAPTER 7

Regulation of Photosynthetic Electron Transport via Supercomplex Formation in the Thylakoid Membrane

Kentaro Ifuku and Toshiharu Shikanai

CONTENTS

ABSTRACT Photosynthetic electron transport in the thylakoid membrane is mediated by two photochemical reactions. In each photosystem, light-harvesting antenna complexes funnel the absorbed light energy into the core complex, resulting in charge separation in the reaction-centre chlorophyll. Because the amount of light energy absorbed often exceeds the capacity of its utilization by CO_2 fixation, phototrophs have multiple mechanisms to regulate the efficiency of light energy utilization. In light of recent progress in research there is a need to reconsider the commonly accepted view of their physiological function. Regulation of light-use efficiency involves dynamic interactions between protein complexes in the thylakoid membrane. The evolutionary strategies for using each process to cope with the natural light environment have diverged between flowering plants and green algae. ΔpH formed across the thylakoid membrane plays a critical role in regulating photosynthetic electron transport. In the photosystem II antenna, ΔpH regulates the photoprotective mechanism underlying non-photochemical quenching of chlorophyll fluorescence. Cyclic electron transport around photosystem I regulates ΔpH formation, which is essential for photosystem I photoprotection. The supercomplexes including photosystem I are the machinery for this electron transport.

7.1 INTRODUCTION

The light reactions of photosynthesis involve electron transport in the thylakoid membrane. This is driven by two photochemical reactions, of photosystem (PS) II and PSI, in O_2-evolving photosynthesis. Linear electron transport (LET) from water to $NADP^+$ is coupled with H^+ translocation across the thylakoid membrane. The resulting ΔpH is used in adenosine triphosphate (ATP) synthesis, as is the membrane potential ($\Delta\psi$). NADPH and ATP are the first stable products of photosynthesis and are used in CO_2 fixation. In LET, electrons are excised from water in PSII and are used to reduce plastoquinone (PQ) in the thylakoid membrane. The cytochrome (Cyt)b_6f complex accepts electrons from PQ and then reduces plastocyanin on the lumen side. The Q cycle is a mechanism for coupling the electron transport through Cytb_6f with H^+ translocation across the thylakoid membrane. PSI mediates light-dependent electron transfer from plastocyanin to ferredoxin (Fd), which is oxidized by Fd-$NADP^+$ reductase (FNR) to produce NADPH.

Because light intensity often limits photosynthesis under natural conditions, plants have to be aggressive to absorb more light energy. Even more frequently, CO_2 fixation limits photosynthesis, especially under stress conditions such as low temperature and drought, in which excessive reducing power generated by LET induces the production of reactive oxygen species (ROS). These problems are serious, because light intensity fluctuates under natural conditions. To sustain maximum photosynthesis performance, the efficiency of light reactions needs to be precisely regulated. To satisfy this flexibility of light reactions, protein complexes in the thylakoid membrane alter their structure and activity dynamically. This is one of topics of this chapter.

Structural changes in the protein complexes in the thylakoid membrane are closely related to the regulation of photosynthetic electron transport. This is another topic of this chapter. To avoid photoinhibition by excessive light energy, the most efficient strategy is the dissipation of absorbed light energy from PSII antennae safely as heat (Niyogi and Truong 2013). This process is regulated by monitoring acidification of the thylakoid lumen. Because ΔpH energizes ATP synthesis, regulation of ΔpH has a central role in orchestrating photosynthesis. Cyclic electron transport around PSI (PSI CET) has a critical role in the regulation of ΔpH, consequently optimizing the trade-off between maximum photosynthesis and photoprotection (Shikanai 2014). State transitions are a process regulating the excitation pressure of both PSs and were originally regarded as a mechanism positively optimizing LET. However, recent studies have clarified its physiological function in the regulation of LET to avoid photodamage to PSI (Grieco et al. 2012). In this chapter, we summarize current progress in research into the regulation of photosynthesis, with a focus on the structure of thylakoid membrane complexes.

7.2 OVERVIEW OF THE PSII SUPERCOMPLEX

7.2.1 Structure and Function of the PSII Core

PSII is one of the key complexes of the redox light reactions in photosynthesis: PSII converts solar energy into the electrochemical potential energy required to split water into H^+, electrons and O_2 (Hankamer et al. 1997). Recent X-ray structural analysis of the cyanobacterial PSII complex to a resolution of 1.9 Å has revealed the location of most subunits, pigments and redox cofactors (Umena et al. 2011). The PSII core is composed of more than 20 subunits, among which CP47, CP43, D1, D2 and the α- and β-subunits of Cytb_{559} are known to be essential for the O_2-evolving reaction in vitro (Satoh 2008). Light excitation of the primary donor P680, a special pair of chlorophyll (Chl)*a* dimers in PSII, leads to electron transfer to a nearby pheophytin. This is followed by electron transfer via PQs (Q_A and Q_B), although the primary charge separation within PSII is suggested to be more complex (Cardona et al. 2012). The resulting primary cation radical of $P680^+$ receives electrons via a redox-active tyrosine of D1 from the Mn_4CaO_5 cluster ligated by D1 and CP43, which accumulates the four oxidizing equivalents required for oxidation of water to O_2 (Vinyard et al. 2013).

The mechanism of water oxidation and the basic subunit structure of the PSII core are highly conserved from cyanobacteria to flowering plants; however, some peripheral subunits differ among the oxygenic photosynthetic organisms. In particular, the composition of the extrinsic subunits of PSII has undergone a great evolutionary change (Ifuku et al. 2011). Therefore, crystallographic information derived from prokaryotic cyanobacterial PSII cannot necessarily be applied in the context of eukaryotes. Green eukaryotes, such as flowering plants, green algae and *Euglena*, have a set of three extrinsic proteins, PsbO, PsbP and PsbQ, binding to the lumenal surface of PSII. In cyanobacteria, however, PsbV and PsbU are present at the expense of PsbP and PsbQ (Bricker et al. 2012). Cyanobacteria have PsbP and PsbQ homologues (CyanoP and CyanoQ, respectively) (De Las Rivas et al. 2004), but they are not yet included in the current structural models. Thus the PsbP and PsbQ proteins in green plants seem to have evolved from their cyanobacterial homologues, although considerable genetic and functional modifications have occurred to generate the present eukaryotic forms (Ifuku 2014). The high-resolution structures of isolated PsbP and PsbQ proteins have been reported (Calderone et al. 2003; Ifuku et al. 2004; Balsera et al. 2005), and their locations and binding topologies in the green plant PSII complex have been proposed (Ido et al. 2012, 2014; Nishimura et al. 2014). It has also been suggested that the PsbP and PsbQ proteins in flowering plants are required for the stable association of intramembranous light-harvesting proteins with PSII (Ido et al. 2009; Allahverdiyeva et al. 2013).

7.2.2 PSII-LHCII Supercomplex in Green Plants

The water-splitting/O_2-evolving reaction in PSII is a multi-electron redox reaction, and photons must incessantly excite the PSII reaction centre (RC). Therefore, an efficient light-harvesting system is crucial for sustaining the photochemical reaction, especially under light-limiting conditions (Horton et al. 1996). In plants and green algae, the PSII core complex is associated with the membrane-embedded light-harvesting antenna complexes (LHCII), forming large macromolecular complexes, namely the PSII-LHCII supercomplexes (Croce and van Amerongen 2014). Figure 7.1a is a schematic model of a PSII-LHCII supercomplex of flowering plants. The organization of the supercomplexes was visualized by electron microscopy and single particle analysis at a resolution of ~12 Å (Dekker and Boekema 2005; Caffarri et al. 2009; Pagliano et al. 2013). The LHCII trimers associated with the PSII core are categorized into three types on the basis of their affinity with the PSII supercomplex, namely those that have a strong (S), moderate (M) or loose (L) association with the PSII core (C). These LHCII trimers consist of different combinations of three LHCII proteins, Lhcb1–3 (Caffarri et al. 2004). In *Arabidopsis*, the $C_2S_2M_2$ supercomplex is

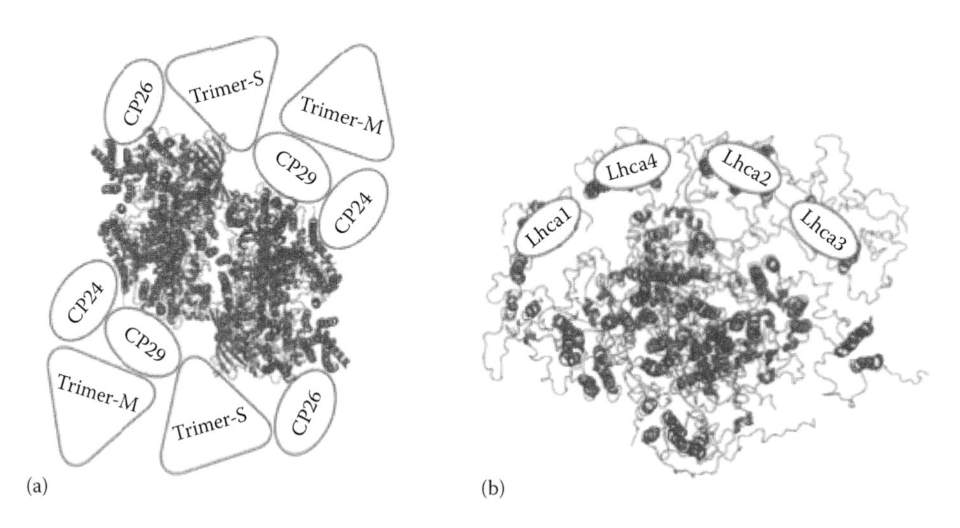

FIGURE 7.1 Organization of the two photosystem core complexes and light-harvesting antenna proteins. (a) Model of the plant PSII-LHCII supercomplex ($C_2S_2M_2$ complex), viewed from the top onto the lumenal surface and utilizing the high-resolution structure of the cyanobacterial PSII core (PDB_ID, 3ARC). (Data from Umena, Y. et al., *Nature*, 473, 55–60, 2011.) The extrinsic PsbV and PsbU known to be absent in plant PSII were excluded from the model. Positioning of the strongly (S) or moderately (M) bound LHC trimers, as well as the minor LHC antennae CP24, CP26, and CP29, is schematically depicted. (Data from Caffarri, S. et al., *EMBO J.*, 28, 3052–3063, 2009.) (b) Crystal structure of the plant PSI-LHCI supercomplex (PDB_ID, 3LW5), viewed from the lumen. (Data from Amunts, A. et al., *J. Biol. Chem.* 285, 3478–3486, 2010.)

the largest PSII-LHCII and is composed of a dimeric core (C_2); two LHCII-S trimers interact directly with the core and with monomeric LHCII CP29 (Lhcb4) and CP26 (Lhcb5), and two M-trimers react with the core via CP29 and CP24 (Lhcb6). In addition, a variable number of LHCII-L trimers supposedly associate with PSII at positions that have been observed in spinach (Boekema et al. 1999).

Recent studies using *Chlamydomonas* have reported a novel complex with $C_2S_2M_2L_2$ organization. Unlike flowering plants, *Chlamydomonas* does not have CP24: the position is replaced by LHCII-L (Tokutsu et al. 2012; Drop et al. 2014a). Furthermore, amino acid sequences are not highly conserved, even between orthologous members of *Chlamydomonas* and *Arabidopsis* (Minagawa and Takahashi 2004). Similar diversification has been observed in LHCI (see Section 7.5.2). Despite the conservation of the core subunits, the light-harvesting systems are rather divergent, and this may be a plant strategy for adapting to different light environments.

7.3 NON-PHOTOCHEMICAL QUENCHING

7.3.1 Components of Non-Photochemical Quenching

Whereas the antenna system increases the effective absorption cross-section of the PSII core, there are often too many excitations to be handled by the downstream photosynthetic reactions under fluctuating light environments (Vass 2012). It has also been suggested that the Mn_4CaO_5 cluster in PSII is intrinsically susceptible to photodamage (Murata et al. 2007). Under such conditions, overexcitation of the PSII RC must be avoided by the regulation of light harvesting in the antenna system (Horton and Ruban 2005). Dissipation of the excess light energy can be monitored as non-photochemical quenching (NPQ) of Chl fluorescence. Traditionally, NPQ is proposed to be the sum of at least three components (Figure 7.2)—qE (energy-dependent quenching), qT (state transitions) and qI (photoinhibitory quenching)—which are distinguished by their relaxation kinetics in the dark (Quick and Stitt 1989; Walters and Horton 1991). These quenching mechanisms have been regarded as independent processes activated in different light conditions; however, recent studies suggest that they may be strongly interconnected, and that each may exploit multiple mechanisms or may even include unknown mechanisms (Tikkanen and Aro 2014). Here, we summarize the recent results regarding each NPQ component. The classification or nomenclature of the NPQ components might be reconsidered in the future.

7.3.2 qE: Energy-Dependent Quenching

In vascular plants, qE-type energy dissipation is the major component of NPQ, which is rapidly induced by ΔpH under excess light and relaxed in the dark (Horton et al. 1996). The acidification of thylakoid lumena activates VDE (violaxanthin de-epoxidase), mediating the conversion of violaxanthin to zeaxanthin via antheraxanthin (xanthophyll cycle) (Demmig-Adams 1990). In addition, a distinct LHC protein, PsbS, senses low pH via the protonation of two glutamic acid residues and modulates qE (Li et al. 2000, 2004). It has been recently suggested that both elevation of zeaxanthin and antheraxanthin contents and protonation of PsbS increase the fluidity of the thylakoid membrane (Johnson et al. 2011a; Goral et al. 2012). This facilitates partial disassembly of the PSII-LHCII supercomplex (Betterle et al. 2009)

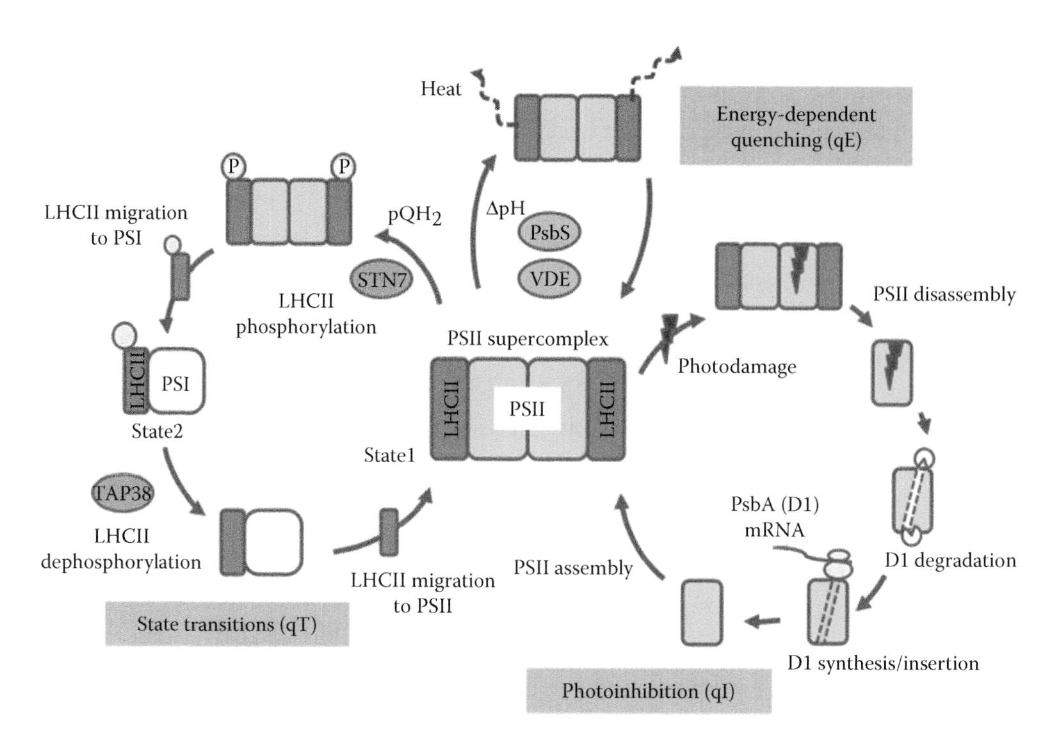

FIGURE 7.2 Classical view of the components of non-photochemical chemical quenching (NPQ) regulating PSII activity. NPQ has been proposed to be the sum of at least three components: energy-dependent quenching (qE), state-transition quenching (qT) and photoinhibitory quenching (qI). For details see Section 7.3.

and causes aggregation of the dissociated LHCII (Kiss et al. 2008; Johnson et al. 2011b), which probably induces conformational change within the major LHCII and minor LHCII towards the quenching state (Ruban et al. 2007; Ahn et al. 2008). The binding of zeaxanthin also induces conformational changes in LHCII proteins (Dall'Osto et al. 2005). It should be noted that the ability to induce the quenching state is an intrinsic property of LHCII proteins (Krüger et al. 2012) and can be induced independently of zeaxanthin or PsbS (Johnson and Ruban 2010, 2011). However, rapid and maximum induction of qE requires both the xanthophyll cycle and PsbS under physiological conditions.

Unlike in vascular plants, in *Chlamydomonas* PsbS is not involved in qE-type quenching (Bonente et al. 2008), whereas the xanthophyll cycle contributes to qE (Niyogi et al. 1997a,b). Instead of PsbS, *Chlamydomonas* has LHCSR protein, another antenna protein specifically involved in quenching modulation (Peers et al. 2009). Unlike the expression of PsbS, expression of LHCSR is not constitutive but is induced under stress conditions, under which qE activation is induced in *Chlamydomonas* (Finazzi et al. 2006). A recent study has demonstrated that LHCSR binds to the PSII-LHCII supercomplex and can switch the light-harvesting state of the purified PSII-LHCII-LHCSR3 supercomplex to an energy-dissipative state in a low-pH-dependent manner (Tokutsu and Minagawa 2013). This suggests that algae have developed a specific mechanism for qE induction under stress conditions. Interestingly, the moss *Physcomitrella patens* can induce high levels of NPQ via both LHCSR- and PsbS-dependent mechanisms (Pinnola et al. 2013). These findings suggest that a gradual shift from an LHCSR-dependent to a PsbS-dependent system has occurred during evolution to adapt to the high light levels of land environments.

7.3.3 qT: State Transitions

qT is related to state transitions and is relaxed within several minutes in the dark. State transitions balance the excitation of PSs by modulating the antenna size of each PS to maximize the efficiency of LET (Allen 1992). Because the light-absorption properties of both PSs are markedly different, state transitions have been suggested to be particularly important for coping with light-quality changes in natural environments (Rochaix 2007). In the classical model, the redox state of the PQ pool controls the activity of the protein kinase Stt7/STN7 (Depège et al. 2003; Bellafiore et al. 2005), which is activated upon binding of plastoquinol (PQH_2) to the Qo site of the Cytb_6f complex and phosphorylates LHCII (Wollman 2001). Phosphorylated LHCII dissociates from PSII and moves to PSI, thereby avoiding over-reduction of the PQ pool (state 2). Oxidation of the PQ pool leads to inactivation of the kinase and consequently dephosphorylation of LHCII by PPH1/TAP38 phosphatase (Pribil et al. 2010; Shapiguzov et al. 2010), resulting in return of the mobile LHCII to PSII (state 1). However, this elegant model may be too simplistic to explain the actual physiological functions of state transitions.

In *Arabidopsis*, only 10%–15% of LHCII is involved in state transitions (Allen 1992). Although the contribution of qT to NPQ is relatively low in plants, recent studies demonstrate that balancing of excitation energy between two PSs is crucial, particularly under fluctuating light (Tikkanen et al. 2011). The mobile pool of LHCs is likely to be part of the LHC-L trimer population (extra LHCII), because remodelling of the PSII-LHCII supercomplex has not been observed in state 2 (Wientjes et al. 2013a). These mobile LHCs transfer energy relatively slowly to PSII in state 1, whereas the energy transfer to PSI is extremely fast and efficient in state 2 (Galka et al. 2012). In fact, the association of extra LHCII with PSI plays an important role in long-term acclimation to different intensities of growth light, indicating that this LHCII serves as an innate antenna of PSI and occasionally transfers energy to PSII in state 2 (Wientjes et al. 2013b). Furthermore, phosphorylation of LHCII does not necessarily cause the dissociation of LHCs from PSII; phosphorylated forms of the LHCII-M and LHCII-S trimers remain bound to PSII, whereas a phosphorylated form of extra LHCII-L would move to PSI (Wientjes et al. 2013a). Differential phosphorylation of Lhcb proteins during state transitions has also been reported (Leoni et al. 2013), and Lhcb2 phosphorylation plays a critical role during state transitions (Pietrzykowska et al. 2014).

In *Chlamydomonas*, it was earlier believed that 80% of LHCII is mobile in the state transitions between PSs (Delosme et al. 1996). However, recent studies suggest that state transitions in *Chlamydomonas* primarily reduce the antenna size of PSII, whereas the effect on PSI is rather small: large parts of the phosphorylated LHCIIs do not bind to PSI, but instead form energetically quenched complexes, which are either associated with PSII supercomplexes or in a free form (Nagy et al. 2014). Furthermore, disassembly of the PSII-LHCII complex, which has been reported to occur in state transitions (Iwai et al. 2008, 2010b), is not necessarily observed during the transition from state1 to state 2, suggesting that only a subpopulation of Lhcb is involved in the process denoted as state transitions (Drop et al. 2014b; Nagy et al. 2014). It has been proposed that qT in algae is an energy-dissipation process, rather than an energy-balancing system between PSs. In fact, state transitions have been shown to have an important role as a short-term response mechanism to increase resistance to high light levels (Allorent et al. 2013). Further research is needed into the exact energy-quenching mechanism related to qT. It should be also noted that some non-phosphorylated Lhcbs can associate with PSI and some phosphorylated forms with PSII, indicating that the original concept is not absolutely true in *Chlamydomonas* (Drop et al. 2014b), as well as in *Arabidopsis*.

In summary, the above-mentioned results disclose the different roles of state transitions in algae and vascular plants. There is a need to revise our concept of state transitions under both mechanisms. Vascular plants use qT to modulate excitation balance between two PSs under fluctuating light; under constant high light levels they immediately switch on photoprotective qE via PsbS. In *Chlamydomonas*, however, qE depends on light-inducible LHCSR. For a more rapid response to high light, *Chlamydomonas* also depends on quenching related to qT. Interplay between qT and qE is also discussed in recent reviews (Tikkanen et al. 2011; Tikkanen and Aro 2014).

7.3.4 Photoinhibition (qI)

The most slowly relaxing component of NPQ is qI, which is generally believed to reflect photoinhibitory quenching. Because the water-splitting reaction via light-excited Chl molecules is extremely oxidizing, photooxidative damage to PSII inevitably occurs. Photoinhibition has been considered as the selective degradation of D1 protein in photodamaged PSII, which is restored by a specific repair system (Aro et al. 1993). However, the quantitative link is unclear between the qI level and the extent of PSII photodamage (Walters and Horton 1993). Therefore, the exact mechanism behind qI-type quenching remains to be elucidated. It has been proposed that this phase includes a long-lasting quenching process, possibly related to conformational changes in LHC proteins upon zeaxanthin binding (qZ) (Dall'Osto et al. 2005; Nilkens et al. 2010). A more recent study has shown that the thioredoxin-like/β-propeller protein SOQ1 could prevent the formation of slowly reversible NPQ (Brooks et al. 2013). Its thioredoxin-like domain on the thylakoid lumenal side may be the site for redox regulation in this novel NPQ mechanism. Interactions among light-harvesting antenna complexes are weakened in the absence of SOQ1 (Onoa et al. 2014). qI-type quenching may also be related to the rearrangement of protein complexes in the thylakoid membrane and might have a mechanism similar to that of qE- or qT-type quenching.

7.4 OVERVIEW OF THE PSI SUPERCOMPLEX

7.4.1 Structure and Function of the PSI Core

PSI accepts electrons from plastocyanin or Cytc_6 in the thylakoid lumen and transfers them to Fd on the stromal side. The quantum yield of PSI is close to 100%, suggesting that PSI is the most efficient machinery for the conversion of light energy (Amunts and

Nelson 2009). Two RC proteins, PsaA and PsaB, form a heterodimer that harbours a series of electron carriers. Light energy absorbed by Chls present in the PSI-LHCI supercomplex is funneled to a special Chl dimer (P700), resulting in its charge separation. The primary electron acceptor is a Chl (A_0), which transfers electrons to phylloquinone (A_1). Three Fe_4S_4 clusters (F_X, F_A and F_B) complete the electron transport in the PSI RC and reduce Fd. The stromal side of the PsaA/PsaB heterodimer is equipped with three subunits (PsaC, PsaD and PsaE), which form the docking site with Fd. The last electron carrier, F_B, is coordinated with PsaC.

Both PSII and PSI RCs form heterodimers considered to have originated from an ancestral homodimer that is probably similar to the homodimeric FeS (PSI)-type RC present in heliobacteria, acidobacteria and green sulphur bacteria (Hohmann-Marriott and Blankenship 2011). Each subunit of the FeS-type RC consists of 11 transmembrane helixes (TMHs). In the PSII complex, however, 11 TMHs are split into RC proteins (D1/D2) containing 5 TMHs and core antenna proteins (CP43/CP47) containing 6 TMHs. Although the amino acid sequence is not conserved, PSII and PSI RCs are structurally closely related.

7.4.2 PSI-LHCI Supercomplex in Angiosperms

In angiosperms, the PSI-LHCI supercomplex consists of a PSI core and one copy each of four major LHCI molecules, Lhca1 to 4 (Figure 7.1b; Amunts et al. 2010). In the supercomplex, a fixed position is assigned for each of the molecules, which form two heterodimers, Lhca1/4 and Lhca2/3 (Amunts et al. 2007; Wientjes et al. 2009). In addition to these four major LHCI molecules, the *Arabidopsis* genome encodes two minor LHCI molecules, Lhca5 and Lhca6, the accumulation levels of which are substoichiometric compared with that of the PSI RC. In a mutant lacking a single gene for a major LHCI molecule the missing subunit was not replaced by others (Wientjes et al. 2009). The only exception is Lhca4, which is partly replaced by Lhca5. Consistently, higher levels of Lhca5 have been detected in mutants defective in Lhca1/4 than in the wild type (Lucinski et al. 2006). In contrast, the levels of Lhca5 were reduced in the mutant defective in Lhca2/3. Lhca5 may interact with the Lhca2/3 dimer, and the Lhca5 homodimer partly substitutes for the Lhca1/4 dimer.

The structure of the PSI-LHCI complex depends on the organism (Busch and Hippler 2011). In *Chlamydomonas*, it is composed of one core complex and nine LLHCI proteins (Lhca1–9). In addition to the four LHCI molecules directly interacting with the PSI core, five LHCI molecules form the outer half-ring (Drop et al. 2011). In *Arabidopsis*, Lhca5 likely localizes to the outside of the inner ring composed of Lhca1–4, as occurs with the outer-ring LHCI molecules in *Chlamydomonas* (Lucinski et al. 2006). However, Lhca5 behaves as a linker between the PSI supercomplex and NDH in *Arabidopsis* (see Section 7.5.3).

7.5 PSI CET

7.5.1 Physiological Function of PSI CET

In PSI CET, electrons are recycled from Fd to PQ, contributing to the formation of ΔpH without accumulation of NADPH (Figure 7.3). It was more than 60 years ago when CET was discovered as cyclic phosphorylation (Arnon et al. 1954), but its physiological significance

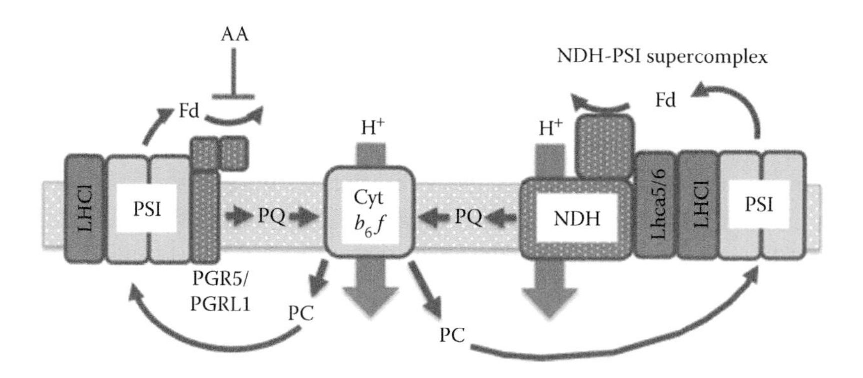

FIGURE 7.3 Machinery of PSI CET in *Arabidopsis*. PGR5/PGRL1 interacts with PSI to mediate AA-sensitive, Fd-dependent PQ reduction. In *Chlamydomonas*, this complex further associates with the Cytb_6f complex and some other factors. Chloroplast NDH interacts with PSI via Lhca5/6 and mediates AA-insensitive Fd-dependent PQ reduction.

has only recently been recognized (Shikanai 2007). In angiosperms, PSI CET consists of two routes of electrons. Arnon's pathway is sensitive to antimycin A (AA) and corresponds to the route depending on the proteins PROTON GRADIENT REGULATION 5 (PGR5) and PGR5-LIKE PHOTOSYNTHETIC PHENOTYPE 1 (PGRL1) (Tagawa et al. 1962; Hertle et al. 2013; Sugimoto et al. 2013). The minor pathway is resistant to AA and is mediated by chloroplast NDH (Shikanai et al. 1998).

The *Arabidopsis pgr5* mutant was identified on the basis of its high Chl fluorescence at high light intensity (Munekage et al. 2002). PGR5-dependent PSI CET acidifies the thylakoid lumen to less than pH 6.0 to induce qE. The *pgr5* mutant grows like the wild type under constant low light, but in double mutants defective in both pathways of PSI CET (e.g. *crr2 pgr5*), plant growth and photosynthesis are severely impaired (Munekage et al. 2004). Because the double mutants are sensitive even to extremely low light, the phenotype cannot be explained by the defect in qE induction. Most likely, PSI CET compensates for the ATP/NADPH production ratio in photosynthesis. Consistently, the contribution of the PGR5 pathway to total ATP synthesis has been estimated to be 13% (Avenson et al. 2005). The size of the proton motive force ($\Delta pH + \Delta\psi$) is reduced in *pgr5* mutants in *Arabidopsis* and rice (Avenson et al. 2005; Nishikawa et al. 2012; Wang et al. 2015).

7.5.2 Machinery of PGR5/PGRL1-Dependent PSI CET

The contribution of state transitions to the regulation of photosynthesis in *Chlamydomonas* is more important than that in angiosperms. In state 2, where more light energy is funnelled to the PSI RC, PSI CET preferentially operates instead of LET (Finazzi et al. 2002). From this perspective, the protein complex specifically accumulating in state 2 has been purified via His-tagged PsaA and shown to include PSI, Cytb_6f, LHCII, FNR, PetO and PGRL1 (Iwai et al. 2010a). The purified complex mediates Fd-dependent Cyt*b* reduction in the light, suggesting that this supercomplex is the machinery of PSI CET. PGRL1 is required for PSI CET in both *Arabidopsis* and *Chlamydomonas* (DalCorso et al. 2008; Petroutsos et al. 2009), suggesting that the machinery of PSI CET is conserved between the two organisms. However, this idea was not supported by the following facts: (1) The supercomplex including PSI and Cytb_6f has not been identified in angiosperms; in *Chlamydomonas*, (2) PGR5 has not been detected in the CET supercomplex and (3) PSI CET activity is resistant to AA (Iwai et al. 2010a). However, the *Chlamydomonas pgr5* mutant was recently shown to have a phenotype similar to that of the *pgrl1* mutant and PGR5 was unstable in the *pgrl1* mutant, suggesting that the function of PGR5 is conserved in *Chlamydomonas* (Johnson et al. 2014). A single amino acid alteration in PGR5 conferred resistance to AA (Sugimoto et al. 2013), suggesting that the sensitivity to AA was affected by a minor alteration in the PGR5 structure. Despite the difference in the supercomplex structure, the core component of PSI CET is likely conserved between *Chlamydomonas* and *Arabidopsis*.

Although operation of PSI CET is linked to state transitions in *Chlamydomonas* (Finazzi et al. 2002), the two regulatory mechanisms are independent in *Arabidopsis*, as evident in the different mutant phenotypes (Munekage et al. 2002; Bellafiore et al. 2005). The difference in dependency on state transitions suggests that the machinery of PSI CET differs at least partly between *Chlamydomonas* and *Arabidopsis*. However, a recent study demonstrated that, in *Chlamydomonas*, operation of PSI CET, as well as formation of the CET supercomplex, could be induced by reducing conditions in chloroplasts independently of the induction of state 2 (Takahashi et al. 2013). Operation of PGR5-dependent PSI CET also requires the accumulation of reducing equivalents in the stroma in *Arabidopsis* (Okegawa et al. 2008), suggesting that the regulatory mechanism of PSI CET is conserved. The critical difference between the two organisms is in the conditions in which state transitions are induced: in angiosperms, state 2 is induced only at low light intensity, because excessive absorbed light energy is dissipated via the qE mechanism in state 1 at high light intensity (Tikkanen et al. 2011). In contrast, induction of state 2 coincides with the operation of PSI CET in *Chlamydomonas*, although the two regulatory events are independent.

7.5.3 Structure of the Chloroplast NDH Complex

Complete sequencing of plastid genomes in tobacco (*Nicotiana tabacum*) and liverwort (*Marchantia polymorpha*) has revealed 11 *ndh* genes that encode proteins homologous to subunits of NADH dehydrogenase (complex I) in mitochondria and bacteria (Matsubayashi et al. 1987). The discovery was unexpected, because complex I functions in respiratory electron transport. As a result of several breakthroughs, we are sure that chloroplast NDH is the machinery of PSI CET (Shikanai 2007). On the basis of differences in structure, function and even enzyme activity, we have distinguished photosynthetic NDH present in cyanobacteria and chloroplasts from respiratory NDH (complex I) present in mitochondria and non-photosynthetic bacteria (Shikanai and Aro 2014). Respiratory NDH consists of three modules. The Q (quinone) and M (membrane) modules are conserved in photosynthetic NDH and are involved in quinone reduction and proton pumping, respectively (Friedrich and Scheide 2000). The N module is the site of NADH oxidation and is absent in photosynthetic NDH (Yamamoto et al. 2011). The electron donor–binding module in photosynthetic NDH, as well as the electron donor, has been long unclear (Friedrich et al. 1995). Although the discrepancy has still not been fully explained by early works, which suggest that chloroplast NDH accepts electrons from NADH (Sazanov et al. 1998), chloroplast NDH most likely accepts electrons from Fd (Yamamoto et al. 2011). Chloroplast NDH is a Fd-dependent PQ reductase (FQR): We have proposed that the NDH complex represents an NADH dehydrogenase–like complex. This conclusion depends mainly on the FQR activity detected in ruptured chloroplasts (Endo et al. 1998; Munekage et al. 2004) and on the discovery of an NDH subunit, NdhS, which forms an Fd-binding site on the stromal side of photosynthetic NDH (Yamamoto et al. 2011; Yamamoto and Shikanai 2013).

In angiosperms, chloroplast NDH forms a supercomplex with PSI (Peng et al. 2008). Lhca5 and Lhca6 (minor LHCI molecules) are linkers that mediate supercomplex formation (Peng et al. 2009). Recently, a single particle analysis visualized this supercomplex consisting of a single copy of NDH sandwiched by two copies of the PSI-LHCI supercomplexes (Kouřil et al. 2014). In *lhca5* or *lhca6* single mutants, NDH subunits were detected in slightly smaller complexes than the intact NDH-PSI supercomplex, whereas in the *lhca5 lhca6* double mutant they were detected in a 700 kDa complex, the molecular weight of which corresponded to that of the NDH monomer without PSI (Peng and Shikanai 2011). These results suggest a model in which Lhca5 and Lhca6 independently contribute to the interaction with a single copy of PSI. The contributions of Lhca5 and Lhca6 are not equal in terms of the stability of the NDH complex. In *lhca6*, NDH is unstable in mature leaves, whereas in *lhca5* NDH stability is unaffected (Peng et al. 2009). However, the *lhca5* defect decreases the stability of NDH in the *lhca6* background, especially at high light intensity (Peng and Shikanai 2011).

In *M. polymorpha*, chloroplast NDH does not form a supercomplex with PSI (Ueda et al. 2012). Although *P. patens* also lacks a key component of supercomplex formation (Lhca6), a small amount NDH subunits has been detected in the high molecular weight complex including PSI, suggesting that a modified form of the supercomplex is present (Armbruster et al. 2013). Supercomplex formation via Lhca6 was likely established during the evolution of land plants. A central question is that of the reason for supercomplex formation. In the absence of Lhca6, chloroplast NDH is unstable in *Arabidopsis*, suggesting that the purpose of supercomplex formation is to give stability (Peng and Shikanai 2011). Chloroplasts may have required additional machinery for the assembly or stability of protein complexes. An example is observed in the evolution of Cpn60β4, which is specifically required for the assembly of NdhH in angiosperms, but this is not the case in *Physcomitrella* (Peng et al. 2011).

7.6 PHOTODAMAGE TO PSI

It is generally accepted that the primary target of photoinhibition is PSII. Plants have a system for rapidly repairing the damaged PSII RC. To make PSII sensitive to light, plants control the rate of electron transport towards PSI. In pioneering work, PSI photodamage has been observed in chilling-sensitive plants at low light intensity (Sonoike 2011). Because of the limitation of electron acceptors from PSI, P700 is reduced by electrons via charge recombination, resulting in ROS generation, which damages PSI irreversibly. Reduction of P700 is not specific to chilling-sensitive plants and has been observed in plants growing under severe light conditions in the field (Endo et al. 2005). In wild-type plants, P700 is oxidized more in the course of an increase in light intensity. However, this is not the case in the *Arabidopsis pgr5* mutant, probably because of limitation of electron acceptors (Munekage et al. 2002) and disturbed regulation of Cytb_6f via high lumen pH (Suorsa et al. 2012). PGR5-dependent PSI CET plays a central role in regulating the redox state of the intersystem to avoid PSI photoinhibition (Shikanai 2014). In angiosperms another mechanism for preventing PSI photoinhibition is state transitions (see Section 7.3.3). The *Arabidopsis stn7* mutant, which is defective in state transitions, is sensitive to fluctuating light (Bellafiore et al. 2005). It is essential to be in state 2 to prevent over-reduction of PSI during a shift from low light to high light (Grieco et al. 2012). In light of recent research progress, we may have to change our view on PSI photoinhibition. PSI photodamage does not happen often, because plants have sophisticated strategies for avoiding it.

7.7 CONCLUDING REMARKS

The current concept of photosynthetic electron transport was established more than half a century ago. At present, researchers are discussing the reactions of each component at scales as small as the atomic level (Umena et al. 2011). In terms of the regulation of electron transport, however, there is a need for substantial reconsideration of the historical concepts. The classical models may have been too static to give us an adequate understanding of the ingenuity of plant strategies for adapting to fluctuating light environments. Technical breakthroughs have enabled us to reconsider the molecular mechanisms of the regulation of photosynthesis as the dynamic interaction of thylakoid membrane complexes.

REFERENCES

Ahn TK, Avenson TJ, Ballottari M et al. (2008). Architecture of a charge-transfer state regulating light harvesting in a plant antenna protein. *Science* 320:794–7.

Allahverdiyeva Y, Suorsa M, Rossi F et al. (2013). Arabidopsis plants lacking PsbQ and PsbR subunits of the oxygen-evolving complex show altered PSII super-complex organization and short-term adaptive mechanisms. *Plant J* 75:671–84.

Allen JF. (1992). Protein phosphorylation in regulation of photosynthesis. *Biochim Biophys Acta—Bioenerg* 1098:275–335.

Allorent G, Tokutsu R, Roach T et al. (2013). A dual strategy to cope with high light in *Chlamydomonas reinhardtii*. *Plant Cell* 25:545–57.

Amunts A, Drory O, Nelson N. (2007). The structure of a plant photosystem I supercomplex at 3.4 A resolution. *Nature* 447:58–63.

Amunts A, Nelson N. (2009). Plant photosystem I design in the light of evolution. *Structure* 17:637–50.

Amunts A, Toporik H, Borovikova A, Nelson N. (2010). Structure determination and improved model of plant photosystem I. *J Biol Chem* 285:3478–86.

Armbruster U, Labs M, Pribil M et al. (2013). Arabidopsis CURVATURE THYLAKOID1 proteins modify thylakoid architecture by inducing membrane curvature. *Plant Cell* 25:2661–78.

Arnon DI, Allen MB, Whatley FR. (1954). Photosynthesis by isolated chloroplasts. *Nature* 174:394–6.

Aro E-M, Virgin I, Andersson B. (1993). Photoinhibition of photosystem II. Inactivation, protein damage and turnover. *Biochim Biophys Acta* 1143:113–34.

Avenson TJ, Cruz JA, Kanazawa A, Kramer DM. (2005). Regulating the proton budget of higher plant photosynthesis. *Proc Natl Acad Sci USA* 56:395–406.

Balsera M, Arellano JB, Revuelta JL et al. (2005). The 1.49 A resolution crystal structure of PsbQ from photosystem II of *Spinacia oleracea* reveals a PPII structure in the N-terminal region. *J Mol Biol* 350:1051–60.

Bellafiore S, Barneche F, Peltier G, Rochaix J-D. (2005). State transitions and light adaptation require chloroplast thylakoid protein kinase STN7. *Nature* 433:892–5.

Betterle N, Ballottari M, Zorzan S et al. (2009). Light-induced dissociation of an antenna hetero-oligomer is needed for non-photochemical quenching induction. *J Biol Chem* 284:15255–66.

Boekema EJ, Van Roon H, Van Breemen JF, Dekker JP. (1999). Supramolecular organization of photosystem II and its light-harvesting antenna in partially solubilized photosystem II membranes. *Eur J Biochem* 266:444–52.

Bonente G, Passarini F, Cazzaniga S et al. (2008). The occurrence of the *psbS* gene product in *Chlamydomonas reinhardtii* and in other photosynthetic organisms and its correlation with energy quenching. *Photochem Photobiol* 84:1359–70.

Bricker TM, Roose JL, Fagerlund RD et al. (2012). The extrinsic proteins of photosystem II. *Biochim Biophys Acta* 1817:121–42.

Brooks MD, Sylak-Glassman EJ, Fleming GR, Niyogi KK. (2013). A thioredoxin-like/β-propeller protein maintains the efficiency of light harvesting in arabidopsis. *Proc Natl Acad Sci USA* 110:E2733–40.

Busch A, Hippler M. (2011). The structure and function of eukaryotic photosystem I. *Biochim Biophys Acta* 1807:864–77.

Caffarri S, Croce R, Cattivelli L, Bassi R. (2004). A look within LHCII: Differential analysis of the Lhcb1-3 complexes building the major trimeric antenna complex of higher-plant photosynthesis. *Biochemistry* 43:9467–76.

Caffarri S, Kouřil R, Kereïche S et al. (2009). Functional architecture of higher plant photosystem II supercomplexes. *EMBO J* 28:3052–63.

Calderone V, Trabucco M, Vujicić A et al. (2003). Crystal structure of the PsbQ protein of photosystem II from higher plants. *EMBO Rep* 4:900–5.

Cardona T, Sedoud A, Cox N, Rutherford AW. (2012). Charge separation in photosystem II: A comparative and evolutionary overview. *Biochim Biophys Acta* 1817:26–43.

Croce R, van Amerongen H. (2014). Natural strategies for photosynthetic light harvesting. *Nat Chem Biol* 10:492–501.

DalCorso G, Pesaresi P, Masiero S et al. (2008). A complex containing PGRL1 and PGR5 is involved in the switch between linear and cyclic electron flow in arabidopsis. *Cell* 132:273–85.

Dall'Osto L, Caffarri S, Bassi R. (2005). A mechanism of nonphotochemical energy dissipation, independent from PsbS, revealed by a conformational change in the antenna protein CP26. *Plant Cell* 17:1217–32.

Dekker JP, Boekema EJ. (2005). Supramolecular organization of thylakoid membrane proteins in green plants. *Biochim Biophys Acta* 1706:12–39.

De Las Rivas J, Balsera M, Barber J. (2004). Evolution of oxygenic photosynthesis: Genome-wide analysis of the OEC extrinsic proteins. *Trends Plant Sci* 9:18–25.

Delosme R, Olive J, Wollman F-A. (1996). Changes in light energy distribution upon state transitions: An in vivo photoacoustic study of the wild type and photosynthesis mutants from *Chlamydomonas reinhardtii*. *Biochim Biophys Acta* 1273:150–8.

Demmig-Adams B. (1990). Carotenoids and photoprotection in plants: A role for the xanthophyll zeaxanthin. *Biochim Biophys Acta* 1020:1–24.

Depège N, Bellafiore S, Rochaix J-D. (2003). Role of chloroplast protein kinase Stt7 in LHCII phosphorylation and state transition in *Chlamydomonas*. *Science* 299:1572–5.

Drop B, Webber-Birungi M, Fusetti F et al. (2011). Photosystem I of *Chlamydomonas reinhardtii* contains nine light-harvesting complexes (Lhca) located on one side of the core. *J Biol Chem.* 286:44878–87.

Drop B, Webber-Birungi M, Yadav SKN et al. (2014a) Light-harvesting complex II (LHCII) and its supramolecular organization in *Chlamydomonas reinhardtii*. *Biochim Biophys Acta* 1837:63–72.

Drop B, Yadav KNS, Boekema EJ, Croce R. (2014b). Consequences of state transitions on the structural and functional organization of Photosystem I in the green alga *Chlamydomonas reinhardtii*. *Plant J* 78:181–91.

Endo T, Kawase D, Sato F. (2005). Stromal over-reduction by high-light stress as measured by decreases in P700 oxidation by far-red light and its physiological relevance. *Plant Cell Physiol* 46:775–81.

Endo T, Shikanai T, Sato F, Asada K. (1998). NAD(P)H dehydrogenase-dependent, antimycin A-sensitive electron donation to plastoquinone in tobacco chloroplasts. *Plant Cell Physiol* 39:1226–31.

Finazzi G, Johnson GN, Dall'Osto L et al. (2006). Nonphotochemical quenching of chlorophyll fluorescence in *Chlamydomonas reinhardtii*. *Biochemistry* 45:1490–8.

Finazzi G, Rappaport F, Furia A et al. (2002). Involvement of state transitions in the switch between linear and cyclic electron flow in *Chlamydomonas reinhardtii*. *EMBO Rep* 3:280–5.

Friedrich T, Scheide D. (2000). The respiratory complex I of bacteria, archaea and eukarya and its module common with membrane-bound multisubunit hydrogenases. *FEBS Lett* 479:1–5

Friedrich T, Steinmüller K, Weiss H. (1995). The proton-pumping respiratory complex I of bacteria and mitochondria and its homologue in chloroplasts. *FEBS Lett* 367:107–11.

Galka P, Santabarbara S, Khuong TTH et al. (2012). Functional analyses of the plant photosystem I-light-harvesting complex II supercomplex reveal that light-harvesting complex II loosely bound to photosystem II is a very efficient antenna for photosystem I in state II. *Plant Cell* 24:2963–78.

Goral TK, Johnson MP, Duffy CDP et al. (2012). Light-harvesting antenna composition controls the macrostructure and dynamics of thylakoid membranes in Arabidopsis. *Plant J* 69:289–301.

Grieco M, Tikkanen M, Paakkarinen V, Kangasjärvi S, Aro E-M. (2012). Steady-state phosphorylation of light-harvesting complex II proteins preserves photosystem I under fluctuating white light. *Plant Physiol* 160:1896–910.

Hankamer B, Barber J, Boekema EJ. (1997). Structure and membrane organization of Photosytem II in green plants. *Annu Rev Plant Physiol Plant Mol Biol* 48:641–71.

Hertle AP, Blunder T, Wunder T et al. (2013). PGRL1 is the elusive ferredoxin-plastoquinone reductase in photosynthetic cyclic electron flow. *Mol Cell* 49:511–23.

Hohmann-Marriott MF, Blankenship RE. (2011). Evolution of photosynthesis. *Annu Rev Plant Biol* 62:515–48.

Horton P, Ruban AV. (2005). Molecular design of the photosystem II light-harvesting antenna: Photosynthesis and photoprotection. *J Exp Bot* 56:365–73.

Horton P, Ruban AV, Walters RG. (1996). Regulation of light harvesting in green plants. *Annu Rev Plant Physiol Plant Mol Biol* 47:655–84.

Ido K, Ifuku K, Yamamoto Y et al. (2009). Knockdown of the PsbP protein does not prevent assembly of the dimeric PSII core complex but impairs accumulation of photosystem II supercomplexes in tobacco. *Biochim Biophys Acta* 1787:873–81.

Ido K, Kakiuchi S, Uno C et al. (2012). The conserved His-144 in the PsbP protein is important for the interaction between the PsbP N-terminus and the Cyt b_{559} subunit of photosystem II. *J Biol Chem* 287:26377–87.

Ido K, Nield J, Fukao Y et al. (2014). Cross-linking evidence for multiple interactions of the PsbP and PsbQ proteins in a higher plant Photosystem II supercomplex. *J Biol Chem* 289:20150–7.

Ifuku K. (2014). The PsbP and PsbQ family proteins in the photosynthetic machinery of chloroplasts. *Plant Physiol Biochem* 81:108–14.

Ifuku K, Ido K, Sato F. (2011). Molecular functions of PsbP and PsbQ proteins in the photosystem II supercomplex. *J Photochem Photobiol B Biol* 104:158–64.

Ifuku K, Nakatsu T, Kato H, Sato F. (2004). Crystal structure of the PsbP protein of photosystem II from *Nicotiana tabacum*. *EMBO Rep* 5:362–7.

Iwai M, Takahashi Y, Minagawa J. (2008). Molecular remodeling of photosystem II during state transitions in *Chlamydomonas reinhardtii*. *Plant Cell* 20:2177–189.

Iwai M, Takizawa K, Tokutsu R et al. (2010a) Isolation of the elusive supercomplex that drives cyclic electron flow in photosynthesis. *Nature* 464:1210–13.

Iwai M, Yokono M, Inada N, Minagawa J. (2010b) Live-cell imaging of photosystem II antenna dissociation during state transitions. *Proc Natl Acad Sci USA* 107:2337–42.

Johnson MP, Brain APR, Ruban AV. (2011a) Changes in thylakoid membrane thickness associated with the reorganization of photosystem II light harvesting complexes during photoprotective energy dissipation. *Plant Signal Behav* 6:1386–90.

Johnson MP, Goral TK, Duffy CDP et al. (2011b) Photoprotective energy dissipation involves the reorganization of photosystem II light-harvesting complexes in the grana membranes of spinach chloroplasts. *Plant Cell* 23:1468–79.

Johnson MP, Ruban AV. (2010). Arabidopsis plants lacking PsbS protein possess photoprotective energy dissipation. *Plant J* 61:283–9.

Johnson MP, Ruban AV. (2011). Restoration of rapidly reversible photoprotective energy dissipation in the absence of PsbS protein by enhanced ΔpH. *J Biol Chem* 286:19973–81.

Johnson X, Steinbeck J, Dent RM et al. (2014). Proton gradient regulation 5-mediated cyclic electron flow under ATP- or redox-limited conditions: A study of Δ*ATpase pgr5* and Δ*rbcL pgr5* mutants in the green alga *Chlamydomonas reinhardtii*. *Plant Physiol* 165:438–52.

Kiss AZ, Ruban AV, Horton P. (2008). The PsbS protein controls the organization of the photosystem II antenna in higher plant thylakoid membranes. *J Biol Chem* 283:3972–8.

Kouřil R, Strouhal O, Nosek L et al. (2014). Structural characterization of a plant photosystem I and NAD(P)H dehydrogenase supercomplex. *Plant J* 77:568–76.

Krüger TPJ, Ilioaia C, Johnson MP et al. (2012). Controlled disorder in plant light-harvesting complex II explains its photoprotective role. *Biophys J* 102:2669–76.

Leoni C, Pietrzykowska M, Kiss AZ et al. (2013). Very rapid phosphorylation kinetics suggest a unique role for Lhcb2 during state transitions in Arabidopsis. *Plant J* 76:236–46.

Li X-P, Björkman O, Shih C et al. (2000). A pigment-binding protein essential for regulation of photosynthetic light harvesting. *Nature* 403:391–5.

Li X-P, Gilmore AM, Caffarri S et al. (2004). Regulation of photosynthetic light harvesting involves intrathylakoid lumen pH sensing by the PsbS protein. *J Biol Chem* 279:22866–74.

Lucinski R, Schmid VHR, Jansson S, Klimmek F. (2006). Lhca5 interaction with plant photosystem I. *FEBS Lett* 580:6485–8.

Matsubayashi T, Wakasugi T, Shinozaki K et al. (1987). Six chloroplast genes (*ndhA-F*) homologous to human mitochondrial genes encoding components of the respiratory chain NADH dehydrogenase are actively expressed: Determination of the splice sites in *ndhA* and *ndhB* pre-mRNAs. *Mol Gen Genet* 210:385–93.

Minagawa J, Takahashi Y. (2004). Structure, function and assembly of Photosystem II and its light-harvesting proteins. *Photosynth Res* 82:241–63.

Munekage Y, Hashimoto M, Miyake C et al. (2004). Cyclic electron flow around photosystem I is essential for photosynthesis. *Nature* 429:579–82.

Munekage Y, Hojo M, Meurer J et al. (2002). PGR5 is involved in cyclic electron flow around photosystem I and is essential for photoprotection in Arabidopsis. *Cell* 110:361–71.

Murata N, Takahashi S, Nishiyama Y, Allakhverdiev SI. (2007). Photoinhibition of photosystem II under environmental stress. *Biochim Biophys Acta* 1767:414–21.

Nagy G, Ünnep R, Zsiros O et al. (2014). Chloroplast remodeling during state transitions in *Chlamydomonas reinhardtii* as revealed by noninvasive techniques in vivo. *Proc Natl Acad Sci USA* 111:5042–7.

Nilkens M, Kress E, Lambrev P et al. (2010). Identification of a slowly inducible zeaxanthin-dependent component of non-photochemical quenching of chlorophyll fluorescence generated under steady-state conditions in Arabidopsis. *Biochim Biophys Acta* 1797:466–75.

Nishikawa Y, Yamamoto H, Okegawa Y et al. (2012). PGR5-dependent cyclic electron transport around PSI contributes to the redox homeostasis in chloroplasts rather than CO_2 fixation and biomass production in rice. *Plant Cell Physiol* 53:2117–26.

Nishimura T, Uno C, Ido K et al. (2014). Identification of the basic amino acid residues on the PsbP protein involved in the electrostatic interaction with photosystem II. *Biochim Biophys Acta* 1837:1447–53.

Niyogi KK, Björkman O, Grossman AR. (1997a) *Chlamydomonas* xanthophyll cycle mutants identified by video imaging of chlorophyll fluorescence quenching. *Plant Cell* 9:1369–80.

Niyogi KK, Björkman O, Grossman AR. (1997b) The roles of specific xanthophylls in photoprotection. *Proc Natl Acad Sci USA* 94:14162–7.

Niyogi KK, Truong TB. (2013). Evolution of flexible non-photochemical quenching mechanisms that regulate light harvesting in oxygenic photosynthesis. *Curr Opin Plant Biol* 16:307–14.

Okegawa Y, Kagawa Y, Kobayashi Y, Shikanai T. (2008). Characterization of factors affecting the activity of photosystem I cyclic electron transport in chloroplasts. *Plant Cell Physiol* 49:825–34.

Onoa B, Schneider AR, Brooks MD et al. (2014). Atomic force microscopy of photosystem II and its unit cell clustering quantitatively delineate the mesoscale variability in Arabidopsis thylakoids. *PLoS One* 9:e101470.

Pagliano C, Nield J, Marsano F et al. (2013). Proteomic characterization and three-dimensional electron microscopy study of PSII-LHCII supercomplexes from higher plants. *Biochim Biophys Acta* 1837:1454–62.

Peers G, Truong TB, Ostendorf E et al. (2009). An ancient light-harvesting protein is critical for the regulation of algal photosynthesis. *Nature* 462:518–21.

Peng L, Fukao Y, Fujiwara M, Takami T, Shikanai T. (2009). Efficient operation of NAD(P)H dehydrogenase requires supercomplex formation with photosystem I via minor LHCI in arabidopsis. *Plant Cell* 21:3623–40.

Peng L, Fukao Y, Myouga F et al. (2011). A chaperonin subunit with unique structures is essential for folding of a specific substrate. *PLoS Biol* 9:e1001040.

Peng L, Shikanai T. (2011). Supercomplex formation with photosystem I is required for the stabilization of the chloroplast NADH dehydrogenase-like complex in arabidopsis. *Plant Physiol* 155:1629–39.

Peng L, Shimizu H, Shikanai T. (2008). The chloroplast NAD(P)H dehydrogenase complex interacts with photosystem I in arabidopsis. *J Biol Chem* 283:34873–9.

Petroutsos D, Terauchi AM, Busch A et al. (2009). PGRL1 participates in iron-induced remodeling of the photosynthetic apparatus and in energy metabolism in *Chlamydomonas reinhardtii*. *J Biol Chem* 284:32770–81.

Pietrzykowska M, Suorsa M, Semchonok DA et al. (2014). The light-harvesting chlorophyll a/b binding proteins Lhcb1 and Lhcb2 play complementary roles during state transitions in Arabidopsis. *Plant Cell* in press. doi:10.1105/tpc.114.127373.

Pinnola A, Dall'Osto L, Gerotto C et al. (2013). Zeaxanthin binds to light-harvesting complex stress-related protein to enhance nonphotochemical quenching in *Physcomitrella patens*. *Plant Cell* 25:3519–34.

Pribil M, Pesaresi P, Hertle A et al. (2010). Role of plastid protein phosphatase TAP38 in LHCII dephosphorylation and thylakoid electron flow. *PLoS Biol* 8:e1000288.

Quick W, Stitt M. (1989). An examination of factors contributing to non-photochemical quenching of chlorophyll fluorescence in barley leaves. *Biochim Biophys Acta* 977:287–96.

Rochaix J-D. (2007). Role of thylakoid protein kinases in photosynthetic acclimation. *FEBS Lett* 581:2768–75.

Ruban AV, Berera R, Ilioaia C et al. (2007). Identification of a mechanism of photoprotective energy dissipation in higher plants. *Nature* 450:575–8.

Satoh K. (2008). Protein-pigments and the photosystem II reaction center: A glimpse into the history of research and reminiscences. *Photosynth Res* 98:33–42.

Sazanov LA, Burrows PA, Nixon PJ. (1998). The plastid *ndh* genes code for an NADH-specific dehydrogenase: Isolation of a complex I analogue from pea thylakoid membranes. *Proc Natl Acad Sci USA* 95:1319–24.

Shapiguzov A, Ingelsson B, Samol I et al. (2010). The PPH1 phosphatase is specifically involved in LHCII dephosphorylation and state transitions in arabidopsis. *Proc Natl Acad Sci USA* 107:4782–7.

Shikanai T. (2007). Cyclic electron transport around photosystem I; genetic approaches. *Annu Rev Plant Biol* 58:199–217.

Shikanai T. (2014). Central role of cyclic electron transport around photosystem I in the regulation of photosynthesis. *Curr Opin Biotech* 26:25–30.

Shikanai T, Aro E-M. (2014). Evolution of photosynthetic NDH-1: structure and physiological function. *Adv Photosynth Respir* in press.

Shikanai T, Endo T, Hashimoto T et al. (1998). Directed disruption of the tobacco *ndhB* gene impairs cyclic electron flow around photosystem I. *Proc Natl Acad Sci USA* 95:9705–9.

Sonoike K. (2011). Photoinhibition of photosystem I. *Physiol Plant* 142:56–64.

Sugimoto K, Okegawa Y, Tohri A. et al. (2013). A single amino acid alteration in PGR5 confers resistance to antimycin A in cyclic electron transport around PSI. *Plant Cell Physiol* 54:1525–34.

Suorsa M, Järvi S, Grieco M et al. (2012). PROTON GRADIENT REGULATION5 is essential for proper acclimation of Arabidopsis photosystem I to naturally and artificially fluctuating light conditions. *Plant Cell* 24:2934–48.

Tagawa K, Tsujimoto HY, Arnon DI. (1963). Role of chloroplast ferredoxin in the energy conversion process of photosynthesis. *Proc Natl Acad Sci USA* 49:567–72.

Takahashi H, Clowez S, Wollman F-A, Vallon O, Rappaport F. (2013). Cyclic electron flow is redox-controlled but independent of state transition. *Nat Commun* 4:1954.

Tikkanen M, Aro E-M. (2014). Integrative regulatory network of plant thylakoid energy transduction. *Trends Plant Sci* 19:10–7.

Tikkanen M, Grieco M, Aro E-M. (2011). Novel insights into plant light-harvesting complex II phosphorylation and "state transitions". *Trends Plant Sci* 16:126–31.

Tokutsu R, Kato N, Bui KH et al. (2012). Revisiting the supramolecular organization of photosystem II in *Chlamydomonas reinhardtii*. *J Biol Chem* 287:31574–81.

Tokutsu R, Minagawa J. (2013). Energy-dissipative supercomplex of photosystem II associated with LHCSR3 in *Chlamydomonas reinhardtii*. *Proc Natl Acad Sci USA* 110:10016–21.

Ueda M, Kuniyoshi T, Yamamoto Y et al. (2012). Composition and physiological function of the chloroplast NADH dehydrogenase-like complex in *Marchantia polymorpha*. *Plant J* 72:683–93.

Umena Y, Kawakami K, Shen J-R, Kamiya N. (2011). Crystal structure of oxygen-evolving photosystem II at a resolution of 1.9 Å. *Nature* 473:55–60.

Vass I. (2012). Molecular mechanisms of photodamage in the Photosystem II complex. *Biochim Biophys Acta* 1817:209–17.

Vinyard DJ, Ananyev GM, Dismukes GC. (2013). Photosystem II: The reaction center of oxygenic photosynthesis. *Annu Rev Biochem* 82:577–606.

Walters RG, Horton P. (1991). Resolution of components of non-photochemical chlorophyll fluorescence quenching in barley leaves. *Photosynth Res* 27:121–33.

Walters RG, Horton P. (1993). Theoretical assessment of alternative mechanisms for non-photochemical quenching of PS II fluorescence in barley leaves. *Photosynth Res* 36:119–39.

Wang C, Yamamoto H, Shikanai T. (2015). Role of cyclic electron transport around photosystem I in regulating proton motive force. *Biochim. Biophys. Acta* 1847:931–8.

Wientjes E, Drop B, Kouřil R et al. (2013a). During state 1 to state 2 transition in *Arabidopsis thaliana*, the photosystem II supercomplex gets phosphorylated but does not disassemble. *J Biol Chem* 288:32821–6.

Wientjes E, Oostergetel GT, Jansson S et al. (2009). The role of Lhca complexes in the supramolecular organization of higher plant photosystem I. *J Biol Chem* 284:7803–10.

Wientjes E, van Amerongen H, Croce R. (2013b). LHCII is an antenna of both photosystems after long-term acclimation. *Biochim Biophys Acta* 1827:420–6.

Wollman F-A. (2001). State transitions reveal the dynamics and flexibility of the photosynthetic apparatus. *EMBO J* 20:3623–30.

Yamamoto H, Peng L, Fukao Y, Shikanai T. (2011). An Src homology 3 domain-like fold protein forms a ferredoxin-binding site for the chloroplast NADH dehydrogenase-like complex in Arabidopsis. *Plant Cell* 23:1480–93.

Yamamoto H, Shikanai T. (2013). In planta mutagenesis of Src homology 3 domain-like fold of NdhS, a ferredoxin-binding subunit of the chloroplast NADH dehydrogenase-like complex in Arabidopsis. A conserved Arg-193 plays a critical role in ferredoxin binding. *J Biol Chem* 288:36328–37.

CHAPTER 8

Microbial Redox Proteins and Protein Complexes for Extracellular Respiration

Liang Shi, Ming Tien, James K. Fredrickson, John M. Zachara and Kevin M. Rosso

CONTENTS

8.1 INTRODUCTION

Prokaryotes exhibit extremely versatile respiration capabilities. Most eukaryotes, including humans, use only molecular oxygen (O_2) as the terminal electron acceptor for aerobic respiration, while some eukaryotes, such as fungi, diatoms and foraminifers, can use nitrate (NO_3^-) and elemental sulphur (S^0) as the terminal electron acceptors for anaerobic respiration (Risgaard-Petersen et al., 2006; Abe et al., 2007; Pina-Ochoa et al., 2010; Kamp et al., 2011; Stief et al., 2014). Thus, to date, O_2, NO_3 and S^0 have been the only known terminal electron acceptors for eukaryotic respiration. In contrast to eukaryotes, prokaryotes use much more diverse terminal electron acceptors for respiration. In addition to O_2, NO_3^- and S^0, other terminal electron acceptors used for prokaryotic respiration include nitrite (NO_2^-), trimethylamine N-oxide [TMAO, $(CH_3)_3NO$], sulphate (SO_4^{2-}), sulphite (SO_3^{2-}), thiosulphate ($S_2O_3^{2-}$), tetrathionate ($S_4O_6^{2-}$), dimethyl sulfoxide [DMSO, $(CH_3)_2SO$], fumarate, electrodes, chlorinated organic compounds, humic substances and different metals and metalloids, such as arsenate [As(V)], cobalt [Co(III)], chromium [(Cr(VI)], ferric iron [Fe(III)]/manganese [Mn(III/IV)] and their oxides, selenate/selenite [Se(VI/IV)], uranium [U(VI)] and vanadium [V(V)] (Lovley et al., 1991, 1996; Caccavo et al., 1994; Stolz and Oremland, 1999; Richardson, 2000; Amend and Shock, 2001; Nealson et al., 2002; Ortiz-Bernad et al., 2004; Torres et al., 2010; Shirodkar et al., 2011; Wagner et al., 2012). In eukaryotes, all the respiratory terminal reductases are located in the mitochondria within the eukaryotic cells (Muller et al., 2012). In prokaryotes, however, the respiratory terminal reductases are found not only inside but also outside the cells.

One reason that certain groups of prokaryotic microorganisms develop the extracellular respiration capabilities is that some terminal electron acceptors exist as solids that are too large to enter the cells. This is especially true for Fe(III) and Mn(III/IV) oxides, which are insoluble in water at circumneutral pH and in the absence of strong complexing ligand. The solid-phase Fe(III) and Mn(III/IV) oxides cannot cross the microbial cell envelope and thus remain external to the microbial cells. Similarly, electrodes and high molecular weight humic substances are external to the microbial cells, and their respirations occur extracellularly. Extracellular respiration is also considered as a detoxification mechanism. After microbial reduction of Cr(VI), Se(VI/IV) and U(VI), their respective reduced products

Cr_2O_3, Se(0) and UO_2 are poorly soluble in water and often form nano-sized particles that may interfere with the normal functions of microbial cells. Thus, extracellular respirations of Cr(VI), Se(VI/IV) and U(VI) could avoid the accumulation of these water-insoluble products inside the microbial cells to mitigate their negative impacts on the microbial functions (Belchik et al., 2011; Cologgi et al., 2011).

Over the past decade, significant progress has been made in identifying and characterizing microbial redox proteins and protein complexes that directly participate in the extracellular electron transfer reactions for respiration of the terminal electron acceptors external to microbial cells. These proteins often form pathways for transferring electrons from the quinone/quinol pool in the cytoplasmic membrane, across the entire cell envelope to the extracellular electron acceptors. It should be noted that no energy-conserving processes have been observed so far during the electron transfer processes mediated by these extracellular pathways that are beyond the quinone/quinol pool in the cytoplasmic membrane. Thus, the function of these microbial extracellular electron transfer pathways is primarily to dissipate the electrons accumulated in the cytoplasmic membrane after oxidation of organic matter or H_2, which is analogous to the electrical ground system.

This chapter examines and reviews first the Mtr (i.e. metal-reducing) and DMSO-respiring pathways of the metal- and DMSO-respiring bacterium *Shewanella oneidensis* MR-1 and then the proteins and protein complexes involved in extracellular respiration by the metal-respiring bacteria *Geobacter* spp. and other bacteria. For brevity, this chapter focuses on the recent progress in identifying and characterizing the proteins and protein complexes that are directly involved in extracellular respiration by these bacteria. The reader is referred to other excellent reviews for the historic details of related topics (Nealson and Saffarini, 1994; Richardson, 2000; Nealson et al., 2002; Lovley et al., 2004, 2011; Gralnick and Newman, 2007; Shi et al., 2007, 2009, 2012a; Fredrickson et al., 2008; Richter et al., 2012).

8.2 METAL-REDUCING AND DMSO-RESPIRING EXTRACELLULAR ELECTRON TRANSFER PATHWAYS OF *SHEWANELLA ONEIDENSIS* MR-1

The Gram-negative and gammaproteobacterium *S. oneidensis* MR-1 is among the first microorganisms discovered to respire on solid-phase Fe(III) and Mn(III/IV) oxides (Myers and Nealson, 1988, 1990). In addition to Fe(III) and Mn(III/IV) oxides, *S. oneidensis* MR-1 also respires on O_2, NO_3^-, NO_2^-, TMAO, S^0, SO_4^{2-}, SO_3^{2-}, $S_2O_3^{2-}$, DMSO, fumarate and electrodes (Myers and Nealson, 1988, 1990; Nealson et al., 2002; Gralnick et al., 2006; Burns and DiChristina, 2009; Shirodkar et al., 2011; Renslow et al., 2013). Thus, *S. oneidensis* MR-1 possesses the most flexible respiration capabilities for a single organism that has been characterized to date. Most of the respiratory terminal reductases of *S. oneidensis* MR-1 are localized in either the cytoplasmic membrane or periplasm, while the terminal reductases for Fe(III) and Mn(III/IV) oxides, electrodes and DMSO are localized on the exterior surface of the bacterial outer membrane. The Mtr pathway and DMSO-reducing pathway of *S. oneidensis* MR-1 transfer electrons from the quinone/quinol pool in the cytoplasmic membrane, across the periplasm and outer membrane to the surfaces of metal oxides and electrodes and DMSO, respectively. The protein components identified to date include five multiheme *c*-type cytochromes (or *c*-Cyts; CymA, Fcc_3, MtrA, MtrC and OmcA) and a porin-like and outer-membrane protein (MtrB) for the Mtr pathway; and two multiheme *c*-Cyts (CymA and DmsE, an MtrA homologue), a porin-like outer-membrane protein (DmsF, an MtrB homologue), an molybdopterin cluster protein (DmsA) and an Fe-S cluster protein (DmsB) for the DMSO-respiring pathway. Because they share a certain degree of similarity in their protein components, both Mtr and DMSO-respiring pathways are described in this section.

8.2.1 CymA: Electron Transfer Hub

CymA is a cytoplasmic membrane *c*-Cyt that belongs to the NapC/NrfH family of quinol dehydrogenases. In addition to respiration of Fe(III) and Mn(III/IV) oxides, electrodes and DMSO, CymA is also required for respiration of NO_3, NO_2 and fumarate by *S. oneidensis* MR-1 and As(V) respiration by *Shewanella* sp. strain ANA-3 and *S. putrefaciens* CN-32 (Myers and Myers, 1997, 2000; Schwalb et al., 2003; Lies et al., 2005; Murphy and Saltikov, 2007). Thus, CymA is an electron transfer hub that is critical to the anaerobic respiration flexibility of *Shewanella* spp. (Marritt et al., 2012a,b).

CymA contains a transmembrane domain that anchors CymA to the cytoplasmic membrane and a periplasmic domain that binds four hemes covalently. Characterizations of purified CymA of *S. oneidensis* MR-1 reveal three low-spin hemes whose Fe atoms are coordinated by two histidine residues. The fourth heme is a high-spin heme whose Fe atom is coordinated by a histidine residue and another axial ligand that is either empty, a H_2O molecule, or a hydroxide. At neutral pH, the measured midpoint redox potentials (E_m) are −110, −190 and −265 mV versus standard hydrogen electrode (SHE) for respective three low-spin hemes and −240 mV versus SHE for the high-spin heme. This high-spin heme is close to the quinone-binding site and is suggested to be the heme I (Marritt et al., 2012a).

Under anaerobic conditions, the quinone/quinol pool in the cytoplasmic membrane of *S. oneidensis* MR-1 consists of mainly menaquinones and methylmenaquinones (Myers and Myers, 1997, 2000). Purified CymA from *S. oneidensis* MR-1 possesses a tightly bound menaquinone-7 as a co-factor that is required for the catalytic activity of CymA. It is proposed that CymA has two quinone-binding sites of which the $Q_{MQ\text{-}7}$ site binds a menaquinone-7 tightly as a co-factor, while the Q_w site weakly binds another menaquinone-7 or a different quinone molecule, such as menaquinone-0 or ubiquinone-10 that is a dominant quinone molecule under aerobic condition. Thus, the reduction of quinone or oxidation of quinol by CymA is believed to occur between the menaquinone-7 at the $Q_{MQ\text{-}7}$ site and the quinone molecule at the Q_w site via a two-electron transfer reaction. McMillan et al. (2012) hypothesized that strong binding of menaquinone-7 at the $Q_{MQ\text{-}7}$ site would stabilize the reaction intermediate, semiquinone, of the bound menaquinone-7 and consequently lower the reactivity of the semiquinone, the side reactions of which could generate detrimental radical oxygen species.

By itself, the purified CymA can only reduce quinone, such as menaquinone-7, and exhibits no quinol oxidation activity, which contrasts with its suggested role in extracellular respiration. However, formation of a stable protein complex with the fumarate reductase Fcc_3, which is one of the native electron transfer partners of CymA in the periplasm, reverses the catalytic activity of CymA towards oxidation of menaquinol-7. This is the first time that protein–protein interaction has been shown to regulate the direction of catalysis and electron transfer of a redox enzyme complex. Association with Fcc_3 is believed to increase redox potential of the solvent-exposed heme of CymA, which alters the direction of catalysis and electron transfer of CymA in favour of oxidation of menaquinol-7 (McMillan et al., 2013). After quinol oxidation at the Q_w site, the released electrons are proposed to be transferred to Fcc_3 via the $Q_{MQ\text{-}7}$ site, heme I, other hemes and the solvent-exposed heme of CymA. Thus, for CymA to oxidize quinol in the cytoplasmic membrane during extracellular respiration, additional periplasmic proteins are required to activate the quinol oxidation activity of CymA via protein–protein interaction.

8.2.2 CymA-Associated Periplasmic Proteins

In addition to Fcc_3, CymA interacts with other periplasmic redox proteins. Transient protein–protein interaction is observed between purified CymA and purified small tetraheme cytochrome (STC) (a periplasmic 4-heme *c*-Cyt) (Firer-Sherwood et al., 2011; Louro and Paquete, 2012; Fonseca et al., 2013). CymA is also believed to transfer electrons directly to nitrate reductase NapAB and nitrite reductase NrfA in the periplasm. The periplasmic *c*-Cyt MtrA was originally thought to receive electrons directly from CymA (Shi et al., 2007, 2009). However, MtrA was later found to be an integral part of the MtrABC trans-outer membrane complex in which MtrA is believed to be inserted into MtrB (Ross et al., 2007; Hartshorne et al., 2009). Given that the length of MtrA is ~104 Å and the periplasmic width of *S. oneidensis* MR-1 is 235 ± 37 Å (Dohnalkova et al., 2011; Firer-Sherwood et al., 2011), it is unlikely that CymA in the cytoplasmic membrane directly interacts with MtrA in the outer membrane.

Protein–protein interactions are also observed between purified MtrA and purified STC and Fcc_3 (Firer-Sherwood et al., 2011; Fonseca et al., 2013). Thus, STC or Fcc_3 is proposed to bridge the periplasmic gap between CymA in the cytoplasmic membrane and MtrA in the outer membrane. However, deletion of the gene for STC or Fcc_3 has no impact on the reduction of Fe(III) oxide by *S. oneidensis* MR-1 (Schuetz et al., 2009; Coursolle and Gralnick, 2010). These results suggest that (1) the periplasmic proteins other than STC and Fcc_3 are involved in electron transfer between CymA and MtrA, (2) both STC and Fcc_3 are involved in electron transfer between CymA and MtrA or both (1) and (2). Because it interacts with CymA homologue NrfH and MtrA homologue NrfB, NrfA is suggested to mediate electron transfer between CymA and MtrA (Shi et al., 2007, 2012a). Consistent with this suggestion, MtrA expressed in *Escherichia coli* receives electrons from NrfA (Pitts et al., 2003).

8.2.3 MtrABC: First Characterized Trans-Outer Membrane Porin-Cytochrome Electron Conduit

MtrA is a periplasmic 10-heme *c*-Cyt (Pitts et al., 2003; Shi et al., 2005), while MtrB is a large, porin-like, outer-membrane protein with 28 predicted trans-outer membrane motifs (White et al., 2013), and MtrC is an outer-membrane 10-heme *c*-Cyt that is cell surface-exposed

(Myers and Myers, 2003; Shi et al., 2006, 2008; Hartshorne et al., 2007; Lower et al., 2009; Reardon et al., 2010; Wang et al., 2013). MtrAB can be chemically cross-linked *in vivo,* and MtrABC can be co-purified from the outer membrane fraction as a protein complex (Ross et al., 2007; Hartshorne et al., 2009). The isolated MtrABC complex is a 20-heme heterotrimer (Hartshorne et al., 2009). When it is inserted into the lipid bilayer of proteoliposomes, the topology of MtrABC complex is identical to that in the bacterial outer membrane. MtrABC reconstituted in the proteoliposomes can transfer electrons across the lipid bilayer of the liposomes to the surfaces of Fe(III) oxides with a rate that is high enough to support *in vivo* respiration of Fe(III) oxides by *S. oneidensis* MR-1 (White et al., 2012, 2013). It is hypothesized that MtrB serves as a scaffold through which MtrA and MtrC are inserted. After receiving electrons from the periplasmic *c*-Cyts, such as Fcc_3, MtrA transfers electrons across the outer membrane to the surface-exposed MtrC that then transfers electrons to the surfaces of metal oxides and electrodes, small and water-soluble electron carriers (i.e. flavins) and other proteins, such as OmcA (Hartshorne et al., 2009; Richardson et al., 2012; Shi et al., 2012a; White et al., 2013). Thus, the MtrABC porin-cytochrome complex is the first characterized extracellular electron conduit in the bacterial outer membrane.

8.2.4 OmcA and the Outer-Membrane *c*-Cyts MtrF and UndA of *Shewanella*

8.2.4.1 Association of OmcA with MtrC

OmcA is involved in Fe(III) oxide reduction by *S. oneidensis* MR-1 (Gorby et al., 2006; Mitchell et al., 2012). Like MtrC, OmcA is an outer-membrane 10-heme *c*-Cyt that is cell surface-exposed (Myers and Myers, 2003; Shi et al., 2006, 2008; Lower et al., 2009; Reardon et al., 2010). Both MtrC and OmcA are translocated across the outer membrane by the bacterial type II secretion system (DiChristina et al., 2002; Donald et al., 2008; Shi et al., 2008). *In vivo,* MtrC and OmcA can be chemically cross-linked on the bacterial cell surface and can also be co-purified even without chemical cross-linking (Shi et al., 2006; Ross et al., 2007; Tang et al., 2007; Zhang et al., 2008, 2009). *In vitro,* the purified MtrC and OmcA form a stable complex with a $K_d < 500$ nM and a stoichiometry of 1:2 (Shi et al., 2006). Purified OmcA can transfer electrons to hematite (an Fe(III) oxide) via direct binding (Xiong et al., 2006; Lower et al., 2007; Meitl et al., 2009; Johs et al., 2010). Moreover, the voltammograms of purified MtrC and OmcA on the hematite electrodes are nearly identical to those of *S. oneidensis* MR-1 with only MtrC and OmcA, respectively, and both MtrC and OmcA possess a hematite-binding motif (i.e. Ser/Thr-Pro-Ser/Thr) (Lower et al., 2008; Meitl et al., 2009; Edwards et al., 2014), showing that like MtrC, OmcA can also function as a terminal reductase for Fe(III) oxides.

8.2.4.2 Molecular Structures of OmcA, MtrF and UndA

OmcA structure determined at a resolution of 2.7 Å supports the role of OmcA as a terminal reductase for Fe(III) oxide. The overall structure of OmcA is very similar to that of MtrF and UndA, a 10-heme and an 11-heme outer-membrane *c*-Cyt of *Shewanella,* respectively, despite the fact that they share low sequence identity (Clarke et al., 2011; Shi et al., 2011; Edwards et al., 2012). They are all folded into four distinct domains.

Domains I and III each contains a Greek key split-barrel structure, while domains II and IV each binds five hemes covalently, with the exception of UndA whose domain IV binds 6 hemes. The four domains fold together so that domains II and IV are packed to form a central core with domains I and III flanking either side (Figure 8.1a). The heme groups are organized into a unique *staggered cross* of short and long heme chains in which heme 5 and hemes 10/11 comprise the termini of the long chain (Clarke et al., 2011; Edwards et al., 2012, 2014). All heme groups are closely packed, each within 7 Å of its nearest neighbour, enabling the prospect of rapid intra-molecular electron transfer (Figure 8.1b) (Edwards et al., 2014).

8.2.4.3 Molecular Simulations of MtrF

Molecular dynamics simulations show that the calculated free energy landscape of MtrF is nearly symmetric in terms of a heme chain from heme 5 to heme 10 (Figure 8.1b), which suggests that intra-molecular electron transfer between heme 5 and heme 10 is bidirectional (Breuer et al., 2012a,b). Indeed, MtrABC transfer electrons across the lipid bilayer of liposomes both outside-in and inside-out and the Mtr pathway of *S. oneidensis* MR-1 can also transfer electrons from electrodes to CymA in the cytoplasmic membrane (Hartshorne et al., 2009; Ross et al., 2011; White et al., 2013). Further quantum mechanics/molecular mechanics simulations show that the calculated maximum one-electron transfer rate along heme chains in MtrF is $10^4 - 10^5\ s^{-1}$ (Breuer et al., 2014), which is consistent with the rates measured with the MtrABC proteoliposome system (White et al., 2013). This rapid electron transfer is achieved in part by stacking the heme pairs, which increases the electronic coupling and thus the electron transfer probability, in regions of the heme chains for which electron transfer is thermodynamically uphill. This design principle that balances the redox potentials and strengths of electronic coupling between heme pairs is believed to be applicable to OmcA and UndA, which share similar structural folds and heme arrangements with MtrF (Breuer et al., 2014).

8.2.4.4 OmcA Dimer and Hematite-Binding Motif

Unlike MtrF and UndA, OmcA crystallizes as a dimer in a way that is often observed for other protein homodimers, in which two OmcA monomers interact with each other at the area close to heme 5 of domain II. The inter-molecular distance between two heme 5s is 9 Å. Thus, formation of an OmcA dimer results in a branched, 20-heme network of 169 Å long (Figure 8.2a and b) (Edwards et al., 2014). Formation of an OmcA dimer is also consistent with the results showing that one MtrC interacts with two OmcAs (Shi et al., 2006). Structural results reveal that the hematite-binding motif (i.e. Thr-Pro-Ser) of OmcA is on the OmcA surface and close to the solvent-exposed heme 10 of OmcA (Figure 8.3). Thus, the OmcA dimer may receive electrons from MtrC at the heme 5s of the OmcA dimer. The electrons then flow along the heme network to the heme 10s of the OmcA dimer. The hematite-binding motifs adjacent to the heme 10s bind to the surfaces of Fe(III) oxides, which could bring the heme 10s close to the Fe(III) oxides for rapid electron transfer from the heme 10s to the oxide surface (Figure 8.3) (Edwards et al., 2014).

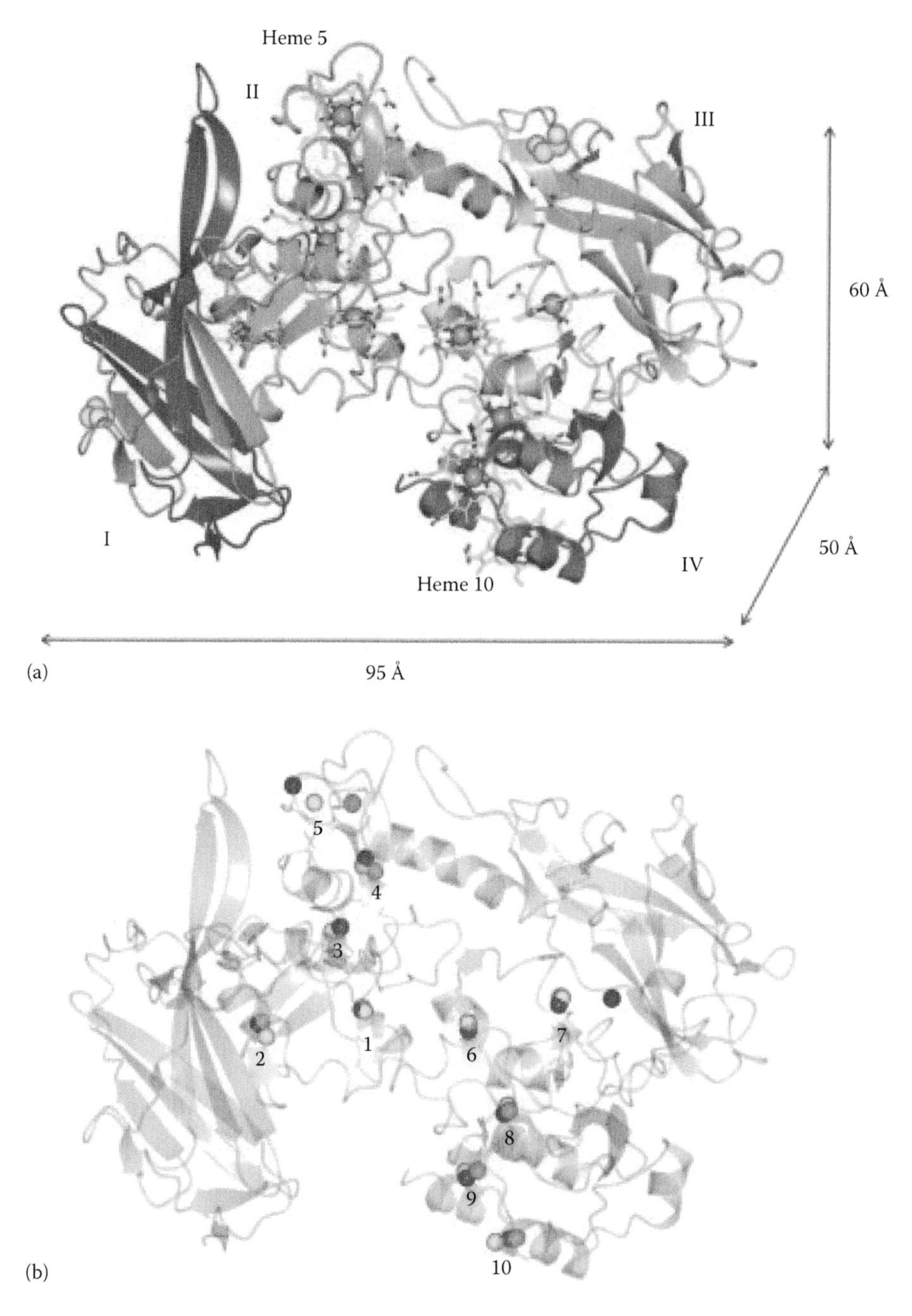

FIGURE 8.1 OmcA structure. Crystal structure of OmcA at 2.7 Å resolution (PDB ID: 4LMH). (a) The polypeptide chains are shown in ribbon representation. The domains are indicated by Roman numerals. The Fe atoms of the hemes are represented as grey spheres, and the porphyrin rings of the hemes are shown as sticks. The crystal structure disulphide bonds are represented as grey spheres. (b) Superposition of Fe atoms from the structures of OmcA (light grey), MtrF (medium grey) and UndA (dark grey). The polypeptide chain of OmcA is shown as a transparent carton, and the Fe atoms are shown as spheres. (Reprinted from Edwards, M.J. et al., *FEBS Lett.*, 588, 1886–1890, 2014. With permission.)

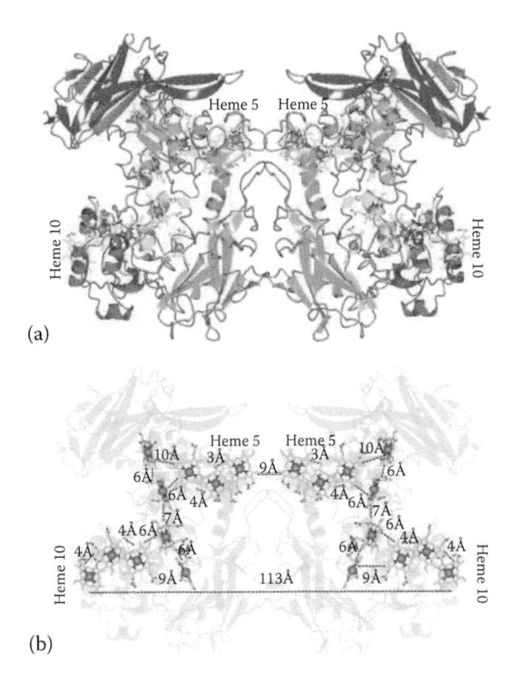

FIGURE 8.2 Dimer formation and heme arrangement within OmcA. (a) Arrangement of monomers in the potential OmcA dimer within the crystallographic asymmetric unit. The polypeptide chain of the OmcA molecules are shown in the same way as in Figure 8.1. (b) Heme-heme distance of potential OmcA dimers found within the asymmetric unit. The polypeptide chains are shown as a transparent grey background in cartoon representation. Hemes of both OmcA molecules are shown as sticks with Fe atoms shown as sphere. (Reprinted from Edwards, M.J. et al., *FEBS Lett.*, 588, 1886–1890, 2014. With permission.)

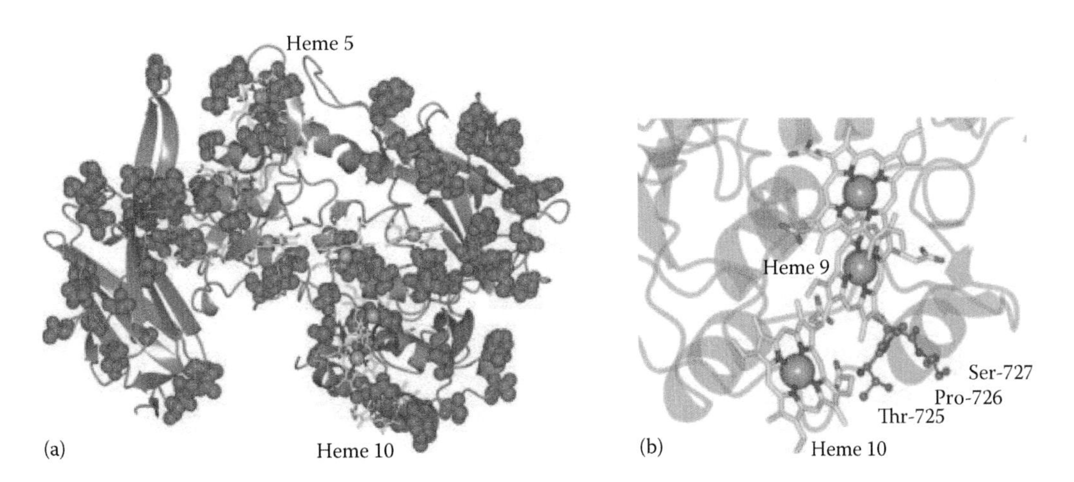

FIGURE 8.3 Hydroxylated surface of OmcA showing motif. (a) Distribution of serine and threonine residues of OmcA. Spheres indicate the positions of atoms of serine/threonine residues. (b) Residues of the proposed Fe(III) oxide binding motif of OmcA. (Data from Lower, B.H. et al., *Environ. Sci. Technol.*, 42, 3821–3827, 2008.) The residues of the Fe(III) oxide binding motif are shown in ball and stick representation. (Reprinted from Edwards, M.J. et al., *FEBS Lett.*, 588, 1886–1890, 2014. With permission.)

8.2.5 Proposed Mtr Extracellular Electron Transfer Pathway

The proposed Mtr extracellular electron transfer pathway in *S. oneidensis* MR-1 begins at the cytoplasmic membrane where *c*-Cyt CymA oxidizes quinol and then transfers the electrons to the periplasmic *c*-Cyt Fcc_3. Fcc_3 relays the electrons to *c*-Cyt MtrA either directly or indirectly via other periplasmic redox proteins. MtrA is believed to be inserted into the porin-like outer-membrane protein MtrB. Together, MtrAB transfer electrons across the outer membrane to the cell surface-exposed *c*-Cyt MtrC. On the bacterial cell surface, MtrC and the outer-membrane *c*-Cyt OmcA form a protein complex that transfers electrons directly to the surfaces of Fe(III)/Mn(III, IV) oxides and electrodes via their solvent-exposed hemes, such as heme 10 of OmcA (Figure 8.4). Although our understanding of the

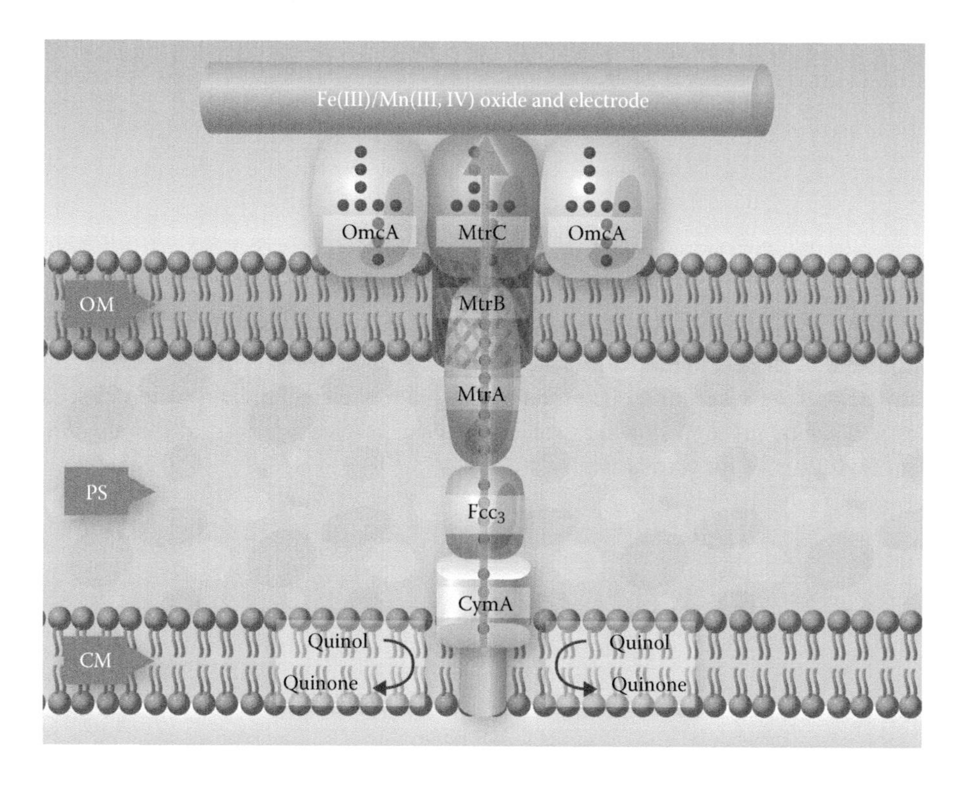

FIGURE 8.4 The proposed Mtr extracellular electron transfer pathway of *S. oneidensis* MR-1. The proposed pathway includes a 4-heme *c*-Cyt CymA; a periplasmic 10-heme *c*-Cyt MtrA; a porin-like outer-membrane (OM) protein MtrB; two OM 10-heme *c*-Cyts MtrC and OmcA; and probably a periplasmic 4-heme *c*-Cyt Fcc_3. Association with Fcc_3 activates the quinol oxidase activity of CymA. After oxidation of quinol in the cytoplasmic membrane (CM), CymA transfers electrons to Fcc_3 that then relays the electrons across the periplasm (PS) to MtrA either directly or indirectly via other periplasmic redox proteins. MtrABC form a trans-OM protein complex for mediating electron transfer across the OM. On the outmost of the OM, MtrC also interacts with OmcA and both of them transfer electrons directly to the surfaces of Fe(III)/Mn(III, IV) oxides and electrodes via their solvent-exposed hemes, such as heme 10 of OmcA. The arrow indicates the electron transfer path during extracellular respiration of Fe(III)/Mn(III, IV) oxide and electrode. The non-protein components flavins whose roles in the pathway remain unclear are not included.

Mtr pathway in *S. oneidensis* MR-1 is far from complete, it is the best characterized electron transfer pathway used for microbial extracellular respiration.

8.2.6 Flavins: Electron Shuttles or Co-Factors?

In addition to the protein components, flavins (e.g. flavin mononucleotide [FMN] and riboflavin) secreted by *S. oneidensis* MR-1 are involved in extracellular electron transfer to the surfaces of Fe(III) oxides and electrodes (Marsili et al., 2008; von Canstein et al., 2008; Ross et al., 2009; Coursolle et al., 2010; Kotloski and Gralnick, 2013). Because chemically reduced FMN and riboflavin transfer electrons to Fe(III) oxides directly (Marsili et al., 2008; von Canstein et al., 2008; Shi et al., 2012c, 2013) and oxidized FMN transiently interact with MtrC and OmcA (Paquete et al., 2014), they are proposed to function as diffusible electron carriers that shuttle electrons between MtrC/OmcA and surfaces of Fe(III) oxides and electrodes (Brutinel and Gralnick, 2012). In this case, MtrC and OmcA are proposed to reduce flavins via a 2-electron transfer reaction, and, in turn, the reduced flavins then transfer two electrons to Fe(III) oxides and electrodes. Under this paradigm, MtrC and OmcA would function as flavin reductases rather than Fe(III) oxide reductases. However, the E_m of free FMN is −220 mM versus SHE, which is lower than that of MtrC and OmcA, making it thermodynamically unfavourable to reduce free FMN by MtrC and OmcA (Okamoto et al., 2013).

Alternative to the *shuttle* paradigm, FMN and riboflavin are proposed to be bound by MtrC and OmcA as co-factors, respectively. This *co-factor* paradigm is based on the observations of flavin semiquinones in the monolayer biofilm of *S. oneidensis* MR-1 that is formed on the electrode surface. In that biofilm, MtrC binds FMN semiquinone, while OmcA binds riboflavin semiquinone. The observed peak potential in differential pulse voltammogram (E_p) of flavin semiquinones in the biofilms is −145 mV versus SHE that is 115 mV higher than the E_p of free flavins, which is −260 mV versus SHE. This positively shifted E_p makes it thermodynamically more favourable for MtrC and OmcA to reduce flavin semiquinones than free flavins. Calculated electron transfer rate of reducing flavin semiquinones by MtrC and OmcA is $10^3 - 10^5$ fold faster than that of reducing free flavins. Thus, MtrC and OmcA are proposed to be the terminal reductases of Fe(III) oxides with bound flavin semiquinones as co-factors for mediating two 1-electron transfers to Fe(III) oxides and electrodes (Okamoto et al., 2013, 2014; Richardson et al., 2013). Consistent with the *co-factor* paradigm, the electron transfer rates by MtrABC in proteoliposomes to Fe(III) oxides in the absence of flavins are much faster than those mediated by chemically reduced flavins (Shi et al., 2012c, 2013; White et al., 2013). However, it should be pointed out that the determined molecular structure of OmcA contains no bound flavin molecule (Figure 8.1) (Edwards et al., 2014) and the precise locations of MtrC and OmcA for interacting with flavins remain undermined. Thus, the exact role of flavins in the Mtr pathway is currently unclear.

8.2.7 *Shewanella* Nanowires: Outer Membrane and Periplasmic Extensions with Mtr Proteins

Under the electron-acceptor-limited condition, *S. oneidensis* MR-1 produces extracellular appendages called *nanowires* that are electrically conductive (Gorby et al., 2006; El-Naggar et al., 2008, 2010; Leung et al., 2013). Filamentous proteins, such as pilin protein PilA, were originally thought to be the structural components of *Shewanella* nanowires (Gorby et al., 2006). Deletion of the gene for PilA, however, has no impact on the formation of *Shewanella* nanowires. Moreover, quantitative-PCR analyses show no change of mRNA levels of *pilA* or *pilE* after switching to the electron-acceptor-limited condition. Thus, pili and pilin proteins are not essential components of *S. oneidensis* MR-1 nanowires (Pirbadian et al., 2014).

Outer-membrane vesicles are formed by *Shewanella* cells growing under the electron-acceptor-limited conditions and these vesicles often associate with *Shewanella* nanowires (Gorby et al., 2008). *In vivo* fluorescence measurements show that the nanowires formed on *S. oneidensis* MR-1 cells are labelled by FM 4-64FX, a styryl dye that specifically stains lipid bilayers. This reveals that the *Shewanella* nanowires are the cellular structures enveloped by the outer membrane (Pirbadian et al., 2014). Furthermore, *in vivo* fluorescence measurements show the presence of the periplasmic proteins, but not the cytoplasmic proteins in the *Shewanella* nanowires (Pirbadian et al., 2014). MtrC and OmcA are required for the electrical conductivity of *Shewanella* nanowires as the nanowires isolated from a *S. oneidensis* MR-1 mutant lacking both MtrC and OmcA are not electrically conductive (Gorby et al., 2006). Quantitative-PCR analyses also show significant increases of mRNA abundances of Mtr genes, such as *mtrA*, *mtrC* and *omcA*, in *S. oneidensis* MR-1 after changing to the electron-acceptor-limited condition. Moreover, immunolabelling with MtrC- or OmcA-specific antibodies co-localizes MtrC and OmcA on the surface of *S. oneidensis* MR-1 nanowires. Collectively, these results demonstrate that the nanowires of *S. oneidensis* MR-1 are the extensions of the outer membrane and periplasm with Mtr proteins. It is proposed that, after receiving electrons from the periplasmic MtrA, the surface-exposed MtrC and OmcA mediate electron conductance along the *Shewanella* nanowires (Figure 8.5) (Pirbadian et al., 2014). This hypothesis is consistent with the previous observations that *Shewanella* nanowires transfer electrons via microbial redox chains by a multistep hopping mechanism (Pirbadian and El-Naggar, 2012).

MtrC, OmcA and their homologues are also associated with extracellular polymeric substances (EPS) where they have been implicated in extracellular reduction of Cr(VI), U(VI) and Fe(III) oxides and co-localized with hematite, nanoparticles that contain Cr(IV) and U(IV)O_2, and Fe(II)-containing secondary mineral phases. These EPS-associated MtrC and OmcA may be part of the *Shewanella* nanowires where they participate in extracellular reduction of the Fe(III) oxides, Cr(VI) and U(VI) distant from the surface of *Shewanella* cells (Rosso et al., 2003; Marshall et al., 2006; Bose et al., 2009; Reardon et al., 2010; Belchik et al., 2011; Cao et al., 2011a,b; Shi et al., 2012b; Wang et al., 2013; Pirbadian et al., 2014).

8.2.8 Mtr Homologues: Their Distribution in Different Bacteria and Biological Functions

8.2.8.1 In Other Metal-Reducing Bacteria

Mtr homologues are found in all metal-reducing *Shewanella* spp. analyzed (Fredrickson et al., 2008; Shi et al., 2011). In the genome of *S. oneidensis* MR-1, the genes that encode MtrABC and OmcA are clustered in the same region (i.e. the *mtr* gene clusters) that also includes *mtrD* (an *mtrA* homologue), *mtrE* (an *mtrB* homologue) and *mtrF* (an *mtrC* homologue).

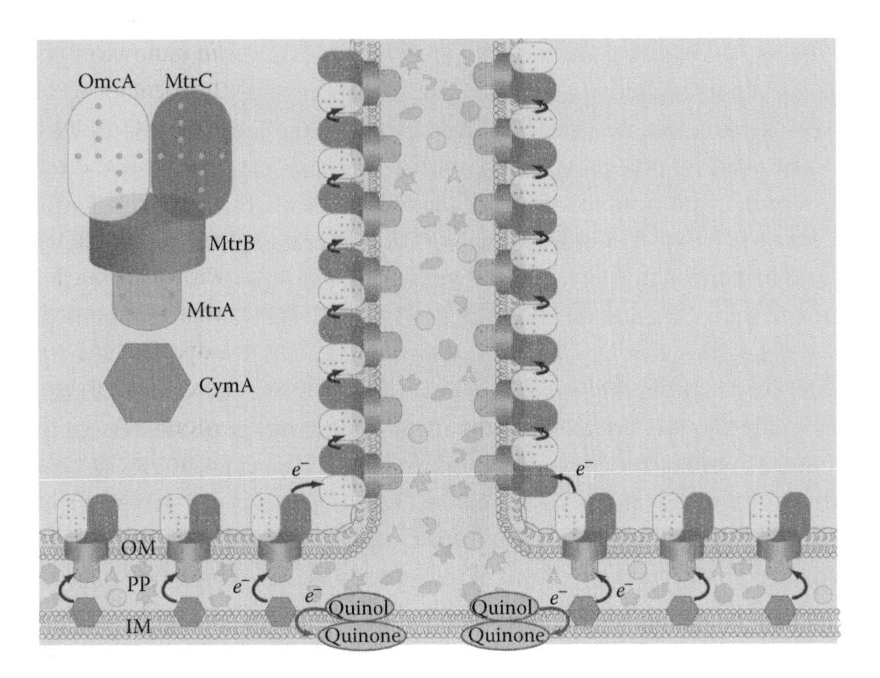

FIGURE 8.5 Proposed structural model of *Shewanella* nanowires. *S. oneidensis* MR-1 nanowires are the outer membrane (OM) and periplasmic (PP) extensions including the multiheme cytochromes responsible for extracellular electron transfer. (Reprinted from Pirbadian, S. et al., *Proc. Natl. Acad. Sci. U. S. A.*, 111, 12883–12888, 2014. With permission.)

The *mtr* gene clusters are found in the genomes of 19 sequenced *Shewanella* spp. where *mtrC-mtrA-mtrB* are conserved. Different from *mtrC-mtrA-mtrB*, *omcB* is often replaced by *undA* or *undA1* that encodes 11-heme *c*-Cyts. The numbers of genes found in *mtr* gene clusters in *Shewanella* genomes also vary, ranging from 4 to 11 (Fredrickson et al., 2008; Shi et al., 2012b).

The *mtr* gene clusters are also found in the genomes of the Fe(III)-reducing bacteria *Aeromonas hydrophila*, *Ferrimonas balearica* and *Rhodoferax ferrireducens*; the soil bacterium *Candidatus Solibacter usitatus* Ellin6076; and the marine bacteria *Vibrio* sp. Ex25, *V. parahaemolyticus*, *V. vulnificus* CMCP6, *V. vulnificus* MO6-24/O and *V. vulnificus* YJ016 (Hartshorne et al., 2009; Shi et al., 2012b). Different from that in the genomes of *Shewanella* spp. in which *cymA* is not part of the *mtr* gene clusters, *cymA* is clustered with the *mtr* genes in the genomes of *F. balearica* and *R. ferrireducens* (Shi et al., 2012b). Presence of the *mtr* gene clusters in the genomes of *Candidatus Solibacter usitatus* Ellin6076, *Vibrio* sp. Ex25, *V. parahaemolyticus*, *V. vulnificus* CMCP6, *V. vulnificus* MO6-24/O and *V. vulnificus* YJ016 suggests that these bacteria have the capacity for extracellular electron transfer and may reduce Fe(III) oxides (Shi et al., 2012b). Indeed, *V. parahaemolyticus*, a human pathogen, reduces Fe(III) and Mn(IV) oxides (Wee et al., 2014). It is suggested that the respiration of Fe(III) and Mn(III/IV) oxides by these *Vibrio* spp. provides an alternative way for energy conservation in the absence of other terminal electron acceptors (Shi et al., 2012b).

8.2.8.2 In the Fe(II)-Oxidizing Bacteria

Mtr homologues are required for extracellular oxidation of Fe(II) by *Rhodopseudomonas palustris* TIE-1 (Gralnick et al., 2006; Jiao and Newman, 2007). In the genome of *R. palustris* TIE-1, *pioA* (an *mtrA* homologue), *pioB* (an *mtrB* homologue) and *pioC* that encode a high potential Fe-S protein are clustered together and are all essential for extracellular oxidation of Fe(II) by *R. palustris* TIE-1 (Jiao and Newman, 2007; Bird et al., 2014). The *pioABC* gene cluster is also involved in electron uptake from electrodes by *R. palustris* TIE-1 (Bose et al., 2014).

In addition to that of *R. palustris* TIE-1, *mtrAB* gene clusters are found in the genomes of the Fe(II)-oxidizing bacteria *Dechloromonas aromatica* RCB, *Gallionella capsiferriformans* ES-2 and *Sideroxydans lithotrophicus* ES-1 (Liu et al., 2012; Shi et al., 2012b; Emerson et al., 2013). These gene clusters are called *mto* (metal-oxidizing) gene clusters to distinguish them from those found in the Fe(III)-reducing bacteria (Liu et al., 2012; Shi et al., 2012b). In these *mto* gene clusters, no *pioC*-like gene is found. Instead, the genes encoding putative 1-heme *c*-Cyts MtoD and the cytoplasmic membrane *c*-Cyts that participate or are predicted to participate in quinone/quinol redox cycle in the cytoplasmic membrane, such as CymA and MtoC, are part of these *mto* gene clusters (Liu et al., 2012; Shi et al., 2012b). Further characterization of MtoA (an MtrA homologue) of *S. lithotrophicus* ES-1 demonstrates its ability to oxidize Fe(II), including the Fe(II)-bearing minerals, which is consistent with its predicted role in Fe(II) oxidation (Liu et al., 2012, 2013). Similar to the Mtr proteins found in the Fe(III)-reducing bacteria, the Mto proteins found in the Fe(II)-oxidizing bacteria are proposed to form pathways that oxidize Fe(II) on the bacterial cell surface and then transfer the electrons into the quinone/quinol pool in the cytoplasmic membrane (Liu et al., 2012; Shi et al., 2012b). Metagenomic sequencing of the microbial community in the subsurface sediments of an aquifer close to the Colorado River, Colorado, identified a novel microorganism, RBG-1, that is dominant in this microbial community. Mto homologues are found in RBG-1 in which they are proposed to participate in Fe(II) oxidation (Castelle et al., 2013).

8.2.9 Proposed DMSO-Respiring Pathway

In the genome of *S. oneidensis* MR-1, two additional *mtrAB* like gene pairs, *dmsEF* and SO4360 (an *mtrA* homologue) and SO4359 (an *mtrB* homologue), are found in addition to *mtrAB* and *mtrDE* gene pairs. Both gene pairs are clustered with four additional genes that putatively encode a molybdopterin protein (DmsA/SO4358), an Fe-S protein (DmsB/SO4357), a chaperone for DMSO reductases (DmsG/SO4361) and an accessory protein for DMSO reductases (DmsH/SO4362). DMSO respiration by *S. oneidensis* MR-1 significantly increases the mRNA abundances of *dmsA*, *dmsE* and *dmsF*, but not those for SO4357, SO4338 or SO4360 (Gralnick et al., 2006). Inactivation of *dmsA*, *dmsB* or *dmsE* lowers the ability of *S. oneidensis* MR-1 to respire on DMSO, while deletion of SO4360 has no impact on DMSO respiration (Gralnick et al., 2006; Coursolle and Gralnick, 2010). Recombinant DmsA is localized on the bacterial surface (Gralnick et al., 2006). Thus, similar to the Mtr pathway described in previous sections, the proposed DMSO-respiring pathway of *S. oneidensis* MR-1 begins with quinol oxidation in the cytoplasmic membrane by CymA. The released electrons are then transferred across the periplasm to DmsE, which is believed to be inserted

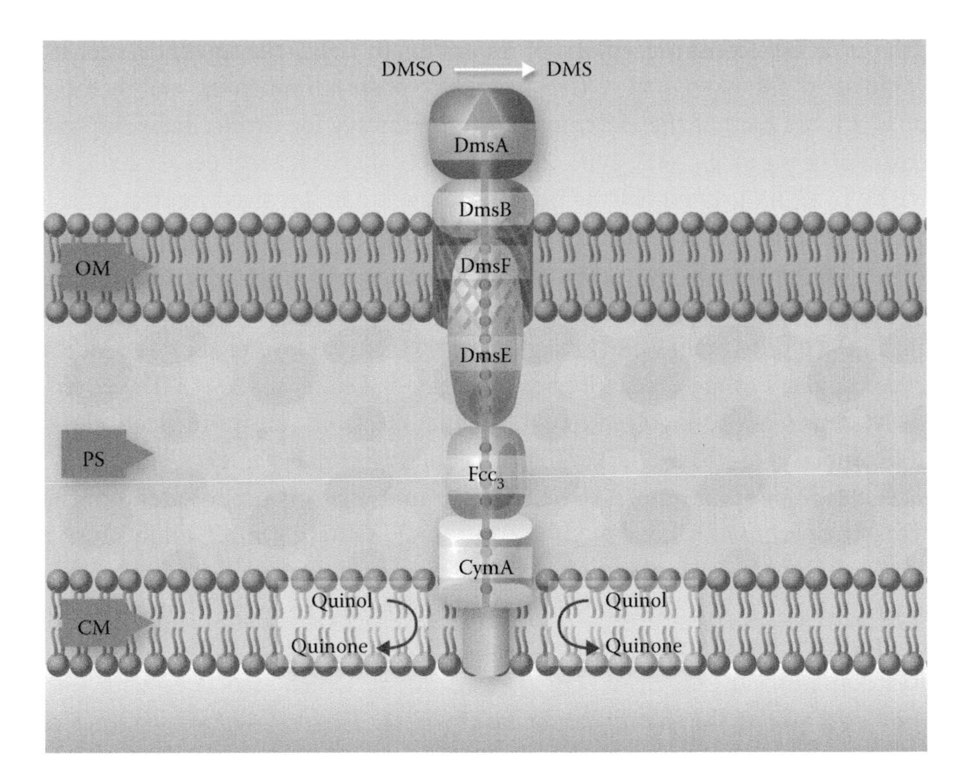

FIGURE 8.6 The proposed DMSO-respiring pathway of *S. oneidensis* MR-1. The proposed pathway consists of CymA; a periplasmic 10-heme *c*-Cyt DmsF (an MtrA homologue); a porin-like outer-membrane (OM) proteins DmsF (an MtrB homologue); an Fe-S protein (DmsB); an molybdopterin protein (DmsA); and probably *c*-Cyt Fcc_3. After quinol oxidation in the cytoplasmic membrane (CM), CymA transfers the released electrons to Fcc_3 that replays electrons across the periplasm (PS) to DmsE either directly or indirectly via other periplasmic redox proteins. DmsE is proposed to be inserted into DmsF. Together, DmsEF transfer electrons across the outer membrane (OM) to DmsAB that reduce DMSO to DMS. The arrow indicates the electron transfer path during extracellular respiration of DMSO.

into DmsF for transferring electrons across the outer membrane. DmsAB are proposed to receive electrons from DmsEF via protein–protein interaction and, in turn, use the electrons to reduce extracellular DMSO (Figure 8.6) (Gralnick et al., 2006; Hartshorne et al., 2009; Coursolle and Gralnick, 2010). The *dmsEF* gene pairs and their associated genes are also identified in the genomes of other *Shewanella* spp. and *F. balearica*, suggesting a possible extracellular respiration of DMSO by these bacteria (Gralnick et al., 2006; Shi et al., 2012b).

8.3 REDOX PROTEINS AND PROTEIN COMPLEXES FOR EXTRACELLULAR ELECTRON TRANSFERS IN *GEOBACTER* SPP.

Similar to *S. oneidensis* MR-1, *Geobacter metallireducens* is among the first microorganisms discovered to use solid-phase Fe(III) and Mn(IV) oxides as the terminal electron acceptors for respiration (Lovley and Phillips, 1988). In addition to Fe(III) and Mn(IV), Co(III), electrodes, chlorinated organic compounds, humic substances, S^0, U(VI) and V(V) can also be used as the terminal electron acceptors for respiration by *Geobacter* spp. (Lovley et al., 1996, 2011). Similar to *Shewanella* spp., *Geobacter* spp. possess numerous multiheme *c*-Cyts, some of which are part of the electron transfer pathways for extracellular respiration of Fe(III) oxides (Leang et al., 2003, 2005, 2010; Lloyd et al., 2003; Mehta et al., 2005; Aklujkar et al., 2013). In *G. sulfurreducens*, the proteins known to be involved in extracellular electron transfer to Fe(III) oxides include three periplasmic *c*-Cyts (PpcA, OmaB and OmaC), two porin-like outer-membrane proteins (OmbB and OmbC), four outer-membrane *c*-Cyts (OmcB, OmcC, OmcE and OmcS), two outer-membrane multicopper proteins (OmpB and OmpC) and the pilin protein PilA (Leang et al., 2003, 2010; Lloyd et al., 2003; Mehta et al., 2005, 2006; Reguera et al., 2005; Holmes et al., 2008; Liu et al., 2014). The outer-membrane *c*-Cyt OmcZ is also involved in extracellular electron transfer to electrodes (Nevin et al., 2009; Inoue et al., 2010, 2011). The exact roles of OmcE, OmcZ, OmpB and OmpC in extracellular respiration remain to be characterized. PpcA is proposed to transfer electrons across the periplasm (Lloyd et al., 2003). OmaB, OmaC, OmbB, OmbC, OmcB and OmcC form the porin-cytochrome (Pcc) complexes for mediating electron transfer across the outer membrane (Liu et al., 2014). PilA is involved in the long-distance conductivity along the *Geobacter* nanowires (Reguera et al., 2005; Leang et al., 2010; Malvankar et al., 2011; Reardon and Mueller, 2013; Vargas et al., 2013). OmcS associated with *Geobacter* nanowires is believed to the terminal reductase for Fe(III) oxides, as purified OmcS reduces Fe(III) oxide (Leang et al., 2010; Qian et al., 2011).

8.3.1 PpcA: A Periplasmic *c*-Cyt

As a member of the cytochrome *c7* family, the 3-heme PpcA is involved in the reduction of Fe(III)-citrate when acetate is used as an electron donor (Lloyd et al., 2003; Pokkuluri et al., 2004). The *c7* *c*-Cyts are the homologues of the 4-heme *c*-Cyts of cytochrome *c3* family. Because the nomenclature for naming the hemes in *c3* *c*-Cyts is also used for that in the *c7* *c*-Cyts, and the *c7* *c*-Cyts lack the heme equivalent to the heme II of *c3* *c*-Cyts, the three hemes of *c7* *c*-Cyts are named as heme I, heme III and heme IV, respectively. PpcA structure shows that the heme III is in the middle of a 3-heme chain. The Fe atoms of all three hemes are coordinated by two histidine residues, and distance between the neighbouring heme Fe atoms is 11.2 and 12.6 Å, respectively. Although they are all exposed to the solvents, heme I is the most solvent-exposed, while heme III is the least (Pokkuluri et al., 2004).

When PpcA is fully reduced and protonated, the redox potentials for heme I, heme III and heme IV are −162, −143 and −133 mV versus SHE, respectively (Pessanha et al., 2006). One of the suggested roles for PpcA in extracellular respiration of Fe(III) oxides is electron transfer across the periplasm (Lloyd et al., 2003). Because protons may be released in the periplasm after electron transfer by PpcA, PpcA is also proposed to play a role in maintaining the membrane potential for ATP synthesis (Mahadevan et al., 2006; Pessanha et al., 2006). Notably, *G. sulfurreducens* PCA possesses additional four PpcA homologues (PpcB-E). Comparative analyses of these homologues reveal different thermodynamic properties among these *c*-Cyts, which suggest that they may have different functions (Morgado et al., 2010).

8.3.2 Pcc Protein Complex: Different Type of Trans-Outer Membrane Porin-Cytochrome Electron Conduit

In *G. sulfurreducens*, the outer-membrane *c*-Cyt OmcB is cell surface exposed. The *omcB* gene is part of two tandem four-gene clusters. Each encodes a transcriptional factor (OrfR/OrfS), a porin-like outer-membrane protein (OmbB/OmbC) with 20 predicted trans-outer membrane motifs, a periplasmic 8-heme *c*-Cyt (OmaB/OmaC), and an outer-membrane 12-heme *c*-Cyt (OmcB/OmcC), respectively (Figure 8.7a) (Leang et al., 2003; Qian et al., 2007; Liu et al., 2014).

Comparison of the *omcB*-associated gene clusters of *G. sulfurreducens* PCA with the *mtr* gene clusters of *S. oneidensis* MR-1 reveals a remarkable case of apparent convergent evolution. Although they share no identity at the amino acid sequence level with exception of heme-binding motifs of the *c*-Cyts, the proposed functions and cellular localizations of the proteins encoded by the *ombB-omaB-omcB/ombC-omaC-omcC* gene clusters of *G. sulfurreducens* PCA are analogous to that of the *mtrC-mtrA-mtrB/mtrD-mtrE-mtrF* gene clusters of *S. oneidensis* MR-1 (Figure 8.7a) (Fredrickson et al., 2008; Liu et al., 2014). As discussed in previous sections, each of the *mtr* gene clusters also encodes a porin-like outer-membrane protein (MtrB/MtrE), a periplasmic *c*-Cyt (MtrA/MtrD) and an outer-membrane *c*-Cyt (MtrC/MtrF).

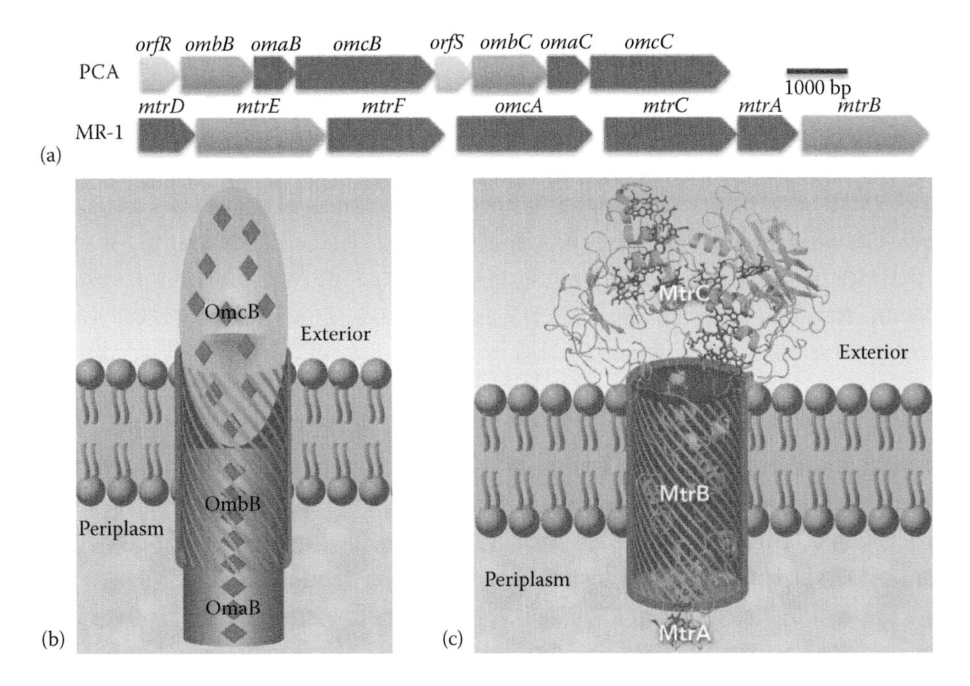

FIGURE 8.7 The proposed *Geobacter* Pcc and *Shewanella* MtrABC trans-outer membrane porin-cytochrome models. (a) Comparison of the *omcB*-associated gene clusters of *G. sulfurreducens* PCA and the *mtr* gene clusters of *S. oneidensis* MR-1. (b) The proposed Pcc model of *G. sulfurreducens* PCA. Only OmaB, OmbB and OmcB are shown. (c) The proposed MtrABC model of *S. oneidensis* MR-1. MtrA and MtrC are modelled based on the structures of NrfB and MtrF, respectively. (Data from Clarke, T.A. et al., *Biochem. J.*, 406, 19–30, 2007; Data from Clarke, T.A. et al., *Proc. Natl. Acad. Sci. U. S. A.*, 108, 9384–9389, 2011; Modified from Liu, Y. et al., *Environ. Microbiol. Rep.*, 6, 776–785, 2014.)

MtrABC of *S. oneidensis* MR-1 form a protein complex for trans-outer membrane electron conductance (Fredrickson et al., 2008; Hartshorne et al., 2009). Similar to the MtrABC proteins, OmbB/OmbC, OmaB/OmaC and OmcB/OmcC (i.e. porin-cytochrome or Pcc) proteins of *G. sulfurreducens* DL-1 associate with the outer membrane (Ding et al., 2006). These observed similarities between the Pcc and the Mtr proteins suggest that the Pcc proteins of *G. sulfurreducens* also form complexes for transferring electrons across the bacterial outer membrane (Liu et al., 2014).

Consistent with this suggestion, the Pcc proteins are isolated as complexes from the membrane fractions of *G. sulfurreducens* PCA that is grown with Fe(III)-citrate as the terminal electron acceptor. Similar to the MtrABC complex, the isolate Pcc complex is a 20-heme heterotrimer. When inserted in the liposome, Pcc complexes transfer electrons from the reduced methyl viologen inside the proteoliposomes across the lipid bilayer to the Fe(III)-citrate and Fe(III) oxides outside the proteoliposomes. Moreover, both *ombB-omaB-omcB* and *ombC-omaC-omcC* clusters are involved in extracellular reduction of Fe(III)-citrate and Fe(III) oxides by *G. sulfurreducens* PCA. Collectively, these results demonstrate that the Pcc proteins of *G. sulfurreducens* PCA form complexes for transferring electrons across the outer membrane. It is proposed that the porin-like outer-membrane proteins OmbB/OmbC function as sheaths through which the *c*-Cyts OmaB/OmaC and OmcB/OmcC are inserted to form a continuous, heme-based electron conduit that spans across the outer membrane, similar to the proposed MtrABC complex (Figure 8.7b and c) (Liu et al., 2014).

The *pcc* gene clusters are found in all sequenced *Geobacter* genomes and other bacterial genomes from six different phyla. Widespread distribution of the Pcc homologues in phylogenetically diverse bacteria suggests a broad application of the Pcc protein complexes in extracellular electron transfer by the Gram-negative bacteria. This finding is very similar to the Mtr proteins whose homologues are found in all characterized metal-reducing *Shewanella* strains and many other bacteria, as discussed in previous sections (Fredrickson et al., 2008; Shi et al., 2012; Liu et al., 2014).

8.3.3 *Geobacter* Nanowires: Proposed Pilin-Based Electron Conduits

Geobacter nanowires appear to be compositionally distinct from those produced by *Shewanella* spp. *Shewanella* nanowires are the outer membrane and periplasm extensions where MtrC and OmcA are the redox proteins responsible for transferring electrons along the nanowires via the heme Fe of MtrC and OmcA by a multistep hopping mechanism (Gorby et al., 2006; Pirbadian et al., 2014). *Geobacter* nanowires, however, consist of mainly pilin protein PilA and exhibit metallic-like conductivity, which is proposed to be attributed to overlapping pi–pi orbitals of aromatic amino acid residues of PilA (Malvankar et al., 2011; Vargas et al., 2013). Comparative analyses of the amino acid sequences of matured *Geobacter* PilA and PilA homologues from the bacteria that do not produce electrically conductive nanowire, such as *Pseudomonas aeruginosa* PAO1, reveal that although the first 22 amino acid residues at their N-terminal regions are well conserved among the PilAs compared, the polypeptide of *Geobacter* PilA (i.e. 61 amino acid residues for the PilA of *G. sulfurreducens*, or GSu PilA) is much shorter than that of *P. aeruginosa* PAO1 (141 amino acid residues).

Sequence analyses also reveal the existence of up to five aromatic amino acid residues outside the conservative N-terminal regions of *Geobacter* PilA (Reardon and Mueller, 2013; Vargas et al., 2013). Mutation of all five of these aromatic amino acid residues to alanine in GSu PilA renders the nanowires nonconductive, even though OmcS is still associated with the nanowire. The *Geobacter* PilA mutant also displays the impaired ability to reduce Fe(III) oxide and to generate electricity on the electrodes (Vargas et al., 2013).

The molecular structure of GSu PilA has been determined at atomic resolution with solution state NMR spectroscopy. Similar to the structure of PilA from *Neisseria gonorrhoeae* (Craig et al., 2006), GSu PilA contains a bent α-helix (residues of 1–52) of ~75 Å long. In contrast to that of PilA from *N. gonorrhoeae*, GSu PilA contains a short and flexible C-terminal region (Figure 8.8a) (Reardon and Mueller, 2013). Based on the GSu PilA structure and pilus model of *N. gonorrhoeae*, a model for *Geobacter* pilus fibre has been constructed (Craig et al., 2006; Reardon and Mueller, 2013). This proposed model indicates that the aromatic acid residues (Phe-2, Phe-24, Tyr-27, Tyr-32, Phe-51 and Tyr-57) may be clustered to form a continuous aromatic band along the proposed structure of *Geobacter* pilus.

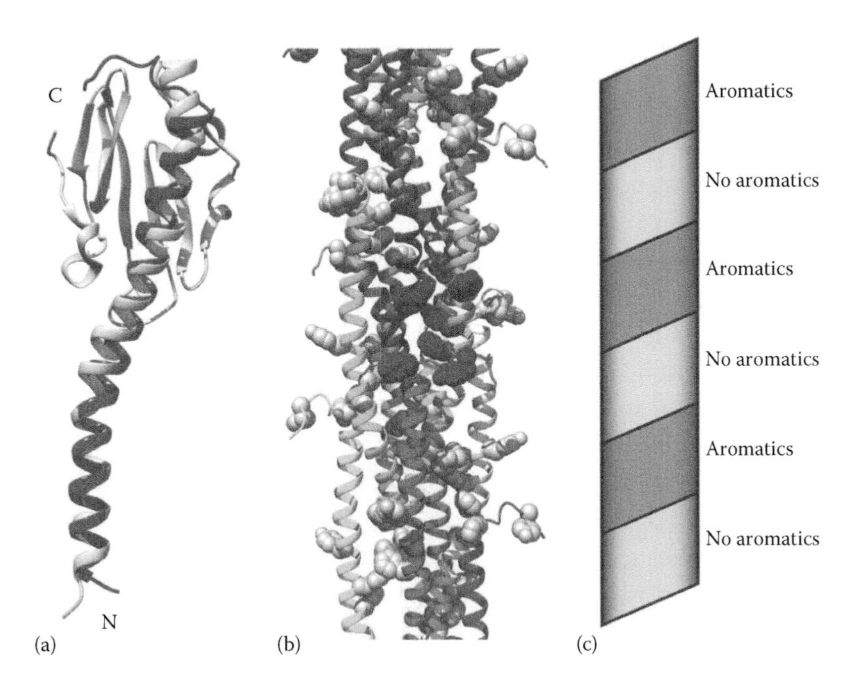

FIGURE 8.8 Model of a nanowire fibre based on the structure of *G. sulfurreducens* PilA. (a) Superimposition of PilA from *G. sulfurreducens* on to the homologous type IV pilin from *Neisseria gonorrhoeae* (Protein Data Bank code 2HIL). (Data from Craig, L. et al., *Mol. Cell,* 23, 651–662, 2006.) The amino terminus and carboxyl terminus are indicated by N and C, respectively. (b) Model of the bacterial nanowire, based on the pilus assembly of *N. gonorrhoeae* (Protein Data Bank code 2HIL). (Data from Craig, L. et al., *Mol. Cell,* 23, 651–662, 2006.) Aromatic side chains are shown in space filling. (c) Schematic diagram showing the progression of the aromatic clusters up the pilus structure. The aromatic and non-aromatic bands are labelled. (Reprinted from Reardon, P.N., and Mueller, K.T., *J. Biol. Chem.,* 288, 29260–29266, 2013. With permission.)

It is suggested that this aromatic band could mediate long-distance electron transfer via delocalized orbitals and/or electron hopping through the aromatic amino acid residues (Figure 8.8b and c) (Reardon and Mueller, 2013).

It should be pointed out that whether GSu PilA itself is a redox protein has never been experimentally verified. Alternative hypotheses and mechanisms have also been suggested for the long-distance electron transfer in *Geobacter* biofilms. For example, it has been proposed that outer-membrane and multiheme *c*-Cyts, rather than PilA, are the redox proteins responsible for mediating long-distance electron transfer in *Geobacter* biofilms by a multistep hopping mechanism (Strycharz-Glaven et al., 2011; Bond et al., 2012; Snider et al., 2012), analogous to the roles of MtrC and OmcA in *Shewanella* nanowires (Pirbadian et al., 2014).

8.4 PUTATIVE ELECTRON TRANSFER PATHWAYS FOR EXTRACELLULAR RESPIRATION IN OTHER MICROORGANISMS

The Gram-positive bacterium *Thermincola potens* strain JR respires on electrodes and Fe(III) oxides via a direct electron transfer mechanism. The cell envelope of *T. potens* strain JR consists of a cytoplasmic membrane, a 16-nm-thick periplasm and a 17-nm-thick cell wall (Wrighton et al., 2011). Analyses of the genome of *T. potens* strain JR identify 35 putative multiheme *c*-Cyts (Byrne-Bailey et al., 2010; Wrighton et al., 2011). Evidence suggests that some of these multiheme *c*-Cyts are involved in transferring electrons from the cytoplasmic membrane, through the periplasm and across the cell wall to the surfaces of electrodes and Fe(III) oxides. Pre-treatment of the cells of *T. potens* strain JR with cyanide that binds to heme Fe prevents accumulation of Ag(0) precipitate on the cell surface after incubation with Ag(I), suggesting cell-surface localization of heme-containing proteins, such as *c*-Cyts. Analyses with surface-enhanced Raman spectroscopy also suggest surface localization of *c*-Cyts on the cells of *T. potens* strain JR. Extraction of the cells of *T. potens* strain JR under denaturing or low pH conditions result in release of TherJR-1122, a 6-heme *c*-Cyt. Proteomic analyses of the proteins released after sonication coupled to trypsin treatment identify four *c*-Cyts (10-heme TherJR_0333, 10-heme TherJR_1117, TherJR_1122 and 9-heme TherJR_2595), three cell surface layer proteins (TherJR_1061, TherJR_2582 and TherJR_2871) and other proteins. Trypsin treatment of the cells of *T. potens* strain JR also negatively impacts the ability of *T. potens* strain JR to reduce Fe(III) oxides. Based on these results, an extracellular electron transfer pathway that consists of four *c*-Cyts (TherJR_0333, TherJR_1117, TherJR_1122 and TherJR_2595) and a hydrogenase is proposed for transferring electrons from the complex III in the cytoplasmic membrane, through the periplasm and across the cell wall to the surfaces of electrode and Fe(III) oxides (Figure 8.9) (Carlson et al., 2012).

The Gram-positive bacterium *Carboxydothermus ferrireducens* and archaeon *Geoglobus ahangari* also respire on Fe(III) oxides via a direct contact mechanism. Cell surface-exposed *c*-Cyts, heme-containing proteins and pilin proteins are all proposed to mediate the extracellular electron transfer reactions in *C. ferrireducens* and *G. ahangari* (Gavrilov et al., 2012; Manzella et al., 2013). However, the proteins and protein complexes that are responsible for electron transfer during extracellular respiration by these two microorganisms remain uninvestigated.

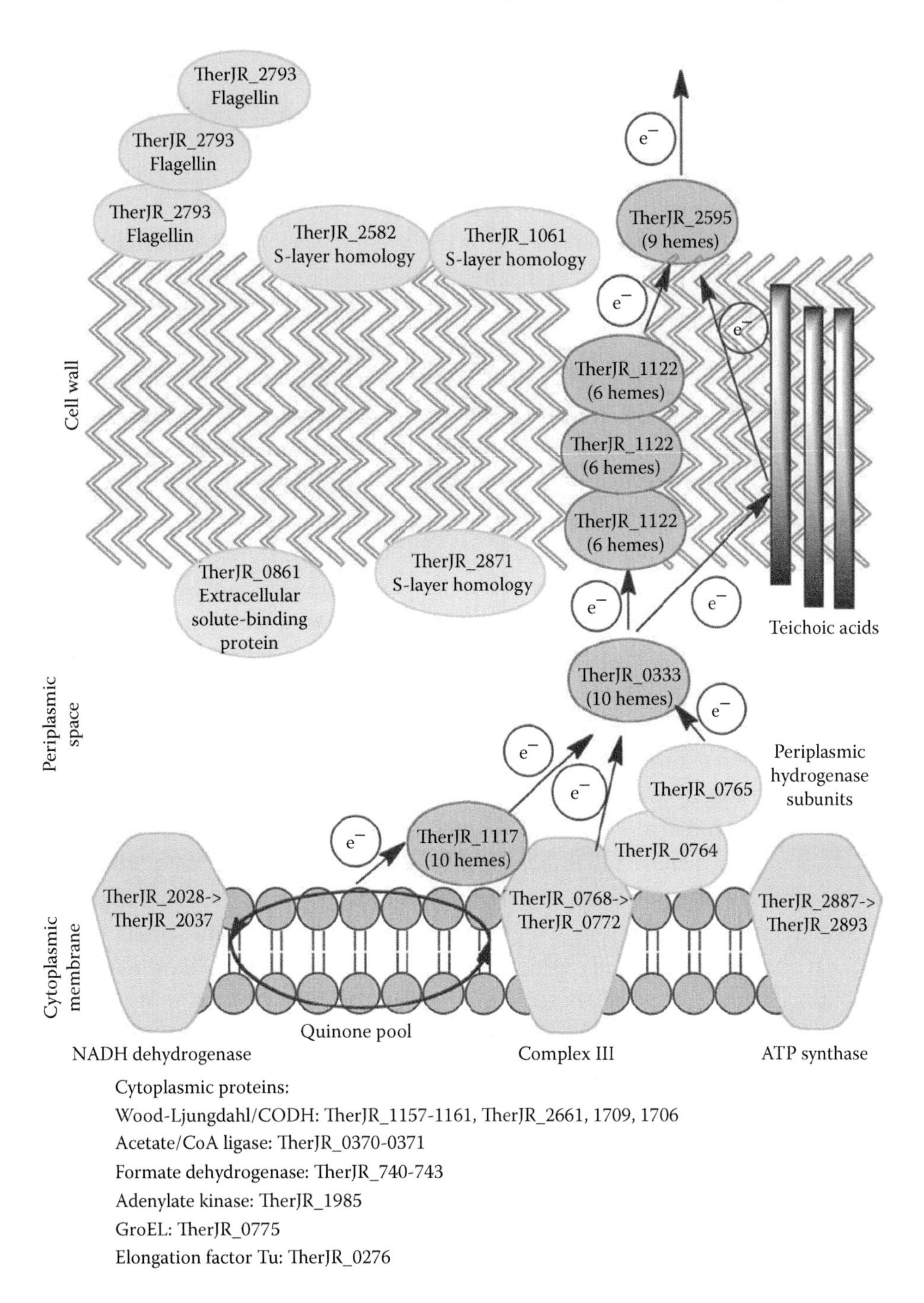

FIGURE 8.9 Model for proposed extracellular electron transfer conduit by *Thermincola potens*. The likely localization of multiheme *c*-type cytochromes and other proteins observed in proteomic experiments is indicated. (Reprinted from Carlson, H.K. et al., *Proc. Natl. Acad. Sci. U. S. A.*, 109, 1702–1707, 2012. With permission.)

8.5 CONCLUSIONS

To respire on extracellular terminal electron acceptors, some microorganisms have developed electron transfer pathways that physically link the quinone/quinol pool in the cytoplasmic membrane to the electron acceptors external to the microbial cells. A common theme emerging from investigations of the microbial redox proteins and protein complexes that are the key components of these extracellular electron transfer pathways is the importance of *c*-Cyts, especially those of multiple hemes, in the microbial extracellular respiratory pathways. They are involved in every step of the electron transfer reactions of the microbial extracellular respiratory pathways, that is, oxidation of quinol in the cytoplasmic membrane, electron transfer across the entire cell envelope, extracellular reduction of terminal electron acceptors and long-distance electron transfer along the microbial nanowires. The extensive use of *c*-Cyts in extracellular respiration can be attributed to their covalent binding of hemes; each heme is typically bound via two thioether bonds to the proximal cysteines in the polypeptides. The covalent binding of heme permits efficient packing of multiple hemes by proteins, as many as 58 hemes can be present in a single polypeptide (Carlson et al., 2012). These closely packed hemes in conjunction with physical association among the multiheme *c*-Cyts form the continuous, heme-based conduits for rapid electron transfer over the distances that are long enough to span across the entire width of microbial cell envelope as well as to connect microbial cells to distant electron acceptors via the nanowires.

Another common theme emerging from these investigations is the formation of protein complexes that consist of porin-like outer-membrane proteins and multiheme *c*-Cyts for transferring electrons across the outer membrane of the Gram-negative bacteria. Although the molecular structures for these trans-outer membrane porin-cytochrome complexes remain unsolved, it is believed that the cytochromes are inserted into the porin-like outer-membrane proteins. Together they mediate electron transfer across the outer membrane. To date, two different types of the trans-outer membrane porin-cytochrome complexes (i.e. Mtr and Pcc) have been identified and characterized. As research on microbial extracellular respiration continues, additional trans-outer membrane extracellular electron transfer complexes with designs similar to those of the Mtr and Pcc complexes are expected to be discovered.

Identification and characterization of microbial redox proteins and proteins complexes directly involved in extracellular electron transfer have contributed significantly to the molecular understanding of the electron transfer processes involved in microbial extracellular respiration. This in-depth understanding will help develop biotechnological applications of these extracellular electron transfer processes in bioremediation of contaminants in the subsurface sediments, bioenergy production and electrobiosynthesis of advanced biofuels and other valuable chemicals.

8.6 ACKNOWLEDGEMENTS

We acknowledge the support from the Subsurface Biogeochemical Research program (SBR) via the Pacific Northwest National Laboratory Scientific Focus Area and the Genome Science Program (GSP) (DE-SC0007229)/Office of Biological and Environmental Research (BER), U.S. Department of Energy (DOE). Pacific Northwest National Laboratory is operated for the DOE by Battelle Memorial Institute under contract DE-AC05-76RLO 1830.

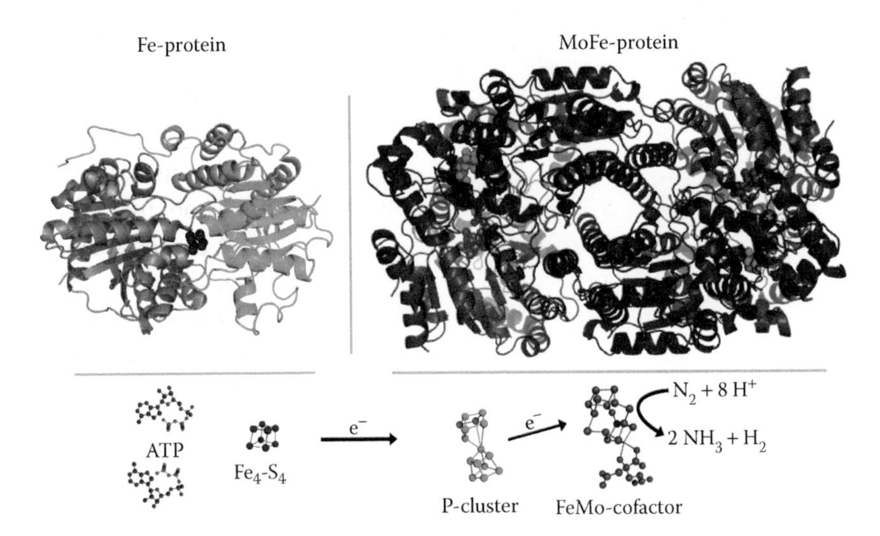

FIGURE 1.11 Mo-Nitrogenase with cofactors. The top shows ribbon diagrams of the homodimeric Fe-protein component (right) and the $\alpha_2\beta_2$ heterodimeric MoFe-protein component (left). The two subunits of the Fe-protein are shown in different shades of green, while the [Fe_4-S_4] cluster is represented in red. The α- and β-subunits of the MoFe-protein are shown in pink and blue, respectively, while the P-cluster and the FeMo-cofactor are represented in green and orange, respectively. Representations were made with PyMol, using the nitrogenase structure from *Azotobacter vinelandii* (PDB code 1M1Y). Represented below are the ATP molecules and the metal clusters involved in the catalysis process.

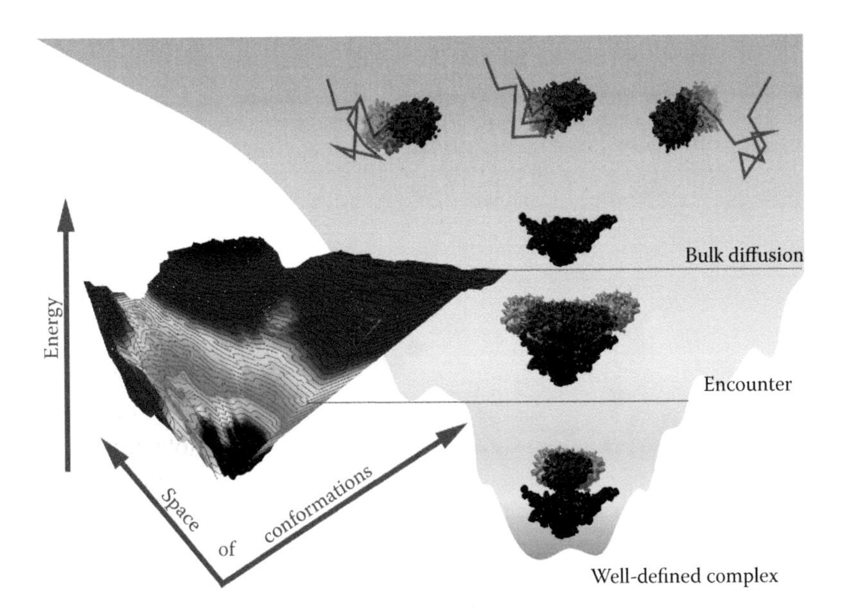

FIGURE 2.2 Energetic landscape of a transient complex. *Left:* Contour levels correspond to the energy of each point of the 2D projection of the coordinate space. The representation is coloured from low energy level (blue) to high (red) of each contour. *Right:* Funnel cross-section showing the three major processes involved in the binding event. The proteins participating in the complex are represented by low-resolution surfaces. Partner 1 (purple) remains at a fixed position, whereas partner 2 (magenta) is represented as a mobile partner.

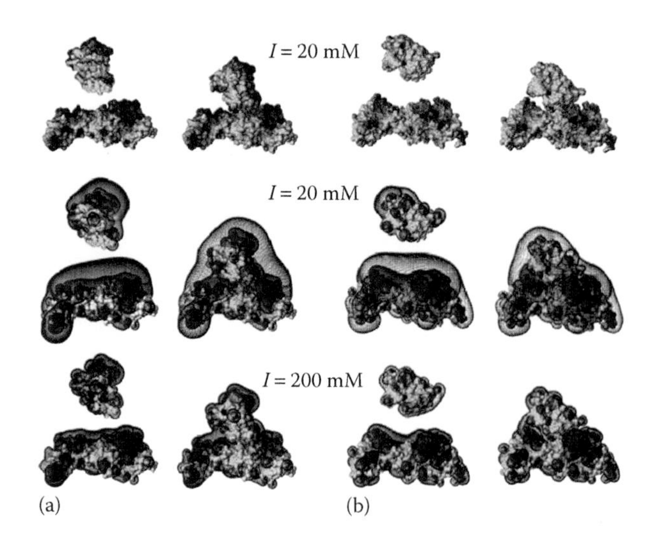

FIGURE 2.3 Electrostatics of complexes between plastocyanin (Pc) and cytochrome *f* (C*f*) from the cyanobacteria (a) *Phormidium* and (b) *Nostoc*. The electrostatic potentials of Pc and C*f* are represented for both the free proteins and the complex composed by them. *Upper:* Electrostatic surface potential (ESP) calculated with DelPhi at an ionic strength of 20 mM with full scale ranging from $-5\ k_BT/e$ (blue) to $+\ 5\ k_BT/e$ (red). *Middle* and *lower:* Isopotential surfaces at $\pm 1\ k_BT/e$ (mesh) and $\pm 2.5\ k_BT/e$ (solid) superimposed to the ESP representations at ionic strength of 20 mM or 200 mM. Blue and red surfaces correspond to positive and negative potential values, respectively.

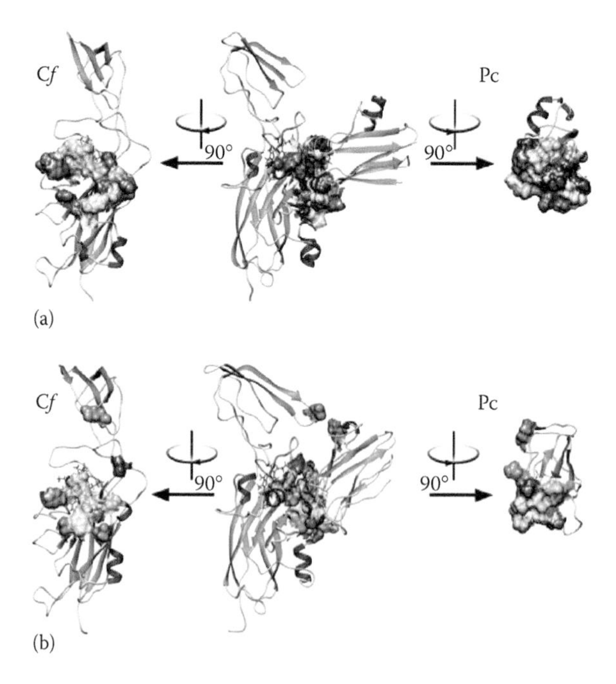

FIGURE 2.4 Interfaces of complexes between plastocyanin (Pc) and cytochrome *f* (C*f*) from the cyanobacteria (a) *Phormidium* and (b) *Nostoc*. Surface residues of each partner closer than 4 Å to the surface of the other partner were chosen for the representations. The surfaces are coloured according to the Doodlittle scale of hydrophobicity from −4.5 (green) to +4.5 (brown). The Pc-C*f* complex is in the middle, with free Pc at the right and free C*f* at the left following a 90° rotation to better show the residues at the interface.

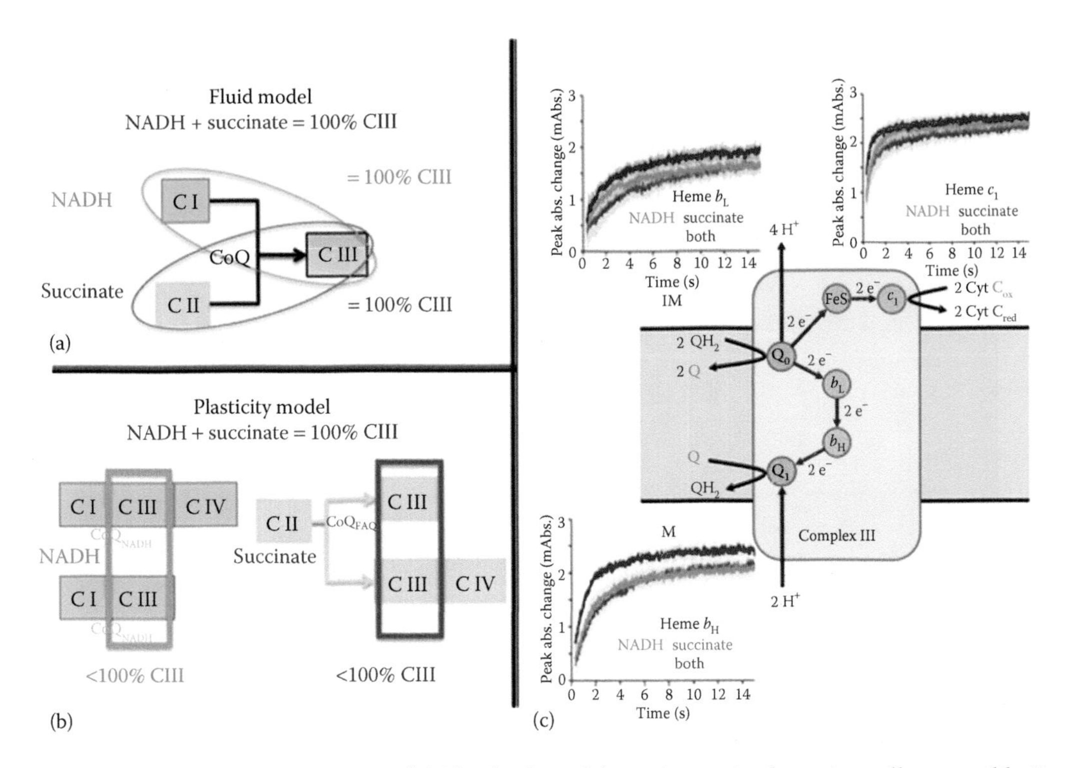

FIGURE 3.1 One or more CoQ pools? The fluid model requires a single, universally accessible Q pool that is not partitioned and the plasticity model proposes that different forms of superassembly define functionally dedicated CoQ pools (*cf.* Lapuente-Brun et al., 2013). Expected versus observed maximum reduction of CIII containing cytochromes as predicted from: (a) the fluid model: NADH or Succinate alone should be able to reduce the bulk of cytochromes in CIII; or (b) the plasticity model: since the CoQ pool is functionally segmented, to be able to reduce the bulk of CIII containing cytochromes both substrates (NADH and succinate) are required simultaneously. (c) Scheme representing CIII cytochrome reduction (bH, bL and c_1) showing the results obtained by Blaza et al. monitoring the reduction of CIII cytochromes and clearly showing that none of them can be fully reduced by NADH or succinate alone and that the simultaneous presence of both is required to reach complete reduction. (Reproduced from Figure 3A, 3C and 3D in Blaza, J.N. et al., *Proc. Nat. Acad. Sci. U.S.A.*, 111, 15735–15740, 2014. With permission.)

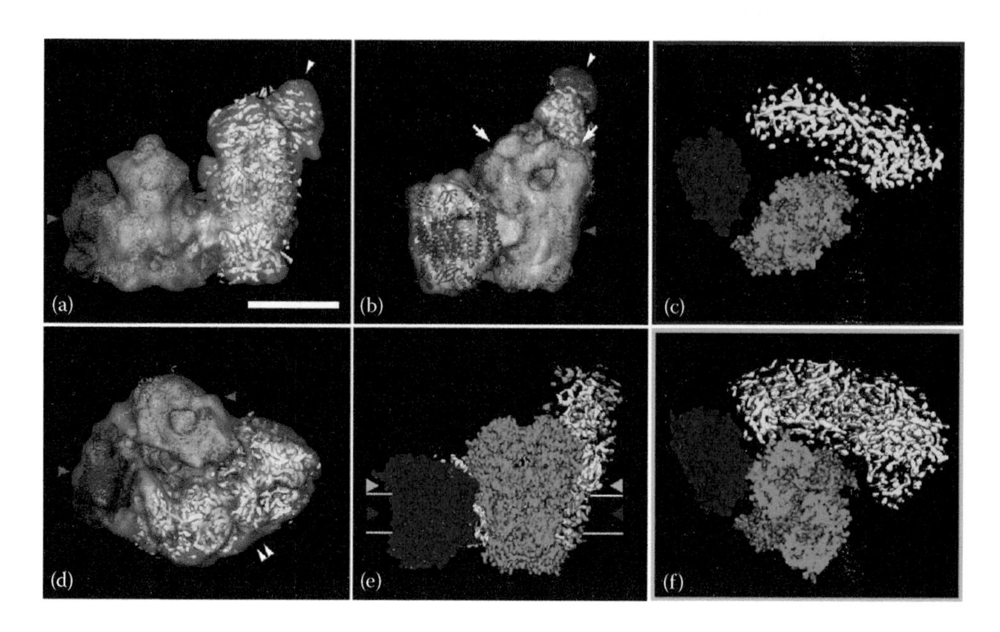

FIGURE 5.1 Fitting of the high- and medium-resolution structures of complexes I, III_2, and IV to the 3D cryo-EM map of $I_1III_2IV_1$ supercomplex: (a) side view, arrowhead points to flavoproteins; (b) side view from the membrane, arrows point to core I and II subunits of complex III_2, arrowhead to flavoproteins; (c) section through the space-filling model of respirasome on the level of membrane, demonstrating gaps between complexes within the supercomplex; (d) top view from the intermembrane space, double arrowhead points to the bend of complex I in membrane; (e) space-filling model of respirasome seen from the membrane, red and light-blue arrowheads show the level of sections in *C* and *F* and (f) section through the space-filling model of respirasome on the level of matrix. In green, X-ray structure of the bovine dimeric complex III; in purple, X-ray structure of bovine monomeric complex IV; in yellow, the density map of complex I from *Yarrowia lipolytica.* Horizontal lines on (e) indicate the position of the membrane. Orange arrowheads on (a), (b) and (d) point to the position of detergent micelles (scale bar: 10 nm). (Reprinted from Dudkina, N.V. et al., Interaction of complexes I, III, and IV within the bovine respirasome by single particle cryoelectron tomography, *Proc. Natl. Acad. Sci. U.S.A*, 108, 15196–15200. Copyright 2011 National Academy of Sciences. With permission.)

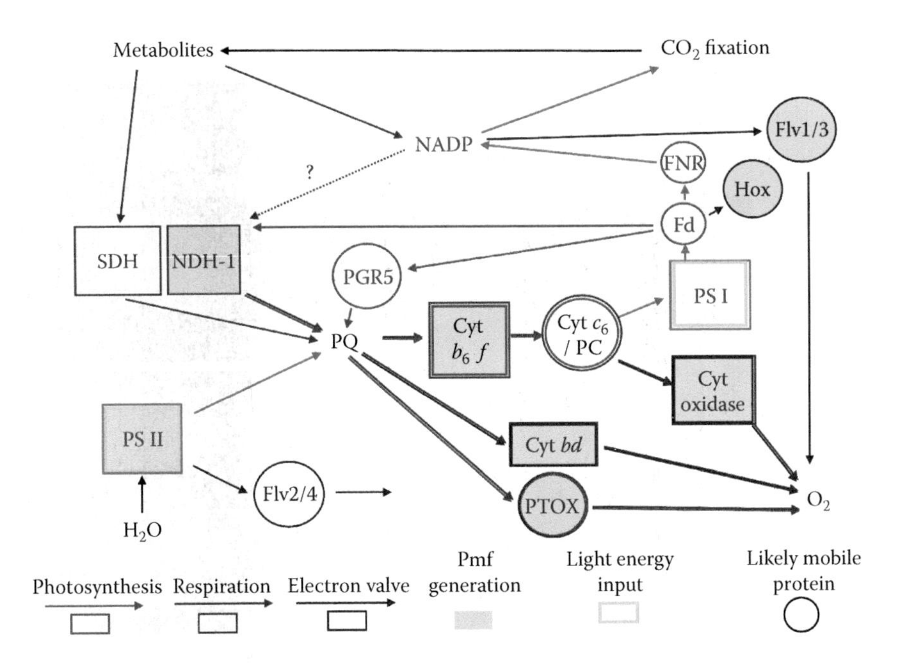

FIGURE 10.2 Pathways of electron transport in and around cyanobacterial thylakoid membranes. Arrows indicate directions of electron flow. Proteins and protein complexes are indicated by frames: small molecules are shown without frames. Frame colour indicates involvement in photosynthesis, respiration or electron dissipation as indicated in the key. Additionally, yellow boxes indicate sites of light energy input and light blue backgrounds show sites of proton-motive force generation (by proton pumping and/or proton consumption/release on a particular side of the membrane). Cytoplasmic, lumenal or peripheral membrane proteins are likely to be rapidly mobile (therefore able to redistribute on timescales of seconds for fast switching of electron transport) and are given circular frames. Membrane-integral protein complexes are likely to show only very restricted mobility and are given square frames. (Adapted from Mullineaux, C.W., *Biochim. Biophys. Acta [BBA-Bioenergetics]*, 1837, 503–511, 2014.) Abbreviations: Cyt, cytochrome; Fd, ferredoxin; Flv, flavodiiron protein; FNR, ferredoxin-NADP oxidoreductase; Hox, bidirectional NiFe hydrogenase; NDH-1, respiratory complex I (NAD[P]H dehydrogenase); PC, plastocyanin; PGR5, proton gradient regulation protein 5; PQ, plastoquinone; PS, photosystem; PTOX, quinol oxidase (plastid terminal oxidase); SDH, respiratory complex II (succinate dehydrogenase).

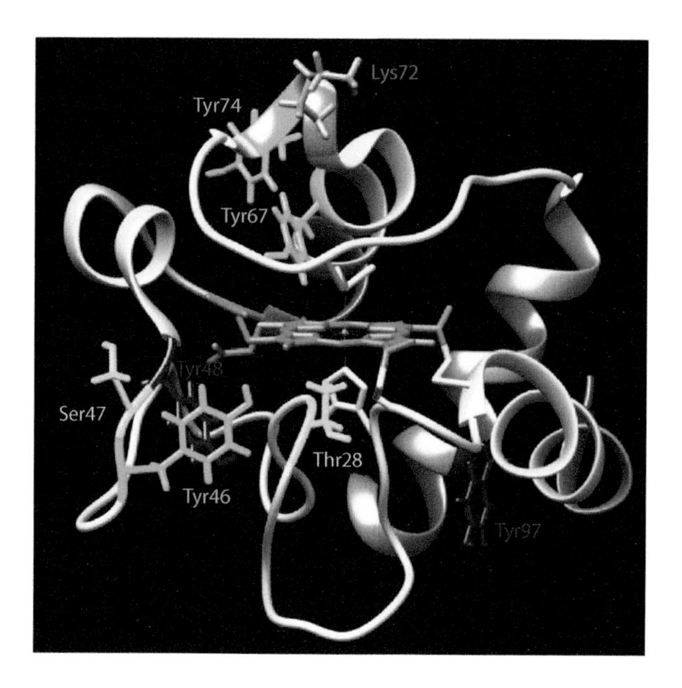

FIGURE 12.1 Post-translational modifications of cytochrome *c*. Human C*c* structure is shown with ribbon diagram (PDB: 1J3S). (Data from Jeng, W.-Y. et al., *J. Bioenerg. Biomembr.,* 34, 423–431, 2002.) Residues Thr28 and Ser47 (in yellow) can be phosphorylated in mammals. Tyr46, Tyr67 and Tyr74 (in blue) reportedly undergo nitration. Tyr48 and Tyr97 (in red) can be nitrated or phosphorylated. Lys72 (in magenta) can be (tri)methylated.

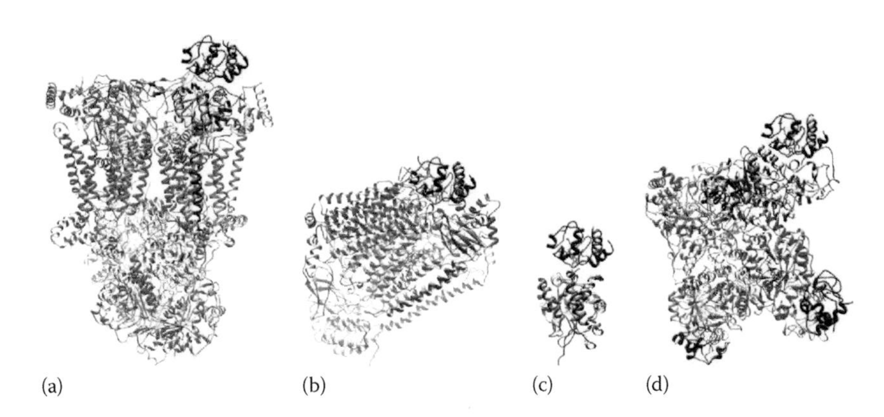

FIGURE 12.3 Structural models for eukaryotic cytochrome *c* complexes in homeostatic conditions. Protein subunits are shown with ribbon diagram. C*c* (red) is presented in the same orientation in all panels. Panel a: Overall crystallographic structure of the adduct formed between C*c* and Complex III in yeast. Note that C*c* binds to only one Cc_1 subunit (blue) of Complex III. Other subunits of Complex III are shown (light grey), as are hemes (green). Panel b: Docking model between C*c* and Complex IV in bovine heart. Also represented is subunit 4 in Complex IV (blue), other subunits (light grey), copper atoms (yellow spheres) and hemes (green). Panel c: Solution structure of C*c*–C*c*P complex in yeast as determined by paramagnetic NMR, showing C*c*P (blue) and heme groups (green). Panel d: Computational model of C*c*–FCb_2 complex in yeast, with homotetrameric structure of FCb_2 represented with one C*c* molecule docked onto each FCb_2 subunit (one in blue and the others in light grey). The heme groups (green) and flavine mononucleotide cofactors (purple) are also represented.

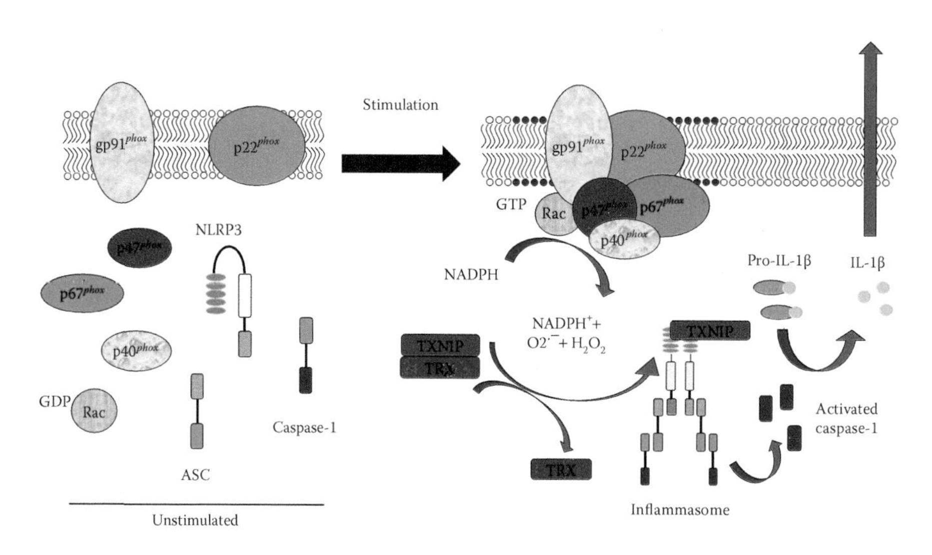

FIGURE 13.3 Activation of NLRP3 inflammasomes linking to MR redox signalosomes formation. NLRP3 inflammasomes are activated by NADPH oxidase-derived ROS through lipid raft clustering. ROS derived from MR redox signalosomes activate NLRP3 inflammasomes through the binding of thioredoxin-interacting protein (TXNIP) to NLRP3. TXNIP is a binding partner to NLRP3, which dissociates from TRX and then binds with NLRP3 when ROS is increased.

REFERENCES

Abe, T., Hoshino, T., Nakamura, A., and Takaya, N. (2007). Anaerobic elemental sulfur reduction by fungus *Fusarium oxysporum. Biosci Biotechnol Biochem* **71**: 2402–2407.

Aklujkar, M., Coppi, M. V., Leang, C., Kim, B. C., Chavan, M. A., Perpetua, L. A. et al. (2013). Proteins involved in electron transfer to Fe(III) and Mn(IV) oxides by *Geobacter sulfurreducens* and *Geobacter uraniireducens. Microbiology* **159**: 515–535.

Amend, J. P., and Shock, E. L. (2001). Energetics of overall metabolic reactions of thermophilic and hyperthermophilic Archaea and bacteria. *FEMS Microbiol Rev* **25**: 175–243.

Belchik, S. M., Kennedy, D. W., Dohnalkova, A. C., Wang, Y., Sevinc, P. C., Wu, H. et al. (2011). Extracellular reduction of hexavalant chromium by cytochromes MtrC and OmcA of *Shewanella oneidensis* MR-1. *Appl Environ Microbiol* **77**: 4035–4041.

Bird, L. J., Saraiva, I. H., Park, S., Calcada, E. O., Salgueiro, C. A., Nitschke, W. et al. (2014). Nonredundant roles for cytochrome c2 and two high-potential iron-sulfur proteins in the photoferrotroph *Rhodopseudomonas palustris* TIE-1. *J Bacteriol* **196**: 850–858.

Bond, D. R., Strycharz-Glaven, S. M., Tender, L. M., and Torres, C. I. (2012). On electron transport through *Geobacter* biofilms. *ChemSusChem* **5**: 1099–1105.

Bose, A., Gardel, E. J., Vidoudez, C., Parra, E. A., and Girguis, P. R. (2014). Electron uptake by iron-oxidizing phototrophic bacteria. *Nat Commun* **5**: 3391.

Bose, S., Hochella, M. F., Gorby, Y., Kennedy, D. W., McCready, D. E., Madded, A. S., and Lower, B. H. (2009). Bioreduction of hematite nanoparticles by the dissimilatory iron reducing bacterium *Shewanella oneidensis* MR-1. *Geochim Cosmochim Acta* **73**: 962–976.

Breuer, M., Rosso, K. M., and Blumberger, J. (2014). Electron flow in multiheme bacterial cytochromes is a balancing act between heme electronic interaction and redox potentials. *Proc Natl Acad Sci U S A* **111**: 611–616.

Breuer, M., Zarzycki, P., Blumberger, J., and Rosso, K. M. (2012a). Thermodynamics of electron flow in the bacterial deca-heme cytochrome MtrF. *J Am Chem Soc* **134**: 9868–9871.

Breuer, M., Zarzycki, P., Shi, L., Clarke, T. A., Edwards, M. J., Butt, J. N. et al. (2012b). Molecular structure and free energy landscape for electron transport in the decahaem cytochrome MtrF. *Biochem Soc Trans* **40**: 1198–1203.

Brutinel, E. D., and Gralnick, J. A. (2012). Shuttling happens: Soluble flavin mediators of extracellular electron transfer in *Shewanella. Appl Microbiol Biotechnol* **93**: 41–48.

Burns, J. L., and DiChristina, T. J. (2009). Anaerobic respiration of elemental sulfur and thiosulfate by *Shewanella oneidensis* MR-1 requires *psrA*, a homolog of the *phsA* gene of *Salmonella enterica* serovar typhimurium LT2. *Appl Environ Microbiol* **75**: 5209–5217.

Byrne-Bailey, K. G., Wrighton, K. C., Melnyk, R. A., Agbo, P., Hazen, T. C., and Coates, J. D. (2010). Complete genome sequence of the electricity-producing "Thermincola potens" strain JR. *J Bacteriol* **192**: 4078–4079.

Caccavo, F., Jr., Lonergan, D. J., Lovley, D. R., Davis, M., Stolz, J. F., and McInerney, M. J. (1994). *Geobacter sulfurreducens* sp. nov., a hydrogen- and acetate-oxidizing dissimilatory metal-reducing microorganism. *Appl Environ Microbiol* **60**: 3752–3759.

Cao, B., Ahmed, B., Kennedy, D. W., Wang, Z., Shi, L., Marshall, M. J. et al. (2011a). Contribution of extracellular polymeric substances from *Shewanella* sp. HRCR-1 biofilms to U(VI) immobilization. *Environ Sci Technol* **45**: 5483–5490.

Cao, B., Shi, L., Brown, R. N., Xiong, Y., Fredrickson, J. K., Romine, M. F. et al. (2011b). Extracellular polymeric substances from *Shewanella* sp. HRCR-1 biofilms: Characterization by infrared spectroscopy and proteomics. *Environ Microbiol* **13**: 1018–1031.

Carlson, H. K., Iavarone, A. T., Gorur, A., Yeo, B. S., Tran, R., Melnyk, R. A. et al. (2012). Surface multiheme *c*-type cytochromes from *Thermincola* potens and implications for respiratory metal reduction by Gram-positive bacteria. *Proc Natl Acad Sci U S A* **109**: 1702–1707.

Castelle, C. J., Hug, L. A., Wrighton, K. C., Thomas, B. C., Williams, K. H., Wu, D. et al. (2013). Extraordinary phylogenetic diversity and metabolic versatility in aquifer sediment. *Nat Commun* **4**: 2120.

Clarke, T. A., Cole, J. A., Richardson, D. J., and Hemmings, A. M. (2007). The crystal structure of the pentahaem *c*-type cytochrome NrfB and characterization of its solution-state interaction with the pentahaem nitrite reductase NrfA. *Biochem J* **406**: 19–30.

Clarke, T. A., Edwards, M. J., Gates, A. J., Hall, A., White, G. F., Bradley, J. et al. (2011). Structure of a bacterial cell surface decaheme electron conduit. *Proc Natl Acad Sci U S A* **108**: 9384–9389.

Cologgi, D. L., Lampa-Pastirk, S., Speers, A. M., Kelly, S. D., and Reguera, G. (2011). Extracellular reduction of uranium via *Geobacter* conductive pili as a protective cellular mechanism. *Proc Natl Acad Sci U S A* **108**: 15248–15252.

Coursolle, D., Baron, D. B., Bond, D. R., and Gralnick, J. A. (2010). The Mtr respiratory pathway is essential for reducing flavins and electrodes in *Shewanella oneidensis*. *J Bacteriol* **192**: 467–474.

Coursolle, D., and Gralnick, J. A. (2010). Modularity of the Mtr respiratory pathway of *Shewanella oneidensis* strain MR-1. *Mol Microbiol* **77**: 995–1008.

Craig, L., Volkmann, N., Arvai, A. S., Pique, M. E., Yeager, M., Egelman, E. H. and Tainer, J. A. (2006). Type IV pilus structure by cryo-electron microscopy and crystallography: Implications for pilus assembly and functions. *Mol Cell* **23**: 651–662.

DiChristina, T. J., Moore, C. M., and Haller, C. A. (2002). Dissimilatory Fe(III) and Mn(IV) reduction by *Shewanella putrefaciens* requires *ferE*, a homolog of the *pulE* (*gspE*) type II protein secretion gene. *J Bacteriol* **184**: 142–151.

Ding, Y. H., Hixson, K. K., Giometti, C. S., Stanley, A., Esteve-Nunez, A., Khare, T. et al. (2006). The proteome of dissimilatory metal-reducing microorganism *Geobacter sulfurreducens* under various growth conditions. *Biochim Biophys Acta* **1764**: 1198–1206.

Dohnalkova, A. C., Marshall, M. J., Arey, B. W., Williams, K. H., Buck, E. C., and Fredrickson, J. K. (2011). Imaging hydrated microbial extracellular polymers: Comparative analysis by electron microscopy. *Appl Environ Microbiol* **77**: 1254–1262.

Donald, J. W., Hicks, M. G., Richardson, D. J., and Palmer, T. (2008). The *c*-type cytochrome OmcA localizes to the outer membrane upon heterologous expression in *Escherichia coli*. *J Bacteriol* **190**: 5127–5131.

Edwards, M. J., Baiden, N. A., Johs, A., Tomanicek, S. J., Liang, L., Shi, L. et al. (2014). The X-ray crystal structure of *Shewanella oneidensis* OmcA reveals new insight at the microbe-mineral interface. *FEBS Lett* **588**: 1886–1890.

Edwards, M. J., Hall, A., Shi, L., Fredrickson, J. K., Zachara, J. M., Butt, J. N. et al. (2012). The crystal structure of the extracellular 11-heme cytochrome UndA reveals a conserved 10-heme motif and defined binding site for soluble iron chelates. *Structure* **20**: 1275–1284.

El-Naggar, M. Y., Gorby, Y. A., Xia, W., and Nealson, K. H. (2008). The molecular density of states in bacterial nanowires. *Biophys J* **95**: L10–L12.

El-Naggar, M. Y., Wanger, G., Leung, K. M., Yuzvinsky, T. D., Southam, G., Yang, J. et al. (2010). Electrical transport along bacterial nanowires from *Shewanella oneidensis* MR-1. *Proc Natl Acad Sci U S A* **107**: 18127–18131.

Emerson, D., Field, E. K., Chertkov, O., Davenport, K. W., Goodwin, L., Munk, C. et al. (2013). Comparative genomics of freshwater Fe-oxidizing bacteria: Implications for physiology, ecology, and systematics. *Front Microbiol* **4**: 254.

Firer-Sherwood, M. A., Ando, N., Drennan, C. L., and Elliott, S. J. (2011). Solution-based structural analysis of the decaheme cytochrome, MtrA, by small angle X-ray scattering and analytical ultracentrifugation. *J Phys Chem B* **115**: 11208–11214.

Firer-Sherwood, M. A., Bewley, K. D., Mock, J. Y., and Elliott, S. J. (2011). Tools for resolving complexity in the electron transfer networks of multiheme cytochromes c. *Metallomics* **3**: 344–348.

Fonseca, B. M., Paquete, C. M., Neto, S. E., Pacheco, I., Soares, C. M., and Louro, R. O. (2013). Mind the gap: Cytochrome interactions reveal electron pathways across the periplasm of *Shewanella oneidensis* MR-1. *Biochem J* **449**: 101–108.

Fredrickson, J. K., Romine, M. F., Beliaev, A. S., Auchtung, J. M., Driscoll, M. E., Gardner, T. S. et al. (2008). Towards environmental systems biology of *Shewanella*. *Nat Rev Microbiol* **6**: 592–603.

Gavrilov, S. N., Lloyd, J. R., Kostrikina, N. A., and Slobodkin, A. I. (2012). Fe(III) oxide reduction a Gram-positive thermophile: Physiological mechanisms for dissimilatory reduction of poorly crystalline Fe(III) oxide by a thermophilic Gram-positive bacterium *Carboxydothermus ferrireducens*. *Geomicrobiol J* **29**: 804–819.

Gorby, Y., McLean, J., Korenevsky, A., Rosso, K., El-Naggar, M. Y., and Beveridge, T. J. (2008). Redox-reactive membrane vesicles produced by *Shewanella*. *Geobiology* **6**: 232–241.

Gorby, Y. A., Yanina, S., McLean, J. S., Rosso, K. M., Moyles, D., Dohnalkova, A. et al. (2006). Electrically conductive bacterial nanowires produced by *Shewanella oneidensis* strain MR-1 and other microorganisms. *Proc Natl Acad Sci U S A* **103**: 11358–11363.

Gralnick, J. A., and Newman, D. K. (2007). Extracellular respiration. *Mol Microbiol* **65**: 1–11.

Gralnick, J. A., Vali, H., Lies, D. P., and Newman, D. K. (2006). Extracellular respiration of dimethyl sulfoxide by *Shewanella oneidensis* strain MR-1. *Proc Natl Acad Sci U S A* **103**: 4669–4674.

Hartshorne, R. S., Jepson, B. N., Clarke, T. A., Field, S. J., Fredrickson, J., Zachara, J. et al. (2007). Characterization of *Shewanella oneidensis* MtrC: A cell-surface decaheme cytochrome involved in respiratory electron transport to extracellular electron acceptors. *J Biol Inorg Chem* **12**: 1083–1094.

Hartshorne, R. S., Reardon, C. L., Ross, D., Nuester, J., Clarke, T. A., Gates, A. J. et al. (2009). Characterization of an electron conduit between bacteria and the extracellular environment. *Proc Natl Acad Sci U S A* **106**: 22169–22174.

Holmes, D. E., Mester, T., O'Neil, R. A., Perpetua, L. A., Larrahondo, M. J., Glaven, R. et al. (2008). Genes for two multicopper proteins required for Fe(III) oxide reduction in *Geobacter sulfurreducens* have different expression patterns both in the subsurface and on energy-harvesting electrodes. *Microbiology* **154**: 1422–1435.

Inoue, K., Leang, C., Franks, A. E., Woodard, T. L., Nevin, K. P., and Lovley, D. R. (2011). Specific localization of the *c*-type cytochrome OmcZ at the anode surface in current-producing biofilms of *Geobacter sulfurreducens*. *Environ Microbiol Rep* **3**: 211–217.

Inoue, K., Qian, X., Morgado, L., Kim, B. C., Mester, T., Izallalen, M. et al. (2010). Purification and characterization of OmcZ, an outer-surface, octaheme *c*-type cytochrome essential for optimal current production by *Geobacter sulfurreducens*. *Appl Environ Microbiol* **76**: 3999–4007.

Jiao, Y., and Newman, D. K. (2007). The *pio* operon is essential for phototrophic Fe(II) oxidation in *Rhodopseudomonas palustris* TIE-1. *J Bacteriol* **189**: 1765–1773.

Johs, A., Shi, L., Droubay, T., Ankner, J. F., and Liang, L. (2010). Characterization of the decaheme *c*-type cytochrome OmcA in solution and on hematite surfaces by small angle x-ray scattering and neutron reflectometry. *Biophys J* **98**: 3035–3043.

Kamp, A., de Beer, D., Nitsch, J. L., Lavik, G., and Stief, P. (2011). Diatoms respire nitrate to survive dark and anoxic conditions. *Proc Natl Acad Sci U S A* **108**: 5649–5654.

Kotloski, N. J., and Gralnick, J. A. (2013). Flavin electron shuttles dominate extracellular electron transfer by *Shewanella oneidensis*. *mBio* **4**: e00553-12.

Leang, C., Adams, L. A., Chin, K. J., Nevin, K. P., Methe, B. A., Webster, J. et al. (2005). Adaptation to disruption of the electron transfer pathway for Fe(III) reduction in *Geobacter sulfurreducens*. *J Bacteriol* **187**: 5918–5926.

Leang, C., Coppi, M. V., and Lovley, D. R. (2003). OmcB, a *c*-type polyheme cytochrome, involved in Fe(III) reduction in *Geobacter sulfurreducens*. *J Bacteriol* **185**: 2096–2103.

Leang, C., Qian, X., Mester, T., and Lovley, D. R. (2010). Alignment of the *c*-type cytochrome OmcS along pili of *Geobacter sulfurreducens*. *Appl Environ Microbiol* **76**: 4080–4084.

Leung, K. M., Wanger, G., El-Naggar, M. Y., Gorby, Y., Southam, G., Lau, W. M. and Yang, J. (2013). *Shewanella oneidensis* MR-1 bacterial nanowires exhibit p-type, tunable electronic behavior. *Nano Lett* **13**: 2407–2411.

Lies, D. P., Hernandez, M. E., Kappler, A., Mielke, R. E., Gralnick, J. A., and Newman, D. K. (2005). *Shewanella oneidensis* MR-1 uses overlapping pathways for iron reduction at a distance and by direct contact under conditions relevant for biofilms. *Appl Environ Microbiol* **71**: 4414–4426.

Liu, J., Pearce, C. I., Liu, C., Wang, Z., Shi, L., Arenholz, E., and Rosso, K. M. (2013). $Fe_{(3-x)}Ti_{(x)}O_4$ nanoparticles as tunable probes of microbial metal oxidation. *J Am Chem Soc* **135**: 8896–8907.

Liu, J., Wang, Z., Belchik, S. M., Edwards, M. J., Liu, C., Kennedy, D. W. et al. (2012). Identification and characterization of MtoA: A decaheme *c*-type cytochrome of the neutrophilic Fe(II)-oxidizing bacterium *Sideroxydans lithotrophicus* ES-1. *Front Microbiol* **3**: 37.

Liu, Y., Wang, Z., Liu, J., Levar, C., Edwards, M. J., Babauta, J. T. et al. (2014). A trans-outer membrane porin-cytochrome protein complex for extracellular electron transfer by *Geobacter sulfurreducens* PCA. *Environ Microbiol Rep* **6**: 776–785.

Lloyd, J. R., Leang, C., Hodges Myerson, A. L., Coppi, M. V., Cuifo, S., Methe, B. et al. (2003). Biochemical and genetic characterization of PpcA, a periplasmic *c*-type cytochrome in *Geobacter sulfurreducens*. *Biochem J* **369**: 153–161.

Louro, R. O., and Paquete, C. M. (2012). The quest to achieve the detailed structural and functional characterization of CymA. *Biochem Soc Trans* **40**: 1291–1294.

Lovley, D. R., Coates, J. D., Blunt-Harris, E. L., Phillips, E. J. P., and Woodward, J. C. (1996). Humic substances as electron acceptors for microbial respiration. *Nature* **382**: 445–448.

Lovley, D. R., Holmes, D. E., and Nevin, K. P. (2004). Dissimilatory Fe(III) and Mn(IV) reduction. *Adv Microb Physiol* **49**: 219–286.

Lovley, D. R., and Phillips, E. J. (1988). Novel mode of microbial energy metabolism: Organic carbon oxidation coupled to dissimilatory reduction of iron or manganese. *Appl Environ Microbiol* **54**: 1472–1480.

Lovley, D. R., Phillips, E. J. P., Gorby, Y. A., and Landa, E. R. (1991). Microbial reduction of uranium. *Nature* **350**: 413–416.

Lovley, D. R., Ueki, T., Zhang, T., Malvankar, N. S., Shrestha, P. M., Flanagan, K. A. et al. (2011). *Geobacter*: The microbe electric's physiology, ecology, and practical applications. *Adv Microb Physiol* **59**: 1–100.

Lower, B. H., Lins, R. D., Oestreicher, Z., Straatsma, T. P., Hochella, M. F., Jr., Shi, L., and Lower, S. K. (2008). In vitro evolution of a peptide with a hematite binding motif that may constitute a natural metal-oxide binding archetype. *Environ Sci Technol* **42**: 3821–3827.

Lower, B. H., Shi, L., Yongsunthon, R., Droubay, T. C., McCready, D. E., and Lower, S. K. (2007). Specific bonds between an iron oxide surface and outer membrane cytochromes MtrC and OmcA from *Shewanella oneidensis* MR-1. *J Bacteriol* **189**: 4944–4952.

Lower, B. H., Yongsunthon, R., Shi, L., Wildling, L., Gruber, H. J., Wigginton, N. S., Reardon, C. L., Pinchuk, G. E., Droubay, T. C., Boily J. F., and Lower S. K. (2009). Antibody recognition force microscopy shows that outer membrane cytochromes OmcA and MtrC are expressed on the exterior surface of Shewanella oneidensis MR-1. *Appl Environ Microbiol* **75**: 2931–2935.

Mahadevan, R., Bond, D. R., Butler, J. E., Esteve-Nunez, A., Coppi, M. V., Palsson, B. O. et al. (2006). Characterization of metabolism in the Fe(III)-reducing organism *Geobacter sulfurreducens* by constraint-based modeling. *Appl Environ Microbiol* **72**: 1558–1568.

Malvankar, N. S., Vargas, M., Nevin, K. P., Franks, A. E., Leang, C., Kim, B. C. et al. (2011). Tunable metallic-like conductivity in microbial nanowire networks. *Nat Nanotechnol* **6**: 573–579.

Manzella, M. P., Reguera, G., and Kashefi, K. (2013). Extracellular electron transfer to Fe(III) oxides by the hyperthermophilic archaeon *Geoglobus ahangari* via a direct contact mechanism. *Appl Environ Microbiol* **79**: 4694–4700.

Marritt, S. J., Lowe, T. G., Bye, J., McMillan, D. G., Shi, L., Fredrickson, J. et al. (2012a). A functional description of CymA, an electron-transfer hub supporting anaerobic respiratory flexibility in *Shewanella*. *Biochem J* **444**: 465–474.

Marritt, S. J., McMillan, D. G., Shi, L., Fredrickson, J. K., Zachara, J. M., Richardson, D. J. et al. (2012b). The roles of CymA in support of the respiratory flexibility of *Shewanella oneidensis* MR-1. *Biochem Soc Trans* **40**: 1217–1221.

Marshall, M. J., Beliaev, A. S., Dohnalkova, A. C., Kennedy, D. W., Shi, L., Wang, Z. et al. (2006). *c*-Type cytochrome-dependent formation of U(IV) nanoparticles by *Shewanella oneidensis. PLoS Biol* **4**: e268.

Marsili, E., Baron, D. B., Shikhare, I. D., Coursolle, D., Gralnick, J. A., and Bond, D. R. (2008). *Shewanella* secretes flavins that mediate extracellular electron transfer. *Proc Natl Acad Sci U S A* **105**: 3968–3973.

McMillan, D. G., Marritt, S. J., Butt, J. N., and Jeuken, L. J. (2012). Menaquinone-7 is specific cofactor in tetraheme quinol dehydrogenase CymA. *J Biol Chem* **287**: 14215–14225.

McMillan, D. G., Marritt, S. J., Firer-Sherwood, M. A., Shi, L., Richardson, D. J., Evans, S. D. et al. (2013). Protein-protein interaction regulates the direction of catalysis and electron transfer in a redox enzyme complex. *J Am Chem Soc* **135**: 10550–10556.

Mehta, T., Childers, S. E., Glaven, R., Lovley, D. R., and Mester, T. (2006). A putative multicopper protein secreted by an atypical type II secretion system involved in the reduction of insoluble electron acceptors in *Geobacter sulfurreducens. Microbiology* **152**: 2257–2264.

Mehta, T., Coppi, M. V., Childers, S. E., and Lovley, D. R. (2005). Outer membrane *c*-type cytochromes required for Fe(III) and Mn(IV) oxide reduction in *Geobacter sulfurreducens. Appl Environ Microbiol* **71**: 8634–8641.

Meitl, L. A., Eggleston, C. M., Colberg, P. J. S., Khare, N., Reardon, C. L., and Shi, L. (2009). Electrochemical interaction of *Shewanella oneidensis* MR-1 and its outer membrane cytochromes OmcA and MtrC with hematite electrodes. *Geochim Cosmochim Acta* **2009**: 5292–5307.

Mitchell, A. C., Peterson, L., Reardon, C. L., Reed, S. B., Culley, D. E., Romine, M. R., and Geesey, G. G. (2012). Role of outer membrane c-type cytochromes MtrC and OmcA in *Shewanella oneidensis* MR-1 cell production, accumulation, and detachment during respiration on hematite. *Geobiology* **10**: 355–370.

Morgado, L., Bruix, M., Pessanha, M., Londer, Y. Y., and Salgueiro, C. A. (2010). Thermodynamic characterization of a triheme cytochrome family from *Geobacter sulfurreducens* reveals mechanistic and functional diversity. *Biophys J* **99**: 293–301.

Muller, M., Mentel, M., van Hellemond, J. J., Henze, K., Woehle, C., Gould, S. B. et al. (2012). Biochemistry and evolution of anaerobic energy metabolism in eukaryotes. *Microbiol Mol Biol Rev* **76**: 444–495.

Murphy, J. N., and Saltikov, C. W. (2007). The *cymA* gene, encoding a tetraheme *c*-type cytochrome, is required for arsenate respiration in *Shewanella* species. *J Bacteriol* **189**: 2283–2290.

Myers, C. R., and Myers, J. M. (1997). Cloning and sequence of *cym*A, a gene encoding a tetraheme cytochrome c required for reduction of iron(III), fumarate, and nitrate by *Shewanella putrefaciens* MR-1. *J Bacteriol* **179**: 1143–1152.

Myers, C. R., and Myers, J. M. (2003). Cell surface exposure of the outer membrane cytochromes of *Shewanella oneidensis* MR-1. *Lett Appl Microbiol* **37**: 254–258.

Myers, C. R., and Nealson, K. H. (1988). Bacterial manganese reduction and growth with manganese oxide as the sole electron acceptor. *Science* **240**: 1319–1321.

Myers, C. R., and Nealson, K. H. (1990). Respiration-linked proton translocation coupled to anaerobic reduction of manganese(IV) and iron(III) in *Shewanella putrefaciens* MR-1. *J Bacteriol* **172**: 6232–6238.

Myers, J. M., and Myers, C. R. (2000). Role of the tetraheme cytochrome CymA in anaerobic electron transport in cells of *Shewanella putrefaciens* MR-1 with normal levels of menaquinone. *J Bacteriol* **182**: 67–75.

Nealson, K. H., Belz, A., and McKee, B. (2002). Breathing metals as a way of life: Geobiology in action. *Antonie Van Leeuwenhoek* **81**: 215–222.

Nealson, K. H., and Saffarini, D. (1994). Iron and manganese in anaerobic respiration: Environmental significance, physiology, and regulation. *Annu Rev Microbiol* **48**: 311–343.

Nevin, K. P., Kim, B. C., Glaven, R. H., Johnson, J. P., Woodard, T. L., Methe, B. A. et al. (2009). Anode biofilm transcriptomics reveals outer surface components essential for high density current production in *Geobacter sulfurreducens* fuel cells. *PLoS One* **4**: e5628.

Okamoto, A., Hashimoto, K., Nealson, K. H., and Nakamura, R. (2013). Rate enhancement of bacterial extracellular electron transport involves bound flavin semiquinones. *Proc Natl Acad Sci U S A* **110**: 7856–7861.

Okamoto, A., Kalathil, S., Deng, X., Hashimoto, K., Nakamura, R., and Nealson, K. H. (2014). Cell-secreted flavins bound to membrane cytochromes dictate electron transfer reactions to surfaces with diverse charge and pH. *Sci Rep* **4**: 5628.

Ortiz-Bernad, I., Anderson, R. T., Vrionis, H. A., and Lovley, D. R. (2004). Vanadium respiration by *Geobacter metallireducens*: Novel strategy for in situ removal of vanadium from groundwater. *Appl Environ Microbiol* **70**: 3091–3095.

Paquete, C. M., Fonseca, B. M., Cruz, D. R., Pereira, T. M., Pacheco, I., Soares, C. M., and Louro, R. O. (2014). Exploring the molecular mechanisms of electron shuttling across the microbe/metal space. *Front Microbiol* **5**: 318.

Pessanha, M., Morgado, L., Louro, R. O., Londer, Y. Y., Pokkuluri, P. R., Schiffer, M., and Salgueiro, C. A. (2006). Thermodynamic characterization of triheme cytochrome PpcA from *Geobacter sulfurreducens*: Evidence for a role played in e-/H+ energy transduction. *Biochemistry* **45**: 13910–13917.

Pina-Ochoa, E., Hogslund, S., Geslin, E., Cedhagen, T., Revsbech, N. P., Nielsen, L. P. et al. (2010). Widespread occurrence of nitrate storage and denitrification among *Foraminifera* and *Gromiida*. *Proc Natl Acad Sci U S A* **107**: 1148–1153.

Pirbadian, S., Barchinger, S. E., Leung, K. M., Byun, H. S., Jangir, Y., Bouhenni, R. A. et al. (2014). *Shewanella oneidensis* MR-1 nanowires are outer membrane and periplasmic extensions of the extracellular electron transport components. *Proc Natl Acad Sci U S A* **111**: 12883–12888.

Pirbadian, S., and El-Naggar, M. Y. (2012). Multistep hopping and extracellular charge transfer in microbial redox chains. *Phys Chem Chem Phys* **14**: 13802–13808.

Pitts, K. E., Dobbin, P. S., Reyes-Ramirez, F., Thomson, A. J., Richardson, D. J., and Seward, H. E. (2003). Characterization of the *Shewanella oneidensis* MR-1 decaheme cytochrome MtrA: Expression in *Escherichia coli* confers the ability to reduce soluble Fe(III) chelates. *J Biol Chem* **278**: 27758–27765.

Pokkuluri, P. R., Londer, Y. Y., Duke, N. E., Long, W. C., and Schiffer, M. (2004). Family of cytochrome *c7*-type proteins from *Geobacter sulfurreducens*: Structure of one cytochrome *c7* at 1.45 A resolution. *Biochemistry* **43**: 849–859.

Qian, X., Mester, T., Morgado, L., Arakawa, T., Sharma, M. L., Inoue, K. et al. (2011). Biochemical characterization of purified OmcS, a c-type cytochrome required for insoluble Fe(III) reduction in *Geobacter sulfurreducens*. *Biochim Biophys Acta* **1807**: 404–412.

Qian, X., Reguera, G., Mester, T., and Lovley, D. R. (2007). Evidence that OmcB and OmpB of *Geobacter sulfurreducens* are outer membrane surface proteins. *FEMS Microbiol Lett* **277**: 21–27.

Reardon, C. L., Dohnalkova, A. C., Nachimuthu, P., Kennedy, D. W., Saffarini, D. A., Arey, B. W. et al. (2010). Role of outer-membrane cytochromes MtrC and OmcA in the biomineralization of ferrihydrite by *Shewanella oneidensis* MR-1. *Geobiology* **8**: 56–68.

Reardon, P. N., and Mueller, K. T. (2013). Structure of the type IV a major pilin from the electrically conductive bacterial nanowires of *Geobacter sulfurreducens*. *J Biol Chem* **288**: 29260–29266.

Reguera, G., McCarthy, K. D., Mehta, T., Nicoll, J. S., Tuominen, M. T., and Lovley, D. R. (2005). Extracellular electron transfer via microbial nanowires. *Nature* **435**: 1098–1101.

Renslow, R., Babauta, J., Majors, P., and Beyenal, H. (2013). Diffusion in biofilms respiring on electrodes. *Energy Environ Sci* **6**: 595–607.

Richardson, D. J. (2000). Bacterial respiration: A flexible process for a changing environment. *Microbiology* **146 (Pt 3)**: 551–571.

Richardson, D. J., Butt, J. N., and Clarke, T. A. (2013). Controlling electron transfer at the microbe-mineral interface. *Proc Natl Acad Sci U S A* **110**: 7537–7538.

Richardson, D. J., Butt, J. N., Fredrickson, J. K., Zachara, J. M., Shi, L., Edwards, M. J. et al. (2012). The "porin-cytochrome" model for microbe-to-mineral electron transfer. *Mol Microbiol* **85**: 201–212.

Richter, K., Schicklberger, M., and Gescher, J. (2012). Dissimilatory reduction of extracellular electron acceptors in anaerobic respiration. *Appl Environ Microbiol* **78**: 913–921.

Risgaard-Petersen, N., Langezaal, A. M., Ingvardsen, S., Schmid, M. C., Jetten, M. S., Op den Camp, H. J. et al. (2006). Evidence for complete denitrification in a benthic foraminifer. *Nature* **443**: 93–96.

Ross, D. E., Brantley, S. L., and Tien, M. (2009). Kinetic characterization of OmcA and MtrC, terminal reductases involved in respiratory electron transfer for dissimilatory iron reduction in *Shewanella oneidensis* MR-1. *Appl Environ Microbiol* **75**: 5218–5226.

Ross, D. E., Flynn, J. M., Baron, D. B., Gralnick, J. A., and Bond, D. R. (2011). Towards electrosynthesis in *Shewanella*: Energetics of reversing the *mtr* pathway for reductive metabolism. *PLoS One* **6**: e16649.

Ross, D. E., Ruebush, S. S., Brantley, S. L., Hartshorne, R. S., Clarke, T. A., Richardson, D. J., and Tien, M. (2007). Characterization of protein-protein interactions involved in iron reduction by *Shewanella oneidensis* MR-1. *Appl Environ Microbiol* **73**: 5797–5808.

Rosso, K. M., Zachara, J. M., Fredrickson, J. K., Gorby, Y. A., and Smith, S. C. (2003). Nonlocal bacterial electron transfer to hematite surface. *Geochim Cosmochim Acta* **67**: 1081–1087.

Schuetz, B., Schicklberger, M., Kuermann, J., Spormann, A. M., and Gescher, J. (2009). Periplasmic electron transfer via the *c*-type cytochromes MtrA and FccA of *Shewanella oneidensis* MR-1. *Appl Environ Microbiol* **75**: 7789–7796.

Schwalb, C., Chapman, S. K., and Reid, G. A. (2003). The tetraheme cytochrome CymA is required for anaerobic respiration with dimethyl sulfoxide and nitrite in *Shewanella oneidensis*. *Biochemistry* **42**: 9491–9497.

Shi, L., Belchik, S. M., Wang, Z., Kennedy, D. W., Dohnalkova, A. C., Marshall, M. J. et al. (2011). Identification and characterization of $\text{UndA}_{\text{HRCR-6}}$, an outer membrane endecaheme *c*-type cytochrome of *Shewanella* sp. strain HRCR-6. *Appl Environ Microbiol* **77**: 5521–5523.

Shi, L., Chen, B., Wang, Z., Elias, D. A., Mayer, M. U., Gorby, Y. A. et al. (2006). Isolation of a high-affinity functional protein complex between OmcA and MtrC: two outer membrane decaheme *c*-type cytochromes of *Shewanella oneidensis* MR-1. *J Bacteriol* **188**: 4705–4714.

Shi, L., Deng, S., Marshall, M. J., Wang, Z., Kennedy, D. W., Dohnalkova, A. C. et al. (2008). Direct involvement of type II secretion system in extracellular translocation of *Shewanella oneidensis* outer membrane cytochromes MtrC and OmcA. *J Bacteriol* **190**: 5512–5516.

Shi, L., Lin, J. T., Markillie, L. M., Squier, T. C., and Hooker, B. S. (2005). Overexpression of multi-heme C-type cytochromes. *Biotechniques* **38**: 297–299.

Shi, L., Richardson, D. J., Wang, Z., Kerisit, S. N., Rosso, K. M., Zachara, J. M., and Fredrickson, J. K. (2009). The roles of outer membrane cytochromes of *Shewanella* and *Geobacter* in extracellular electron transfer. *Environ Microbiol Rep* **1**: 220–227.

Shi, L., Rosso, K. M., Clarke, T. A., Richardson, D. J., Zachara, J. M., and Fredrickson, J. K. (2012a). Molecular underpinnings of Fe(III) oxide reduction by *Shewanella oneidensis* MR-1. *Front Microbiol* **3**: 50.

Shi, L., Rosso, K. M., Zachara, J. M., and Fredrickson, J. K. (2012b). Mtr extracellular electron-transfer pathways in Fe(III)-reducing or Fe(II)-oxidizing bacteria: A genomic perspective. *Biochem Soc Trans* **40**: 1261–1267.

Shi, L., Squier, T. C., Zachara, J. M., and Fredrickson, J. K. (2007). Respiration of metal (hydr)oxides by *Shewanella* and *Geobacter*: A key role for multihaem *c*-type cytochromes. *Mol Microbiol* **65**: 12–20.

Shi, Z., Zachara, J., Wang, Z., Shi, L., and Fredrickson, J. (2013). Reductive dissolution of goethite and hematite by reduced flavins. *Geochim. Cosmochim. Acta* **121**: 139–154.

Shi, Z., Zachara, J. M., Shi, L., Wang, Z., Moore, D. A., Kennedy, D. W., and Fredrickson, J. K. (2012c). Redox reactions of reduced flavin mononucleotide (FMN), riboflavin (RBF), and anthraquinone-2,6-disulfonate (AQDS) with ferrihydrite and lepidocrocite. *Environ Sci Technol* **46**: 11644–11652.

Shirodkar, S., Reed, S., Romine, M., and Saffarini, D. (2011). The octahaem SirA catalyses dissimilatory sulfite reduction in *Shewanella oneidensis* MR-1. *Environ Microbiol* **13**: 108–115.

Snider, R. M., Strycharz-Glaven, S. M., Tsoi, S. D., Erickson, J. S., and Tender, L. M. (2012). Long-range electron transport in *Geobacter sulfurreducens* biofilms is redox gradient-driven. *Proc Natl Acad Sci U S A* **109**: 15467–15472.

Stief, P., Fuchs-Ocklenburg, S., Kamp, A., Manohar, C. S., Houbraken, J., Boekhout, T. et al. (2014). Dissimilatory nitrate reduction by *Aspergillus terreus* isolated from the seasonal oxygen minimum zone in the Arabian Sea. *BMC Microbiol* **14**: 35.

Stolz, J. F., and Oremland, R. S. (1999). Bacterial respiration of arsenic and selenium. *FEMS Microbiol Rev* **23**: 615–627.

Strycharz-Glaven, S. M., Snider, R. M., Guiseppi-Elite, A., and Tender, L. M. (2011). On the electrical conductivity of microbial nanowires and biofilms. *Energy Environ Sci* **4**: 4366–4379.

Tang, X., Yi, W., Munske, G. R., Adhikari, D. P., Zakharova, N. L., and Bruce, J. E. (2007). Profiling the membrane proteome of *Shewanella oneidensis* MR-1 with new affinity labeling probes. *J Proteome Res* **6**: 724–734.

Torres, C. I., Marcus, A. K., Lee, H. S., Parameswaran, P., Krajmalnik-Brown, R., and Rittmann, B. E. (2010). A kinetic perspective on extracellular electron transfer by anode-respiring bacteria. *FEMS Microbiol Rev* **34**: 3–17.

Vargas, M., Malvankar, N. S., Tremblay, P. L., Leang, C., Smith, J. A., Patel, P. et al. (2013). Aromatic amino acids required for pili conductivity and long-range extracellular electron transport in *Geobacter sulfurreducens*. *mBio* **4**: e00105–e00113.

von Canstein, H., Ogawa, J., Shimizu, S., and Lloyd, J. R. (2008). Secretion of flavins by *Shewanella* species and their role in extracellular electron transfer. *Appl Environ Microbiol* **74**: 615–623.

Wagner, D. D., Hug, L. A., Hatt, J. K., Spitzmiller, M. R., Padilla-Crespo, E., Ritalahti, K. M. et al. (2012). Genomic determinants of organohalide-respiration in *Geobacter lovleyi*, an unusual member of the *Geobacteraceae*. *BMC Genomics* **13**: 200.

Wang, Y., Sevinc, P. C., Belchik, S. M., Fredrickson, J., Shi, L., and Lu, H. P. (2013). Single-cell imaging and spectroscopic analyses of Cr(VI) reduction on the surface of bacterial cells. *Langmuir* **29**: 950–956.

Wee, S. K., Burns, J. L., and DiChristina, T. J. (2014). Identification of a molecular signature unique to metal-reducing Gammaproteobacteria. *FEMS Microbiol Lett* **350**: 90–99.

White, G. F., Shi, Z., Shi, L., Dohnalkova, A. C., Fredrickson, J. K., Zachara, J. M. et al. (2012). Development of a proteoliposome model to probe transmembrane electron-transfer reactions. *Biochem Soc Trans* **40**: 1257–1260.

White, G. F., Shi, Z., Shi, L., Wang, Z., Dohnalkova, A. C., Marshall, M. J. et al. (2013). Rapid electron exchange between surface-exposed bacterial cytochromes and Fe(III) minerals. *Proc Natl Acad Sci U S A* **110**: 6346–6351.

Wrighton, K. C., Thrash, J. C., Melnyk, R. A., Bigi, J. P., Byrne-Bailey, K. G., Remis, J. P. et al. (2011). Evidence for direct electron transfer by a gram-positive bacterium isolated from a microbial fuel cell. *Appl Environ Microbiol* **77**: 7633–7639.

Xiong, Y., Shi, L., Chen, B., Mayer, M. U., Lower, B. H., Londer, Y. et al. (2006). High-affinity binding and direct electron transfer to solid metals by the *Shewanella oneidensis* MR-1 outer membrane *c*-type cytochrome OmcA. *J Am Chem Soc* **128**: 13978–13979.

Zhang, H., Tang, X., Munske, G. R., Tolic, N., Anderson, G. A., and Bruce, J. E. (2009). Identification of protein-protein interactions and topologies in living cells with chemical cross-linking and mass spectrometry. *Mol Cell Proteomic* **8**: 409–420.

Zhang, H., Tang, X., Munske, G. R., Zakharova, N., Yang, L., Zheng, C. et al. (2008). In vivo identification of the outer membrane protein OmcA-MtrC interaction network in *Shewanella oneidensis* MR-1 cells using novel hydrophobic chemical cross-linkers. *J Proteome Res* **7**: 1712–1720.

CHAPTER 9

Unravelling New Metabolic Pathways

Supramolecular Organization of Aerobic Bacteria Respiratory Chains

Ana M.P. Melo, Emma B. Gutiérrez-Cirlos and Miguel Teixeira

CONTENTS

9.1 PROTOTYPE: MAMMALIAN MITOCHONDRIA

Oxidative phosphorylation is the process by which aerobic respiratory chains use the energy of the oxidation of reducing equivalents, conserved in a transmembrane ion-motive force, to synthesize adenosine triphosphate (ATP). In eukaryotes, this process occurs in the inner membrane of mitochondria, and in prokaryotes in the membrane.

Mammalian mitochondria are the paradigm of the aerobic respiratory chains, which comprise a set of four major protein complexes, nicotinamide adenine dinucleotide (NADH):ubiquinone oxidoreductase (complex I), succinate:ubiquinone oxidoreductase (complex II), ubiquinol:cytochrome *c* oxidoreductase (complex III) and cytochrome *c*:oxygen oxidoreductase (complex IV), linked by lipophilic and soluble electron carriers, ubiquinone

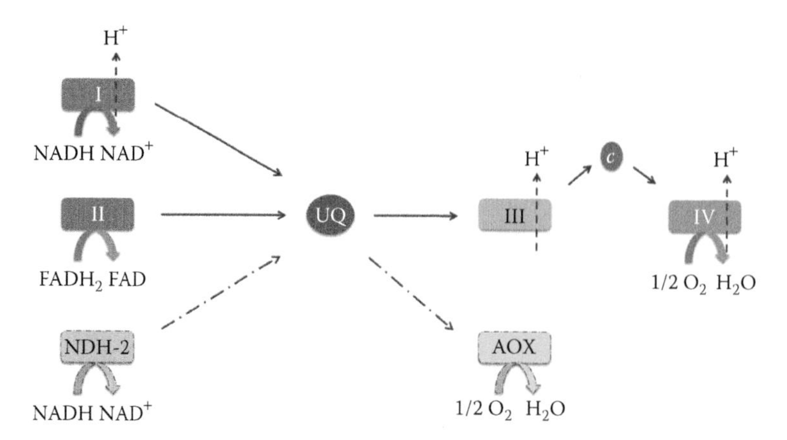

FIGURE 9.1 Mammalian mitochondria electron transfer chain. In dashed boxes, the alternative enzymes present in the mitochondria of plants and fungi. I, II, III and IV; NADH:ubiquinone oxidoreductase (complex I); succinate:ubiquinone oxidoreductase (complex II); ubiquinol:cytochrome *c* oxidoreductase (complex III) and cytochrome *c*:oxygen oxidoreductase (complex IV), respectively. NDH-2, type II or alternative NADH dehydrogenase, AOX, alternative oxidase. Solid arrows represent the direction of electron flow; dashed when following the alternative pathways.

and cytochrome *c* (Figure 9.1). Concomitantly with the downhill transfer of electrons through the respiratory chain, there is the uphill proton uptake from the matrix to the intermembrane space via complexes I, III and IV, generating the proton motive force that drives ATP synthase (complex V) towards the synthesis of ATP (Hatefi, 1985), or is directly used for motility and ion transport.

Mitochondria from plants, fungi and protists present alternative enzymes to the mammalian oxidative phosphorylation pathway. Different types of alternative rotenone-insensitive NAD(PH):quinone oxidoreductases, also designated type II NAD(P)H dehydrogenases, have been identified (e.g. Melo et al., 1996, 1999), some of which are sensitive to calcium (e.g. Melo et al., 2001). Unlike complex I, these enzymes are encoded by a single gene and form homo-dimers attached to the membrane via amphipathic helices. The so-called alternative oxidases (AOXs), absent from mammalian mitochondria, are the homodimeric quinol:oxygen oxidoreductases resistant to cyanide. Like type II NAD(P)H dehydrogenases, AOX do not translocate protons, thus uncoupling oxygen reduction from ATP synthesis, the energy of the proton-motive force being dissipated as heat. The latter may be used for instance in the volatilization of the cell odours used to attract insects in plant pollination (e.g. *Arum maculatum*, Wagner et al., 2008). The AOX active site of oxygen reduction is a non-heme di-iron carboxylate centre, which is buried in a four-helix bundle (Shiba et al., 2013).

9.1.1 Models of Organization of Mitochondrial Respiratory Chains

The organization of respiratory chains within the mitochondrial and cytoplasmic membranes has been a topic of intense debate since decades, which received recently major advances, taking advantage from the large development of techniques like electron microscopy and single particle analysis, and the strong improvements in membrane protein solubilization and analysis techniques such as blue native polyacrylamide gels (BN-PAGE), *in gel* activity detection and in mass spectrometry analysis, namely liquid chromatography coupled with tandem mass spectrometry (LC-MS/MS).

By the middle of the twentieth century, the solid state model was proposed to explain the structural and functional organization of the respiratory chains in the inner mitochondrial membrane, based on the knowledge that the transfer of electrons from reducing substrates to oxygen occurred via a sequentially organized cytochrome pathway, that included cytochromes *b*, *c*, *a* and a_3. These enzymes would be tightly packed in a spatial arrangement ensuring substrate accessibility and a high catalytic rate (Chance and Williams, 1955). The recent isolation of respirasomes in mitochondria from plants, mammals and fungi, and from bacteria further supported this model.

Nonetheless, by the sixties, the isolation and reconstitution of functionally active individual respiratory complexes (Hatefi et al., 1962), and the demonstration that the electron transfer from different electron donors mediated by ubiquinone and cytochrome *c* required multiple collisions and occurred across long range distances led Hackenbrock and co-workers to propose the random collision model. In accordance with this model, the respiratory enzymes are not in permanent contact with each other, but diffuse laterally and independently of each other, and the electron transfer is multi-collisional and directly dependent on the diffusion rates of the redox components (Hackenbrock et al., 1986). Later on, fluorescence recovery after photobleaching (FRAP) experiments supplied further evidences for this model and, in particular, determined diffusion coefficients of complexes III and IV, and of cytochrome *c* and ubiquinone (Gupte et al., 1984).

The so-called plasticity model reconciles the former proposals postulating the coexistence of populations of individual respiratory components with supramolecular assemblies of these components, in a dynamic model where individual components assemble and disassemble, presumably, in response to cell types and metabolic requirements (Acín-Pérez and Enriquez, 2014). This model was further supported by experiments in potato mitochondria, where variations in the amount and composition of supercomplexes within different cell types were observed (Acín-Pérez et al., 2008), and the supercomplex composition varied with pH and oxygen availability (Ramírez-Aguilar et al., 2011). More recently, a human complex I intermediate that acted as a scaffold for the assembly of complexes III and IV, in a biogenesis process concluded by the assembly of complex I catalytic subunit was proposed (Moreno-Lastres et al., 2012).

In this chapter we will focus on the respiratory chain supercomplexes of prokaryotic organisms. Nevertheless, it is worth mentioning that the supramolecular organization of mitochondrial respiratory chains was observed in all domains of Eukaryotic life, particularly in the form of respirasomes (supramolecular assembly of complexes I, III and IV) and multiple oligomerizations of ATP synthase.

The supramolecular organization of respiratory chains is acknowledged to optimize oxidative phosphorylation, namely by allowing substrate channelling, individual complexes stability (Acín-Pérez et al., 2004) and tight regulation of reactive oxygen species (ROS) production (Diaz et al., 2012).

9.2 PROKARYOTIC AEROBIC RESPIRATORY CHAINS

In this section, general features of aerobic respiratory pathways of the Bacteria and Archaea domains will be considered and examples of their respiratory supercomplexes reported to date will be described, with special focus on the bacterial supramolecular structures.

The different number of membranes and the characteristics of prokaryotic cell walls are also distinctive among Archaea and Bacteria and among types of Bacteria. The most well studied Bacteria belong to Gram-positive and Gram-negative types. The first group is characterized by having a single membrane, the cytoplasmic membrane, and several peptidoglycan layers, the latter reacting positively to the Gram test. In contrast, Gram-negative bacteria contain a cytoplasmic membrane surrounded by a single peptidoglycan layer, in turn surrounded by an outer membrane, and these microorganisms test negatively to the Gram staining. Archaea, as Gram-positive bacteria, contain only the cytoplasmic membrane, but in contrast to Bacteria, their cell wall lacks peptidoglycans, and S-layers are the only cell wall component external to the cytoplasmic membrane (Sleytr and Sára, 1997).

The prokaryotic counterparts of mitochondrial respiratory chains, located in the cytoplasmic membranes of Bacteria and Archaea, are more flexible, including not only functional and structural variants of mammalian mitochondria complexes, but also alternative pathways, whose expression is essentially regulated by environmental conditions like oxygen tension and substrate availability. As mentioned above, the main focus of this section will be Bacteria.

Aerobic bacteria respiratory chains have three main distinct types of NADH:quinone oxidoreductases and all possible combinations of coexistence of these enzymes can be observed (Melo et al., 2004). NDH-1, the bacterial homologue of complex I, is composed by 13 or 14 subunits disposed in two perpendicular arms. The soluble arm is projected into the cytoplasm, where NADH oxidation occurs and through which the electrons are transferred to ubiquinone via a set of redox co-factors, namely a non-covalently bound FMN molecule and several $[2Fe_2S]^{2+/1+}$ and $[4Fe_4S]^{2+/1+}$ clusters, while the other arm, responsible for ion translocation, is embedded in the membrane (Yagi et al., 2001). Although there are reports suggesting that NDH-1 from *Escherichia coli* (Steuber et al., 2000; Stolpe and Friedrich, 2004), *Rhodothermus marinus* (Batista et al., 2010) and *Klebsiella pneumoniae* (Gemperli et al., 2002) may translocate sodium ions, protons are the ions most commonly translocated by NDH-1. Type II NAD(P)H:quinone oxidoreductases, in prokaryotes also called NDH-2, like their eukaryotic homologues are soluble single polypeptides interacting with membrane surfaces via amphipathic helices and have often been described as homo-dimers. The third type of NADH:quinone oxidoreductase is a sodium translocating membranous enzyme (Na^+-NQR), often present in marine and pathogenic bacteria. It is composed by six subunits and harbours a $[2Fe\text{-}2S]^{2+/1+}$ cluster, a non-covalently bound FAD, two covalently bound FMN and one non-covalently bound riboflavin; none of these subunits displays significant similarity with NDH-1 and NDH-2 (Dimroth and Thomer, 1989; Nakayama et al., 1998; Verkhovsky and Bogachev, 2010).

Succinate:quinone oxidoreductase, which besides the respiratory chain also takes part in the Krebs cycle, is a transmembrane enzyme that catalyzes the transfer of two electrons from succinate to quinone, without ion translocation across the membrane (Lancaster, 2002). In prokaryotes, it is composed by 3 or 4 subunits, depending on the number of hydrophobic anchors of the enzyme. The soluble subunits face the cytoplasm, one carrying a FAD co-factor and the other assembling three iron-sulphur clusters, generally one of each type, $[2Fe_2S]^{2+/1+}$, $[4Fe_4S]^{2+/1+}$ and $[3Fe_4S]^{1+/0}$. The hydrophobic subunits of succinate:quinone oxidoreductases may contain one, two or no *b*-type hemes.

Other types of quinone reductases have been reported in prokaryotes, some of which will be addressed in later sections.

Depending on the environmental conditions, prokaryotes synthesize distinct types of quinones with different reduction potentials (e.g. ubiquinone, menaquinone and caldariella quinone) (Pereira et al., 2004). The nature of the present quinone is also related to the downstream pathways that the electrons will follow, though several different pathways may coexist in a single microorganism. These may comprise a single quinol:oxygen reductase, like the cytochromes bo_3, aa_3, *bd* or AOX, or include a set of proteins organized in a similar way to the mitochondrial cytochrome pathway. In the latter case, such pathways contain a cytochrome bc_1 or analogue enzyme that accepts electrons from the quinol pool to reduce cytochrome *c* (e.g. Stelter et al., 2008), high potential iron-sulphur protein (HiPIP) (e.g. Stelter et al., 2010), which in turn will reduce oxygen via oxygen reductases that can be of aa_3, caa_3, cbb_3 and ba_3 types (see Figure 9.2). Proton translocation is performed by all these enzymes, except cytochrome *bd* and AOX.

The bc_1 complex present in many species of bacteria is similar in multiple aspects to the cytochrome b_6f (Crofts, 2004). In bacteria, it is generally composed by three subunits, the Rieske Fe-S protein, the cytochrome *b* subunit and the c_1 subunit (Berry et al., 2004). The Rieske and the cytochrome *b* subunits are conserved between different species, while the catalytic subunit varies between phyla. In fact, the cytochrome c_1-like subunit is only present in α-, β- and γ-proteobacteria (Schütz et al., 2000).

As mentioned above, prokaryotes express a wide variety of oxygen reductases, which are classified into three main groups: (1) heme-copper oxygen reductases, (2) cytochrome *bd* oxygen reductases and (3) AOXs.

The superfamily of heme-copper oxygen reductases was divided into three main types, namely A, B and C, according to the characteristics of intra-protein proton channels (Pereira et al., 2001). Mitochondrial cytochrome aa_3 (complex IV) is the prototype of the A family, where the closely related aa_3 enzymes from α-proteobacteria belong. Heme-copper oxygen reductases translocate protons across the membrane concomitantly with oxygen reduction. Generally, these contain four redox groups, located in subunits I and II. Subunit I has most commonly 12-transmembrane helices, which assemble a heme of type *a* or *b*, and a binuclear heme-copper centre complex, where oxygen reduction takes place. The second heme may be of the *a*, *b* or *o* type and carries the suffix 3 that distinguishes the high-spin heme from the former low-spin heme. Generally, subunits II have two transmembrane helices and a globular domain projected into the periplasmic side of the membrane, where a binuclear Cu_A centre, absent in quinol oxygen reductases, accepts electrons from cytochrome *c* or analogous electron carriers (Pereira et al., 2001). These enzymes are designated as aa_3, caa_3,

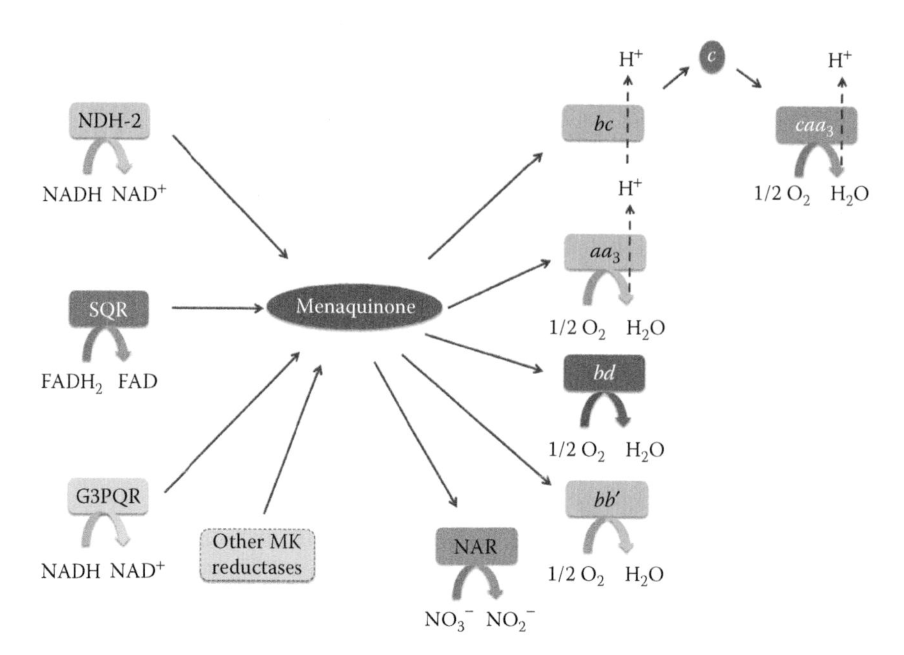

FIGURE 9.2 Aerobic respiratory chain of *B. subtilis*. G3PQR, glycerol-3-phosphate:menaquinone oxidoreductase; MK, menaquinone; SQR, succinate:menaquinone oxidoreductase; aa_3, *bb*', and *bd*, menaquinol:oxygen oxidoreductases; *bc*, menaquinol:cytochrome *c* oxidoreductase; caa_3, cytochrome *c*:oxygen oxidoreductase. Although not part of the aerobic respiratory chain, nitrate reductase (NAR) is included in the present scheme since it was identified in *B. subtilis* membranes grown aerobically. Dashed arrows represent the proton translocation. Solid arrows represent the electron flow.

bo_3, ba_3 and cbb_3, irrespective of their family. The oxygen reductases of the C family, all cbb_3 enzymes, have as subunits 2 and 3 a mono and a di-heme cytochrome *c*, respectively (Buschmann et al., 2010).

Oxygen reductases of the *bd* type are non-proton-translocating two-subunit membrane enzymes that transfer electrons from quinol to oxygen. They accommodate a low-spin heme b_{558} in subunit I, and two high-spin hemes, b_{595} and *d*, in subunit II, forming the catalytic site. Although there is no proton translocation in this reaction, it is electrogenic, involving the uptake of four protons from the cytoplasmic side and the release of two protons in the periplasmic side of the membrane (Mogi et al., 2008).

The available information regarding prokaryotic AOXs is still scarce. As well as in mitochondria, F_0-F_1 ATP synthase couples the energy of the ion-motive force, conserved from the electron transfer reactions, to the phosphorylation of ADP into ATP. In *E. coli*, this enzyme is composed by eight different subunits: ab_2c_{10} subunits compose the oligomycin-sensitive membrane-anchored F_0 domain, and $\alpha_3\beta_3\gamma\delta\varepsilon$ subunits form the peripheric catalytic F_1 domain of the complex. F_0 couples proton translocation to ATP synthesis or hydrolysis by F_1, via a rotary mechanism. To date, and in contrast with mitochondrial ATP synthases, the bacterial counterparts have not been described as dimers, maybe in agreement with the straight shape of the cytoplasmic membrane.

9.2.1 Bacterial Respiratory Chain Supercomplexes

Escherichia coli and *Bacillus subtilis* are examples of Gram-negative and Gram-positive bacteria, respectively. These are often considered as paradigms for each group, given that both genomes have been completely sequenced. Moreover, both have been extensively studied in many aspects of their development, cell biology, molecular biology, and also with respect to their respiratory chains. As mentioned above, while Gram-positive bacteria have only one cytoplasmic membrane, the Gram-negative ones have two membranes, with separate types of protein pathways in each membrane, which leads to some metabolic *compartmentalization* (Lai et al., 2004).

9.2.1.1 Respiratory Supercomplexes of B. subtilis

B. subtilis lives in the soil and forms an endospore when the surroundings change unfavourably for growth. The genome of this bacterium was published in 1997 and its analysis showed 78 genes coding for respiratory chain complexes, and the ATP synthase (Kunst et al., 1997). Menaquinone-7 is the lipophilic electron carrier synthesized for *B. subtilis* electron transfer chain (Figure 9.2).

The main electron-entry points that reduce menaquinone in the *B. subtilis* aerobic respiratory chain are the *yjlD*-encoded NDH-2 (Bergsma et al., 1982) and the succinate: menaquinol oxidoreductase, whose transmembrane subunit harbours two *b*-type hemes (Hägerhäll et al., 1992). There are three other genes that code for NADH dehydrogenases (*yutJ, yumB and ydgI*) but, to date, there are no reports that relate these enzymes to the respiratory chain. Downstream menaquinone, the electrons may follow several pathways. One of these includes the unique *bc* complex, a menadiol:cytochrome *c* oxidoreductase, analogous to the mitochondrial bc_1 complex. It is composed of three subunits, QcrA, similar to Rieske-type iron-sulphur proteins; QcrB, similar to the cytochrome *b* subunit of the b_6f complex; and QcrC, resembling a fusion of subunit IV from b_6f complex with a *c*-type cytochrome. In *B. subtilis*, cytochromes c_{550} and c_{551}, differing from their soluble counterparts by displaying a membrane attaching element, may accomplish electron transfer from the *bc* complex to a caa_3 oxygen reductase (Bengtsson et al., 1999). In cytochrome c_{550}, this attaching element is a transmembrane domain in the N-terminus of the protein, while in cytochrome c_{551} the N-terminus of the protein has a Leu-Ala-Ala-Cys consensus sequence, distinctive of lipoproteins and binds a diacylglycerol lipid. *In vitro* experiments suggest that cytochrome c_{550} is the preferred substrate of the caa_3 complex (David et al., 1999).

Three O_2 reductases, menadiol:O_2 oxidoreductases may conclude the electron transfer pathway in *B. subtilis*, a proton-pumping aa_3 O_2 reductase and cytochromes *bd* and *bb*' (Azarkina et al., 1999; Lauraeus and Wikström, 1993; Winstedt and von Wachenfeldt, 2000).

B. subtilis aerobic respiratory chain relies solely on a menaquinone-dependent electron transfer to operate. In fact, this electron transfer was proposed to be kinetically controlled by membrane energization at the level of menaquinone reduction by the dehydrogenases (Azarkina and Konstantinov, 2002). Due to the negative reduction potential of the *B. subtilis* menaquinone/menaquinol couple (−80 mV), free menaquinol is prone to rapid autoxidation (Azarkina and Konstantinov, 2002).

Considering the lack of membrane *compartmentalization* of Gram-positive bacteria, *B. subtilis* respiratory chain proteins may interact with a variety of proteins with other functions, which in some cases may facilitate cell adaptation to new environmental conditions. For instance, glucose limitation, in the growth and adaptation to stationary phase of *B. subtilis* strain IS58, triggers the presence of alternative carbon source transporters like mannitol, lactate and glycerol. The same condition promotes an increase in respiratory chain proteins like the *bc* complex, quinol oxygen reductase aa_3 and ATP synthase, and also TCA cycle proteins like CitZ, Icd and SdhABC (Dreisbach et al., 2008). The expression of genes involved in cellular energy pathways was affected by oxygen supply. Namely, ATP synthase, *bd*, caa_3 and aa_3 oxygen reductases are induced by low oxygen supply, showing a demand for ATP. *nasDE* (soluble nitrate reductase) and *narGHI* (nitrate reductase) were also induced by low oxygen supply (Yu et al., 2011).

Taking advantage of digitonin solubilization techniques, the availability of respiratory chain mutant strains, BN- and CN-PAGE among other meanwhile improved techniques, the supramolecular organization of *B. subtilis* was recently readdressed, confirming the existence of the previously suggested $bc{:}caa_3$ supercomplex and of an unexpected supercomplex composed by succinate:quinone oxidoreductase (SQR) and nitrate reductase (NAR) (García Montes de Oca et al., 2012; Sousa et al., 2013b) (Figure 9.3a, Table 9.1). Moreover, the dimeric assemblies of SQR and cytochrome caa_3 oxygen reductase were also observed. It is worth mentioning that these results were obtained from membranes of *B. subtilis* grown in distinct media, the conventional Luria Bertani broth and in 3% succinate super rich medium (Henning et al., 1995).

Three different stoichiometries of the $bc{:}caa_3$ supercomplex were retrieved, distinct with respect to their molecular masses, namely $(bc)_4{:}(caa_3)_2$, $(bc)_2{:}(caa_3)_4$ and $[(bc)_2(caa_3)_4]_2$ supercomplexes. This led Sousa et al. to propose that the largest form represents a megacomplex (Sousa et al., 2013b), possibly corresponding to the binding of two $(bc)_2{:}(caa3)_4$ building blocks, resulting the remaining forms from incomplete solubilization of this megacomplex. Such string-like organization was previously suggested for plant, mammalian and fungi mitochondria (Bultema et al., 2009; Heinemeyer et al., 2007; Wittig et al., 2006), but never previously observed in prokaryotic respiratory chains. Given that the CtaC (subunit II in type A oxygen reductases) (Saraste et al., 1991) subunit of *B. subtilis* caa_3 oxygen reductase has an additional covalently bound cytochrome *c*, it is possible that a functional supramolecular association of these enzymes does not require an independent cytochrome *c* to accomplish electron transfer. Nevertheless, García Montes de Oca et al. (2012) identified cytochrome c_{550} on the basis of TMBZ (3,3′,5,5′-tetramethylbenzidine) stained 2D-SDS PAGE of the supercomplex followed by mass spectrometry analysis. This finding could be attributed to the fact that membranes were obtained by lysozyme treatment, whereas Sousa et al. (2013b) obtained their membranes by French press disruption. Noteworthy, a $bc_1{:}aa_3$ supercomplex containing cytochrome c_{552} was found in the respiratory chain of *Paracoccus denitrificans* (Stroh et al., 2004). The widespread nature of this complex III: complex IV supercomplex or supercomplexes involving analogous enzymes, like the $bc{:}caa_3$, suggests that it may be crucial for a tight regulation of the respiratory chain cytochrome pathway.

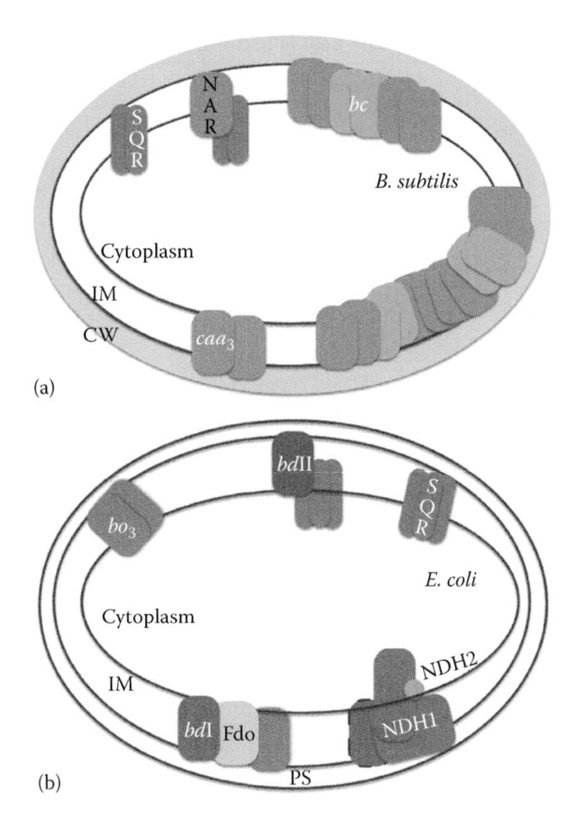

FIGURE 9.3 Supramolecular organization of the respiratory chains from *B. subtilis* (a) and *E. coli* (b), grown in the presence of oxygen. *bc*, menaquinol:cytochrome *c* oxidoreductase; *bd*I, *bd*II and bo_3, ubiquinol:oxygen oxidoreductases; caa_3, cytochrome *c*:oxygen oxidoreductase; CW, cell wall; Fdo, oxygen-induced formate dehydrogenase; IM, inner membrane; NAR, nitrate reductase; PS, periplasmic space and SQR, succinate:quinone oxidoreductase. A dashed border surrounds cytochrome *bd*I in the NADH oxidoreductase supercomplex, since this enzyme was not retrieved by mass spectroscopy or BN-page, although NADH:oxygen oxidoreductase activity was retrieved in solubilized membranes fractionated by sucrose gradient, and a positive correlation between *cydA*, *ndh* and *nuoF* was found in a Kendal correlation analysis. (Data from Sousa, P.M.F. et al., *Biochimie,* **93,** 418–425, 2011; Sousa, P.M.F. et al., *Microbiology,* **158,** 2408–2418, 2012.)

The SQR:NAR oxidoreductase supercomplex from *B. subtilis*, with a stoichiometry of 2:1 (García Montes de Oca et al., 2012; Sousa et al., 2013a), is the first example of a respiratory chain supercomplex that couples the electron transfer from the oxidation of succinate to the reduction of nitrate and raises the question of the association of an enzyme commonly associated with aerobic respiration to another consensually admitted to take part in anaerobic respiration. The presence of this supercomplex was confirmed by the methylviologen:nitrate oxidoreductase activity detected in the BN-PAGE, which was absent from SQR mutant's *B. subtilis* membranes (Sousa et al., 2013b).

It is tempting to speculate that a physiological justification for a $(SQR)_2$:NAR supercomplex may arise from the unique succinate oxidation reaction proposed to occur in

TABLE 9.1 Reported Prokaryotic Respiratory Chain Supercomplexes and Their Phylogenetic Distribution

Supercomplex	Stoichiometry	Organisms	Classification	References
NDH-1:bc_1:aa_3	1:4:4	Pd	B, GN, α-P	Stroh et al., 2004
NDH-1:NDH-2	1:1	Ec	B, GN, γ-P	Sousa et al., 2011
FDH-O:bo_3:*bd*I	1:1:1	Ec	B, GN, γ-P	Sousa et al., 2011, 2013b
SDH:*bd*II	nd	Ec	B, GN, γ-P	Sousa et al., 2012
SDH:aa_3	nd	Bs	B, GP	García Montes de Oca et al., 2012
SDH:NAR	1:1; 2:1	Bs	B, GP	García Montes de Oca et al., 2012; Sousa et al., 2013b
SQR:bc_1:ba_3	nd	Aae	B, Aqu	Prunetti et al., 2010
bc_{1}:aa_3	4:4; 4:2	Cg, Pd	B, GP; B, GN, α-P	Berry and Trumpower, 1985; Niebisch and Bott, 2003; Stroh et al., 2004
bc_1:cbb_3	nd	Bj	B, GN, α-P	Keefe and Maier, 1993
bc_1:caa_3	1:1	PS3	B, GP	Sone et al., 1987
bc:caa_3	2(2:4); 2:4; 4:2	Bs	B, GP	García Montes de Oca et al., 2012; Sousa et al., 2013b
bcc:aa_3	nd	Ms	B, Act	Megehee et al., 2006
ACIII:caa_3	nd	Rm	B, BC	Refojo et al., 2010
caa_3:F_1F_O-ATPase	1:1	Bp	B, GP	Liu et al., 2007
aa_3:F1-ATPase	nd	Bs	B, GP	García Montes de Oca et al., 2012
SoxABCD	nd	Sa	A, C	Lübben et al., 1994
SoxM	nd	Sa	A, C	Komorowski et al., 2002
Cyc2:Rcy:Cyc1:caa_3	nd	Af	B, P	Castelle et al., 2008
HYD:SR	nd	Aae, Aam, Pa, Tn	B, Aqu; A, C	Dirmeier et al., 1998; Guiral et al., 2005; Laska and Kletzin, 2000; Laska et al., 2003

Notes: Pd, *P. denitrificans*; Bj, *B. japonicum*; Rm, *R. marinus*; Aae, *A. aeolicus*; Af, *A. ferrooxidans*; Ec, *E. coli*; Bs, *B. subtilis*; BP, *B. pseudofirmus* OF4; PS3, *Bacillus sp.* PS3; Cg, *C. glutamicum*; Ms, *M. smegmatis*; Sc, *S. acidocaldarius*; Pa, *P. abyssi*; Tn, *T. neutrophilus*; Aam, *A. ambivalens*. ACIII, alternative complex III; FDF-O, oxygen expressed formate dehydrogenase; HYD, hydrogenase; NAR, nitrate reductase; SDH, succinate dehydrogenase; SQR, sulphide:quinone oxidoreductase; SR, sulphur reductase. Rcy, Rusticyanin; A, Archae; Act, Actinobacteria; Aqu, Aquificae; B, Bacteria; BC, Bacteroidetes-Chlorobi; C, Crenarchaeota; E, Euryarchaeota; GN, Gram-negative; GP, Gram-positive; P, Proteobacteria.

menaquinone-dependent SQRs (Lancaster et al., 2005; Schirawski and Unden, 1998) together with the fact that the *B. subtilis* NAR is likely to catalyze the reduction of nitrate to nitrite in the cytoplasm, with the concomitant uptake of two protons from the cytoplasm (Bertero et al., 2003). In fact, the succinate oxidation mediated by menaquinone-dependent SQRs involves the transfer of two electrons to menaquinone in an endergonic process, due to the higher reduction potential of the succinate/fumarate redox pair relatively to the one of menaquinone/menaquinol. This thermodynamic feature was suggested to be circumvented by the concomitant uptake of two protons from the inner wall zone to the menaquinone binding site, which could be performed by the NAR within the supercomplex. It is noteworthy that besides oxygen, the preferred terminal electron acceptor of *B. subtilis* respiratory chain is nitrate.

9.2.1.2 Supercomplexes in Other Gram-Positive Bacteria

A bc_1:caa_3 supercomplex was identified in the thermophilic bacterium PS3, long before the BN- and CN-PAGE development and the use of milder detergents in the membrane solubilization process (Sone et al., 1987) (Table 9.1). A stoichiometry of 1:1 was proposed to this supercomplex based on cross-linking experiments (Tanaka et al., 1996). In *Mycobacterium smegmatis* (Megehee et al., 2006) and *Corynebacterium glutamicum* (Niebisch and Bott, 2003) respiratory chains, supramolecular associations of the *bcc* complex (equivalent to the *B. subtilis bc* complex) and cytochrome aa_3 oxygen reductase were described (Table 9.1). In the alkaliphilic *Bacillus pseudofirmus* OF4, cytochrome caa_3 and F_1F_0-ATP synthase interact directly, suggesting a physical association between these complexes (Liu et al., 2007) (Table 9.1).

9.2.1.3 Respiratory Supercomplexes of E. coli

E. coli is a facultative anaerobe, γ-proteobacterium whose aerobic respiratory chain is located in the cytoplasmic membrane. It is composed by several quinone reductases, namely NADH:, succinate:, lactate:, glycerol-1-phosphate:, and, already as a result of investigating its supercomplexes, formate:quinone oxidoreductases, and three quinol:oxygen oxidoreductases (cytochromes bo_3, *bd*I and *bd*II) (Sousa et al., 2011; Unden and Bongaerts, 1997) (Figure 9.4). This respiratory chain has many different types of NADH:quinone oxidoreductases (e.g. NDH-1, NDH-2, WrbA, YhdH and QOR), of which NDH-1 and NDH-2 are the most important.

The expression of *E. coli* respiratory enzymes, similar to other bacteria, is growth phase dependent and conditioned by substrate availability (Kita et al., 1984a; Unden and Bongaerts, 1997).

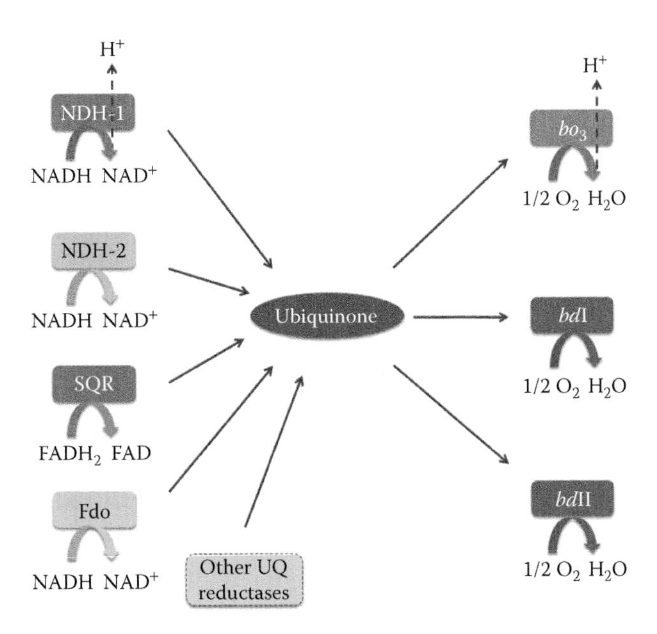

FIGURE 9.4 Aerobic respiratory chain of *E. coli*. *bd*I, *bd*II, bo_3, ubiquinol:oxygen oxidoreductases; FDO, oxygen-induced formate dehydrogenase; SQR, succinate:quinone oxidoreductase; and UQ, ubiquinone. Solid arrows represent the direction of electron flow. Dashed arrows represent the H^+ translocation.

Until recently, the existence of aerobic respiratory supercomplexes in *E. coli* was not reported, and it was generally accepted that, since its respiratory chain was devoid of cytochrome bc_1 and assembled a detergent stable complex I, such supramolecular structure was not needed. However, based on the heterogeneous distribution of cytochrome *bd*I *in vivo*, and the possibility that complexes I, II and V may also share the same distribution pattern, it was speculated that there would be specialized patches with respiratory enzymes within the plasma membrane called *respirazones* (Lenn et al., 2008). Noteworthy, the same group has recently contradicted this hypothesis, claiming that there was no significant fluorescence overlapping between the selected respiratory enzyme pairs analyzed (Llorente-Garcia et al., 2014).

Challenging the state of the art, Melo and co-workers addressed the *in vitro* investigation of supramolecular assemblies in *E. coli* grown under aerobic conditions and detected several homo- and hetero-oligomerizations of respiratory enzymes. The authors found evidences for homo-dimers of cytochrome bo_3, and homo-trimers of succinate:quinone oxidoreductase, in addition to a NADH oxidoreductase supercomplex (Sousa et al., 2011), a formate:oxygen oxidoreductase (FdOx) (Sousa et al., 2011, 2013b) and a succinate:oxygen oxidoreductase supercomplexes (Sousa et al., 2012) (Figure 9.3b), based on the analyses of digitonin-solubilized membranes from wild type and respiratory mutants by BN-PAGE and sucrose gradients. LC/MS-MS was used to corroborate the identity of supercomplex components in BN-PAGE bands detected with NADH oxidoreductase activity.

The NADH oxidoreductase supercomplex is composed by NDH-1 and NDH-2 in a stoichiometry of 1:1, as deduced from comparison with the masses of the individual enzymes with the estimated mass of the supercomplex band detected in the BN-PAGE. The eventual quinol oxidizing partner could be cytochrome *bd*I, relying on the identification of the positive correlation observed between the gene transcription of NDH-1, NDH-2 and cytochrome *bd*I (Sousa et al., 2012). It is worth mentioning that this band has some NADH:O_2 oxidoreductase activity inhibited by potassium cyanide (KCN). Further experiments from different technical approaches are necessary to corroborate this hypothesis, which was discouraged by *in vivo* studies where NDH-1 and cytochrome *bd*I tagged with fluorescent probes did not show significant co-localization (Llorente-Garcia et al., 2014).

Additionally, the NADH:oxidoreductase supercomplex could be involved in the critical balance of the [NADH]/[NAD+] ratio in the cytoplasm, since NDH-2 is able to enhance NADH oxidation flux in the respiratory chain in comparison with NDH-1, and bypassing one proton-translocation site (Melo et al., 2004). The ~10 times higher K_M of NADH oxidation relative to NDH-1 may confer particular relevance to the former when high concentrations of NADH are present (Friedrich et al., 1994). It is also very likely that in *E. coli,* the assembly of a stable complex I occurs only in the presence of NDH-2, given the absence of NDH-1 in digitonin-solubilized membranes from a strain devoid of NDH-2 resolved by BN-PAGE (Sousa et al., 2011).

The FdOx supercomplex assembles cytochromes bo_3 and *bd*I, and the formate dehydrogenase (Fdo), in a 1:1:1 stoichiometry as inspected from its mass in BN-PAGE compared with the masses of the individual enzymes. Moreover, a principal component analysis of *E. coli* aerobic respiratory chain gene transcription and enzyme activity, and growth dynamics correlated formate:oxygen oxidoreductase activity and the transcription of the genes encoding cytochromes *bo3* and *bd*I, corroborating previous evidences that associated these complexes in FdOx. This was the first respirasome reported in *E. coli* and the first evidence of a new oxygen-terminated respiratory pathway involving the aerobic formate dehydrogenase (FDH-O). Noteworthy, this was also the first example of an oxygen-terminated supercomplex that did not include a cytochrome bc_1 or analogous protein, challenging previous theories that predicted the general requirement of its presence for supercomplex assembly.

Besides NADH:oxidoreductase activity, the oxidoreductase activity of formate was also detected in membranes of *E. coli* resolved by BN-PAGE, but compromised in membranes of strains where cytochromes bo_3 or *bd*I were inactivated. Bands excised from BN-PAGE resolved wild type membranes were also capable of cyanide sensitive formate:oxygen and quinol:oxygen oxidoreductase activity. Formate:oxygen oxidoreductase activity was measured in *E. coli* membranes, being KCN sensitive.

Considering that the determined K_M for formate oxidation by the FdOx supercomplex was 169 ± 21 μM as measured in *E. coli* membranes, in the range of that established for the anaerobic FDH-N K_M (120 μM) (Sawers, 1994), and the high similarity between FDH-O and FDH-N amino acid sequences, it can be expected that FDH-O and FDH-N share similar structures and catalytic functions, which are well established for the latter (Sawers, 2005). As such, FDH-O would be involved in the oxidation of the periplasmic formate, requiring the uptake of protons from the cytoplasm side of the membrane to the quinone pool, hence contributing to the proton-motive force with a H^+/e^- ratio of 1 (Jormakka et al., 2002). A respiratory pathway involving FDH-O and cytochromes *bd*I or bo_3 allows the drifting between two H^+/e^- ratios, depending on the electron flow to oxygen being accomplished via cytochrome bo_3 (Abramson et al., 2000) or cytochrome *bd*I (Borisov et al., 2011) oxygen reductases, enabling this supercomplex to modulate the magnitude of the proton-motive force. The distinct affinities of cytochromes *bd*I or bo_3 towards oxygen (Kita et al., 1984a,b) could be the oxygen reductase selection factor that permits the supercomplex to be active in a wide range of oxygen concentrations.

The succinate:oxygen oxidoreductase supercomplex, containing SQR and cytochrome *bd*II in an unknown stoichiometry, was the second oxygen-terminated supercomplex identified in the *E. coli* respiratory chain. It was suggested by a global quantitative analysis of the *E. coli* respiratory components and further corroborated by means of a mutant devoid of NDH-1 and cytochromes *bd*I and bo_3. The levels of expression of cytochrome *bd*II in wild type cells are so low that this supercomplex should be hard to observe. Nevertheless, in an *E. coli* strain where cytochrome *bd*II was disrupted, oxygen consumption due to succinate oxidation decreased by half in comparison to wild type. With respect to energy conservation by the succinate:oxygen oxidoreductase supercomplex, it will contribute exclusively via cytochrome *bd*II (Borisov et al., 2011).

In 2014, Friedrich's group detected fluorescent patches, larger than the respiratory supercomplexes identified so far (Erhardt et al., 2014). This observation is compatible not only with patches composed by oligomers of the same enzyme, like dimers of cytochrome bo_3 (Sousa et al., 2011; Stenberg et al., 2007) and trimers of SQR (Sousa et al., 2011;

Yankovskaya et al., 2003), but also with megacomplex organization of the respiratory chains as was proposed for the *bc*:*caa*$_3$ supercomplexes in *B. subtilis* (Sousa et al., 2013a) and to the respirasomes of mammalian (Wittig et al., 2006) and potato (Bultema et al., 2009) mitochondria. Unfortunately, and since the pairs of already identified supercomplex components were not tested in the above-mentioned study, nothing can be said about the *in vivo* assembly of the supramolecular structures already identified in *E. coli in vitro*.

Comprehensive quantitative multi-level analysis of the aerobic respiratory chain of *E. coli* included the monitorization of the transcription and activity of its components during growth, and the use of Kendall's rank coefficient and the principal component analysis (Sousa et al., 2012, 2013a). This quantitative study showed a tightly regulated respiratory chain, with selective correlations between its individual components, which may favour the assembly of specific supercomplexes. It is worth mentioning that the supercomplexes whose activity was observed in BN-PAGE were present throughout growth phases.

9.2.1.4 Supercomplexes in Other Gram-Negative Bacteria

In *P. denitrificans*, a fully functional $I_1 + III_4 + IV_4$ supercomplex, resembling closely its mitochondrial counterpart, was described, along with $III_4 + IV_4$ and $III_4 + IV_2$ assemblies (Berry and Trumpower, 1985; Stroh et al., 2004) (Table 9.1). In *Bradyrhizobium japonicum*, functional associations between *bc*$_1$ complex and cytochrome *cbb*$_3$ oxygen reductase were reported (Keefe and Maier, 1993) (Table 9.1). In *R. marinus*, a supercomplex of an alternative complex III and *caa*$_3$ oxygen reductase was identified (Refojo et al., 2010) (Table 9.1).

Two respiratory supercomplexes were characterized in *Aquifex aeolicus*. One is composed by a sulphide:quinone reductase, a dimeric *bc*$_1$ complex and a cytochrome *ba*$_3$ oxygen reductase, and was proposed to catalyze the reduction of oxygen and the oxidation of hydrogen sulphide (Gao et al., 2012; Prunetti et al., 2011) (Table 9.1). The other supercomplex is not related to the aerobic metabolism and assembles a hydrogenase and a sulphur reductase, coupling the reduction of sulphur with the periplasmic oxidation of molecular hydrogen (Guiral et al., 2005) (Table 9.1).

A respirasome with iron oxidase and oxygen reductase activity was characterized in the respiratory chain of *Acidithiobacillus ferrooxidans* (Castelle et al., 2008) (Table 9.1). It is composed by a *bc* complex and a cytochrome *c* (Cyc) besides the outer-membrane bound cytochrome *c* (Cyc2) and an oxidoreductase of the *aa*$_3$ type (Table 9.1).

9.2.2 Supercomplexes from Archaea

SoxABCD and SoxM are two supercomplexes with quinol:oxygen oxidoreductase activity, similar to the III–IV oligomerizations, described in the respiratory chains of *Sulfolobus acidocaldarius* and *Sulfolobus tokodai* strain 7 (Lübben et al., 1992; Iwasaki et al., 1995; Komorowski et al., 2002). In *S. acidocaldarius*, SoxABC is encoded by a single operon. SoxA and B are similar to subunits II and I of heme-copper oxygen oxidoreductases. SoxC is an analogue of complex III cytochrome *b*, where the hemes are of *a* type (Lübben et al., 1992). SoxM supercomplex is composed of six subunits (SoxEFGHIM) arranged in two functional subcomplexes, which are similar to the respiratory complexes III and IV, the electron transfer between them being mediated by sulfocyanin (SoxE subunit), a copper protein.

Subcomplex SoxM-SoxH contains two *b*-type hemes (*b* and b_3), a Cu_B centre and a binuclear Cu_A centre, being homologous to type A heme-copper oxygen reductases. The other subcomplex is composed of subunits SoxF, a homologue of complex III Rieske Fe-S protein subunit also harbouring a binuclear [Fe-S] cluster, and SoxG, a structural analogue to complex III cytochrome *b* subunit, which, like SoxC from the SoxABCD supercomplex, contains two *a*-type hemes (Komorowski et al., 2001, 2002). In the hyperthermophilic archaeons, *Pyrodictium abyssi* (Dirmeier et al., 1998), *Thermoproteus neutrophilus* (Laska and Kletzin, 2000) and *Acidianus ambivalens* (Laska et al., 2003), hydrogenase:sulphur reductase supercomplexes were also identified.

9.3 FINAL REMARKS

In the last 10 years, there has been accumulating evidence of respiratory supercomplexes in all life domains. Their constitution differs in the type and stoichiometry of each individual complex. Most important have been the structures modelled by electron microscopy and single particle analysis of mitochondrial supercomplexes and also the determination of other electron:oxygen oxidoreductases.

Mitochondrial-like bacteria, as *P. denitrificans*, demonstrated to have the NADH:O_2 oxidoreductase respirasome, like mitochondria from most eukaryotic organisms investigated, as well as sub-oligomerizations of complexes III and IV homologues. This finding opened the sights to move forward to investigate the supramolecular organization of the respiratory chains of other types of bacteria and also in Archaea. To date, such organization of respiratory chains was identified in a wide variety of bacterial organisms, from proteobacteria to mycobacteria and Gram-positive bacteria (Table 9.1). Respirasomes of different compositions were identified, according with the genome of different bacteria, involving both unforeseen quinone reducing and oxidizing enzymes. These achievements contributed to demonstrate that the supramolecular organization of respiratory chains is an ancient metabolic strategy widespread in nature, possibly to overcome the poor compartmentalization of bacterial enzymes. In addition, the study of prokaryotic respiratory chain supercomplexes unveiled unexpected pathways, opening new fronts of research for microbiologists and biochemists. Interestingly, to date, no homo-oligomers of ATP synthase have been observed in prokaryotes. This fact may be related to the smooth nature of prokaryotic membranes, in contrast with the invaginated inner mitochondria membranes (cristae). It is tempting to speculate that the dimerization of ATP synthase and the profuse oligomerization of its dimers are a mark of evolution of prokaryotes to eukaryotes, envisaging a more effective energy production network to attend the new energetic demands of the highly specialized eukaryotic cell.

In spite of the enormous technological progress of the electron microscope that coupled to single particle analyses may achieve up to 3 Å resolution and cryo-electron tomography that allows 10 Å resolution, to date, it was not possible to obtain structures of prokaryotic respiratory supercomplexes. In this case, the limitation relies on the small amounts of supercomplexes isolated. As already mentioned, the cytoplasmic membranes of prokaryotes contain the whole set of membrane proteins of these organisms, raising the need of more complex purification procedures. Moreover, the coexistence of several respiratory pathways in the same cell also makes it difficult to get enriched samples of the proteins we aim to purify. An alternative technique that is being used to overcome the limitation of supercomplex amounts is fluorescence co-localization, which this far was not able to support the BN-PAGE/MS-based results. The bacterial world offers, thus, a wide challenge for supercomplex researchers from microbiology to microscopy and fluorescence, very likely with strong support of molecular biology.

Many issues remain to be answered, the first being the observation and characterization of respiratory supercomplexes by *in vivo* techniques. Others rely on the confirmation of the physiological advantages associated with these supramolecular structures: substrate channelling, faster intra-molecular electron exchange, protein stability and avoidance of ROS formation. Answers to these and other questions will certainly come by in the forthcoming years.

REFERENCES

Abramson J, Riistama S, Larsson G, Jasaitis A, Svensson-Ek M, Laakkonen L, Puustinen A, Iwata S, and Wikström M. (2000). The structure of the ubiquinol oxidase from *Escherichia coli* and its ubiquinone binding site. *Nat Struct Biol* 7: 910–917.

Acín-Pérez R, Bayona-Bafaluy MP, Fernández-Silva P, Moreno-Loshuertos R, Pérez-Martos A, Bruno C, Moraes CT, and Enríquez JA. (2004). Respiratory complex III is required to maintain complex I in mammalian mitochondria. *Mol Cell* **13**: 805–815.

Acín-Pérez R, and Enriquez JA. (2014). The function of the respiratory supercomplexes: The plasticity model. *Biochim Biophys Acta* **1837**: 444–450.

Acín-Pérez R, Fernández-Silva P, Peleato ML, Pérez-Martos A, and Enriquez JA. (2008). Respiratory active mitochondrial supercomplexes. *Mol Cell* **32**: 529–539.

Azarkina N, and Konstantinov AA. (2002). Stimulation of menaquinone-dependent electron transfer in the respiratory chain of *Bacillus subtilis* by membrane energization. *J Bacteriol* **184**: 5339–5347.

Azarkina N, Siletsky S, Borisov V, von Wachenfeldt C, Hederstedt L, and Konstantinov AA. (1999). A cytochrome *bb*'-type quinol oxidase in *Bacillus subtilis* strain 168. *J Biol Chem* **274**: 32810–32817.

Batista AP, Fernandes AS, Louro RO, Steuber J, and Pereira MM. (2010). Energy conservation by *Rhodothermus marinus* respiratory complex I. *Biochim Biophys Acta* **1797**: 509–515.

Bengtsson J, Rivolta C, Hederstedt L, and Karamata D. (1999). *Bacillus subtilis* contains two small *c*-type cytochromes with homologous heme domains but different types of membrane anchors. *J Biol Chem* **274**: 26179–26184.

Bergsma J, Van Dongen MB, and Konings WN. (1982). Purification and characterization of NADH dehydrogenase from *Bacillus subtilis*. *Eur J Biochem* **128**: 151–157.

Berry EA, Huang LS, Saechao LK, Pon NG, Valkova-Valchanova M, and Daldal F. (2004). X-Ray structure of rhodobacter capsulatus cytochrome $bc_{(1)}$: Comparison with its mitochondrial and chloroplast counterparts. *Photosynth Res* **81**: 251–275.

Berry EA, and Trumpower BL. (1985). Isolation of ubiquinol oxidase from Paracoccus denitrificans and resolution into cytochrome bc_1 and cytochrome c-aa_3 complexes. *J Biol Chem* **260**: 2458–2467.

Bertero MG, Rothery RA, Palak M, Hou C, Lim D, Blasco F, Weiner JH, and Strynadka NC. (2003). Insights into the respiratory electron transfer pathway from the structure of nitrate reductase A. *Nat Struct Biol* **10**: 681–687.

Borisov VB, Murali R, Verkhovskaya ML, Bloch DA, Han H, Gennis RB, and Verkhovsky MI. (2011). Aerobic respiratory chain of *Escherichia coli* is not allowed to work in fully uncoupled mode. *Proc Natl Acad Sci U S A* **108**: 17320–17324.

Bultema JB, Braun HP, Boekema EJ, and Kouril R. (2009). Megacomplex organization of the oxidative phosphorylation system by structural analysis of respiratory supercomplexes from potato. *Biochim Biophys Acta* **1787**: 60–67.

Buschmann S, Warkentin E, Xie H, Langer JD, Ermler U, and Michel H. (2010). The structure of cbb_3 cytochrome oxidase provides insights into proton pumping. *Science* **329**: 327–330.

Castelle C, Guiral M, Malarte G, Ledgham F, Leroy G, Brugna M, and Giudici-Orticoni MT. (2008). A new iron-oxidizing/O_2-reducing supercomplex spanning both inner and outer membranes, isolated from the extreme acidophile *Acidithiobacillus ferrooxidans*. *J Biol Chem* **283**: 25803–25811.

Chance B, and Williams GR. (1955). A method for the localization of sites for oxidative phosphorylation. *Nature* **176**: 250–254.

Crofts AR, (2004). The cytochrome bc_1 complex: Function in the context of structure. *Annu Rev Physiol* **66**: 689–733.

David PS, Morrison MR, Wong SL, and Hill BC. (1999). Expression, purification, and characterization of recombinant forms of membrane-bound cytochrome *c*-550 nm from *Bacillus subtilis*. *Protein Expr Purif* **15**: 69–76.

Diaz F, Enríquez JA, and Moraes CT. (2012). Cells lacking Rieske iron-sulfur protein have a reactive oxygen species-associated decrease in respiratory complexes I and IV. *Mol Cell Biol* **32**: 415–429.

Dimroth P, and Thomer A. (1989). A primary respiratory Na+ pump of an anaerobic bacterium: The Na+-dependent NADH:quinone oxidoreductase of *Klebsiella pneumoniae*. *Arch Microbiol* **151**: 439–444.

Dirmeier R, Keller M, Frey G, Huber H, and Stetter KO. (1998). Purification and properties of an extremely thermostable membrane-bound sulfur-reducing complex from the hyperthermophilic *Pyrodictium abyssi*. *Eur J Biochem* **252**: 486–491.

Dreisbach A, Otto A, Becher D, Hammer E, Teumer A, Gouw JW, Hecker M, and Völker U. (2008). Monitoring of changes in the membrane proteome during stationary phase adaptation of *Bacillus subtilis* using *in vivo* labeling techniques. *Proteomics* **8**: 2062–2076.

Erhardt H, Dempwolff F, Pfreundschuh M, Riehle M, Schäfer C, Pohl T, Graumann P and Friedrich T. (2014). Organization of the *Escherichia coli* aerobic enzyme complexes of oxidative phosphorylation in dynamic domains within the cytoplasmic membrane. *Microbiologyopen* **3**: 316–326.

Friedrich T, van Heek P, Leif H, Ohnishi T, Forche E, Kunze B, Jansen R, Trowitzsch-Kienast W, Höfle G, and Reichenbach H. (1994). Two binding sites of inhibitors in NADH: Ubiquinone oxidoreductase (complex I). Relationship of one site with the ubiquinone-binding site of bacterial glucose:ubiquinone oxidoreductase. *Eur J Biochem* **219**: 691–698.

Gao Y, Meyer B, Sokolova L, Zwicker K, Karas M, Brutschy B, Peng G, and Michel H. (2012). Heme-copper terminal oxidase using both cytochrome *c* and ubiquinol as electron donors. *Proc Natl Acad Sci U S A* **109**: 3275–3280.

García Montes de Oca LY, Chagolla-López A, González de la Vara L, Cabellos-Avelar T, Gómez-Lojero C, and Gutiérrez Cirlos EB. (2012). The composition of the *Bacillus subtilis* aerobic respiratory chain supercomplexes. *J Bioenerg Biomembr* **44**: 473–486.

Gemperli AC, Dimroth P, and Steuber J. (2002). The respiratory complex I (NDH I) from *Klebsiella pneumoniae*, a sodium pump. *J Biol Chem* **277**: 33811–33817.

Guiral M, Tron P, Aubert C, Gloter A, Iobbi-Nivol C, and Giudici-Orticoni MT. (2005). A membrane-bound multienzyme, hydrogen-oxidizing, and sulfur-reducing complex from the hyperthermophilic bacterium *Aquifex aeolicus*. *J Biol Chem* **280**: 42004–42015.

Gupte S, Wu ES, Hoechli L, Hoechli M, Jacobson K, Sowers AE, and Hackenbrock CR. (1984). Relationship between lateral diffusion, collision frequency, and electron transfer of mitochondrial inner membrane oxidation-reduction components. *Proc Natl Acad Sci U S A* **81**: 2606–2610.

Hackenbrock CR, Chazotte B, and Gupte SS. (1986). The random collision model and a critical assessment of diffusion and collision in mitochondrial electron transport. *J Bioenerg Biomembr* **18**: 331–368.

Hägerhäll C, Aasa R, von Wachenfeldt C, and Hederstedt L. (1992). Two hemes in *Bacillus subtilis* succinate:menaquinone oxidoreductase (complex II). *Biochemistry* **31**: 7411–7421.

Hatefi Y. (1985). The mitochondrial electron transport and oxidative phosphorylation system. *Annu Rev Biochem* **54**: 1015–1069.

Hatefi Y, Haavik AG, Fowler LR, and Griffiths DE. (1962). Studies on the electron transfer system. XLII. Reconstitution of the electron transfer system. *J Biol Chem* **237**: 2661–2669.

Heinemeyer J, Braun HP, Boekema EJ, and Kouril R. (2007). A structural model of the cytochrome *c* reductase/oxidase supercomplex from yeast mitochondria. *J Biol Chem* **282**: 12240–12248.

Henning W, Vo L, Albanese J, and Hill BC. (1995). High-yield purification of cytochrome aa_3 and cytochrome caa_3 oxidases from *Bacillus subtilis* plasma membranes. *Biochem J* **309 (Pt 1)**: 279–283.

Iwasaki T, Matsuura K, and Oshima T. (1995). Resolution of the aerobic respiratory system of the thermoacidophilic archaeon, *Sulfolobus* sp. strain 7. I. The archaeal terminal oxidase supercomplex is a functional fusion of respiratory complexes III and IV with no *c*-type cytochromes. *J Biol Chem* **270**: 30881–30892.

Jormakka M, Törnroth S, Byrne B, and Iwata S. (2002). Molecular basis of proton motive force generation: Structure of formate dehydrogenase-N. *Science* **295**: 1863–1868.

Keefe RG, and Maier RJ. (1993). Purification and characterization of an O_2-utilizing cytochrome-c oxidase complex from *Bradyrhizobium japonicum* bacteroid membranes. *Biochim Biophys Acta* **1183**: 91–104.

Kita K, Konishi K, and Anraku Y. (1984a). Terminal oxidases of *Escherichia coli* aerobic respiratory chain. I. Purification and properties of cytochrome b_{562}-*o* complex from cells in the early exponential phase of aerobic growth. *J Biol Chem* **259**: 3368–3374.

Kita K, Konishi K, and Anraku Y. (1984b). Terminal oxidases of *Escherichia coli* aerobic respiratory chain. II. Purification and properties of cytochrome b_{558}-*d* complex from cells grown with limited oxygen and evidence of branched electron-carrying systems. *J Biol Chem* **259**: 3375–3381.

Komorowski L, Anemüller S, and Schäfer G. (2001). First expression and characterization of a recombinant CuA-containing subunit II from an archaeal terminal oxidase complex. *J Bioenerg Biomembr* **33**: 27–34.

Komorowski L, Verheyen W, and Schäfer G. (2002). The archaeal respiratory supercomplex SoxM from *S. acidocaldarius* combines features of quinol and cytochrome *c* oxidases. *Biol Chem* **383**: 1791–1799.

Kunst F, Ogasawara N, Moszer I et al. (1997). The complete genome sequence of the Gram-positive bacterium *Bacillus subtilis*. *Nature* **390**: 249–256.

Lai EM, Nair U, Phadke ND, and Maddock JR. (2004). Proteomic screening and identification of differentially distributed membrane proteins in Escherichia coli. *Mol Microbiol* **52**: 1029–1044.

Lancaster CR. (2002). Succinate:quinone oxidoreductases: An overview. *Biochim Biophys Acta* **1553**: 1–6.

Lancaster CR, Sauer US, Gross R, Haas AH, Graf J, Schwalbe H, Mäntele W, Simon J, and Madej MG. (2005). Experimental support for the "E pathway hypothesis" of coupled transmembrane e- and H+ transfer in dihemic quinol:fumarate reductase. *Proc Natl Acad Sci U S A* **102**: 18860–18865.

Laska S, and Kletzin A. (2000). Improved purification of the membrane-bound hydrogenase-sulfur-reductase complex from thermophilic archaea using epsilon-aminocaproic acid-containing chromatography buffers. *J Chromatogr B Biomed Sci Appl* **737**: 151–160.

Laska S, Lottspeich F, and Kletzin A. (2003). Membrane-bound hydrogenase and sulfur reductase of the hyperthermophilic and acidophilic archaeon *Acidianus ambivalens*. *Microbiology* **149**: 2357–2371.

Lauraeus M, and Wikström M. (1993). The terminal quinol oxidases of *Bacillus subtilis* have different energy conservation properties. *J Biol Chem* **268**: 11470–11473.

Lenn T, Leake MC, and Mullineaux CW. (2008). Clustering and dynamics of cytochrome bd-I complexes in the Escherichia coli plasma membrane in vivo. *Mol Microbiol* **70**: 1397–1407.

Liu X, Gong X, Hicks DB, Krulwich TA, Yu L, and Yu CA. (2007). Interaction between cytochrome caa3 and F1F0-ATP synthase of alkaliphilic *Bacillus pseudofirmus* OF4 is demonstrated by saturation transfer electron paramagnetic resonance and differential scanning calorimetry assays. *Biochemistry* **46**: 306–313.

Llorente-Garcia I, Lenn T, Erhardt H et al. (2014). Single-molecule *in vivo* imaging of bacterial respiratory complexes indicates delocalized oxidative phosphorylation. *Biochim Biophys Acta* **1837**: 811–824.

Lübben M, Arnaud S, Castresana J, Warne A, Albracht SP, and Saraste M. (1994). A second terminal oxidase in *Sulfolobus acidocaldarius*. *Eur J Biochem* **224**: 151–159.

Lübben M, Kolmerer B, and Saraste M. (1992). An archaebacterial terminal oxidase combines core structures of two mitochondrial respiratory complexes. *EMBO J* **11**: 805–812.

Megehee JA, Hosler JP, and Lundrigan MD. (2006). Evidence for a cytochrome *bcc-aa*$_3$ interaction in the respiratory chain of *Mycobacterium smegmatis*. *Microbiology* **152**: 823–829.

Melo AMP, Bandeiras TM, and Teixeira M. (2004). New insights into Type II NAD(P)H:quinone oxidoreductases. *Microbiol Mol Biol Rev* **68**: 603–616.

Melo AMP, Duarte M, Moller IM, Prokisch H, Dolan PL, Pinto L, Nelson MA, and Videira A. (2001). The external calcium-dependent NADPH dehydrogenase from *Neurospora crassa* mitochondria. *J Biol Chem* **276**: 3947–3951.

Melo AMP, Duarte M, and Videira A. (1999). Primary structure and characterisation of a 64 kDa NADH dehydrogenase from the inner membrane of *Neurospora crassa* mitochondria. *Biochim Biophys Acta—Bioenerg* **1412**: 282–287.

Melo AMP, Roberts TH, and Moller IM. (1996). Evidence for the presence of two rotenone-insensitive NAD(P)H dehydrogenases on the inner surface of the inner membrane of potato tuber mitochondria. *Biochim Biophys Acta—Bioenerg* **1276**: 133–139.

Mogi T, Ui H, Shiomi K, Omura S, and Kita K. (2008). Gramicidin S identified as a potent inhibitor for cytochrome *bd*-type quinol oxidase. *FEBS Lett* **582**: 2299–2302.

Moreno-Lastres D, Fontanesi F, García-Consuegra I, Martín MA, Arenas J, Barrientos A, and Ugalde C. (2012). Mitochondrial complex I plays an essential role in human respirasome assembly. *Cell Metab* **15**: 324–335.

Nakayama Y, Hayashi M, and Unemoto T. (1998). Identification of six subunits constituting Na+-translocating NADH-quinone reductase from the marine *Vibrio alginolyticus*. *FEBS Lett* **422**: 240–242.

Niebisch A, and Bott M. (2003). Purification of a cytochrome bc-aa3 supercomplex with quinol oxidase activity from *Corynebacterium glutamicum*. Identification of a fourth subunity of cytochrome aa3 oxidase and mutational analysis of diheme cytochrome c_1. *J Biol Chem* **278**: 4339–4346.

Pereira MM, Bandeiras TM, Fernandes AS, Lemos RS, Melo AMP, and Teixeira M. (2004). Respiratory chains from aerobic thermophilic prokaryotes. *J Bioenerg Biomembr.* **36**: 93–105.

Pereira MM, Santana M, and Teixeira M. (2001). A novel scenario for the evolution of haem-copper oxygen reductases. *Biochim Biophys Acta* **1505**: 185–208.

Prunetti L, Brugna M, Lebrun R, Giudici-Orticoni MT, and Guiral M. (2011). The elusive third subunit IIa of the bacterial B-type oxidases: The enzyme from the hyperthermophile *Aquifex aeolicus*. *PLoS One* **6**: e21616.

Prunetti L, Infossi P, Brugna M, Ebel C, Giudici-Orticoni MT, and Guiral M. (2010). New functional sulfide oxidase-oxygen reductase supercomplex in the membrane of the hyperthermophilic bacterium *Aquifex aeolicus*. *J Biol Chem* **285**: 41815–41826.

Ramírez-Aguilar SJ, Keuthe M, Rocha M, Fedyaev VV, Kramp K, Gupta KJ, Rasmusson AG, Schulze WX, and van Dongen JT. (2011). The composition of plant mitochondrial supercomplexes changes with oxygen availability. *J Biol Chem* **286**: 43045–43053.

Refojo PN, Teixeira M, and Pereira MM. (2010). The alternative complex III of *Rhodothermus marinus* and its structural and functional association with caa3 oxygen reductase. *Biochim Biophys Acta* **1797**: 1477–1482.

Saraste M, Metso T, Nakari T, Jalli T, Lauraeus M, and Van der Oost J. (1991). The *Bacillus subtilis* cytochrome-*c* oxidase. Variations on a conserved protein theme. *Eur J Biochem* **195**: 517–525.

Sawers G. (1994). The hydrogenases and formate dehydrogenases of *Escherichia coli*. *Antonie Van Leeuwenhoek* **66**: 57–88.

Sawers RG. (2005). Formate and its role in hydrogen production in *Escherichia coli*. *Biochem Soc Trans* **33**: 42–46.

Schirawski J, and Unden G. (1998). Menaquinone-dependent succinate dehydrogenase of bacteria catalyzes reversed electron transport driven by the proton potential. *Eur J Biochem* **257**: 210–215.

Schütz M, Brugna M, Lebrun E et al. (2000). Early evolution of cytochrome *bc* complexes. *J Mol Biol* **300**: 663–675.

Shiba T, Kido Y, Sakamoto K et al. (2013). Structure of the trypanosome cyanide-insensitive alternative oxidase. *Proc Natl Acad Sci U S A* **110**: 4580–4585.

Sleytr UB, and Sára M. (1997). Bacterial and archaeal S-layer proteins: Structure-function relationships and their biotechnological applications. *Trends Biotechnol* **15**: 20–26.

Sone N, Sekimachi M, and Kutoh E. (1987). Identification and properties of a quinol oxidase supercomplex composed of a bc_1 complex and cytochrome oxidase in the thermophilic bacterium PS3. *J Biol Chem* **262**: 15386–15391.

Sousa PMF, Silva STN, Hood BL, Charro N, Carita JN, Vaz F, Penque D, Conrads TP, and Melo AMP. (2011). Supramolecular organizations in the aerobic respiratory chain of *Escherichia coli*. *Biochimie* **93**: 418–425.

Sousa PMF, Videira MA, and Melo AM. (2013b). The formate:oxygen oxidoreductase supercomplex of Escherichia coli aerobic respiratory chain. *FEBS Lett* **587**: 2559–2564.

Sousa PMF, Videira MAM, Bohn A, Hood BL, Conrads TP, Goulao LF, and Melo AMP. (2012). The aerobic respiratory chain of *Escherichia coli*: From genes to supercomplexes. *Microbiology* **158**: 2408–2418.

Sousa PMF, Videira MAM, Santos FAS, Hood BL, Conrads TP, and Melo AMP. (2013a). The *bc:caa*$_3$ supercomplexes from the Gram positive bacterium *Bacillus subtilis* respiratory chain: A megacomplex organization? *Arch Biochem Biophys* **537**: 153–160.

Stelter M, Melo AMP, Hreggvidsson GO, Hjorleifsdottir S, Saraiva LM, Teixeira M, and Archer M. (2010). Structure at 1.0 resolution of a high-potential iron-sulfur protein involved in the aerobic respiratory chain of *Rhodothermus marinus*. *J Biol Inorg Chem* **15**: 303–313.

Stelter M, Melo AMP, Pereira MM, Gomes CM, Hreggvidsson GO, Hjorleifsdottir S, Saraiva LM, Teixeira M, and Archer M. (2008). A novel type of monoheme cytochrome *c*: Biochemical and structural characterization at 1.23 angstrom resolution of *Rhodothermus marinus* Cytochrome *c*. *Biochemistry* **47**: 11953–11963.

Stenberg F, von Heijne G, and Daley DO. (2007). Assembly of the cytochrome *bo*$_3$ complex. *J Mol Biol* **371**: 765–773.

Steuber J, Schmid C, Rufibach M, and Dimroth P. (2000). Na^+ translocation by complex I (NADH:quinone oxidoreductase) of *Escherichia coli*. *Mol Microbiol* **35**: 428–434.

Stolpe S, and Friedrich T. (2004). The *Escherichia coli* NADH:ubiquinone oxidoreductase (complex I) is a primary proton pump but may be capable of secondary sodium antiport. *J Biol Chem* **279**: 18377–18383.

Stroh A, Anderka O, Pfeiffer K, Yagi T, Finel M, Ludwig B, and Schägger H. (2004). Assembly of respiratory complexes I, III, and IV into NADH oxidase supercomplex stabilizes complex I in *Paracoccus denitrificans*. *J Biol Chem* **279**: 5000–5007.

Tanaka T, Inoue M, Sakamoto J, and Sone N. (1996). Intra- and inter-complex cross-linking of subunits in the quinol oxidase super-complex from thermophilic *Bacillus* PS3. *J Biochem* **119**: 482–486.

Unden G, and Bongaerts J. (1997). Alternative respiratory pathways of *Escherichia coli*: Energetics and transcriptional regulation in response to electron acceptors. *Biochim Biophys Acta* **1320**: 217–234.

Verkhovsky MI, and Bogachev AV. (2010). Sodium-translocating NADH:quinone oxidoreductase as a redox-driven ion pump. *Biochim Biophys Acta* **1797**: 738–746.

Wagner AM, Krab K, Wagner MJ, and Moore AL. (2008). Regulation of thermogenesis in flowering Araceae: The role of the alternative oxidase. *Biochim Biophys Acta* **1777**: 993–1000.

Winstedt L, and von Wachenfeldt C. (2000). Terminal oxidases of *Bacillus subtilis* strain 168: One quinol oxidase, cytochrome $aa_{(3)}$ or cytochrome *bd*, is required for aerobic growth. *J Bacteriol* **182**: 6557–6564.

Wittig I, Carrozzo R, Santorelli FM, and Schägger H. (2006). Supercomplexes and subcomplexes of mitochondrial oxidative phosphorylation. *Biochim Biophys Acta* **1757**: 1066–1072.

Yagi T, Seo BB, Di Bernardo S, Nakamaru-Ogiso E, Kao MC, and Matsuno-Yagi A. (2001). NADH dehydrogenases: from basic science to biomedicine. *J Bioenerg Biomembr* **33**: 233–242.

Yankovskaya V, Horsefield R, Törnroth S, Luna-Chavez C, Miyoshi H, Léger C, Byrne B, Cecchini G, and Iwata S. (2003). Architecture of succinate dehydrogenase and reactive oxygen species generation. *Science* **299**: 700–704.

Yu WB, Gao SH, Yin CY, Zhou Y, and Ye BC. (2011). Comparative transcriptome analysis of *Bacillus subtilis* responding to dissolved oxygen in adenosine fermentation. *PLoS One* **6**: e20092.

CHAPTER 10

Membrane Organisation and Electron Transport Switches in Cyanobacteria

Conrad W. Mullineaux and Tchern Lenn

CONTENTS

10.1 ELECTRON TRANSPORT PATHWAYS IN CYANOBACTERIAL THYLAKOID MEMBRANES: INTERACTION BETWEEN THE PHOTOSYNTHETIC AND RESPIRATORY ELECTRON TRANSPORT CHAINS

Cyanobacteria are phototrophic prokaryotes with the ability to extract electrons from water and to use the reducing power generated to fix carbon dioxide. They generate oxygen as a waste product from solar-powered water-splitting. Cyanobacteria are an ancient, widespread and diverse group of bacteria that played a crucial role in the history of the biosphere by generating the atmospheric oxygen needed for aerobic heterotrophic life. Cyanobacteria today remain abundant in the oceans and in a wide variety of other habitats, and they still generate a significant proportion of the oxygen that we breathe. The remainder of the oxygen in the atmosphere is generated by the chloroplasts of plants and algae, organelles that are themselves the descendents of a free-living cyanobacterium. Chloroplasts carry out their photosynthetic light reactions in an internal membrane system called the *thylakoid membranes*, a feature that they have inherited from their cyanobacterial ancestors. Nearly all cyanobacteria contain thylakoid membranes: the single known exception is the atypical cyanobacterium *Gloeobacter violaceus*, which lacks internal membranes and houses its photosynthetic apparatus in dedicated zones of the cytoplasmic membrane (Rexroth et al., 2011). In all other known cyanobacteria, the photosynthetic apparatus is housed in the thylakoids. In both cyanobacteria and chloroplasts, the thylakoids are a topologically complex system of internal membranes. Connections with the surrounding envelope membrane (the inner envelope in chloroplasts or the cytoplasmic membrane in cyanobacteria) appear absent in chloroplasts (Vothknecht and Westhoff, 2001) and are ill-defined in cyanobacteria (Liberton et al., 2006; van de Meene et al., 2006; Nevo et al., 2007). The thylakoids serve as a chemiosmotic barrier between the cytoplasm (or the stroma in chloroplasts) and a separate aqueous compartment, the thylakoid lumen. Ultrastructural studies from some cyanobacteria suggest that the entire thylakoid membrane system in the cell may be interconnected, with a continuous membrane surface enclosing a single continuous lumen (Nevo et al., 2007).

Electron transport in the thylakoids results in the acidification of the lumen, and the resulting proton gradient across the thylakoid membrane is used to power the synthesis of ATP by the proton-translocating ATPase. Cyanobacterial thylakoids are topologically rather different from those of green plant chloroplasts, which contain prominent structures known as *grana*, containing multiple stacked layers of thylakoid membrane linked by an unstacked membrane network known as the *stroma lamellae*. The grana membranes of chloroplasts are highly enriched in photosystem II, and thus, higher plant chloroplasts have a very obvious lateral heterogeneity in their protein composition (Dekker and Boekema, 2005). Cyanobacterial thylakoid membranes lack grana, and at the ultrastructural level, they appear much more homogeneous than chloroplast thylakoids (Figure 10.1). Nevertheless, there is now strong evidence for patchy localisation of specific complexes in the membrane, the existence of functionally specialised zones in the membrane and lateral heterogeneity on scales of around 100–300 nm. This will be discussed in detail in Section 10.3.

Plants and algae have adopted a division of labour in electron transport, with all photosynthetic electron transport occurring in the chloroplasts and the great majority of respiratory electron transport occurring in the mitochondria. Cyanobacteria carry out respiration as well as photosynthesis. In the dark, the respiratory consumption of stored photosynthate is important for maintaining cell functions. However, in contrast to plants and algae, the majority of respiratory electron transport in cyanobacteria occurs in the thylakoid membrane, in proximity to the photosynthetic apparatus (Lea-Smith et al., 2013; Mullineaux, 2014). This proximity of the photosynthetic and respiratory systems opens up numerous possibilities for *hybrid* modes of electron transport involving electron exchange between photosynthetic and respiratory complexes (Figure 10.2), and indeed, many such

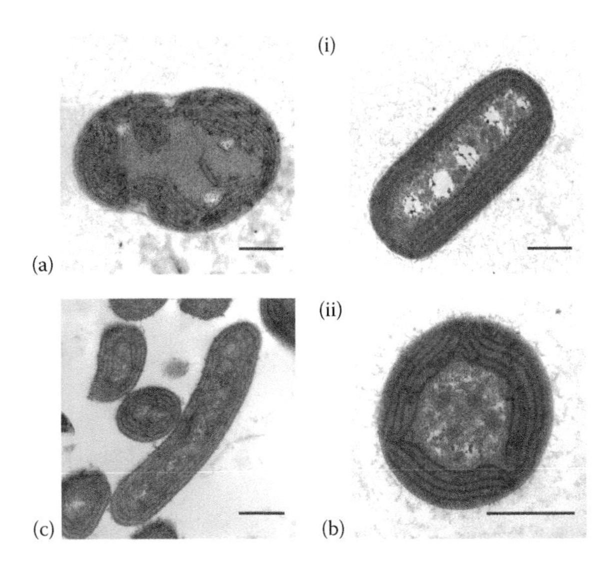

FIGURE 10.1 Thin-section electron micrographs showing the conformation of thylakoid membranes in a range of unicellular cyanobacteria. (a) *Synechocystis* sp. PCC6803; (b) *Synechococcus* sp. PCC7002, with longitudinal (i) and transverse (ii) sections and (c) *Synechococcus* sp. PCC7942, with longitudinal and transverse sections visible. All scale bars 500 nm. (Courtesy of Giulia Mastroianni, Queen Mary University of London, London.)

electron transport activities have been detected *in vivo* (Vermaas, 1996; Bailey et al., 2008). These activities can strongly influence the outputs of photosynthesis, specifically the ratio of reductant (NADPH) to free energy source (ATP) (Figure 10.2). This ratio is of critical importance for the physiology of the cell, and therefore, it must be important for cells to regulate the pathways of electron transport. Precedents from other bioenergetic membranes suggest that the distribution and supramolecular interactions of the electron transport complexes could be crucial for controlling the pathways of electron transport (Iwai et al., 2010; Lapuente-Brun et al., 2013). Therefore, the location of the electron transport complexes is an important experimental question.

10.2 NANOMETRE-SCALE ORGANISATION OF CYANOBACTERIAL ELECTRON TRANSPORT COMPLEXES

In other bioenergetic systems, the local interactions of electron transport complexes have been probed with both biochemical and ultrastructural techniques. Biochemical isolation has yielded evidence for supercomplexes, with structural association of multiple electron transport complexes (Iwai et al., 2010; Dudkina et al., 2011). Ultrastructural techniques such as cryo-electron tomography have given information on the organisation of complexes in bioenergetic membranes, in some cases confirming the presence of bioenergetic supercomplexes in the membrane (Davies et al., 2011). Cyanobacterial thylakoid membranes contain a rich variety of electron transport complexes (Figure 10.2), but information on the location and interactions of many of these complexes is sparse. Quantitatively, the predominant complexes in the membrane are the photosynthetic reaction centres, photosystem I and photosystem II. In most cyanobacteria, and under most conditions, photosystem I reaction centres

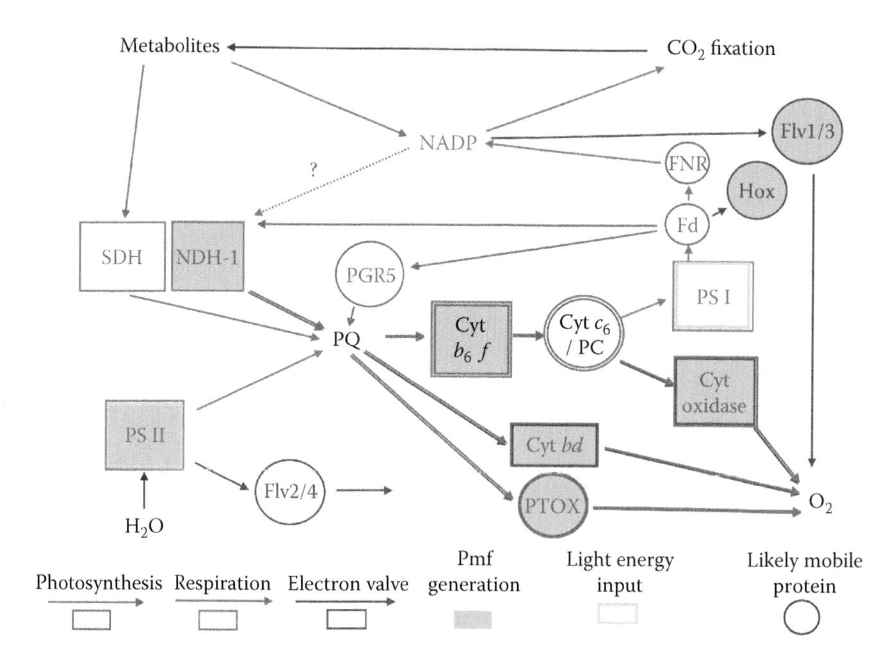

FIGURE 10.2 **(See colour insert.)** Pathways of electron transport in and around cyanobacterial thylakoid membranes. Arrows indicate directions of electron flow. Proteins and protein complexes are indicated by frames: small molecules are shown without frames. Frame colour indicates involvement in photosynthesis, respiration or electron dissipation as indicated in the key. Additionally, yellow boxes indicate sites of light energy input and light blue backgrounds show sites of proton-motive force generation (by proton pumping and/or proton consumption/release on a particular side of the membrane). Cytoplasmic, lumenal or peripheral membrane proteins are likely to be rapidly mobile (therefore able to redistribute on timescales of seconds for fast switching of electron transport) and are given circular frames. Membrane-integral protein complexes are likely to show only very restricted mobility and are given square frames. (Adapted from Mullineaux, C.W., *Biochim. Biophys. Acta [BBA-Bioenergetics]*, 1837, 503–511, 2014.) Abbreviations: Cyt, cytochrome; Fd, ferredoxin; Flv, flavodiiron protein; FNR, ferredoxin-NADP oxidoreductase; Hox, bidirectional NiFe hydrogenase; NDH-1, respiratory complex I (NAD[P]H dehydrogenase); PC, plastocyanin; PGR5, proton gradient regulation protein 5; PQ, plastoquinone; PS, photosystem; PTOX, quinol oxidase (plastid terminal oxidase); SDH, respiratory complex II (succinate dehydrogenase).

greatly outnumber photosystem II reaction centres. Photosystem I is predominantly trimeric, while photosystem II is predominantly dimeric (Rakhimberdieva et al., 2001). Freeze-fracture electron microscopy has shown that photosystem II dimers are sometimes arranged in rows in the membrane, with photosystem I complexes generally assumed to be packed into the spaces between the rows of photosystem II (Moerschel and Schatz, 1987; Vernotte et al., 1990). The phycobilisome light-harvesting complexes are large extrinsic complexes that can cover much of the cytoplasmic surface of the membrane, and there is circumstantial evidence that the phycobilisomes can be arranged in rows on the surface corresponding to rows of photosystem II complexes within the membrane (Olive et al., 1997). However, the ultrastructural organisation of the reaction centres is probably quite variable. Electron microscopy on partially solubilised thylakoid membrane fragments has provided evidence for two-dimensional arrays of photosystem II, in contrast to the one-dimensional rows of PSII dimers observed by freeze-fracture electron microscopy (Folea et al., 2008). There is strong evidence for supramolecular association of reaction centres with light-harvesting complexes, the phycobilisomes on the membrane surface (Liu et al., 2013) and the membrane integral IsiA chlorophyll-binding protein, which is induced under some stress conditions (Boekema et al., 2001). By contrast, evidence for supramolecular association of reaction centres with other electron transport components is very sparse. Of course, this does not necessarily mean that electron transport supercomplexes are absent: it may just be that no one has yet found the right way to look for them. A combination of affinity pull-downs with chemical cross-linking was recently used to isolate a cyanobacterial thylakoid membrane *megacomplex* containing phycobilisomes, photosystem II and photosystem I (Liu et al., 2013). However, although the structural organisation of this megacomplex may well be very significant for photosynthetic light harvesting, it does not constitute a functional electron transport unit. Electrons have to travel from photosystem II to photosystem I via the mobile electron carriers plastoquinone and plastocyanin (or cytochrome c_6) and the cytochrome b_6f complex (Figure 10.2). However, the cytochrome b_6f complex is notably absent from the isolated phycobilisome-photosystem II-photosystem I megacomplex, which means that this megacomplex is not an electron transport unit. Plastoquinol molecules carrying electrons from photosystem II would have to leave the megacomplex to find cytochrome b_6f, and plastocyanin/cytochrome c_6 molecules carrying electrons from cytochrome b_6f would have to find their way back to the megacomplex to complete the electron transport chain.

To date, the location of cytochrome b_6f in cyanobacterial thylakoid membranes has proved very elusive, which means that we have a very incomplete picture of the physical layout of electron transport in the membrane. This topic will be discussed further in Section 10.6. In the chloroplast of the green alga *Chlamydomonas reinhardtii*, there is good evidence for the structural association of photosystem I with cytochrome b_6f in a supercomplex that appears functional in cyclic electron transport: formation of this complex is stimulated by particular physiological conditions (Iwai et al., 2010). Chloroplasts are, of course, related to cyanobacteria, so this study should act as a stimulus to look for similar supercomplexes in cyanobacterial membranes. To date, however, we are not aware of any similar evidence for such an electron transport supercomplex in cyanobacteria. One structural association that is likely to be significant for electron transport can be inferred indirectly, since the soluble cytoplasmic electron carrier ferredoxin:NADP oxidoreductase (FNR) has been shown to be structurally associated with the rod subunits of the phycobilisomes (van Thor et al., 1999), probably close to the interface between the phycobilisome rods and the core (Arteni et al., 2009). If photosystem I is associated with phycobilisomes (e.g. as part of a phycobilisome-photosystem II:photosystem I megacomplex), then the binding of FNR to the phycobilisome rods should bring FNR into proximity with the acceptor side of photosystem I, which may facilitate electron transfer from photosystem I to NADP. This could influence the ratio of linear electron transfer to photosystem I cyclic electron transfer, since a likely pathway for cyclic electron transfer is from photosystem I to respiratory complex I via ferredoxin (Figure 10.2; Battchikova et al., 2011). Proximity of FNR to the acceptor side of photosystem I might tend to steer electrons into the linear route to NADP reduction, the Calvin cycle and CO_2 fixation rather than into this cyclic pathway, which will generate ATP but not reducing power.

Respiratory complexes such complex I, complex II and terminal oxidases are quantitatively minor components of the thylakoid membrane, but potentially important players in determining the balance of electron transport pathways in the membrane (Figure 10.2; Howitt and Vermaas, 1998; Cooley and Vermaas, 2001; Battchikova et al., 2011; Liu et al., 2012; Lea-Smith et al., 2013). Possible supramolecular associations of these complexes have yet to be explored, and there is only limited information on their location in the membrane. The information that we have on the location of these complexes, and the roles that their location could play in controlling the pathways of electron transport, will be discussed in Sections 10.3 and 10.5.

10.3 SUBMICRON-SCALE ORGANISATION AND DYNAMICS OF CYANOBACTERIAL THYLAKOID MEMBRANES

Despite the rather homogeneous appearance of cyanobacterial thylakoid membranes (Figure 10.1), there are numerous indications that the composition and function of the membrane is not laterally homogeneous. The distribution of chlorophyll fluorescence in the membrane appears rather uniform, suggesting that the majority of the membrane area contains photosynthetic complexes and is devoted to photosynthetic electron transport. This is unsurprising, as the photosynthetic reaction centres are far more abundant than the other membrane proteins. In contrast to the photosynthetic reaction centres, other membrane components often appear very patchily distributed in the membrane. Fluorescent protein tagging has shown localised patches of the FtsH proteases, which are involved in the repair of photodamaged photosystem II reaction centres. Immunogold electron microscopy suggests the presence of specialised photosystem II assembly zones in close proximity to the cytoplasmic membrane: these membrane areas can be separated from the bulk thylakoid membrane by density gradient centrifugation (Stengel et al., 2012). A cryo-electron tomographic study of *Synechocystis* sp. PCC6803 also suggested the presence of specialised protein assembly zones, but at the opposite side of the thylakoid membrane system. Ribosome-rich areas, most likely containing membranes and connected to the thylakoids, were observed protruding into the central cytoplasm of the cell (van de Meene et al., 2006). These studies suggest the presence of a variety of specialised membrane regions devoted to the assembly and repair of photosynthetic complexes, contiguous with, but distinct from the bulk membrane devoted to photosynthetic electron transport.

There are also indications that the electron transport complexes are themselves not uniformly distributed in the thylakoid membrane, and this could have strong implications for electron transport and its regulation. A study of the larger-scale distribution of photosystem I and photosystem II in *Synechocystis* sp. PCC6803 by hyperspectral fluorescence imaging indicated that photosystem I is concentrated in the inner thylakoid membrane sacs adjacent to the central cytoplasm, while the outer thylakoid membrane sacs closer to the cytoplasmic membrane have a higher photosystem II/photosystem I ratio (Vermaas et al., 2008). This would imply that linear photosynthetic electron transport (involving both photosystem I and photosystem II) and cyclic photosynthetic electron transport (involving photosystem I but not photosystem II) are concentrated in distinct, well-separated regions of the membrane. A study using immunogold electron microscopy in *Synechococcus* sp. PCC7942 came to a similar overall conclusion, although in this case it appears that the linear electron transport regions are adjacent to the central cytoplasm and the cyclic electron transport regions are adjacent to the cytoplasmic membrane (Sherman et al., 1994). Studies of the membrane localisation of minor electron transport components give indications of a more extreme heterogeneity in their distribution in the membrane. Fluorescent protein tagging of respiratory electron donor complexes I and II (NDH-1 and SDH) in *Synechococcus* sp. PCC7942 indicates that under some conditions these complexes are strongly concentrated in very localised membrane zones around 100–300 nm in diameter (Liu et al., 2012). The bidirectional hydrogenase, which acts as an electron acceptor from photosystem I under anaerobic conditions (Appel et al., 2000), is an extrinsic protein complex in the cytoplasm. However, fluorescent protein tagging in *Synechocystis* sp. PCC6803 indicates that the hydrogenase is strongly thylakoid membrane-associated and heterogeneously distributed, with a proportion of the complex concentrated in foci towards the distal side of the thylakoid membrane system, adjacent to the cytoplasmic membrane (Burroughs et al., 2014). The implications of heterogeneous distribution of these electron transport complexes for the regulation of electron transport are explored further in Section 10.5.

Cyanobacterial thylakoid membranes appear to be a very crowded environment with dense packing of protein complexes (Folea et al., 2008). Probably as a consequence of this dense packing of complexes, the lateral diffusion of membrane-intrinsic protein complexes is greatly restricted. Fluorescence recovery after photobleaching (FRAP) measurements showed that photosystem II is almost immobile in the membrane under normal conditions, although the membrane-intrinsic chlorophyll-binding protein IsiA can diffuse somewhat faster (Sarcina and Mullineaux, 2004). Respiratory complex I (NDH-1) redistributes only very slowly in the thylakoid membrane, on a timescale of about 30 min (Liu et al., 2012), and this is probably a consequence of membrane crowding leading to very slow diffusion. The situation in cyanobacterial thylakoids is in striking contrast to other bacterial membranes such the cytoplasmic membrane of *Escherichia coli*, where even large protein assemblages are very mobile in the membrane (Llorente-Garcia et al., 2014). The slow diffusion of membrane-intrinsic complexes in cyanobacterial thylakoids must restrict the rate at which these complexes can re-organise in the membrane, and it is very likely to contribute to the maintenance of a patchy and heterogeneous distribution of complexes.

A key question for the function of electron transport in cyanobacterial thylakoids is the mobility of small electron carriers such as plastoquinone: how rapidly can plastoquinone diffuse and are there barriers to plastoquinone diffusion in the membrane? These are also important considerations for electron transport in chloroplasts. Unfortunately, we have no direct information on the mobility of plastoquinone in cyanobacterial thylakoids. A fluorescence recovery after photobleaching (FRAP) study using a fluorescent ubiquinone analogue showed that this molecule is extremely mobile in the *E. coli* cytoplasmic membrane, giving the potential for rapid, long-range electron transport in this organism (Llorente-Garcia et al., 2014). However, cyanobacterial thylakoid membranes show much more restricted protein diffusion, so it is not guaranteed that plastoquinone diffusion in cyanobacteria will be as unrestricted as ubiquinone diffusion in *E. coli*. BODIPY FL C_{12}, an artificial fluorescent lipid, shows a diffusion coefficient of about 0.3 $\mu m^2\ s^{-1}$ in cyanobacterial thylakoid membranes (Sarcina et al., 2003). This is about six times slower than the diffusion of the fluorescent ubiquinone analogue in *E. coli* (Llorente-Garcia et al., 2014). The comparison suggests that the diffusion of small molecules may be generally somewhat slower in cyanobacterial thylakoid membranes than in *E. coli*; however, there do not appear to be any strong barriers to long-range diffusion. Long-range electron transport mediated by plastoquinol may therefore be possible in cyanobacterial thylakoid membranes, although in practice electrons are more likely to be delivered to nearby complexes simply because the membrane is crowded with electron transport complexes and the plastoquinol molecule is unlikely to get very far before it encounters an electron acceptor. This point will be considered further in Section 10.6.

10.4 CROWDING AND FLUIDITY IN THE AQUEOUS COMPARTMENTS ON EITHER SIDE OF THE MEMBRANE

In contrast to the membrane-integral protein complexes, extrinsic complexes on the cytoplasmic side of the membrane appear freely mobile, suggesting that the cytoplasm in the region of the thylakoids is a relatively fluid environment. Examples include the phycobilisomes, large extrinsic light-harvesting complexes anchored to the cytoplasmic side of the thylakoid membrane (Watanabe and Ikeuchi, 2013). Despite their size and bulk, the phycobilisomes can diffuse relatively rapidly, as indicated by FRAP measurements (Mullineaux et al., 1997; Sarcina et al., 2001). Fluorescent protein tagging and fluorescence microscopy show that Vipp1, a cytoplasmic protein implicated in the biogenesis of thylakoid membranes and the photosynthetic complexes, can rapidly and dramatically redistribute in the vicinity of the thylakoid membranes (Bryan et al., 2014). These findings suggest that the diffusion of soluble electron carriers on the cytoplasmic side of the membrane could be rapid and unrestricted, giving the potential for long-range electron transport by carriers such as ferredoxin.

In chloroplasts, there appear to be significant restrictions to the diffusion of plastocyanin on the lumenal side of the thylakoid membrane, although these restrictions can be alleviated by swelling of the lumenal space during illumination (Kirchhoff et al., 2011). To date, we have very little information on the mobility of proteins in the cyanobacterial thylakoid lumen, and more information is needed to determine whether long-range electron transport in cyanobacteria could be mediated by lumenal electron carriers such as plastocyanin and cytochrome c_6. However, we have found that a GFP-tagged version of the processing peptidase CtpA can redistribute relatively rapidly and over long ranges in the thylakoid lumen of *Synechocystis* sp. PCC6803 (Joanna Sacharz et al., unpublished). This suggests that there is potential for long-range diffusion of lumenal electron carriers.

10.5 EXAMPLES OF REGULATION OF MEMBRANE ORGANISATION AND ELECTRON TRANSPORT: COMPLEX I, COMPLEX II AND THE BIDIRECTIONAL HYDROGENASE

Fluorescent protein tagging and fluorescence microscopy give a convenient way to monitor the larger-scale distribution of electron transport complexes in bioenergetic membranes. Ideally, the tagged protein subunit should be expressed from the native chromosomal locus, with the modified gene replacing the wild-type gene. This helps to ensure that the tagged protein is expressed in context and at native levels and that the distribution of fluorescence in the membrane gives a complete picture of the distribution of the electron transport complex. As always with fluorescent protein tagging, it is necessary to check that the protein is properly and completely tagged, is present at native levels, is correctly incorporated into its complex and is functional. An advantage with electron transport complexes is that electron transport can easily be quantified (e.g. by measuring the rate of respiratory oxygen consumption under conditions in which the complex of interest should be an active part of the electron transport chain). This allows a quantitative assessment of any perturbation of function by the fluorescent protein tag (Lenn et al., 2008).

The use of fluorescent protein tagging and fluorescence microscopy to probe the membrane distribution of electron transport complexes has several advantages. The tagging gives a very specific picture of the localisation of a particular protein and also a very complete picture of the distribution of the protein within the cell. The localisation of the protein can be monitored *in vivo*, with measurement of its mobility in the membrane as well as its distribution. Redistribution of the complex in response to changing conditions can be observed in real time. Similarly, the accumulation or turnover of the complex resulting from changes in its expression can be followed in real time, although allowance must be made for the time needed for the fluorescent protein chromophore to mature. The drawback of conventional fluorescence microscopy is the low spatial resolution, limited to around 200 nm in the *xy*-plane at the wavelengths used to visualise green fluorescent protein, for example. Higher resolution can be obtained using a variety of super-resolution techniques, although generally at the cost of losing dynamic information and accurate quantitation of the level of the fluorophore. In our experience, super-resolution imaging in cyanobacteria and other photosynthetic systems is particularly challenging because of the complex photophysics of the native photosynthetic pigments. The approach of using fluorescent protein tagging and conventional fluorescence microscopy gives most information when the distribution of complexes in the membrane is patchy on length scales of 100 nm, which can be visualised by conventional fluorescence microscopy. Since cyanobacterial thylakoid membranes do appear to have a rather patchy organisation (as discussed in Section 10.3), the approach can give useful information on the layout of electron transport complexes in the membrane. The best model organisms for the purpose have smooth and regular thylakoid membranes: complex folding and curvature of the membrane will inevitably complicate the interpretation of fluorescence images. *Synechococcus* sp. PCC7942 and *Synechococcus* sp. PCC7002 are examples of cyanobacteria with suitably regular thylakoid membrane organisation (Figure 10.1).

To date, the localisation of three minor electron transport complexes in cyanobacterial thylakoids has been probed with GFP-tagging and fluorescence microscopy: respiratory complex I (NDH-1), respiratory complex II (SDH) (Liu et al., 2012) and the bidirectional NiFe hydrogenase (Hox) (Burroughs et al., 2014). All three complexes prove to have a patchy distribution in the membrane under some conditions. When *Synechococcus* sp PCC7942 cells are grown under very low light, both NDH-1 and SDH are localised in very distinct 100–300 nm patches in the thylakoid membrane, with a distribution that is

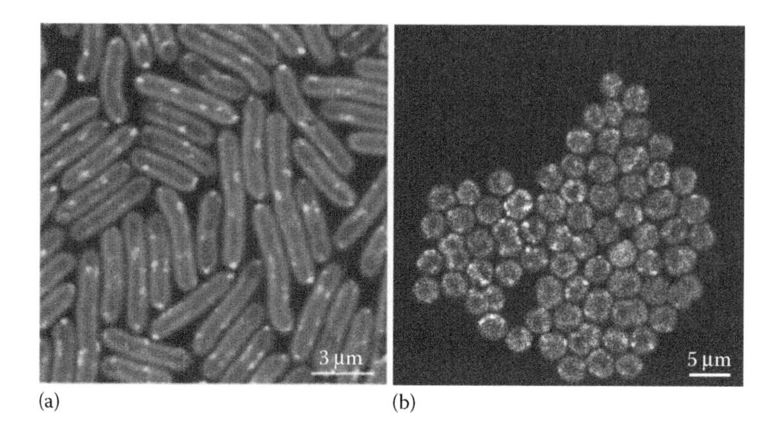

FIGURE 10.3 Two examples of patchy distribution of minor electron transport complexes in cyanobacterial thylakoid membranes, visualised by green fluorescent protein (GFP) tagging and confocal fluorescence microscopy. (a) Respiratory complex I (NDH-I) in *Synechococcus* sp. PCC7942, visualised through a GFP tag on the NdhM subunit. GFP fluorescence is concentrated in bright spots, seen here against a background of chlorophyll fluorescence from the thylakoid membranes. Cells were grown under very low light: a similar patchy distribution is induced whenever the plastoquinone pool is predominantly oxidised. (b) The bidirectional NiFe hydrogenase (Hox) in *Synechocystis* sp. PCC6803, visualised through a GFP tag on the HoxF subunit. The image shows GFP fluorescence in grey scale. Cells were grown in low light and then made anoxic. Hox is predominantly localised to the thylakoid membranes, with local concentrations of Hox visible as bright foci. These foci become more prominent under anoxia, indicating a rapid redistribution of Hox within the thylakoid system. (a: Reproduced from Liu, L-N. et al., *Proc. Natl. Acad. Sci.*, 109, 11431–11436, 2012. With permission; b: Reproduced from Bryan, S.J. et al., *Mol. Microbiol.*, 94, 1194–1195, 2014. With permission.)

very clearly distinct from the chlorophyll fluorescence that indicates the location of the photosynthetic reaction centres (Figure 10.3; Liu et al., 2012). Under moderate light intensities, both complexes redistribute within the membrane, with a more even distribution and much greater co-localisation with chlorophyll fluorescence. The transition between the two states takes about 30 min, and experiments with electron transport blockers indicate that it is triggered by changes in the redox state of plastoquinone or a closely coupled electron carrier: an oxidised plastoquinone pool results in patchy distribution of the complexes, while a reduced plastoquinone pool results in even distribution of the complexes. The transition does not depend on the synthesis of new complexes, since the change in distribution still occurs when all *de novo* protein synthesis is blocked by an antibiotic. The redistribution has very clear functional effects, since it correlates with a major switch in electron transport. In the *patchy* condition, electrons from the respiratory complexes are more likely to go to a terminal oxidase rather than to photosystem I, while in the *even* condition there is an approximately threefold increase in the probability that electrons go from the respiratory complexes to photosystem I (Liu et al., 2012). This constitutes a crucial stage in one of the major pathways for cyclic electron transport around photosystem I (Figure 10.2). Presumably, the electron transport switch happens because respiratory transport complexes come into closer proximity with photosystem I in the *even* condition. A likely physiological consequence is that in the *even* condition, cyclic electron flow around photosystem I will be favoured over linear electron transport involving both photosystems. Consequently, the *even* condition will lead to production of less reducing power (NADPH) relative to ATP, since cyclic electron transport produces ATP only whereas linear electron transport produces not only ATP but also reducing equivalents generated from the extraction of electrons from water by photosystem II (Figure 10.2). The balance of ATP to NADPH production is crucial to the physiology of oxygenic phototrophic organisms. Since the *even* condition is induced by a reduced plastoquinone pool (which can indicate a sufficiency of reducing power in the cell), the redistribution of the respiratory complexes should act as a feedback mechanism to control the redox balance of the cell. A number of key questions remain to be answered, including the nature of the biochemical signal transduction pathway that links plastoquinone redox state to the distribution of the respiratory complexes in the membrane. Crucially, we also need higher resolution information on the organisation of the respiratory complexes, both within the localised patches and in the *dispersed* state. It will obviously be important to find out whether there are direct structural interactions with other electron transport complexes in either state (Figure 10.3).

The bidirectional NiFe hydrogenase Hox provides a further example of a cyanobacterial electron transport complex whose distribution on the thylakoid membrane is patchy and under physiological control. Hox appears to be active only under anoxic conditions, and its most clearly established function is to act as an *electron valve* under these conditions (Appel et al., 2000). Illumination of previously dark-adapted cells under anoxic conditions results in a brief burst of hydrogen production, driven by electrons from the acceptor side of photosystem I (Appel et al., 2000; Cournac et al., 2004). It is likely that flavodoxin and ferredoxin are the direct electron donors to Hox (Gutekunst et al., 2014). GFP-tagging of the HoxF subunit shows that Hox is strongly thylakoid-membrane localised, although the complex itself has no membrane-integral domains (Burroughs et al., 2014). Thylakoid membrane association depends on the HoxE subunit, but the interaction partner in the membrane remains to be determined. NDH-1 has been postulated to be a Hox interaction partner; however, membrane localisation appears unaffected in a mutant background lacking NDH-1. Fluorescence microscopy shows that a proportion of Hox is concentrated in localised patches, and the distribution of Hox becomes significantly patchier under the anoxic conditions in which Hox is active for hydrogen production, suggesting that the patches could be important for physiological electron supply to Hox (Figure 10.3). The transition to a patchier distribution is rapid, within 10 min (Burroughs et al., 2014). In contrast to the redistribution of respiratory electron donors discussed above, the distribution of Hox is not influenced by electron transport donors, suggesting that the triggering signal may be a direct response to anoxia rather than the redox state of an electron transfer component (Burroughs et al., 2014). A working model would be that an oxygen sensor in the cell triggers redistribution of Hox, possibly by promoting direct interactions between Hox complexes. This shift in distribution has the effect of increasing the coupling between Hox and its electron supply from the acceptor side of photosystem I, promoting a transient burst of hydrogen production, which ceases when the oxygen level increases due to photosystem II activity.

Both the respiratory electron donors and the hydrogenase provide examples of correlation between electron transport function and the membrane distribution of electron transport complexes, at scales of 100 nm and above.

10.6 LENGTH SCALES OF ELECTRON TRANSPORT: CYANOBACTERIAL THYLAKOIDS IN COMPARISON TO OTHER BIOENERGETIC MEMBRANES

A crucial question for the *wiring* of electron transport pathways in cyanobacteria is the length scale of electron transport: how far, on average, does a mobile electron carrier diffuse between the donor complex that provided the electron(s) and the acceptor complex that takes them? Other bioenergetic membranes give precedents for a whole range of length scales. At one extreme, electron transport can be confined within supercomplexes at nanometre scales, for example, in mitochondrial respirasomes (Lapuente-Brun et al., 2013) and within the cyclic electron transport supercomplex that has been detected in algal chloroplasts (Iwai et al., 2010). At the other extreme, some modes of electron transport in the cytoplasmic membrane of *E. coli* appear fully delocalised, with ubiquinol molecules potentially travelling microns in the membrane to take electrons from donors to terminal acceptors (Llorente-Garcia et al., 2014). At intermediate scales, chromatophores in purple photosynthetic bacteria are typically spherical membrane vesicles about 45 nm in diameter and therefore with a membrane area of about 6000 nm^2. Electron transport is probably confined within an individual chromatophore (Cartron et al., 2014). Linear electron transport in green plant chloroplasts probably occurs on length scales comparable to the diameter of the grana, about 400 nm (Kirchhoff et al., 2011). We have no definitive answers for cyanobacteria, since, as discussed above, information on the spatial distribution and mobility of some key electron transport components is lacking. The range of electron transport could be restricted in two ways. Firstly, there could be strict barriers confining the diffusion of mobile electron carriers within defined zones. So far, however, there is no indication that the diffusion of small molecules and membrane-extrinsic proteins is confined in this way in cyanobacteria (see Section 10.3). The second way that the range of electron transport could be restricted is through the probability that a mobile electron carrier will encounter and interact with an acceptor complex within a given distance from its starting point. Since the thylakoid membranes of cyanobacteria appear particularly densely packed with electron transport complexes, this may well be the major factor in determining the effective length scale of electron transport. Redistribution of electron transport complexes on length scales of 100–300 nm appears to be a very effective way to control the pathways of electron flow, suggesting that the effective length scale of electron transport in cyanobacterial thylakoids is also within this range.

ACKNOWLEDGEMENTS

Relevant work in the authors' laboratory was supported by Biotechnology and Biological Sciences Research Council grants BB/G021856 and BB/J016985/1 and a Marie Curie Fellowship from the European Commission (contract number: FP7-PEOPLE-2009-IEF 254575) to Dr. Lu-Ning Liu. We thank especially Lu-Ning Liu and Samantha J. Bryan for their contributions to work discussed here, and we thank Giulia Mastroianni for providing electron micrographs.

REFERENCES

Appel J, Phunpruch S, Steinmüller K, Schulz R. (2000). The bidirectional hydrogenase of *Synechocystis* sp. PCC6803 works as an electron valve during photosynthesis. *Arch. Microbiol.* **173**, 333–338.

Arteni AA, Ajlani G, Boekema EJ. (2009). Structural organisation of phycobilisomes from *Synechocystis* sp PCC6803 and their interaction with the membrane. *BBA-Bioenergetics* **1787**, 272–279.

Bailey S, Melis A, Mackey KR, Cardol P, Finazzi G, van Dijken G, Berg GM, Arrigo K, Shrager J, Grossman AR. (2008). Alternative photosynthetic electron flow to oxygen in marine *Synechococcus*. *Biochim. Biophys. Acta* **1777**, 269–276.

Battchikova N, Wei L, Du L, Bersanini L, Aro E-M, Ma W. (2011). Identification of a novel ssl0352 protein, NdhS, essential for efficient operation of cyclic electron transport around Photosystem I in NADPH:Plastoquinone oxidoreducatase (NDH-1) complexes of *Synechocystis* sp. PCC6803. *J. Biol. Chem.* **286**, 36992–37001.

Boekema EJ, Hifney A, Yakushevska AE, Piotrowski M, Keegstra W, Berry S, Michel KP, Pistorius EK, Kruip J. (2001). A giant chlorophyll-protein complex induced by iron-deficiency in cyanobacteria. *Nature* **412**, 745–748.

Bryan SJ, Burroughs NJ, Shevela D, Yu J, Liu L-N, Mastroianni G, Xue Q et al. (2014). Localisation and interactions of the Vipp1 protein in cyanobacteria. *Mol. Microbiol.* **94**, 1194–1195.

Burroughs NJ, Boehm M, Eckert C, Mastroianni G, Spence EM, Yu J, Nixon PJ, Appel J, Mullineaux CW, Bryan SJ. (2014). Solar powered biohydrogen production requires specific localization of the hydrogenase. *Energ. Environ. Sci.* **7**, 3791–3800.

Cartron ML, Olsen JD, Sener M, Jackson PJ, Brindley AA, Qian P, Dickman MJ, Leggett GJ, Schulten K, Hunter CN. (2014). Integration of energy and electron transfer processes in the photosynthetic membrane of *Rhodobacter sphaeroides*. *BBA-Bioenergetics* **1837**, 1769–1780.

Cooley JW, Vermaas WFJ. (2001). Succinate dehydrogenase and other respiratory pathways in thylakoid membranes of *Synechocystis* sp. strain PCC6803: Capacity comparisons and physiological function. *J. Bacteriol.* **183**, 4251–4258.

Cournac L, Guedeney G, Peltier G, Vignais PM. (2004). Sustained photoevolution of molecular hydrogen in a mutant of *Synechocystis* sp. PCC6803 deficient in the Type I NADPH-dehydrogenase complex. *J. Bacteriol.* **186**, 1737–1746.

Davies KM, Strauss M, Daum B, Kief JH, Osiewacz HD, Rycovska A, Zickermann V, Kühlbrandt W. (2011). Macromolecular organization of ATP synthase and complex I in whole mitochondria. *Proc. Natl. Acad. Sci. U. S. A.* **108**, 14121–14126.

Dekker JP, Boekema EJ. (2005). Supramolecular organization of thylakoid membrane proteins in green plants. *Biochim. Biophys. Acta* **1706**, 12–39.

Dudkina NV, Kudryashev M, Stahlberg H, Boekema EJ. (2011). Interaction of complexes I, III and IV within the bovine respirasome by single particle cryo-electron tomography. *Proc. Natl. Acad. Sci. U. S. A.* **108**, 15196–15200.

Folea IM, Zhang P, Aro E-M, Boekema EJ. (2008). Domain organization of Photosystem II in membranes of the cyanobacterium *Synechocystis* sp. PCC6803 investigated by electron microscopy. *FEBS Lett.* **582**, 1749–1754.

Gutekunst K, Chen X, Schreiber K, Kaspar S, Makam S, Appel J. (2014). The bidirectional NiFe hydrogenase in *Synechocystis* sp. PCC6803 is reduced by flavodoxin and ferredoxin and is essential under mixotrophic, nitrate-limiting conditions. *J. Biol. Chem.* **289**, 1930–1937.

Howitt CA, Vermaas WFJ. (1998). Quinol and cytochrome oxidases in the cyanobacterium *Synechocystis sp.* PCC6803. *Biochemistry* **37**, 17944–17951.

Iwai M, Takizawa K, Tokutsu R, Okamuro A, Takahashi Y, Minagawa J. (2010). Isolation of the elusive supercomplex that drives cyclic electron flow in photosynthesis. *Nature* **464**, 1210–1213.

Kirchhoff H, Hall C, Wood M, Herbstova M, Tsabari O, Nevo R, Charuvi D, Shimoni E, Reich Z. (2011). Dynamic control of protein diffusion within the granal thylakoid lumen. *Proc. Natl. Acad. Sci. U. S. A.* **108**, 20248–20253.

Lapuente-Brun E, Moreno-Loshuertos R, Acín-Pérez R, Latorre-Pellicer A, Colás C, Balsa E, Perales-Clemente E et al. (2013). Supercomplex assembly determines electron flux in the mitochondrial electron transport chain. *Science* **340**, 1567–1570.

Lea-Smith DJ, Ross N, Zori M, Bendall DS, Dennis JS, Scott SA, Smith AG, Howe CJ. (2013). Thylakoid terminal oxidases are essential for the cyanobacterium *Synechocystis* sp. PCC6803 to survive rapidly changing light intensities. *Plant Physiol.* **162**, 484–495.

Lenn T, Leake MC, Mullineaux CW. (2008). Clustering and dynamics of cytochrome *bd*-I complexes in the *Escherichia coli* plasma membrane *in vivo*. *Mol. Microbiol.* **70**, 1397–1407.

Liberton M, Berg RH, Heuser J, Roth R, Pakrasi HB. (2006). Ultrastructure of the membrane systems in the unicellular cyanobacterium *Synechocystis* sp. Strain PCC6803. *Protoplasma* **227**, 129–138.

Liu H, Zhang H, Niedzwiedzki DM, Prado M, He G, Gross ML, Blankenship RE. (2013). Phycobilisomes supply excitations to both photosystems in a megacomplex in cyanobacteria. *Science* **342**, 1104–1107.

Liu L-N, Bryan SJ, Huang F, Yu J, Nixon PJ, Rich PR, Mullineaux CW. (2012). Control of electron transport routes through redox-regulated redistribution of respiratory complexes. *Proc. Natl. Acad. Sci. U.S.A.* **109**, 11431–11436.

Llorente-Garcia I, Lenn T, Erhardt H, Harriman O, Liu L-N, Robson A, Chiu S-W et al. (2014). Single-molecule *in vivo* imaging of bacterial respiratory complexes indicates delocalized oxidative phosphorylation. *Biochim. Biophys. Acta (BBA-Bioenergetics)* **1837**, 811–824.

Moerschel E, Schatz GH. (1987). Correlation of photosystem II complexes with exoplasmatic freeze-fracture particles of thylakoids of the cyanobacterium *Synechococcus* sp. *Planta* **172**, 145–154.

Mullineaux CW. (2014). Co-existence of photosynthetic and respiratory activities in cyanobacterial thylakoid membranes. *Biochim. Biophys. Acta (BBA-Bioenergetics)* **1837**, 503–511.

Mullineaux CW, Tobin MJ, Jones GR. (1997). Mobility of photosynthetic complexes in thylakoid membranes. *Nature* **390**, 421–424.

Nevo R, Charuvi D, Shimoni E, Schwarz R, Kaplan A, Ohad I, Reich Z. (2007). Thylakoid membrane perforations and connectivity enable intracellular traffic in cyanobacteria. *EMBO J.* **26**, 1467–1473.

Olive J, Ajlani G, Astier C, Recouvreur M, Vernotte C. (1997). Ultrastructure and light adaptation of phycobilisome mutants of *Synechocystis* sp PCC6803. *Biochim. Biophys. Acta* **1319**, 275–282.

Rakhimberdieva MG, Boichenko VA, Karapetyan NV, Stadnichuk IN. (2001). Interaction of phycobilisomes with photosystem II dimers and photosystem I monomers and trimers in the cyanobacterium *Spirulina platensis. Biochemistry* **40**, 15780–15788.

Rexroth S, Mullineaux CW, Ellinger D, Sendtko E, Rögner M, Koenig F. (2011). The plasma membrane of the cyanobacterium *Gloeobacter violaceus* contains segregated bioenergetic domains. *Plant Cell* **23**, 2379–2390.

Sacharz J, Yu J, Escobar E, Mastroianni G, Nixon PJ, Mullineaux CW. (2015) Sub-cellular localisation of the CtpA protease in the cyanobacterium *Synechocystis sp.* PCC6803 suggests multiple sites of Photosystem II biogenesis and repair. *In preparation.*

Sarcina M, Mullineaux CW. (2004). Mobility of the IsiA chlorophyll-binding protein in cyanobacterial thylakoid membranes. *J. Biol. Chem.* **279**, 36514–36518.

Sarcina M, Murata N, Tobin MJ, Mullineaux CW. (2003). Lipid diffusion in the thylakoid membranes of the cyanobacterium *Synechococcus* sp.: Effect of fatty acid desaturation. *FEBS Lett.* **553**, 295–298.

Sarcina M, Tobin MJ, Mullineaux CW. (2001). Diffusion of phycobilisomes on the thylakoid membranes of the cyanobacterium *Synechococcus* 7942: Effects of phycobilisome size, temperature and membrane lipid composition. *J. Biol. Chem.* **276**, 46830–46834.

Sherman DM, Troyan TA, Sherman LA. (1994). Localization of membrane proteins in the cyanobacterium *Synechococcus* sp, PCC7942. Radial asymmetry in the photosynthetic complexes. *Plant Physiol.* **106**, 251–262.

Stengel A, Gügel IL, Hilger D, Rengstl B, Jung H, Nickelsen J. (2012). Initial steps of Photosystem II de novo assembly and pre-loading with manganese take place in biogenesis centers in *Synechocystis. Plant Cell* **24**, 660–675.

van de Meene AML, Hohmann-Mariott MF, Vermaas WFJ, Roberson RW. (2006). The three-dimensional structure of the cyanobacterium *Synechocystis* sp. PCC6803. *Arch. Microbiol.* **184**, 259–270.

van Thor JJ, Gruters OWM, Matthijs HCP, Hellingwerf KJ. (1999). Localization and function of ferredoxin:NADP+ reductase bound to the phycobilisomes of *Synechocystis. EMBO J.* **18**, 4128–4136.

Vermaas WFJ. (1996). Molecular genetics of the cyanobacterium *Synechocystis* sp. PCC6803: Principles and possible biotechnology applications. *J. Appl. Phycol.* **8**, 263–273.

Vermaas WFJ, Timlin JA, Jones HDT, Sinclair MB, Nieman LT, Hamad SW, Melgaard DK, Haaland DM. (2008). *In vivo* hyperspectral confocal fluorescence imaging to determine pigment localization and distribution in cyanobacterial cells. *Proc. Natl. Acad. Sci. U. S. A.* **105**, 4050–4055.

Vernotte C, Astier C, Olive J. (1990). State 1-state 2 adaptation in the cyanobacteria *Synechocystis* PCC6714 and *Synechocystis* PCC6803 wild-type and phycocyanin-less mutant. *Photosynth. Res.* **26**, 203–212.

Vothknecht UC, Westhoff P. (2001). Biogenesis and origin of thylakoid membranes. *Biochim. Biophys. Acta* **1541**, 91–101.

Watanabe M, Ikeuchi M. (2013). Phycobilisome: Architecture of a light-harvesting supercomplex. *Photosynth. Res.* **116**, 265–276.

CHAPTER 11

Regulation of Cellular Signalling by Thioredoxin

Toshiya Machida, Hidenori Ichijo and Isao Naguro

CONTENTS

ABSTRACT Reactive oxygen species (ROS) mediate various cellular signal transduction as second messengers. However, excess ROS disrupts cellular functions and homeostasis by oxidizing intracellular molecules. This ROS toxicity is referred to as "oxidative stress". To control the intracellular oxidation-reduction (redox) status and endure oxidative stress, organisms are equipped with various antioxidant systems. Among these, the redox protein thioredoxin (Trx), together with thioredoxin reductase (TrxR) and NADPH, constitutes the Trx system in mammalian cells. Trx reduces the disulphide bonds formed in intracellular proteins by oxidation and plays a significant role in maintaining proper cellular functions by retaining the function of proteins altered by ROS. Trx also participates in cellular signal transduction by forming protein complexes called Trx signalosomes with intracellular proteins. The redox status of Trx is altered by various cell stimuli. Trx associates and dissociates with proteins depending on its redox status, switching the intracellular signallling to induce proper cellular responses. In this chapter, we introduce a variety of Trx signalosomes, focussing on the mechanisms of signal transduction depending on the cellular context. Moreover, accumulating data suggest that Trx signalosomes play pivotal roles in diseases related to oxidative stress. We also describe the roles of the Trx signalosome in stressed cells and disease.

11.1 INTRODUCTION—ROS AS COMPONENTS OF SIGNALLING MOLECULES

In aerobic respiration, molecular oxygen is partially reduced, generating ROS. The one-electron reduction of O_2 generates a superoxide anion (O_2^-), and this highly reactive radical is a progenitor of various ROS in cells. Dismutation of O_2^-, mediated either by the catalytic function of superoxide dismutases (SODs) or by a spontaneous reaction, creates hydrogen peroxide (H_2O_2). H_2O_2 is converted to the hydroxyl radical ($OH^\bullet$) by further partial reduction.[1]

Excess ROS, including O_2^-, H_2O_2, and $OH^\bullet$, can irreversibly damage DNA, protein and lipids,[2] causing cellular dysfunction. This disturbance elicited by ROS is called "oxidative stress". To counteract oxidative stress, organisms have evolutionarily developed antioxidant systems. Peroxiredoxins (Prxs) catalyse the reduction of H_2O_2 to H_2O. To recycle these peroxidases, electrons from NADPH are transferred by the Trx system to reduce the disulphide bonds generated in Prx.[3] Glutaredoxins (Grxs) also convert H_2O_2 to H_2O, and the recycling of these peroxidases is mediated by glutathione, a small molecule containing a reactive thiol group.[3] Cells maintain redox homeostasis by using a series of redox reactions.

Despite their toxicity, cells actively generate ROS in some contexts. When stimulated by pathogens or inflammatory cytokines, phagocytes upregulate NADPH oxidase (NOX), the enzyme that converts O_2 to O_2^-, and produce ROS to damage bacteria.[4] In addition to the injurious role to bacteria, ROS regulate intracellular signalling systems and induce cellular responses.[5] Signalling pathways regulated by ROS mediate the cellular responses to oxidative stress.

In ROS-mediated signalling pathways, H_2O_2 oxidatively modifies various proteins for its stability, powerful oxidative potential and specificity of modification.[6] H_2O_2 oxidizes and modifies the thiol group in specific Cys protein residues. The Cys residues specifically implicated in intracellular signalling pathways have a lower pKa than other Cys residues due to the stabilization of negative charge on the sulphurs of these specific Cys residues by hydrogen bonds with adjacent residues.[7,8] Thus, Cys residues can stably exist in thiolate anions, ensuring their nucleophilicity and reactivity with H_2O_2.[9] Such features of the cysteine guarantee that a specific cysteine can uniquely sense the intracellular redox condition.

Nanomolar concentrations of H_2O_2 can oxidize the thiol group of specific Cys residues to sulphenic acid (—SOH). As sulphenic acid is highly reactive, this group can be attacked by an adjacent thiol group or nitrogen, forming disulphide bonds (—S—S—) and sulphenic amides (—S—N—), respectively.[2] Higher concentrations of H_2O_2 oxidize —SOH to sulphinic acid (—SO_2H) or to sulphonic acid (—SO_3H).[5] Nucleophilic Cys residues also can attack disulphide bonds in proteins. Through disulphide exchange reactions, the disulphide bonds can be transferred between proteins.[3]

Oxidative modifications in the Cys residues alter protein structure, functioning as a trigger for signal transduction. Intracellular signalling systems regulated by cellular redox state are called "redox signalling". The Cys modification mediated by H_2O_2 is a trigger for redox signalling. To date, it has been shown that the modification of Cys residues regulates signalling pathways such as the mitogen-activated protein kinase (MAPK), phosphatidylinositol 3-kinase (PI3K), cytokine receptor and DNA damage response signalling pathways.[5,10]

Emerging evidence indicates that Trx, primarily implicated in the antioxidant defence of the cells, also functions as a regulator of these ROS-sensitive signalling pathways. Trx regulates the function of other proteins by changing the redox states of Cys residues in both Trx itself and in other signalling proteins.[3] Here, we discuss the various modes of signal transduction regulated by Trx.

ROS have been implicated in many types of human disease, including cancer, cardiovascular disease, neurodegenerative diseases and inflammatory disease by causing oxidative stress.[11–15] However, ROS also lead to disease by manipulating and disrupting cellular signalling systems.[5] Therefore, the comprehensive elucidation of redox regulation in signal transduction is important to understand pathogenesis and to develop effective therapies for ROS-related diseases. In this review, we also describe the physiological functions and relevance to pathogenesis of Trx-mediated cellular signal transduction.

11.2 THIOREDOXIN—ITS ROLE IN ANTIOXIDANT SYSTEM AND SIGNALLING

Both prokaryotic and eukaryotic cells are equipped with two distinct antioxidant systems for the reduction of protein disulphide bonds generated by ROS: the Trx system and the glutathione/glutaredoxin system. Trx is one of the constituents of the Trx system. It is a small protein of approximately 12 kDa with a Trx-fold structural motif.[3] The basic structure of the Trx-fold is three α-helices surrounding a central core of four-stranded β-sheets (the structure of human Trx1 is shown in Figure 11.1a).[3] Many Trx-fold family proteins with redox reactivity share the active site motif Cys-X-X-Cys, in which the N-terminal Cys has a low pKa value and high nucleophilicity.[3] Trx was originally identified as a dithiol electron donor for a disulphide bond generated in ribonucleotide reductase in *Escherichia coli*,[16] and mammalian Trx also serves as an electron donor for protein disulphide bonds.[17,18] The two mammalian Trxs, cytosolic Trx (Trx1) and mitochondrial Trx (Trx2), are encoded by distinct genes.

Trx contains a highly conserved Cys-Gly-Pro-Cys sequence in its active site (Figure 11.1b).[3] Because of its low pKa value, the N-terminal Cys (Cys32 of human Trx1) is highly nucleophilic. Therefore, the Cys can attack the target disulphide bond and form a mixed disulphide bond with the target protein. Subsequently, the C-terminal Cys (Cys35 of human Trx1) attacks and resolves the mixed disulphide bond, resulting in reduced dithiols in the substrate protein and the formation of a disulphide bond in Trx.[3] The Trx system has three components: Trx, TrxR and NADPH. Oxidized Trx is reduced by the head-to-tail homodimer of TrxR, a seleno-flavoprotein.[19] The FAD domain of TrxR binds to NADPH, the electron donor, and TrxR transfers electrons from NADPH to Trx.[20]

Major targets of the reduction mediated by the Trx system are the Trx-dependent peroxidases, peroxiredoxins (Prxs).[21,22] 2-Cys Prx, which contains two catalytic Cys residues, forms a head-to-tail homodimer.[23] Prx scavenges H_2O_2 by reducing it to H_2O. During this H_2O_2 detoxification, two oxidized catalytic Cys residues form an intermolecular disulphide bond. Prx is recycled through reduction by Trx. Therefore, Trx counteracts oxidative stress through the recycling of Prxs.

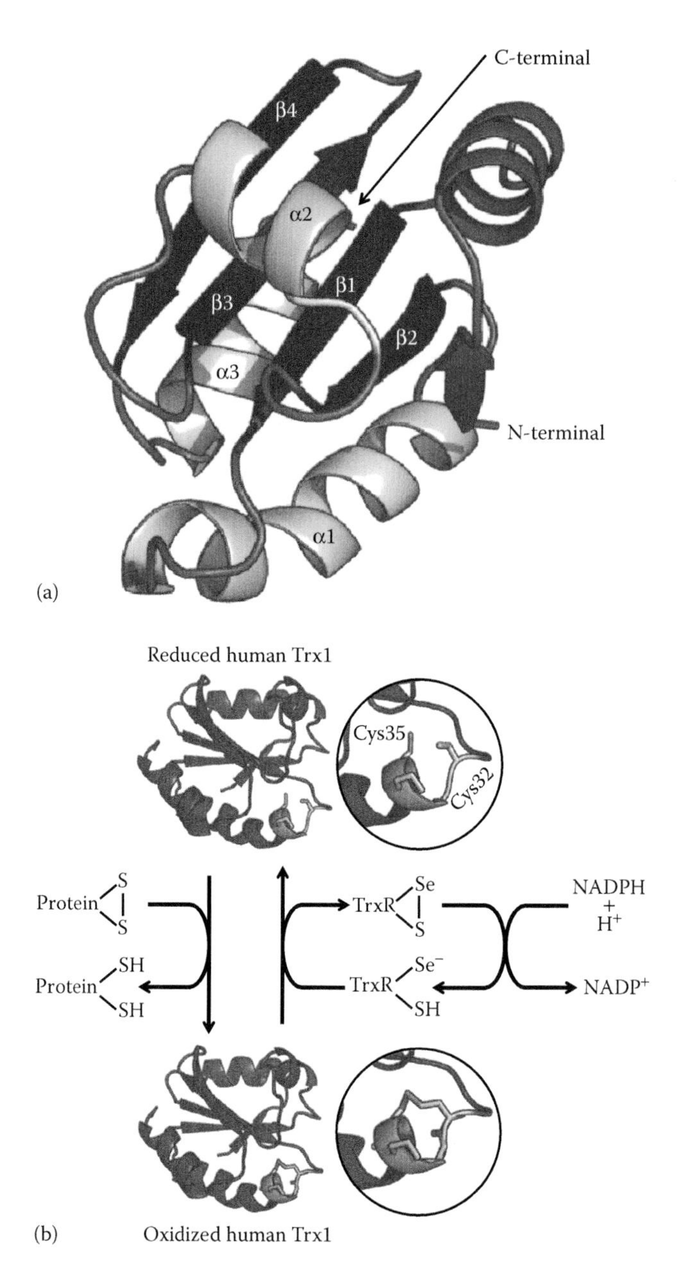

FIGURE 11.1 Molecular structure of Trx and the Trx system. (a) Three structural model of human Trx1 (protein database entry 1ERT). Four α-helices (α1–4) and four β-sheets (β1–4) that comprise the Trx-fold are shown. (b) The structural models of human Trx1 in the reduced form (protein data base entry 1ERT) and the oxidized form (protein database entry 1ERU). A disulphide bond is generated between Cys32 and Cys35 of Trx1 in the process of the reduction of disulphide bonds in other proteins. This disulphide bond is reduced by thioredoxin reductase (TrxR), which transfers electrons from the electron donor NADPH. All structural models were drawn using the Pymol program.

In addition to its role as an antioxidant protein, accumulating evidence suggests that Trx has non-canonical roles in the regulation of cellular signalling. Trx participates in cellular signal transduction by forming protein complexes called thioredoxin signalosomes with intracellular proteins. The redox status of Trx is altered by various cell stimuli. Trx associates and dissociates with proteins depending on its redox status, switching the intracellular signalling to induce proper cellular responses. In a non-stressed state, Trx binds to apoptosis signal-regulating kinase 1 (ASK1), a MAP kinase kinase kinase (MAP3K),[24] and inhibits ASK1 kinase activity through a direct interaction.[24,25] Intracellular redox conditions alter the interaction between Trx and ASK1 by changing the oxidative state of the Cys residues of Trx.[24] Concomitant with the dissociation of Trx, ASK1 autophosphorylates and activates downstream MAPK signalling.[25,26] In this context, Trx alters its structure by oxidation and acts as a redox sensor for the ROS-activated ASK1-MAPK signalling pathway. We will discuss the detailed molecular mechanisms regulating this protein complex in (Section 11.3.1). Trx-interacting protein (TXNIP) also forms a protein complex with Trx.[27] The interaction of Trx and TXNIP is altered by intracellular redox states,[28,29] and regulates various intracellular signalling pathways. The regulatory mechanisms and functions of the Trx-TXNIP protein complex are described in (Section 11.3.2).

Besides, Trx regulates signalling pathways through its catalytic activity of protein disulphide reduction. Signalling proteins are oxidized at specific Cys residues, altering their structures.[30] These structural changes regulate intracellular signalling pathways. The reduction of these proteins by Trx reverses the structure, restoring the altered signalling pathways. We also discuss this different mode of signalling regulation mediated by Trx. Various signalling molecules, such as protein tyrosine phosphatases (PTPs) and phosphatase and tensin homologue (PTEN) are direct substrates of Trx.[31–35] Moreover, new Trx substrates have been recently identified using the Trx1 Cys35Ser mutant, which traps the substrates of Trx1.[36] We will focus on these signalling proteins that are directly reduced by Trx in (Section 11.3.3).

11.3 THIOREDOXIN SIGNALOSOME AND ITS PHYSIOLOGICAL ROLE

11.3.1 Thioredoxin—Apoptosis Signal-Regulating Kinase 1 (ASK1)

In eukaryotic cells, the MAPK pathway is activated and induces proper cellular responses to various stresses, including inflammatory cytokines, oxidative stress, osmotic stress and UV irradiation.[37] Once MAP kinase kinase kinase (MAP3K) is activated, downstream MAP kinase kinase (MAP2K) and MAPK are activated by sequential phosphorylation.[37]

Apoptosis signal-regulating kinase 1 (ASK1), or MAP3K5, was identified as one of the MAP3Ks.[38] ASK1 is activated by oxidative stress,[39] ER stress,[40] Ca^{2+} influx,[41] inflammatory cytokines[38,39] and some pathogen-associated molecular patterns (PAMPs),[42] and subsequently activates the downstream c-jun N-terminal kinase (JNK) and p38 pathways through mitogen-acitvated protein kinase kinase 4/7 (MKK4/7) and MKK3/6, respectively.[43]

Activation of ASK1 can be monitored by phosphorylation of the Thr residue in the activation loop (Thr838 of human ASK1 and Thr845 of mouse ASK1, respectively).[26] This phosphorylation is essential for the kinase activity of ASK1.[26] Homo-oligomerization and autophosphorylation of ASK1 are important for the activation of ASK1.[26,39]

The C-terminal fragment of ASK1 (1108–1374 a.a.) was identified as a binding partner of ASK1 itself by yeast two-hybrid screening.[26] ASK1 ΔCCC, which lacks C-terminal coiled-coil (CCC; 1236–1293 a.a.) domain, is deficient in the formation of the constitutive homo-oligomer and has much lower kinase activity, suggesting that ASK1 homo-oligomerizes through the CCC domain to maintain its kinase activity.[26] Overexpression of kinase deficient ASK1, whose Lys716 residue is mutated to Arg (ASK1 KR), results in much lower Thr838 phosphorylation levels compared to WT ASK1.[26] These data suggest that the activation of ASK1 occurs through the intermolecular autophosphorylation of the ASK1 homo-oligomer.

Using yeast two-hybrid screening, Trx1 was identified as a binding partner of ASK1.[24] ASK1 kinase activity was inhibited when co-expressed with Trx1 in HEK293 cells,[24] suggesting that Trx1 inhibits ASK1 activity. ASK1 associates with Trx1 through the 1–648 a.a. in the N-terminal region,[24] and N-terminal deficient ASK1 (ASK1 ΔN; lacking 1–648 a.a.) has higher kinase activity compared to wild-type ASK1,[24] suggesting that Trx1 inhibits ASK1 activation through a direct interaction.

Treatment of cells with H_2O_2 activates ASK1.[39] Moreover, antioxidant *N*-acetyl-L-cysteine (NAC) treatment attenuates the TNF-α mediated activation of ASK1.[39] These data imply that the mechanism of Trx1 inhibition of ASK1 is redox-dependent. In HEK293 cells, H_2O_2 treatment attenuated the binding between ASK1 and Trx1,[24] indicating that Trx1 binding is redox-dependent. An in vitro pull-down assay revealed that the binding of purified Trx1 to ASK1 was significantly increased by the preceding reduction of Trx1, and this association was disrupted by the addition of H_2O_2.[24] These data suggest that the redox state of Trx1 regulates the binding of Trx1 to ASK1. In the oxidative condition, a disulphide bond is formed in two Cys residues of Trx (Cys32 and Cys35 of human Trx1). The Trx1 double mutant Cys32Ser/Cys35Ser no longer binds to ASK1,[24] suggesting that reduced free Cys32 and Cys35 are essential for the interaction between Trx1 and ASK1. Either Cys32Ser or Cys35Ser mutant of Trx1 remains associated with ASK1 in vitro, even in the presence of H_2O_2,[44] suggesting that the formation of the disulphide bond is critical for the dissociation of Trx1 from ASK1.

TNF-α also activates ASK1.[38,39] ROS are generated after the binding of TNF-α to the TNF receptor.[45] The inhibitory activity of NAC on TNF-α-mediated activation of ASK1 suggests that TNF-α activates ASK1 through a ROS dependent redox mechanism. In RAW264.7 murine macrophage cells, lipopolysaccharide (LPS) activated ASK1.[42] LPS-dependent activation of ASK1 was suppressed by treatment with the antioxidant propyl gallate (PG) or NAC,[42] suggesting the involvement of ROS in LPS-mediated ASK1 activation. In response to LPS, toll-like receptor 4 (TLR4) directly interacts with NADPH oxidase and enhances its activity in HEK 293 cells.[46] These data imply that the ROS-mediated dissociation of Trx1 is also involved in the activation of ASK1 by TNF-α and LPS.

The detailed molecular mechanism of Trx1-mediated inhibition of ASK1 has been investigated using ASK1 deletion mutants. In non-stressed cells, ASK1 forms an inactive homo-oligomer through its CCC domain (Figure 11.2).[26] Although ASK1 ΔCCC is deficient in forming basal homo-oligomers, the mutant retains the ability to form homo-oligomers and undergoes weak activation in response to ROS.[26] This observation suggests that another interface, other than the CCC domain, may form homo-oligomers in response to ROS and

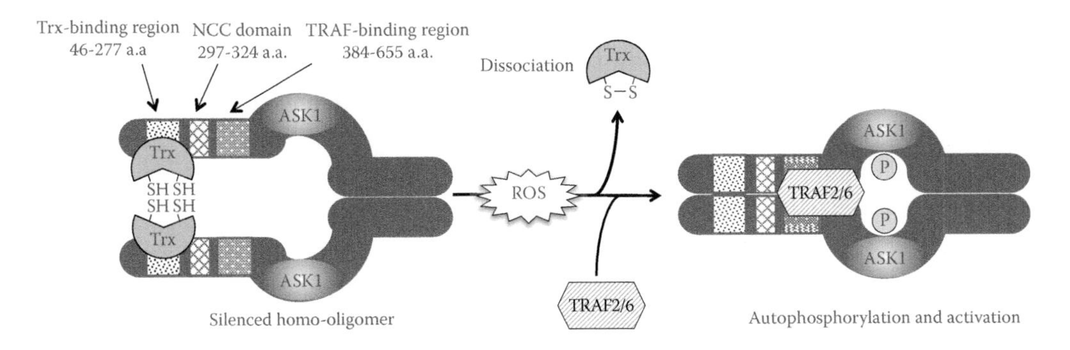

FIGURE 11.2 Activation mechanism of ASK1. In non-stressed cells, ASK1 forms a silent homo-oligomer through the C-terminal coiled-coil (CCC) domain. Trx1 binds to the N-terminal region of ASK1 to inhibit homo-oligomerization of its N-terminal coiled-coil (NCC) domain. Upon the ROS-dependent dissociation of Trx1, the ASK1 homo-oligomer forms another interface through NCC. TRAF2/6 promote this interaction of the NCC through direct binding to the N-terminal region of ASK1. Amino acid numbers are indicated based on human ASK1.

function in ASK1 activation. ASK1 constitutively forms a high molecular mass (approximately 1500–2000 kDa) protein complex, including Trx1.[47] This megadalton ASK1 complex (ASK1-signalosome) increases in size after H_2O_2 treatment to approximately 3000 kDa,[47] suggesting that H_2O_2 alters the components of the ASK1-signalosome. These data suggest that Trx1 may affect the homo-oligomerization of ASK1 and the binding of other molecules to the ASK1 signalosome.

TNF receptor-associated factor (TRAF) family proteins are known for their roles in signalling downstream of the TNF receptor.[48] Since it was first elucidated that the overexpression of TRAF2, TRAF5 and TRAF6 activates ASK1 and its downstream JNK,[49,50] the roles of the TRAF family proteins in ASK1 activation have been extensively investigated. TRAF2 interacts with both the N-terminal and C-terminal regions of ASK1 through its TRAF domain.[51] Moreover, TRAF2 and TRAF6 bind to ASK1 in response to ROS,[47,51] suggesting their role in ROS-mediated activation of ASK1. Mouse embryonic fibroblasts (MEFs) from TRAF-deficient mice revealed that TRAF2 and TRAF6 are required for the ROS-dependent activation of ASK1.[47] In MEFs, TRAF6 was involved in the formation of high molecular mass ASK1 protein complexes by H_2O_2. In TRAF6-deficient MEFs, the ASK1-signalosome no longer increased in size upon H_2O_2 treatment.[47] These data suggest that TRAF6, and presumably TRAF2, is recruited to the ASK1-signalosome and activates ASK1 in response to ROS.

Trx1 overexpression disrupts the binding of ASK1 and TRAF2.[51] This finding suggests that Trx1 inhibits the binding of TRAF2 and ASK1. One report shows the cooperative regulatory mechanism on ASK1 by Trx1 and TRAF2/6.[25] Using truncated ASK1 mutants, it was revealed that the 46–277 a.a. region and the 384–655 a.a. region of ASK1 bind to Trx1 and TRAF2/6, respectively.[25] ASK1 Δ277 (278–1380 a.a.), which lacks the Trx1 binding domain, showed higher binding affinity to TRAF2 and TRAF6 than wild-type ASK1 (ASK1 WT) in non-stressed cells, supporting the finding that Trx1 suppresses the binding between ASK1 and

TRAF2/6.[25] ASK1 has another coiled-coil motif in the N-terminal 297–324 a.a. region, called the N-terminal coiled-coil (NCC) domain, in the vicinity of both the Trx1- and TRAF-binding regions.[25] When overexpressed, ASK1 Δ277 showed higher basal activity than ASK1 WT.[25] However, ASK1 Δ384 (ASK1 385–1380 a.a.) showed lower basal activity than ASK1 Δ277, although it lacks the Trx1-binding region and has a higher binding affinity to TRAF2/6 than ASK1 WT.[25] These results suggest the importance of the NCC domain for the activity of ASK1. In accordance with this hypothesis, NCC domain-deficient ASK1 (ASK1 ΔNCC, which lacks the 277–384 a.a. region) showed low activity when stimulated by H_2O_2.[25] C-terminal region truncated ASK1 (ASK1 ΔC; 1–947 a.a.) weakly homo-oligomerizes, presumably through the NCC domain.[25] Overexpression of Trx1 disrupted and overexpression of TRAF2/6 enhanced this homo-oligomerization of ASK1 ΔC.[25] Furthermore, siRNA-mediated knockdown of TRAF2 and TRAF6 attenuated the H_2O_2-dependent homo-oligomerization of ASK1 ΔC.[25] These data suggest that ASK1 activity is regulated by the following model (Figure 11.2). In non-stressed cells, the reduced Trx1 binds to the N-terminal region of ASK1 and hinders both the NCC-domain mediated homo-oligomerization of ASK1 and the binding of TRAF2/6 to ASK1. Upon ROS stimulation, oxidized Trx1 dissociates from ASK1, due to changes in the affinity of Trx1 to ASK1, enabling ASK1 to interact with TRAF2 and/or TRAF6. ASK1 is thus activated by autophosphorylation, which is dependent on the homophilic interaction of the N-terminal region of ASK1. Although the TRAFs appear to promote the homophilic interaction of the N-terminal region of ASK1, the detailed mechanism by which TRAFs promote the activation of ASK1 remains to be elucidated.

Recently, the low-resolution structure of ASK1-Trx1 was estimated using small angle X-ray scattering technology and computational protein structure modelling.[52] This structural model predicted a large interaction interface between reduced Trx1 and ASK1-Trx-binding-domain (ASK1-TBD, 88–302 a.a. of human ASK1), with the absence of a disulphide bond between them. Although several reports suggested that mixed-disulphide bond formation between Cys32 of Trx1 and Cys250 of ASK1 occurs in unstressed cells,[53,54] the structural modelling of reduced Trx1 and ASK1-TBD suggests that Cys250 of ASK1 is required only for the formation of the Trx1-binding surface of ASK1.[52] The role of Cys250 of ASK1 in the regulation of ASK1 activity remains to be determined.

The function of the ASK1-Trx signalosome is well studied in apoptosis. When the *ASK1* gene was identified, it was also reported that the overexpression of human ASK1 in Mv1Lu cells caused apoptosis,[38] suggesting that ASK1 functions in apoptosis. ASK1 is involved in apoptosis caused by various stress stimuli, such as H_2O_2 and TNF-α.[24,39] Expression of ASK1 ΔN, which does not associate with Trx and which shows constitutive kinase activity, facilitates apoptosis much more strongly than ASK1 WT.[24] This finding suggests that Trx attenuates apoptosis through the inhibition of ASK1. Using MEFs from knockout mice, it was revealed that caspase-9, but not caspase-8, is required for the apoptosis mediated by ASK1 ΔN.[55] Furthermore, ASK1 ΔN facilitates cytochrome *c* release from mitochondria.[55] These data suggest that ASK1 executes apoptosis through a mitochondria-dependent pathway. JNK activates the mitochondrial pathway by phosphorylating Bcl-2/Bcl-xL.[56] ASK1-dependent apoptosis may, at least in part, be attributable to JNK function, which is activated downstream of ASK1.

Apoptosis regulated by the ASK1-Trx signalosome has implications in human disease. ROS generation is involved in the pathogenesis of neurodegenerative diseases.[57] Amyloid β (Aβ), which is assumed to be one of the causes of Alzheimer's disease, generates ROS by disrupting both cellular calcium homeostasis and mitochondrial respiration.[57] ASK1-deficient neurons are protected from Aβ-induced neuronal cell death.[58] Furthermore, $ASK1^{-/-}$ mice exhibited milder symptoms in the 1-methyl-4-phenyl-1,2,3,6-tetrahydropyridine (MPTP)-induced experimental model of Parkinson's disease than wild-type mice.[59]

As mentioned earlier, LPS activates ASK1 in immune cells in a ROS-dependent manner.[42] When stimulated by LPS, p38, but not JNK, showed lower activity in dendritic cells and splenocytes derived from $ASK1^{-/-}$ mice compared to those from wild-type mice.[42] TRAF6 bound to ASK1 in LPS-stimulated RAW264.7 macrophage cells.[42] These findings suggest that the TRAF6-ASK1-p38 signalling axis is activated by ROS upon TLR4 stimulation. Furthermore, cytokine production after LPS stimulation was attenuated in $ASK1^{-/-}$ splenocytes,[42] suggesting that the TRAF6-ASK1-p38 pathway has an essential role in innate immunity. Consistent with this hypothesis, $ASK1^{-/-}$ mice showed lower mortality and cytokine production in a LPS-induced septic shock model compared to wild-type mice.[42]

11.3.2 Thioredoxin—Thioredoxin-Interacting Protein (TXNIP)

Trx-interacting protein (TXNIP), also termed Trx-binding protein 2 (TBP2) or vitamin D_3 upregulated protein 1 (VDUP1), was originally identified as a protein that is upregulated in response to vitamin D_3 in HL-60 human leukaemia cell line.[60] TXNIP is one of six members of the α-arrestin family of proteins, which consists of arrestin domain-containing protein (ARRDC) 1–5 and TXNIP.[61] The amino acid sequences of α-arrestins are similar to the β-arrestin proteins that are known for their interaction with various proteins.[61] The α-arrestin family proteins also interact with other proteins and function as adaptor proteins.[61] α-arrestins and β-arrestins share conserved structures in two major domains, the N-domain and the C-domain, that are important for protein–protein binding.[61]

Yeast two-hybrid screening revealed that TXNIP is a binding partner of Trx1.[62] TXNIP does not bind to the Trx Cys32Ser/Cys35Ser mutant, suggesting that the redox active Cys32 and Cys35 of Trx are required for the interaction between Trx1 and TXNIP.[63] The cell extracts from HEK293 or COS7 cells transfected with TXNIP exhibited suppressed reducing activity to insulin.[62] Moreover, the reductase activity of recombinant Trx1 was inhibited by the addition of TXNIP in vitro.[62] The TXNIP Cys247Ser mutant did not bind to Trx1,[28] resulting in the inability to inhibit Trx1.[28] Thus, TXNIP decreases the disulphide reductase activity of Trx1 through a direct interaction.

The binding mechanism of Trx1 and TXNIP has been intensively analysed. Cys residues in both Trx1 and TXNIP (Cys32 and Cys35 in Trx1 and Cys247 in TXNIP) are required for the interaction, which suggests the disulphide bond formation between Trx1 and TXNIP.[28,63] In the reducing condition of the eukaryotic cytosol, the de novo formation of protein disulphide bonds is not likely to occur.[64] Thus, a disulphide exchange reaction between Trx1 and TXNIP seems to be involved in the binding mechanism between Trx1 and TXNIP. Prior oxidation of Trx1 attenuated the interaction between Trx1 and TXNIP in vitro,[28] suggesting a model that disulphide exchange occurs between reduced Trx1 and a disulphide

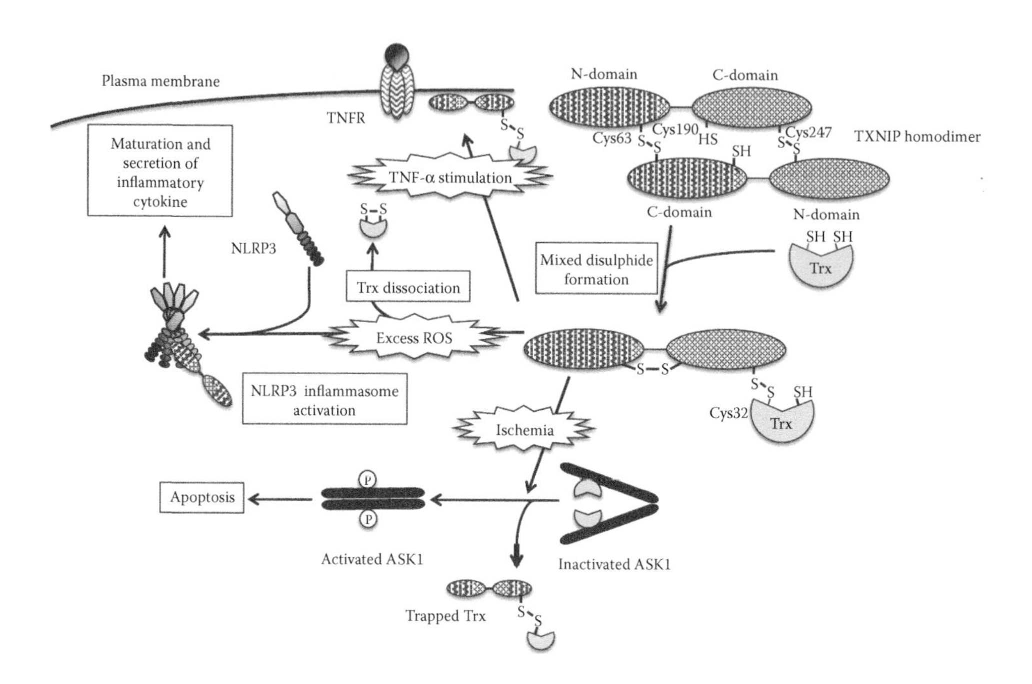

FIGURE 11.3 Functions of Trx-TXNIP complex. TXNIP forms a head-to-tail homodimer by Cys63-Cys190 disulphide bonds in cells. Trx1 binds to TXNIP through mixed-disulphide bond formation, concomitant with the intramolecular disulphide bond formation between Cys63 and Cys190 of TXNIP. In response to excess ROS, Trx1 dissociates from TXNIP and permits the interaction between TXNIP and NLRP3, which is essential for the ROS-dependent activation of the NLRP3 inflammasome. Upon TNF-α stimulation, the Trx-TXNIP signalosome is recruited to the plasma membrane. Ischemia increases TXNIP expression, which traps more Trx1 and leads to the dissociation of Trx1 from ASK1. ASK1 is therefore able to activate and induce apoptosis in cardiac myocytes.

bond in TXNIP. Recently, a redox-dependent mechanism of this disulphide exchange was indicated by crystal structural analysis of the Trx-TXNIP complex.[29] TXNIP forms a head-to-tail homodimer in resting cells, with an intermolecular disulphide bond between Cys63 and Cys247 (Figure 11.3). Upon binding to Trx1, the TXNIP homodimer changes its structure concomitant with disulphide exchange, resulting in a mixed-disulphide bond formation between Cys32 of Trx1 and Cys 247 of TXNIP.[29] Reactive thiol Cys63 freed from the disulphide bond by Trx1 attacks intramolecular Cys190, forming another intramolecular Cys63-Cys190 disulphide bond.[29] The binding of Trx1 and TXNIP was disrupted by the addition of H_2O_2.[29] Furthermore, in THP-1 human macrophage cell lines, Trx1 dissociated from TXNIP after stimulation with an exogenous high concentration (10 mM) of H_2O_2.[65] These data suggest that highly oxidative conditions dissociate Trx from TXNIP. The dissociation of Trx1 from TXNIP is accompanied by intramolecular disulphide formation in Trx,[29] but the detailed mechanism remains unknown.

Although TXNIP was originally described as an intrinsic inhibitor of Trx1 reductase activity, emerging evidence suggests that TXNIP functions as a Trx-TXNIP complex in cellular signalling. One example of the role of the Trx-TXNIP complex in cellular processes is the ROS-mediated activation of the NOD-like receptor, pyrin domain-containing

3 (NLRP3) inflammasome.[65] The NLRP3 inflammasome is activated by stimuli such as ROS, and subsequently activates caspase-1.[66,67] Activated caspase-1 cleaves pro-IL-1β to yield the active form of IL-1β, leading to the final stage of cytokine production in inflammatory cells.[67] NLRP3 associates with the C-domain of TXNIP through the leucine-rich repeat (LRR) domain or NAIP, C2TA, HET-E and TP1 (NACHT) domain.[65] In the resting state, TXNIP binds to Trx1, preventing TXNIP from binding to NLRP3. In human THP-1 macrophages, treatment with high-concentration ROS dissociated TXNIP from Trx1, concomitant with the association of TXNIP to NLRP3 (Figure 11.3).[65] SiRNA knockdown of TXNIP attenuated the activation of caspase-1 and the secretion of mature IL-1β in response to H_2O_2 or other ROS-generating NLRP3 activators, such as monosodium urate and R-837,[65] suggesting a requirement for TXNIP in the activation of the NLRP3 inflammasome. Thus, the Trx-TXNIP complex is a redox-dependent regulator of the cellular inflammatory response.

In addition to the redox state of the cells, the regulation of TXNIP expression is also important to modulate the TXNIP-Trx signalling complex. TXNIP abrogates the inhibitory effect of Trx1 on ASK1 (Section 11.3.1) by disrupting the binding between Trx1 and ASK1 (Figure 11.3).[63] SiRNA knockdown of TXNIP enhanced the binding of Trx1 to ASK1 in human umbilical vein endothelial cells, which coincided with the attenuated JNK and p38 activity and vascular cell adhesion molecule 1 (VCAM1) expression in response to TNF-α.[68] Shear stress, which suppresses TXNIP expression in vascular endothelial cells,[68] attenuates TNF-α-mediated inflammation by inhibiting the activation of the ASK1-JNK/p38 pathway.[69,70] Thus, the inhibitory function of laminar flow in vascular inflammation appears to be due to the suppression of *TXNIP* gene expression.

TXNIP mRNA was upregulated in ischemic myocardium, and the suppression of *TXNIP* gene expression in vivo resulted in attenuated ASK1 activation and the inhibition of apoptosis in the myocardium after acute ischemia.[71] Consistently, when TXNIP expression was downregulated in H9C2 rat cardiomyocytes, ASK1 activity and subsequent apoptosis were suppressed in response to H_2O_2.[71] Thus, the upregulation of *TXNIP* seems to be implicated in ASK1-dependent cardiac apoptosis during cardiac ischemia.

These two examples demonstrate the significance of the regulation of the Trx-TXNIP signalosome by the regulation of *TXNIP* gene expression. However, the example of the ROS-dependent association of TXNIP with NLRP3 exhibits the redox-sensitive regulation of TXNIP by Trx. Trx and TXNIP mutually regulate each other's function in the protein complex through covalent disulphide bonds.

Several reports indicate that TXNIP not only functions as an inhibitor of Trx reductase activity but also mediates the recruitment of Trx to specific cellular fractions. In vascular endothelial cells, cytosolic Trx1 is recruited to the plasma membrane upon TNF-α or H_2O_2 stimulation.[72] This localization of Trx1 coincides with the recruitment of TXNIP to the plasma membrane.[72] Gene silencing of TXNIP by siRNA attenuated the recruitment of Trx1 to the plasma membrane after H_2O_2 stimulation.[72] These observations indicate that TXNIP is essential for the stress-mediated plasma membrane localization of Trx1, although the physiological relevance of this interesting localization of Trx1 under stress conditions remains to be determined.

TXNIP also contributes to the nuclear localization of Trx in coordination with another cofactor. The Trx-TXNIP complex binds to DnaJb5, a DnaJ/HSP40 family protein, through TXNIP in myocytes.[73] Trx1 reduces the Cys274-Cys276 disulphide bond in DnaJb5 generated by hypertrophic stimulus.[73] The Trx-TXNIP-DnaJb5 complex translocates to the nucleus, where the complex associates with HDAC4 through DnaJb5.[73] Trx1 enhances HDAC4 nuclear localization by reducing its Cys667-Cys669 disulphide bond, thus inducing the subsequent HDAC4-mediated arrest in hypertrophic transactivation induced by nuclear factor of activated T-cells.[73,74] Transgenic mice with a dominant-negative Trx1 show both baseline cardiac hypertrophy and augmentation of pressure overload-induced cardiac hypertrophy.[73] Redox regulation of HDAC4 by the Trx-TXNIP-DnaJb5 complex is involved in the inhibitory function of Trx1 on cardiac hypertrophy. As described earlier, TXNIP attenuates the reductase activity of Trx1 by forming disulphide bonds through the redox reactive Cys32 of Trx. Overexpression of DnaJb5 did not interfere with the binding between Trx1 and TXNIP, but restored the reductase activity of Trx1.[73] DnaJb5 therefore mediates the interaction of Trx1 and TXNIP by an unknown mechanism. The localization of the Trx-TXNIP signalosome may be due to the function of TXNIP as an adaptor protein. Further investigation into the relationship between the binding partners of TXNIP and the Trx-TXNIP signalosome would broaden our comprehension of this unique.

11.3.3 Redox Regulation of Signalling Proteins by Trx

Trx also regulates cellular signalling by reducing specific Cys residues in proteins without forming stable complexes. Here, we introduce several noteworthy examples of the signalling regulated by Trx.

As mentioned in Section 11.1, the PI3K pathway is regulated by ROS.[5,75] PTEN and PTPs, negative regulators of the PI3K pathway, are inactivated by ROS generated downstream of growth factors.[31,35,76] When receptor tyrosine kinases (RTKs) are activated, PI3K is recruited to the plasma membrane and phosphorylates phosphatidylinositol (4,5)-bisphosphate ($PI(4,5)P_2$) to phosphatidylinositol (3,4,5)-triphosphate ($PI(3,4,5)P_3$), activating the downstream PI3K pathway. PTEN suppresses PI3K pathway by dephosphorylating PIP_3 to PIP_2.[77] The $PI(3,4,5)P_3$-hydrolizing activity of PTEN was decreased by H_2O_2 treatment in vitro.[31,35] Mass spectrometry analyses revealed the resulting disulphide bond formation between Cys71 and Cys124 of PTEN.[35] Because Cys124 of PTEN is directly involved in the catalytic reaction of PTEN, H_2O_2 inhibits the activity of PTEN through the formation of a disulphide bond in PTEN protein. Oxidation of PTEN was also observed in cells stimulated by the growth factor PDGF,[31] suggesting the physiological incidence of the oxidative inhibition of PTEN. The Trx system reduced oxidized PTEN more efficiently than glutathione.[35] Moreover, Trp14, which is the other member of the Trx family and has similar reduction potential (−0.257 V) as Trx (−0.274 V),[35] shows much lower PTEN reduction efficiency than Trx, suggesting a specific role for Trx in the reduction of PTEN.[35] Silencing of Trx1 results in sustained oxidation and inactivation of PTEN.[33] Trx appropriately reactivates PTEN and terminates the sustained activation of the PI3K pathway.

Inactivation of PTEN permits augmented activation of PI3K-Akt pathway and its downstream mammalian target of rapamycin complex 1 (mTORC1).[78,79] This pathway enhances ribosomal biogenesis and protein synthesis,[79] and has crucial roles in survival and proliferation of cancer cells.[79–81] Redox-dependent PTEN inhibition leads to Akt activation and subsequent cell proliferation, even in normal cells.[31,82] This observation also suggests a role for ROS-mediated PTEN inactivation in the proliferation of cancer cells.

PTPs are the other negative regulators of PI3K pathway. PTPs suppress PI3K signalling by dephosphorylation of phospho-Tyr in proteins involved in the PI3K pathway, such as RTKs.[83] Various PTPs, including PTP1B, Src homology region 2 domain-containing phosphatase 1 (SHP1), SHP2 and CD45, are oxidized and inactivated by ROS.[84,85] Among them, PTP1B is one of the most studied PTPs in ROS-mediated oxidation. ROS, especially H_2O_2, are generated by growth factor stimulation, and reversibly oxidizes PTP1B.[86–89] The catalytic Cys215 of PTP1B is oxidized to sulphenic acid by H_2O_2. Oxidized Cys residues in a protein often form stable disulphide bonds with adjacent Cys residues, which protect the Cys residues from further irreversible oxidation. However, in the case of PTP1B Cys215, there is no Cys residue in its vicinity. Crystal structural analysis revealed that the sulphenic acid formed by PTP1B Cys215 is quickly attacked by the main chain nitrogen of Ser216 and forms -S-N- (sulphenyl amide).[90,91] This intramolecular reaction protects PTP1B from irreversible oxidation to sulphinic acid or sulphonic acid.[90,91] Oxidized PTP1B no longer has catalytic activity or binding potential for RTKs.[90] This characteristic oxidation and inactivation of PTP1B activates the PI3K pathway. PTP1B is reduced and reactivated by the Trx system in vitro.[33,34] In HeLa cells, the ectopically expressed Trx1 Cys35Ser mutant formed a stable disulphide-mediated complex with PTP1B in response to H_2O_2.[34] This finding supports the hypothesis that the reduction of PTP1B is mediated by the Trx system in the physiological condition.

Some reports implicate the redox-dependent regulation of the PI3K pathway in tumourigenesis. PTP1B is inactivated in HepG2 hepatoma cells and A431 epidermoid carcinoma cells, coinciding with the constitutive overproduction of ROS in the cells.[92] Treatment of HepG2 cells with diphenyleniodonium (DPI), an inhibitor of NADPH oxidases, decreased ROS levels in the cells, concomitant with the upregulation of PTP1B activity.[92] Furthermore, DPI treatment attenuated anchorage-independent cell growth,[92] one of the oncogenic characteristics of HepG2 cells. This result suggests that ROS have roles in the abnormal cell growth of some types of cancer cells, presumably through the inhibition of PTP1B.

Recently, several Trx substrates were identified by an interesting method using mutant Trx1. The Trx1 Cys35Ser mutant cannot resolve the disulphide bond between Cys32 and target proteins. Therefore, the Trx1-target complex can be stably preserved in the cells.[36] By utilizing this substrate-trapping mechanism of the Trx1 Cys35Ser mutant, two signalling proteins that undergo redox regulation by Trx have been identified.[93,94]

Collapsin response mediating protein 2 (CRMP2) is an adaptor protein that functions in cytoskeletal regulation in neurons.[95] At steady state, CRMP2 stabilizes microtubules through an interaction with the tubulin homodimer.[95] In response to semaphorin3A (Sema3A), a collapsin, CRMP2 is phosphorylated by both GSK3 and CDK5 and dissociates from the tubulin homodimer, resulting in growth cone collapse.[96–100] Sema3A stimulation generates ROS through molecule interacting with CasL protein (MICAL), a flavoprotein oxidereductase.[93] ROS generated by MICAL facilitate formation of the

CRMP2 homodimer, inducing an intermolecular disulphide bond between two Cys504s, which is subsequently attacked by Trx1.[93] CRMP2, which covalently binds to Trx1, is susceptible to phosphorylation mediated by GSK-3β and CDK5.[93] SiRNA knockdown of MICAL or Trx1 in isolated rat dorsal root ganglia neurons attenuated the growth cone collapse mediated by Sema3A,[93] suggesting the significance of the redox-dependent regulation of CRMP2 phosphorylation in physiological conditions.

The other substrate identified by the substrate-trapping method is AMP-activated kinase (AMPK). When the cellular energy level decreases, the ratio of AMP and ADP to ATP increases. This is sensed by AMPK, leading to the activation of cellular metabolism to retain energy homeostasis.[101] Using transgenic mice with cardiac specific expression of the Trx1 Cys35Ser mutant, the catalytic AMPK α-subunit was identified as a physiological target of Trx1 in cardiomyocytes during ischemia-mediated low glucose.[94] Glucose deprivation caused by ischemia decreases cellular energy levels, leading to subsequent oxidative stress mediated by H_2O_2 generated from Nox4 in the endoplasmic reticulum.[102] AMPK family kinases have two conserved Cys residues in the kinase domain (Cys130 and Cys174 in mouse AMPKα2). H_2O_2 generates an intermolecular disulphide bond between these two Cys residues.[94] Moreover, H_2O_2 pretreatment of AMPKα2 prevents the phosphorylation of AMPKα2 by its upstream kinase LKB1, suggesting that oxidized AMPK is less susceptible to be activated.[94] Substitution of Cys130 to Ser and/or Cys174 to Ser in AMPKα2 also suppressed the binding between AMPKα2 and LKB1.[94] These results suggest that two intact conserved Cys residues are required for the binding between AMPKα2 and LKB1. Trx1 likely retains the susceptibility of AMPK to LKB1-mediated activation by restoring intact Cys130 and Cys174 to AMPKα2. Transgenic mice that harbour the Cys174Ser mutant of AMPKα2 exhibit greater susceptibility to ischemia-mediated cardiac infarction than wild-type mice,[94] suggesting the importance of intact Cys residues of the AMPK α-subunit in the survival of cardiomyocytes in physiological low-glucose conditions. The mice with cardiac-specific overexpression of Trx1 were more resistant, and conversely, the mice with the Trx1 Cys32Ser/Cys35Ser mutant were more susceptible to ischemia-mediated cardiac infarction than wild-type mice.[94] The restoration of intact Cys130 and Cys174 of AMPKα2 may therefore be involved in the inhibitory role of Trx1 on cardiac infarction.

11.4 CONCLUSION

In this chapter, we introduced the various modes of Trx function in signal transduction. Trx stably binds to ASK1 and inhibits ASK1 activity in non-stressed cells. In oxidative stress, Trx dissociates from ASK1 and permits ASK1 activation by enabling its homo-oligomerization via the N-terminus, concomitant with the binding of TRAF2/6. TXNIP forms a covalent protein complex with Trx through a disulphide bond. Redox-dependent regulation of this disulphide bond regulates the binding of Trx and/or TXNIP to other proteins. Furthermore, several studies have suggested the contribution of TXNIP to the cellular localization of Trx. Moreover, Trx also regulates cellular signal transduction, including the PI3K pathway, by reducing specific residues in phosphatases involved in signalling. Substrate-trapping technology using mutant Trx recently revealed new substrates of Trx, CRMP2 and AMPK. Using this system, other signalling proteins could be revealed as Trx substrates.

ROS are involved in various diseases, such as cancer, neurodegenerative disease and cardiovascular disease. Here, we have also mentioned implications of Trx-mediated regulation of signalling in several diseases. Both the characterization of the precise mechanisms of Trx-mediated regulation of cellular signalling and the further elucidation of the physiological roles of cellular signalling regulated by Trx would provide cues for the outstanding cures for diseases caused by disruption of redox homeostasis.

ACKNOWLEDGEMENTS

This work was supported by Grants-in-Aid for Scientific Research (KAKENHI) from JSPS and MEXT, the Strategic Approach to Drug Discovery and Development in Pharmaceutical Sciences, Global Center of Education and Research for Chemical Biology of the Diseases, the "Understanding of molecular and environmental bases for brain health" conducted under the Strategic Research Program for Brain Sciences by MEXT, the Advanced research for medical products Mining Programme of the National Institute of Biomedical Innovation, the Cosmetology Research Foundation and the Tokyo Biochemical Research Foundation.

REFERENCES

1. Imlay JA. Pathways of oxidative damage. *Annu Rev Microbiol.* 2003;57:395–418.
2. Cross CE, Halliwell B, Borish ET et al. Oxygen radicals and human disease. *Ann Intern Med.* 1987;107:526–545.
3. Hanschmann EM, Godoy JR, Berndt C, Hudemann C, Lillig CH. Thioredoxins, glutaredoxins, and peroxiredoxins-molecular mechanisms and health significance: From cofactors to antioxidants to redox signaling. *Antioxid Redox Signal.* 2013;19:1539–1605.
4. Deffert C, Cachat J, Krause KH. Phagocyte NADPH oxidase, chronic granulomatous disease and mycobacterial infections. *Cell Microbiol.* 2014;16:1168–1178.
5. Ray PD, Huang BW, Tsuji Y. Reactive oxygen species (ROS) homeostasis and redox regulation in cellular signaling. *Cell Signal.* 2012;24:981–990.
6. Garcia-Santamarina S, Boronat S, Hidalgo E. Reversible cysteine oxidation in hydrogen peroxide sensing and signal transduction. *Biochemistry.* 2014;53:2560–2580.
7. Roos G, Foloppe N, Messens J. Understanding the pK(a) of redox cysteines: The key role of hydrogen bonding. *Antioxid Redox Signal.* 2013;18:94–127.
8. Ferrer-Sueta G, Manta B, Botti H, Radi R, Trujillo M, Denicola A. Factors affecting protein thiol reactivity and specificity in peroxide reduction. *Chem Res Toxicol.* 2011;24:434–450.
9. Finkel T. From sulfenylation to sulfhydration: What a thiolate needs to tolerate. *Sci Signal.* 2012;5:pe10.
10. West AP, Shadel GS, Ghosh S. Mitochondria in innate immune responses. *Nat Rev Immunol.* 2011;11:389–402.
11. Trachootham D, Alexandre J, Huang P. Targeting cancer cells by ROS-mediated mechanisms: A radical therapeutic approach? *Nat Rev Drug Discov.* 2009;8:579–591.
12. Van Gaal LF, Mertens IL, De Block CE. Mechanisms linking obesity with cardiovascular disease. *Nature.* 2006;444:875–880.
13. Gan L, Johnson JA. Oxidative damage and the Nrf2-ARE pathway in neurodegenerative diseases. *Biochim Biophys Acta.* 2014;1842:1208–1218.
14. Lagan AL, Melley DD, Evans TW, Quinlan GJ. Pathogenesis of the systemic inflammatory syndrome and acute lung injury: Role of iron mobilization and decompartmentalization. *Am J Physiol Lung Cell Mol Physiol.* 2008;294:L161–L174.

15. Harijith A, Ebenezer DL, Natarajan V. Reactive oxygen species at the crossroads of inflammasome and inflammation. *Front Physiol.* 2014;5:352.
16. Laurent TC, Moore EC, Reichard P. Enzymatic synthesis of deoxyribonucleotides. Iv. isolation and characterization of thioredoxin, the hydrogen donor from Escherichia coli B. *J Biol Chem.* 1964;239:3436–3444.
17. Holmgren A. Thioredoxin. *Annu Rev Biochem.* 1985;54:237–271.
18. Eklund H, Gleason FK, Holmgren A. Structural and functional relations among thioredoxins of different species. *Proteins.* 1991;11:13–28.
19. Holmgren A, Bjornstedt M. Thioredoxin and thioredoxin reductase. *Methods Enzymol.* 1995; 252:199–208.
20. Fritz-Wolf K, Kehr S, Stumpf M, Rahlfs S, Becker K. Crystal structure of the human thioredoxin reductase-thioredoxin complex. *Nat Commun.* 2011;2:383.
21. Chae HZ, Chung SJ, Rhee SG. Thioredoxin-dependent peroxide reductase from yeast. *J Biol Chem.* 1994;269:27670–27678.
22. Kim K, Kim IH, Lee KY, Rhee SG, Stadtman ER. The isolation and purification of a specific "protector" protein which inhibits enzyme inactivation by a thiol/Fe(III)/O2 mixed-function oxidation system. *J Biol Chem.* 1988;263:4704–4711.
23. Choi HJ, Kang SW, Yang CH, Rhee SG, Ryu SE. Crystal structure of a novel human peroxidase enzyme at 2.0 A resolution. *Nat Struct Biol.* 1998;5:400–406.
24. Saitoh M, Nishitoh H, Fujii M et al. Mammalian thioredoxin is a direct inhibitor of apoptosis signal-regulating kinase (ASK) 1. *EMBO J.* 1998;17:2596–2606.
25. Fujino G, Noguchi T, Matsuzawa A et al. Thioredoxin and TRAF family proteins regulate reactive oxygen species-dependent activation of ASK1 through reciprocal modulation of the N-terminal homophilic interaction of ASK1. *Mol Cell Biol.* 2007;27:8152–8163.
26. Tobiume K, Saitoh M, Ichijo H. Activation of apoptosis signal-regulating kinase 1 by the stress-induced activating phosphorylation of pre-formed oligomer. *J Cell Physiol.* 2002;191:95–104.
27. Yamanaka H, Maehira F, Oshiro M et al. A possible interaction of thioredoxin with VDUP1 in HeLa cells detected in a yeast two-hybrid system. *Biochem Biophys Res Commun.* 2000; 271:796–800.
28. Patwari P, Higgins LJ, Chutkow WA, Yoshioka J, Lee RT. The interaction of thioredoxin with Txnip. Evidence for formation of a mixed disulfide by disulfide exchange. *J Biol Chem.* 2006;281:21884–21891.
29. Hwang J, Suh HW, Jeon YH et al. The structural basis for the negative regulation of thioredoxin by thioredoxin-interacting protein. *Nat Commun.* 2014;5:2958.
30. O'Brian CA, Chu F. Post-translational disulfide modifications in cell signaling—role of inter-protein, intra-protein, S-glutathionyl, and S-cysteaminyl disulfide modifications in signal transmission. *Free Radic Res.* 2005;39:471–480.
31. Kwon J, Lee SR, Yang KS et al. Reversible oxidation and inactivation of the tumor suppressor PTEN in cells stimulated with peptide growth factors. *Proc Natl Acad Sci U S A.* 2004;101:16419–16424.
32. Sundaresan M, Yu ZX, Ferrans VJ, Irani K, Finkel T. Requirement for generation of H2O2 for platelet-derived growth factor signal transduction. *Science.* 1995;270:296–299.
33. Schwertassek U, Haque A, Krishnan N et al. Reactivation of oxidized PTP1B and PTEN by thioredoxin 1. *FEBS J.* 2014;281:3545–3558.
34. Dagnell M, Frijhoff J, Pader I et al. Selective activation of oxidized PTP1B by the thioredoxin system modulates PDGF-beta receptor tyrosine kinase signaling. *Proc Natl Acad Sci U S A.* 2013;110:13398–13403.
35. Lee SR, Yang KS, Kwon J, Lee C, Jeong W, Rhee SG. Reversible inactivation of the tumor suppressor PTEN by H2O2. *J Biol Chem.* 2002;277:20336–20342.

36. Verdoucq L, Vignols F, Jacquot JP, Chartier Y, Meyer Y. In vivo characterization of a thioredoxin h target protein defines a new peroxiredoxin family. *J Biol Chem*. 1999;274:19714–19722.
37. Tibbles LA, Woodgett JR. The stress-activated protein kinase pathways. *Cell Mol Life Sci*. 1999;55:1230–1254.
38. Ichijo H, Nishida E, Irie K et al. Induction of apoptosis by ASK1, a mammalian MAPKKK that activates SAPK/JNK and p38 signaling pathways. *Science*. 1997;275:90–94.
39. Gotoh Y, Cooper JA. Reactive oxygen species- and dimerization-induced activation of apoptosis signal-regulating kinase 1 in tumor necrosis factor-alpha signal transduction. *J Biol Chem*. 1998;273:17477–17482.
40. Nishitoh H, Kadowaki H, Nagai A et al. ALS-linked mutant SOD1 induces ER stress- and ASK1-dependent motor neuron death by targeting Derlin-1. *Genes Dev*. 2008;22:1451–1464.
41. Takeda K, Matsuzawa A, Nishitoh H et al. Involvement of ASK1 in Ca2+-induced p38 MAP kinase activation. *EMBO Rep*. 2004;5:161–166.
42. Matsuzawa A, Saegusa K, Noguchi T et al. ROS-dependent activation of the TRAF6-ASK1-p38 pathway is selectively required for TLR4-mediated innate immunity. *Nat Immunol*. 2005;6:587–592.
43. Tobiume K, Matsuzawa A, Takahashi T et al. ASK1 is required for sustained activations of JNK/p38 MAP kinases and apoptosis. *EMBO Rep*. 2001;2:222–228.
44. Liu Y, Min W. Thioredoxin promotes ASK1 ubiquitination and degradation to inhibit ASK1-mediated apoptosis in a redox activity-independent manner. *Circ Res*. 2002;90:1259–1266.
45. Han D, Ybanez MD, Ahmadi S, Yeh K, Kaplowitz N. Redox regulation of tumor necrosis factor signaling. *Antioxid Redox Signal*. 2009;11:2245–2263.
46. Park HS, Jung HY, Park EY, Kim J, Lee WJ, Bae YS. Cutting edge: Direct interaction of TLR4 with NAD(P)H oxidase 4 isozyme is essential for lipopolysaccharide-induced production of reactive oxygen species and activation of NF-kappa B. *J Immunol*. 2004;173:3589–3593.
47. Noguchi T, Takeda K, Matsuzawa A et al. Recruitment of tumor necrosis factor receptor-associated factor family proteins to apoptosis signal-regulating kinase 1 signalosome is essential for oxidative stress-induced cell death. *J Biol Chem*. 2005;280:37033–37040.
48. Xie P. TRAF molecules in cell signaling and in human diseases. *J Mol Signal*. 2013;8:7.
49. Nishitoh H, Saitoh M, Mochida Y et al. ASK1 is essential for JNK/SAPK activation by TRAF2. *Mol Cell*. 1998;2:389–395.
50. Song HY, Regnier CH, Kirschning CJ, Goeddel DV, Rothe M. Tumor necrosis factor (TNF)-mediated kinase cascades: Bifurcation of nuclear factor-kappaB and c-jun N-terminal kinase (JNK/SAPK) pathways at TNF receptor-associated factor 2. *Proc Natl Acad Sci U S A*. 1997;94:9792–9796.
51. Liu H, Nishitoh H, Ichijo H, Kyriakis JM. Activation of apoptosis signal-regulating kinase 1 (ASK1) by tumor necrosis factor receptor-associated factor 2 requires prior dissociation of the ASK1 inhibitor thioredoxin. *Mol Cell Biol*. 2000;20:2198–2208.
52. Kosek D, Kylarova S, Psenakova K et al. Biophysical and structural characterization of the thioredoxin-binding domain of protein kinase ASK1 and its interaction with reduced thioredoxin. *J Biol Chem*. 2014;289:24463–24474.
53. Nadeau PJ, Charette SJ, Toledano MB, Landry J. Disulfide bond-mediated multimerization of Ask1 and its reduction by thioredoxin-1 regulate H(2)O(2)-induced c-Jun NH(2)-terminal kinase activation and apoptosis. *Mol Biol Cell*. 2007;18:3903–3913.
54. Nadeau PJ, Charette SJ, Landry J. REDOX reaction at ASK1-Cys250 is essential for activation of JNK and induction of apoptosis. *Mol Biol Cell*. 2009;20:3628–3637.
55. Hatai T, Matsuzawa A, Inoshita S et al. Execution of apoptosis signal-regulating kinase 1 (ASK1)-induced apoptosis by the mitochondria-dependent caspase activation. *J Biol Chem*. 2000;275:26576–26581.
56. Sui X, Kong N, Ye L et al. p38 and JNK MAPK pathways control the balance of apoptosis and autophagy in response to chemotherapeutic agents. *Cancer Lett*. 2014;344:174–179.

57. Tillement L, Lecanu L, Papadopoulos V. Alzheimer's disease: Effects of beta-amyloid on mitochondria. *Mitochondrion*. 2011;11:13–21.
58. Kadowaki H, Nishitoh H, Urano F et al. Amyloid beta induces neuronal cell death through ROS-mediated ASK1 activation. *Cell Death Differ*. 2005;12:19–24.
59. Lee KW, Zhao X, Im JY et al. Apoptosis signal-regulating kinase 1 mediates MPTP toxicity and regulates glial activation. *PLoS One*. 2012;7:e29935.
60. Chen KS, DeLuca HF. Isolation and characterization of a novel cDNA from HL-60 cells treated with 1,25-dihydroxyvitamin D-3. *Biochim Biophys Acta*. 1994;1219:26–32.
61. Kang DS, Tian X, Benovic JL. Role of beta-arrestins and arrestin domain-containing proteins in G protein-coupled receptor trafficking. *Curr Opin Cell Biol*. 2014;27:63–71.
62. Nishiyama A, Matsui M, Iwata S et al. Identification of thioredoxin-binding protein-2/vitamin D(3) up-regulated protein 1 as a negative regulator of thioredoxin function and expression. *J Biol Chem*. 1999;274:21645–21650.
63. Junn E, Han SH, Im JY et al. Vitamin D3 up-regulated protein 1 mediates oxidative stress via suppressing the thioredoxin function. *J Immunol*. 2000;164:6287–6295.
64. Sevier CS, Kaiser CA. Formation and transfer of disulphide bonds in living cells. *Nat Rev Mol Cell Biol*. 2002;3:836–847.
65. Zhou R, Tardivel A, Thorens B, Choi I, Tschopp J. Thioredoxin-interacting protein links oxidative stress to inflammasome activation. *Nat Immunol*. 2010;11:136–140.
66. Dostert C, Petrilli V, Van Bruggen R, Steele C, Mossman BT, Tschopp J. Innate immune activation through Nalp3 inflammasome sensing of asbestos and silica. *Science*. 2008;320:674–677.
67. Schroder K, Tschopp J. The inflammasomes. *Cell*. 2010;140:821–832.
68. Yamawaki H, Pan S, Lee RT, Berk BC. Fluid shear stress inhibits vascular inflammation by decreasing thioredoxin-interacting protein in endothelial cells. *J Clin Invest*. 2005;115:733–738.
69. Gimbrone MA, Jr., Topper JN, Nagel T, Anderson KR, Garcia-Cardena G. Endothelial dysfunction, hemodynamic forces, and atherogenesis. *Ann N Y Acad Sci*. 2000;902:230–239; discussion 239–240.
70. Traub O, Berk BC. Laminar shear stress: Mechanisms by which endothelial cells transduce an atheroprotective force. *Arterioscler Thromb Vasc Biol*. 1998;18:677–685.
71. Xiang G, Seki T, Schuster MD et al. Catalytic degradation of vitamin D up-regulated protein 1 mRNA enhances cardiomyocyte survival and prevents left ventricular remodeling after myocardial ischemia. *J Biol Chem*. 2005;280:39394–39402.
72. World C, Spindel ON, Berk BC. Thioredoxin-interacting protein mediates TRX1 translocation to the plasma membrane in response to tumor necrosis factor-alpha: A key mechanism for vascular endothelial growth factor receptor-2 transactivation by reactive oxygen species. *Arterioscler Thromb Vasc Biol*. 2011;31:1890–1897.
73. Ago T, Liu T, Zhai P et al. A redox-dependent pathway for regulating class II HDACs and cardiac hypertrophy. *Cell*. 2008;133:978–993.
74. Backs J, Olson EN. Control of cardiac growth by histone acetylation/deacetylation. *Circ Res*. 2006;98:15–24.
75. Leslie NR, Bennett D, Lindsay YE, Stewart H, Gray A, Downes CP. Redox regulation of PI 3-kinase signalling via inactivation of PTEN. *EMBO J*. 2003;22:5501–5510.
76. Bae YS, Kang SW, Seo MS et al. Epidermal growth factor (EGF)-induced generation of hydrogen peroxide. Role in EGF receptor-mediated tyrosine phosphorylation. *J Biol Chem*. 1997;272:217–221.
77. Song MS, Salmena L, Pandolfi PP. The functions and regulation of the PTEN tumour suppressor. *Nat Rev Mol Cell Biol*. 2012;13:283–296.
78. Vivanco I, Sawyers CL. The phosphatidylinositol 3-kinase Akt pathway in human cancer. *Nat Rev Cancer*. 2002;2:489–501.
79. Populo H, Lopes JM, Soares P. The mTOR signalling pathway in human cancer. *Int J Mol Sci*. 2012;13:1886–1918.

80. Luo J, Manning BD, Cantley LC. Targeting the PI3K-Akt pathway in human cancer: Rationale and promise. *Cancer Cell.* 2003;4:257–262.
81. Silvera D, Formenti SC, Schneider RJ. Translational control in cancer. *Nat Rev Cancer.* 2010; 10:254–266.
82. Cui W, Matsuno K, Iwata K et al. NOX1/nicotinamide adenine dinucleotide phosphate, reduced form (NADPH) oxidase promotes proliferation of stellate cells and aggravates liver fibrosis induced by bile duct ligation. *Hepatology.* 2011;54:949–958.
83. Tonks NK. Protein tyrosine phosphatases: From genes, to function, to disease. *Nat Rev Mol Cell Biol.* 2006;7:833–846.
84. Chan EC, Jiang F, Peshavariya HM, Dusting GJ. Regulation of cell proliferation by NADPH oxidase-mediated signaling: Potential roles in tissue repair, regenerative medicine and tissue engineering. *Pharmacol Ther.* 2009;122:97–108.
85. Ushio-Fukai M. Localizing NADPH oxidase-derived ROS. *Sci STKE.* 2006;2006:re8.
86. Lee SR, Kwon KS, Kim SR, Rhee SG. Reversible inactivation of protein-tyrosine phosphatase 1B in A431 cells stimulated with epidermal growth factor. *J Biol Chem.* 1998;273:15366–15372.
87. Mahadev K, Zilbering A, Zhu L, Goldstein BJ. Insulin-stimulated hydrogen peroxide reversibly inhibits protein-tyrosine phosphatase 1b in vivo and enhances the early insulin action cascade. *J Biol Chem.* 2001;276:21938–21942.
88. Meng TC, Fukada T, Tonks NK. Reversible oxidation and inactivation of protein tyrosine phosphatases in vivo. *Mol Cell.* 2002;9:387–399.
89. Chiarugi P, Cirri P. Redox regulation of protein tyrosine phosphatases during receptor tyrosine kinase signal transduction. *Trends Biochem Sci.* 2003;28:509–514.
90. Salmeen A, Andersen JN, Myers MP et al. Redox regulation of protein tyrosine phosphatase 1B involves a sulphenyl-amide intermediate. *Nature.* 2003;423:769–773.
91. van Montfort RL, Congreve M, Tisi D, Carr R, Jhoti H. Oxidation state of the active-site cysteine in protein tyrosine phosphatase 1B. *Nature.* 2003;423:773–777.
92. Lou YW, Chen YY, Hsu SF et al. Redox regulation of the protein tyrosine phosphatase PTP1B in cancer cells. *FEBS J.* 2008;275:69–88.
93. Morinaka A, Yamada M, Itofusa R et al. Thioredoxin mediates oxidation-dependent phosphorylation of CRMP2 and growth cone collapse. *Sci Signal.* 2011;4:ra26.
94. Shao D, Oka S, Liu T et al. A redox-dependent mechanism for regulation of AMPK activation by thioredoxin1 during energy starvation. *Cell Metab.* 2014;19:232–245.
95. Hensley K, Venkova K, Christov A, Gunning W, Park J. Collapsin response mediator protein-2: An emerging pathologic feature and therapeutic target for neurodisease indications. *Mol Neurobiol.* 2011;43:180–191.
96. Dickson BJ. Molecular mechanisms of axon guidance. *Science.* 2002;298:1959–1964.
97. Nakamura F, Kalb RG, Strittmatter SM. Molecular basis of semaphorin-mediated axon guidance. *J Neurobiol.* 2000;44:219–229.
98. Goshima Y, Nakamura F, Strittmatter P, Strittmatter SM. Collapsin-induced growth cone collapse mediated by an intracellular protein related to UNC-33. *Nature.* 1995;376:509–514.
99. Yoshimura T, Kawano Y, Arimura N, Kawabata S, Kikuchi A, Kaibuchi K. GSK-3beta regulates phosphorylation of CRMP-2 and neuronal polarity. *Cell.* 2005;120:137–149.
100. Brown M, Jacobs T, Eickholt B et al. Alpha2-chimaerin, cyclin-dependent Kinase 5/p35, and its target collapsin response mediator protein-2 are essential components in semaphorin 3A-induced growth-cone collapse. *J Neurosci.* 2004;24:8994–9004.
101. Hardie DG, Ross FA, Hawley SA. AMPK: A nutrient and energy sensor that maintains energy homeostasis. *Nat Rev Mol Cell Biol.* 2012;13:251–262.
102. Sciarretta S, Zhai P, Shao D et al. Activation of NADPH oxidase 4 in the endoplasmic reticulum promotes cardiomyocyte autophagy and survival during energy stress through the protein kinase RNA-activated-like endoplasmic reticulum kinase/eukaryotic initiation factor 2alpha/activating transcription factor 4 pathway. *Circ Res.* 2013;113:1253–1264.

CHAPTER 12

Cytochrome *c*–Based Signalosome

Katiuska González-Arzola, Blas Moreno-Beltrán, Jonathan Martínez-Fábregas, Miguel A. De la Rosa and Irene Díaz-Moreno

CONTENTS

12.1 CYTOCHROME *c*: MULTI-TASKING POST-TRANSLATIONALLY MODIFIED PROTEIN

Cytochrome *c* (C*c*) is an evolutionarily conserved mitochondrial protein involved in cell life and death decisions. Similar to the photosynthetic soluble metalloproteins (Medina et al., 1992; Díaz et al., 1994a,b; Navarro et al., 1995; Molina-Heredia et al., 1999; Sun et al., 1999; Casaus et al., 2002; Crowley et al., 2002; Díaz-Moreno et al., 2005a,b,c), C*c* participates in electron transfer (ET) as part of the mitochondrial respiratory chain, which is indispensable for energy production. Under apoptotic conditions, C*c* is essential for the formation of the apoptosome (see Section 12.4.1) and triggering of cell death. This dual role of C*c* is regulated by post-translational modifications—namely, phosphorylation and nitration of tyrosine residues—that affect the binding of C*c* to its physiological counterparts, either in the mitochondria or in the nucleus and/or cytoplasm.

C*c* phosphorylation was first found to inhibit ET between C*c* and Complex IV (Lee et al., 2006; Yu et al., 2008), whereas specific phosphorylation of Tyr48 was further reported to disrupt apoptosome activation (Pecina et al., 2010; García-Heredia et al., 2011). Like Tyr48, residues Thr28, Ser47 and Tyr 97 can also become phosphorylated in mammals (Zhao et al., 2011; Figure 12.1). In addition, nitration of Tyr46 and Tyr48 triggers the specific degradation of C*c* and directs it to assemble a non-functional apoptosome (Díaz-Moreno et al., 2011; García-Heredia et al., 2012). Nitration of Tyr74 blocks the ability of C*c* to activate caspase-9, thereby blocking the apoptosis-signalling pathway (García-Heredia et al., 2010; Ly et al., 2012). Interestingly, methodological advances have been recently reported for the analysis of tyrosine nitration (Díaz-Moreno et al., 2012, 2013). Moreover, the post-translational modification of C*c* by methylation has also been reported. In fact, the (tri)methylation of Lys72 in yeast C*c* was found to be related to its apoptosis-related activity (Pollock et al., 1998; Kluck et al., 2000).

12.2 MITOCHONDRIAL-DEPENDENT PATHWAYS IN HOMEOSTASIS

12.2.1 Cytochrome *c* and Complex III

Currently, there are two different models being actively debated for the organization and dynamics of mitochondrial respiratory complexes. According to one of these, the so-called random collision model, membrane components are described as being in constant and independent diffusional motion (Hackenbrock et al., 1986). According to the other, a so-called supercomplex-based model, a stable supramolecular organization of the membrane based on specific interactions across individual complexes has been proposed. In this model, the respirasome, a supercomplex composed of dimer of Complex III and single copies of Complexes I and IV, could steer electrons from NADH to oxygen in the presence of redox carriers (Lenaz and Genova, 2010). Within this framework, the channelling of C*c* molecules between Complexes III and IV has been proposed as occurring in the respirasome in plants, but not in mammals (Genova and Lenaz, 2013; Moreno-Beltrán et al., 2014; Figure 12.2).

The interaction between C*c* and Complex III is highly transient, thereby allowing a high turnover, which is crucial for an efficient electron flow (Sarewicz et al., 2008; Pietras et al., 2014). Complex III (or the cytochrome bc_1 complex) is an integral membrane protein complex that catalyses ET from ubiquinol to C*c* coupled with proton

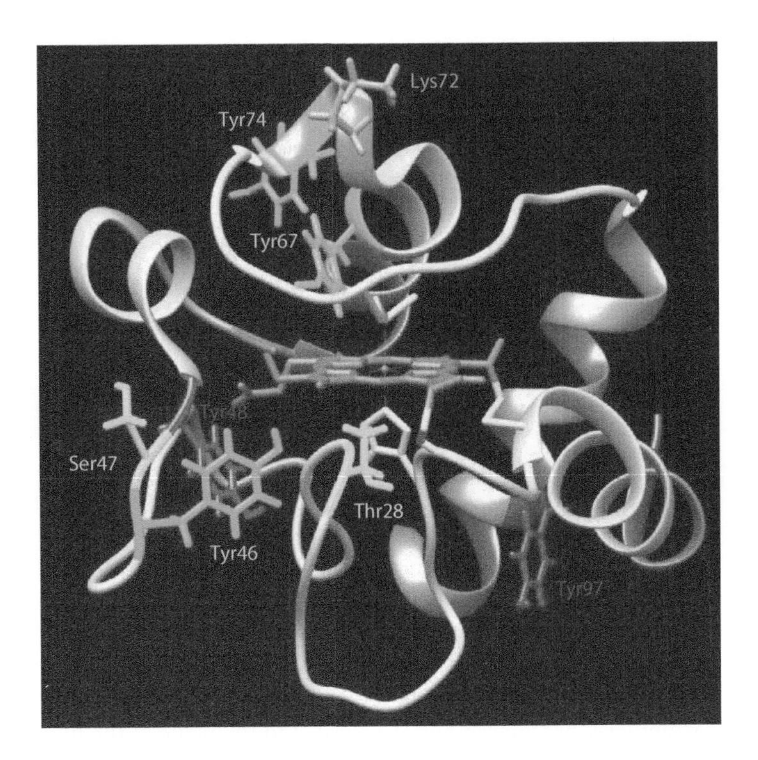

FIGURE 12.1 **(See colour insert.)** Post-translational modifications of cytochrome *c*. Human C*c* structure is shown with ribbon diagram (PDB: 1J3S). (Data from Jeng, W.-Y. et al., *J. Bioenerg. Biomembr.*, 34, 423–431, 2002.) Residues Thr28 and Ser47 (in yellow) can be phosphorylated in mammals. Tyr46, Tyr67 and Tyr74 (in blue) reportedly undergo nitration. Tyr48 and Tyr97 (in red) can be nitrated or phosphorylated. Lys72 (in magenta) can be (tri)methylated.

translocation (Saraste, 1999). Complex III contains a catalytic core formed by three proteins—cytochrome c_1 (Cc_1), cytochrome *b* and Rieske iron-sulphur protein (Berry et al., 2000). It is the soluble domain of Cc_1 that interacts with C*c* during ET reactions. Notably, a small non-polar surface (ca. 957 A^2) constitutes the binding interface in the Complex III–C*c* interaction, as revealed by X-ray diffraction (Lange and Hunte, 2002; Figure 12.3a). In fact, hydrophobic and cation–π contact pairs define the area for ET reactions as running from the Cc_1 to C*c* hemes. The two hemes, which face each other, are bordered by charged residues (Nyola and Hunte, 2008; Solmaz and Hunte, 2008). The Complex III–C*c* interaction is not only steered by non-polar interactions, but also by electrostatic forces (Kokhan et al., 2010; Moreno-Beltrán et al., 2014), a finding that is consistent with the two-step model of complex formation for ET discussed in Chapter 3. According to this model, the C*c* molecules are pre-oriented by long-range electrostatic forces responsible for the initial encounter complex. Hydrophobic interactions, when acting across short distances, account for the more specific complex upon spatial rearrangement of the two adducts (Prudêncio and Ubbink, 2004).

The relatively low surface complementary, the lack of salt bridges and hydrogen bonds and the high solvation of the interface account for the relatively short lifespan and transience

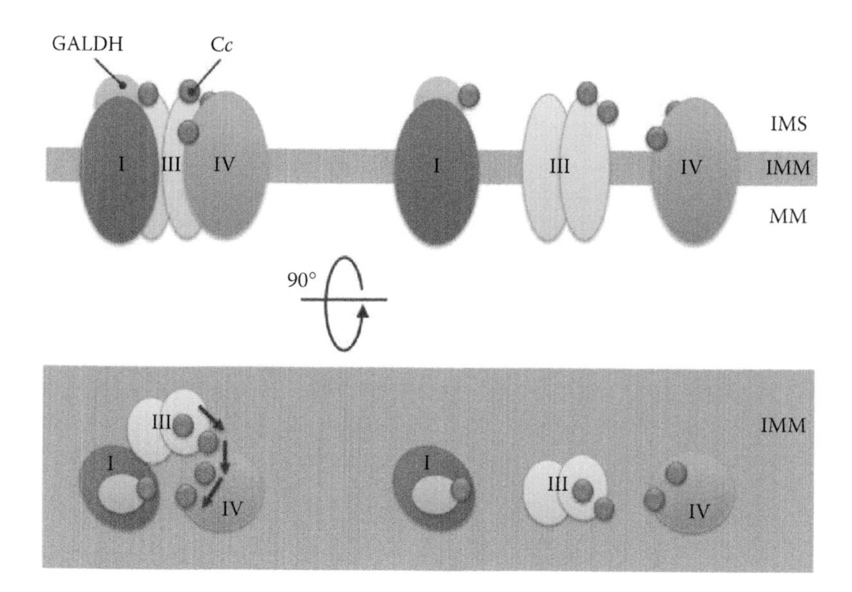

FIGURE 12.2 Role of cytochrome *c* in respirasome in plant mitochondria. Schematic representation of the plant mitochondrial respiratory chain, including two different scenarios: (i) complexes in constant and independent diffusional motion (*right*) and (ii) complexes grouped to form the respirasome (*left*). Complexes are viewed both (i) parallel to the plane of the inner mitochondrial membrane (IMM) with the mitochondrial intermembrane space (IMS) oriented towards the top (*upper*) and (ii) perpendicular to the plane of the IMM from the IMS (*lower*). C*c* is represented as a small sphere at binding sites on GALDH, Complex III (III) and Complex IV (IV), on the latter two of which two binding sites are represented. The channelling of C*c* molecules through the respirasome between Complexes III and IV is indicated by arrows. The mitochondrial matrix (MM) is also indicated.

of the Cc_1–C*c* interaction (Solmaz and Hunte, 2008), the latter of which qualities being inferable from stopped-flow measurements (Saraste, 1999; Yu *et al.*, 2002; Swierczek et al., 2010) that yield ET reaction rates similar to those determined with ruthenium-labelled C*c* derivatives (Tian et al., 2000; Engstrom et al., 2003; Janzon et al., 2008; Millett et al., 2013). These ET rates are dependent on ionic strength since, while the first-order rate constant does not change at low ionic strength, it nevertheless becomes concentration-dependent at high ionic strength (Millett et al., 2013). Furthermore, ionic strength is also the determining factor in the conformational rearrangement of both partners in the final productive complex with respect to ET (Pietras et al., 2014). It is also interesting to note that C*c* binds to Cc_1 at just one of the two possible binding sites on the dimer of Complex III (Lange and Hunte, 2002; Figure 12.3a). This is consistent with cryoelectron tomographies showing a single site for C*c* per dimer (Althoff et al., 2011), but in contrast to classic redox experiments showing C*c* to dock to the surface of Complex III at more than one site. Recently, the existence of two C*c*-binding sites on Cc_1 has been shown in plants, which could be compatible not only with the channelling described for the respirasome in plants, but also with the random collision model (Moreno-Beltrán et al., 2014).

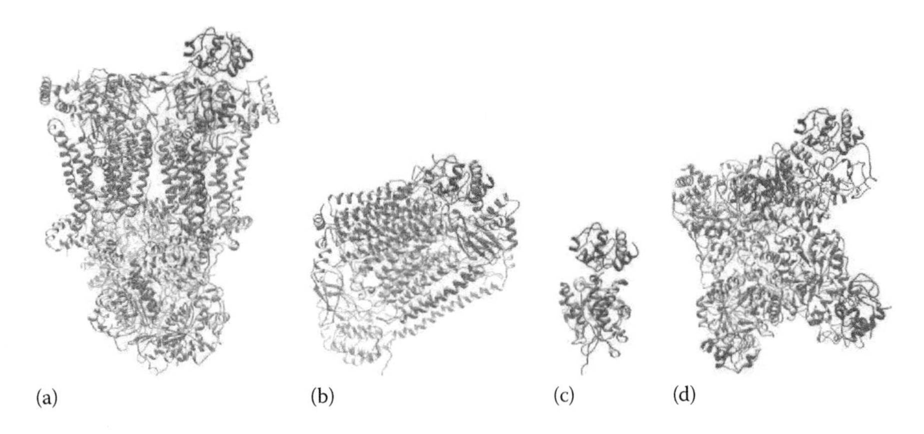

FIGURE 12.3 **(See colour insert.)** Structural models for eukaryotic cytochrome *c* complexes in homeostatic conditions. Protein subunits are shown with ribbon diagram. C*c* (red) is presented in the same orientation in all panels. Panel a: Overall crystallographic structure of the adduct formed between C*c* and Complex III in yeast. Note that C*c* binds to only one Cc_1 subunit (blue) of Complex III. Other subunits of Complex III are shown (light grey), as are hemes (green). Panel b: Docking model between C*c* and Complex IV in bovine heart. Also represented is subunit 4 in Complex IV (blue), other subunits (light grey), copper atoms (yellow spheres) and hemes (green). Panel c: Solution structure of C*c*–C*c*P complex in yeast as determined by paramagnetic NMR, showing C*c*P (blue) and heme groups (green). Panel d: Computational model of C*c*–FCb_2 complex in yeast, with homotetrameric structure of FCb_2 represented with one C*c* molecule docked onto each FCb_2 subunit (one in blue and the others in light grey). The heme groups (green) and flavine mononucleotide cofactors (purple) are also represented.

12.2.2 Cytochrome *c* and Complex IV

The interaction between C*c* and Complex IV is highly dynamic and transient so as to ensure the rapid turnover needed for ET (Prudêncio and Ubbink, 2004), qualities that match well with the two-step docking mechanism proposed previously for the C*c*–Complex III interaction. Complex IV, or the cytochrome *c* oxidase complex, is the final acceptor along the electron transport chain and catalyses the reduction of dioxygen to water coupled with trans-membrane proton translocation (Papa et al., 2004; Heinemeyer et al., 2007). The catalytic core of the complex contains three subunits—COX-1, COX-2 and COX-3. COX-1 contains a mononuclear copper centre (Cu_B), whereas the copper centre of COX-2 (Cu_A) is binuclear (Tsukihara et al., 1996). To determine the C*c* solvent-exposed residues that contact Complex IV, as well as binding affinity, nuclear magnetic resonance (NMR) has been used (Sakamoto et al., 2011). Moreover, to characterize the transient docked complex in bacteria and eukaryotes, several computational dockings have also been performed (Roberts and Pique, 1999; Bertini et al., 2005; Figure 12.3b).

The C*c*-Complex IV adduct is mainly stabilized by hydrophobic forces, which are mediated, on the one hand, by the heme edge and adjacent hydrophobic residues of C*c* (Sakamoto et al., 2011) and, on the other, by the hydrophobic cluster surrounding the Cu_A centre of Complex IV (Yoshikawa et al., 1998; Roberts and Pique, 1999; van Dijk et al., 2007).

The charged residues near the hydrophobic cores could allow long-range electrostatic forces to attract and pre-orient the partners, with the respective redox centres in close proximity to facilitate the final arrangements driven by hydrophobic interactions (Sakamoto et al., 2011). In addition, predicted docking complexes match previous mutagenesis, binding and time-resolved kinetic studies (Roberts and Pique, 1999; Bertini et al., 2005; van Dijk et al., 2007; Figure 12.3b).

It is worth noting that the ET rate reaction constant of the complex plateaus at a particular concentration of *Cc*, implying that the *Cc*–Complex IV binding mechanism includes the formation of a transient complex prior to ET (Rodríguez-Roldán et al., 2008). Actually, the multi-phasic kinetics observed in the oxidation of *Cc* by Complex IV can be fit to a model with just one catalytic site. Such a simple model includes alternative binding conformations of the transient complex, with some of which being unable to transfer electrons while nevertheless affecting the ET reaction rate at the catalytic site (Speck et al., 1984; Garber and Margoliash, 1990). As a single, static structural model of the *Cc*-Complex IV adduct could not simultaneously merge all experimental data collected (Bertini et al., 2005), the scenario in which an encounter complex is formed becomes plausible.

12.2.3 Cytochrome *c* and Cytochrome *c* Peroxidase

The *Cc*–cytochrome *c* peroxidase (*Cc*P) complex serves as a reactive (nitrogen) oxide species (R[N]OS) scavenger in the intermembrane mitochondrial space (IMS) (Giles et al., 2005; Jiang and English, 2006; Bihlmaier et al., 2007). *Cc*P is predominantly an α-helical molecule (Poulos et al., 1980; Volkov et al., 2013), catalysing the reduction of hydrogen peroxidase to water (Yonetani and Ohnishi, 1966). It contains two domains and a heme C, with the heme crevice being located between both domains. Notably, the heme is penta-coordinated with a sixth coordination position remaining available for peroxide binding (Wang et al., 1996).

Cc samples multiple conformations on *Cc*P in the dynamic encounter complex, which, as inferred from paramagnetic NMR spectroscopy and Monte Carlo simulations, is mainly propelled by electrostatic forces (Bashir et al., 2010; Volkov et al., 2010; Schilder et al., 2014). During the encounter, low levels of ET activity are observed due to the large distance between redox centres. However, the well-defined conformation, achieved through electrostatic steering, is, on the contrary, highly productive with regard to ET and could represent up to 70% of the entire lifespan of the complex (Bashir et al., 2010; Volkov et al., 2010; Figures 12.3c and 12.4). To measure ET reaction rates, ruthenium-*Cc* derivatives were used (Liu et al., 1995). The X-ray structure reported for the *Cc*–*Cc*P adduct could correspond to the well-defined, productive conformation in the two-step docking mechanism (Pelletier and Kraut, 1992). Although the binding stoichiometry of the interaction is 1:1, as supported by experimentation with multiple techniques (Pelletier and Kraut, 1992; Pearl et al., 2007, 2008; Volkov et al., 2009, 2010; Bashir et al., 2010), several studies indicate that at low ionic strength the interaction could occur with a stoichiometry of 2:1 and a different binding constant for each site (Zhou and Hoffmann, 1994; Leesch et al., 2000; Mei et al., 2002). The 2:1 complex is also consistent with *Cc* sampling on the surface of *Cc*P (Volkov et al., 2006; Bashir et al., 2010).

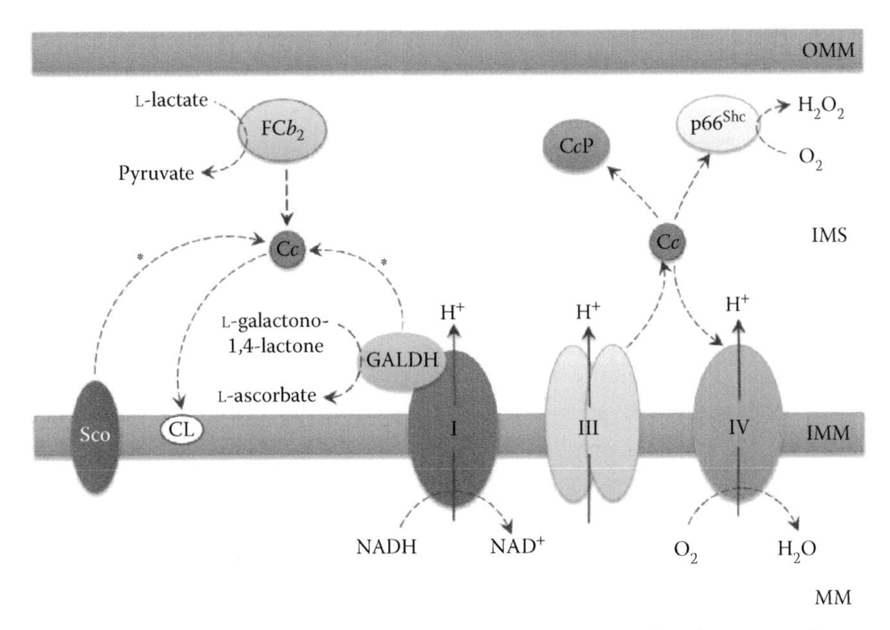

FIGURE 12.4 Cytochrome *c* signalosome in mitochondria. Principal pathways involving C*c* in the IMS are shown, including interactions with Complexes III and IV, C*c*P, SCO, FCb_2, GALDH, p66^{Shc} and CL. Asterisks highlight electron flow between SCO and C*c* reported exclusively in bacteria and between GALDH and C*c* reported exclusively in plants. The outer mitochondrial membrane (OMM) is also represented.

12.2.4 Cytochrome *c* and Flavocytochrome b_2

The functional enzyme flavocytochrome b_2 (FCb_2) is a homotetramer responsible for catalysing the oxidation of L-lactate to pyruvate and is directly coupled with C*c* reduction (Tegoni et al., 1983; Figures 12.3d and 12.4). In FCb_2, each monomer contains two domains—namely, cytochrome b_2 and flavodehydrogenase (Xia and Mathews, 1990). The rapid ET reaction mechanism between FCb_2 and C*c* that occurs within an encounter complex is influenced by ionic strength (Janot et al., 1990; Capeillere-Blandin, 1995) and, as thermodynamic analyses suggest, hydrophobic interactions (Capeillere-Blandin, 1995). Four C*c* molecules docked per FCb_2 tetramer have been devised by computational models (Tegoni et al., 1993; Short et al., 1998; Figure 12.3d), according to which each C*c* molecule interacts with an FCb_2 monomer. It has been postulated that electrostatic interactions could stabilize the binding of C*c* molecules to FCb_2, whereas the intermolecular distances in the models are compatible with those necessary for ET (Tegoni et al., 1993; Short et al., 1998).

12.2.5 Cytochrome *c* and SCO Proteins

Synthesis of C*c* oxidase (SCO) proteins are copper-binding proteins located on the inner mitochondrial membrane (IMM), which is essential for the correct assembly of cytochrome *c* oxidase (Saenkham et al., 2009). SCO proteins are present in all types of organisms, including the vast majority of eukaryotes and many prokaryotes, and play a key role in redox signalling and the maintenance of cellular copper homeostasis (Williams et al., 2005; Leary et al., 2007). Additionally, the absence of SCO in cyanobacteria suggests that the mechanisms ensuring the delivery of copper to cytochrome *c* oxidase in respiration might involve a series of as yet unidentified metallochaperones (Banci et al., 2007). In bacteria, it has been proposed that SCO plays a role in the regulation of the genes related to photosynthetic pathways (Saenkham et al., 2009) and in defence against oxidative stress (Seib et al., 2003). In a recent study of SCO in *Pseudomonas putida*, the interaction between its soluble domain and C*c* (Figure 12.4) has been described as being highly transient (Banci et al., 2011). Like other bacterial signal transduction proteins, both SCO and C*c* could act as sensor domains to perceive changes in periplasmic redox potential, which are essential for signal transmission to the cytoplasm and activation of cellular responses (Banci et al., 2011). The CXXXC motif found in SCO, which is involved in the binding of Cu(I) and, likely, in reduction of C*c*, is highly conserved in all mitochondrial and bacterial SCO proteins. This suggests that ET between SCO and C*c* could be an evolutionarily conserved function in living organisms. Interestingly, human and yeast SCO proteins exhibit a hydrophobic patch in close proximity to the CXXXC motif, which could contribute to tuning different copper binding affinities observed in organisms from bacteria to eukaryotes (Banci et al., 2011).

12.2.6 Cytochrome *c* and L-Galactono-1,4-Lactone Dehydrogenase

L-Galactono-1,4-lactone dehydrogenase (GALDH) is an oxidoreductase catalysing the final step in vitamin C (L-ascorbate) biosynthesis in plant mitochondria (Leferink et al., 2009) and, along with related aldonolactone oxidoreductases, belongs to the vanillyl-alcohol oxidase family of flavoproteins. Members of this family share a two-domain folding, formed by a conserved N-terminal flavin adenine dinucleotide binding domain and a less conserved C-terminal domain responsible for substrate specificity (Fraaije et al., 1998; Leferink et al., 2008). The active site of the protein is located at the interface of the domains, with Glu386 and Arg388 being essential for intermediate stabilization and productive substrate binding, respectively (Leferink et al., 2009).

Although GALDH had been presumed to be an integral protein of the IMM (Bartoli et al., 2000), there is increasing evidence that GALDH represents an NADH dehydrogenase Complex I assembly factor by specifically binding to several of its assembly intermediates (Schertl et al., 2012). However, GALDH was found to be attached only to a slightly smaller version of Complex I (Heazlewood et al., 2003; Millar et al., 2003; Pineau et al., 2008). This flavoenzyme is also important for the central metabolism of plant mitochondria, since clearly retarded growth and production of smaller fruits have been observed in tomato plants upon the silencing of the gene-encoding GALDH (Alhagdow et al., 2007). Moreover, an *Arabidopsis thaliana* knockout lacking the *GALDH* gene was found to have drastically reduced amounts of Complex I (Pineau et al., 2008). Therefore, it seems that GALDH is essential not only for L-ascorbate formation (Schertl et al., 2012), but also for the proper functioning of plant mitochondria.

The catalytic cycle of the flavoenzyme consists of the acceptance of two electrons from L-galactono-1,4-lactone to reduce the flavin cofactor with the concomitant reduction of two C*c* molecules (Figure 12.4). Oxidation of reduced GALDH by C*c* occurs in two successive single steps during ET and involves the formation of an anionic flavin semiquinone as an intermediate (Leferink et al., 2008). As found in laser flash spectroscopy studies, the reaction between GALDH and C*c* is quite fast and the overall rate of GALDH-mediated C*c* reduction is limited by the reduction of GALDH by its carbohydrate (Hervás et al., 2013). Interestingly, C*c* reduction resulted in the formation of a complex with GALDH with low binding affinity and a lifespan in the range of milliseconds. NMR studies revealed residues surrounding the heme edge of C*c* as having taken part in the interaction with the flavoenzyme, in agreement with other ET reaction complexes involving C*c* (Hervás et al., 2013).

12.3 MITOCHONDRIAL-DEPENDENT PATHWAYS IN APOPTOSIS

12.3.1 Cytochrome *c*–Mediated Peroxidation of Cardiolipin

About 15% of C*c* is tightly bound to membrane lipids in the IMM (Rytömaa et al., 1992). Recently, it was reported that the mitochondrial membrane lipid cardiolipin (CL) avidly binds to C*c* and brings about an extraordinary increase in C*c* peroxidase activity (Kagan et al., 2004). Under homeostatic conditions, CL, which represents approximately 25% of all mitochondrial lipids (Vik et al., 1981), is located predominantly in the inner leaflet of the IMM associated with Complexes III and IV, thereby enabling the two to retain their functionality (Haines and Dencher, 2002). In the early stages following the triggering of apoptosis, however, the percentage of CL in the outer mitochondrial membrane (OMM) and the outer leaflet of the IMM increases markedly, reaching up to 40% and 70%, respectively (Kagan et al., 2006). At a similar point in apoptosis, activated caspase-8 results in the cleavage of the pro-apoptotic Bcl-2 family member, Bid, so as to yield a truncated fragment (tBid) that is translocated to the mitochondria and interacts with CL, with the concomitant changes in CL transmembrane distribution (Schug and Gottlieb, 2009). Consequently, significant amounts of CL become available for eventual interactions with C*c* (Figure 12.4).

For the binding of anionic phospholipids, two sites on the surface of C*c*—the so-called A-site and C-site—have been proposed (Rytömaa et al., 1992). The A-site facilitates electrostatic interactions with the negatively charged CL headgroup, whereas the C-site involves hydrophobic forces with the fatty acyl chains of CL. According to the most widely accepted model, the interaction of C*c* with mitochondrial membranes results in significant conformational changes in the former, including the disturbance of bonds with hexacoordinate heme iron, the widening of the heme crevice and the partial unfolding of the protein (Kagan et al., 2004). Binding of C*c* to CL thus causes profound structural changes, which are induced by both electrostatic interactions mediated by a network of positively charged residues at the A-site—mainly conserved Lys72 and Lys73 with negatively charged phospholipids—and by simultaneous hydrophobic interactions at the C-site that facilitate hydrogen bonding of Asn52 with protonated CL molecules (Rytömaa et al., 1992; Kagan et al., 2004; Sinibaldi et al., 2008). Such striking changes in C*c* structure due to its tight association with membranes, especially after disrupting the Fe-S_δ(Met80) bond, give H_2O_2 access to the iron atom and bring about peroxidase activity (Kagan et al., 2004; Belikova et al., 2006). Under homeostatic conditions, the amount of H_2O_2 is the limiting factor in CL peroxidation, as it is regulated by R(N)OS scavengers maintaining cellular metabolism. Upon apoptotic stimuli, however, H_2O_2 ceases to be a limiting factor, insofar as its generation by the p66^{Shc} protein after binding to *Cc* (see Section 12.3.2) contributes to an increase in mitochondrial R(N)OS levels (Giorgio et al., 2005). Once CL is oxidized, its affinity for *Cc* greatly decreases, thereby allowing CL redistribution and providing an abundance of free *Cc*, which in turn is released into the cytosol and directs the activation of the caspase cascade (Kagan et al., 2005; Hüttemann et al., 2011).

CL peroxidation by *Cc* occurs through the formation of tyrosyl radical intermediates resulting from the partial CL-induced unfolding of the heme protein (Kagan et al., 2006). Then, the tyrosyl radical abstracts hydrogen from one of the CL polyunsaturated lipid acyl chains so as to yield a lipid radical that is further converted into CL hydroperoxide (Kagan et al., 2006), which is required for the release of pro-apoptotic factors from mitochondria into the cytosol (Kagan et al., 2005; Bayir et al., 2006). The peroxidase cycle leading to CL peroxidation by *Cc* depends on the generation of tyrosyl radicals, among which the highly conserved Tyr67 is most likely to be involved in oxygenation reactions (Kapralov et al., 2011). In this regard, CL peroxidation can be efficiently triggered by post-translational modifications of *Cc* tyrosine residues. Thus, phosphorylation or nitration of *Cc* tyrosine residues can tune the affinity of *Cc* to CL (Lee et al., 2006; Yu et al., 2008; García-Heredia et al., 2010, 2011; Díaz-Moreno et al., 2011), thereby regulating the first steps in apoptosis.

12.3.2 Reactive Oxygen Species Generation by p66^{Shc}-Mediated Oxidation of Cytochrome *c*

The p66^{Shc} protein is the largest of three isoforms—the other two being p46^{Shc} and p52^{Shc}—encoded by the ShcA locus through two different promoters, one for the former isoform and the other for the latter two (Trinei et al., 2013). p66^{Shc} has a domain organization typical of the Shc family adaptor proteins, with a Src homology 2 domain, two proline- and glycine-rich regions abundant in collagen (CH1 and CH2) and a phosphotyrosine-binding domain (Trinei et al., 2013). p66^{Shc} is regulated via reversible phosphorylation involving the pathways related to protein kinase C β (PKCβ) upon oxidative stress (Wang et al., 2014). Thus, phosphorylation of Ser36, located at the CH2 domain, turns p66^{Shc} into a substrate for prolyl isomerase 1 (Pin1), which in turn induces isomerization of pSer-Pro bonds, leading to mitochondrial accumulation of p66^{Shc} (Pinton et al., 2007). p66^{Shc} then causes alterations of mitochondrial Ca^{2+} responses and triggers apoptosis (Pinton et al., 2007).

Mitochondrial R(N)OS concentration increases markedly following pro-apoptotic signals. In this regard, p66^{Shc} acts as a downstream target of the p53 tumour suppressor and is responsible for the increase in intracellular oxidants during p53-induced apoptosis (Trinei et al., 2002). As demonstrated in a previous study, p66^{Shc} is a redox enzyme that generates mitochondrial R(N)OS—mainly H_2O_2—through the oxidation of *Cc* in response to stress conditions (Giorgio et al., 2005; Figure 12.4). Therefore, pro-apoptotic signals trigger the release of p66^{Shc} from a putative inhibitory complex including members of the translocase of the inner/outer membrane (TIM-TOM) import complex and Hsp70 (Orsini et al., 2004). Active p66^{Shc} then oxidizes *Cc* in the IMS and catalyses the reduction of O_2 to H_2O_2, with the subsequent opening of the mitochondrial permeability transition pore (MPTP)

(Giorgio et al., 2005), leading to the swelling of the mitochondria and, finally, to cell death (Savino et al., 2013). $p66^{Shc}$ uses an N-terminal sequence of 52 amino acids, a so-called cytochrome-binding (CB) domain, to bind to *Cc*. Interestingly, the CB domain is highly conserved among $p66^{Shc}$ sequences in vertebrates and contains glutamate (E125, E132, E133) and tryptophan (W134 and W148) residues that are essential for ET reactions with *Cc*. Despite the CB domain being essential for *Cc* binding, the CH2 domain of $p66^{Shc}$ also contributes to complex formation (Giorgio et al., 2005). Notably, the $p66^{Shc}$ N-terminus forms a redox module responsible for triggering apoptosis, which can be activated through reversible interconversions of tetrameric $p66^{Shc}$ by the formation of two disulphide bonds (Gertz et al., 2008).

12.3.3 Inhibition of Cytochrome *c* Translocation to Cytosol by Bcl-x_L Protein

Bcl-x_L is an anti-apoptotic member of the Bcl-2 protein family located in the membranes of mitochondria and the endoplasmic reticulum, as well as in the nuclear envelope (Carthy et al., 2003). Bcl-x_L, like all Bcl-2 proteins, preserves mitochondrial integrity and prevents the subsequent release of apoptotic molecules. Among its other functions in the intrinsic apoptotic pathway, Bcl-x_L intercepts *Cc* in the cytosol after its early translocation from mitochondria, thereby preventing the assembly of the apoptosome (Figure 12.5). *Cc* was found to interact specifically with Bcl-x_L both *in vivo* and *in vitro* with a binding affinity as high as those reported for other regulatory protein–protein interactions during apoptosis (Yadaiah et al., 2007). Strikingly, an NMR-based structural model reveals the residues of *Cc* involved in the *Cc*-Bcl-x_L adduct to be different from those described for other ET complexes (Bertini et al., 2011). Consequently, the *Cc*-Bcl-x_L ensemble is formed through electrostatic forces and does not require the contribution of the *Cc* heme edge.

12.4 MITOCHONDRIAL-INDEPENDENT PATHWAYS IN APOPTOSIS

12.4.1 Cytochrome *c* as Part of the Apoptosome

The apoptosome—a highly conserved macromolecular platform present from *Caenorhabditis elegans* to *Drosophila melanogaster* to *Homo sapiens*—is essential in the activation of the caspase cascade, which in turn is responsible for the initiation of apoptosis. The caspase cascade is formed by a subfamily of cysteine proteinases that trigger protein degradation (Nuñez et al., 1998) and can be activated through two major stimuli. The first is an extrinsic signal from death receptors, which depends on the interaction between extracellular ligands and transmembrane receptors (Krammer, 2000; Locksley et al., 2001). The other is an intrinsic signal from mitochondria, which is mainly activated in response to DNA damage, oxidative stress or growth factor deprivation (Chipuk and Green, 2005; Suen et al., 2008) and involves the permeabilization of the OMM and release of pro-apoptotic factors such as second mitochondria-derived activator of caspase/direct IAP-binding protein with low pi (Smac/DIABLO), high temperature requirement protein A2/Omi stress-regulated endoprotease (HtrA2/Omi), apoptosis-inducing factor (AIF), endonuclease G (ENDOG), caspase-activated DNase (CAD) and *Cc*. Indeed, *Cc* is essential for the apoptosome assembly in many organisms.

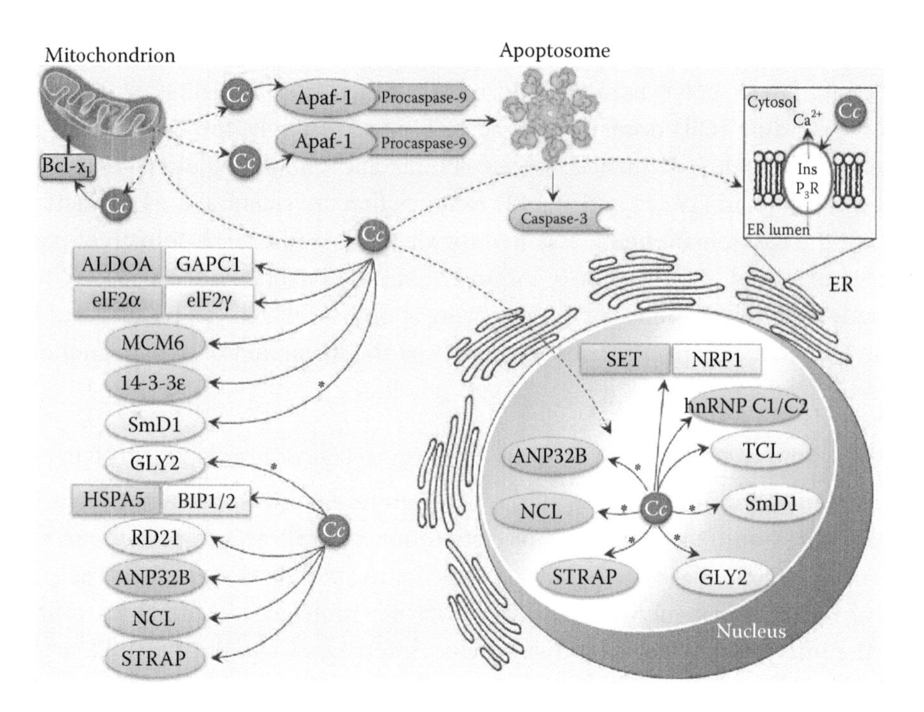

FIGURE 12.5 Nucleocytoplasmic cytochrome *c* signalosome. Binding events of C*c* upon its release from mitochondria (continuous line arrows) with: APAF1, InsP3R at the ER membrane, Bcl-x_L attached to the OMM and recently discovered PCD C*c*-targets (e.g., ALDOA/GAPC1, eIF2α/eIF2γ, MCM6, 14-3-3ε, GLY2, HSPA5/BIP1-BIP2 and RD21 found exclusively in the cytoplasm, and SET/NRP1, hnRNP C1/C2 and TCL found exclusively in the nucleus). Asterisks represent nucleocytoplasmic counterparts of C*c* (e.g., SmD1, GLY2, ANP32B, NCL and STRAP). The diagram differentiates between recently discovered PCD C*c*-targets in human cells (light grey) and in plant cells (white), with analogous proteins in human and plant cells (rectangular boxes) also shown. The diagram also indicates mitochondrial C*c* translocation to the cytoplasm and nucleus (dashed line arrows) upon initiation of PCD.

The structure of the apoptosome has been elucidated using different techniques, such as X-ray diffraction and electron microscopy, thereby providing further insight on the working of this molecular machine. In the particular case of mammals, cytoplasmic C*c* and dATP bind to apoptotic protease-activating factor-1 (APAF1) to form a C*c*/APAF1/caspase-9 complex, the so-called apoptosome. The essential role of C*c* in the assembly of this macromolecular complex was first described biochemically (Li et al., 1997) and then confirmed when its atomic-resolution structure was reported (Acehan et al., 2002; Yu et al., 2005). Biochemical and structural data have revealed that C*c* triggers apoptosome formation by interacting with WD40 repeats of APAF1. These WD40 repeats form two β-propellers, where the caspase activation and recruitment domain (CARD) can be found under normal conditions (Yu et al., 2005). Upon translocation to the cytoplasm, C*c* interacts with the WD40 domains and triggers the conformational change of CARD, thus allowing the oligomerization of APAF1 and the assembly of the apoptosome.

In other organisms, such as *C. elegans* or *D. melanogaster*, C*c* is not essential for the assembly and activation of the apoptosome, as inferred from the 3D-structure of the apoptosome (Qi et al., 2010; Yuan et al., 2011), although protein homologues, such as cell-death abnormality-4 (CED-4) in the former organism and APAF1-related killer (Dark) in the latter, have been found to be essential for caspase cascade activation. In some other organisms, such as *A. thaliana*, neither APAF1 nor any similar proteins are detected (van Nocker and Ludwig, 2003). In fact, only evolutionarily distant caspase homologues (metacaspases) and serine proteases exhibiting caspase-like activity (saspases) have been found in plants (Vartapetian et al., 2011; Minina et al., 2014). However, the nature of their targets and whether they are activated by C*c* remains unclear.

Nevertheless, even in those organisms where APAF1 does not exist or C*c* is not required for apoptosome assembly, the release of C*c* into the cytosol during programmed cell death (PCD) is an evolutionarily conserved event among a wide variety of organisms including yeasts (Giannattasio et al., 2008), plants (Balk et al., 1999), flies (Arama et al., 2006) and mammals (Bossy-Wetzel et al., 1998). That said, only in mammals has a well-established role for C*c* during programmed cell death been reported (Li et al., 1997).

12.4.2 Recently Reported Cytochrome *c* Partners in the Nucleus and/or Cytoplasm

The principal role of C*c* in mammalian apoptosis has been well established *in vitro* although possible new functions for the heme protein are also beginning to emerge. For instance, recent data suggest that C*c* may act in the nucleus (Ruíz-Vela et al., 2002; Nur-E-Kamal et al., 2004) or the endoplasmic reticulum upon apoptotic stimuli (Boehning et al., 2003, 2005; Szado et al., 2008). In the initial stages of apoptosis, C*c* has been shown to amplify its own release from the mitochondria in a coordinated way by interacting with inositol 1,4,5-triphosphate receptor ($InsP_3R$) (Boehning et al., 2003). This interaction between C*c* and $InsP_3R$ leads to a sustained release of calcium from the endoplasmic reticulum, which in turn results in an increased amount of C*c* released to guarantee the rapid execution of apoptosis. All of these novel functions of C*c*, unrelated to the assembly of the apoptosome, highlight the putative multi-functional role of C*c* upon the onset of apoptosis.

The role of C*c* in apoptotic signalling pathways has not been extensively elucidated *in vivo* due to the difficulty of obtaining C*c* knockouts. Recently, however, a successful attempt at a C*c* knockout was reported (Vempati et al., 2007) and was observed as being resistant to factors inducing apoptosis through not only intrinsic, but also extrinsic, stimuli. Intriguingly, apoptosis is triggered in knockouts lacking APAF1—the only apoptotic partner of C*c* described to date *in vitro*—when the death-receptor-initiated pathway is stimulated (Meier et al., 2000; Marsden et al., 2002; Adams, 2003; Shawgo et al., 2009). Besides, the *D. melanogaster* C*c* knockout displays a profound delay in apoptosis, though the heme protein is not essential for the assembly of an apoptosome-like structure (Arama et al., 2006; Mendes et al., 2006).

In two recent papers (Martínez-Fábregas et al., 2013, 2014a), newly discovered C*c* targets have been reported in both human and plant cells under apoptotic conditions. The finding raises the possibility that C*c* may perform additional roles at the onset of PCD, in addition to its well-established role in the assembly of the apoptosome. The new data may help not only to explain the highly conserved nature of the apoptotic release of *Cc*, but also to identify *Cc* nuclear partners and explain the translocation of *Cc* to nucleus described in earlier works (Ruíz-Vela et al., 2002; Nur-E-Kamal et al., 2004).

In humans, *Cc* interacts with proteins controlling pro-survival pathways (Martínez-Fábregas et al., 2014a) such as serine-threonine kinase receptor-associated proteins (STRAP), 14-3-3 protein ε (14-3-3ε), heat shock 70 kDa protein 5 (HSPA5) and nucleolin (NCL) (Figure 12.5). *Cc* also interacts with targets essential for protein synthesis (e.g., eukaryotic translation initiation factor 2 subunit α [eIF2α], energetic metabolism (e.g., fructose-bisphosphate ALDOase A [ALDOA]) and DNA metabolism (e.g., minichromosome maintenance complex 6 [MCM6]). In the nucleus, *Cc* interacts with partners related to transcriptional regulation (e.g., acidic leucine-rich nuclear phosphoprotein 32 family member B [ANP32B]) and DNA damage response (e.g., SET oncoprotein and heterogeneous nuclear ribonucleoproteins C1/C2 [hnRNP C1/C2]).

In plants, *Cc* seems to interfere with biological processes that are essential for cell life (García-Heredia et al., 2008; Martínez-Fábregas et al., 2013), such as protein synthesis (e.g., eukaryotic translation initiation factor 2 γ [eIF2γ]), energetic metabolism (e.g., glyceraldehyde-3-phosphate dehydrogenase C subunit 1 [GAPC1]), DNA metabolism (e.g., nucleosome assembly related protein 1-like [NRP1], transcriptional coactivator-like [TCL] and small nuclear ribonucleoprotein D1 [Sm/D1]) and protein folding and stability (e.g., luminal-binding protein 1 and 2 [BiP1 and BiP2]). Furthermore, *Cc* interacts with proteins playing crucial roles during PCD, such as cysteine proteinase RD21 and hydroxyacylglutathione hydrolase (GLY2) (Figure 12.5).

Interestingly, some of these novel *Cc* partners of identified in independent proteomic experiments in human and plant cells (Martínez-Fábregas et al., 2013, 2014a) are related functionally. ALDOA and GAPC1 are both glycolytic enzymes involved in the synthesis and degradation, respectively, of glyceraldehyde-3 phosphate. NRP1 and its analogous protein SET, a well-known inhibitor of p53, are able to block cell cycle arrest and apoptosis. Moreover, under apoptosis the phosphorylation of eIF2α, which, along with eIF2γ, are two of the three components of eIF2, prevents the assembly of eIF2, thereby blocking protein transcription. Finally, while the human homologue can block eIF2α phosphorylation, HSPA5 and BiP1/BiP2 act as inhibitors of protein kinase R (PKR)-like endoplasmic reticulum kinase (Martinez-Fabregas et al., 2014b). Three of these pairs (i.e., SET/NRP1, eIF2α/eIF2γ and HSPA5/BiP1-BiP2) are involved in convergent pathways that regulate apoptosis and macroautophagy in human cells by affecting eIF2 trimerization (Martínez-Fábregas et al., 2014b).

The existence of *Cc* partners that are functionally related in such evolutionarily distant groups of organisms—that is, plants and humans—raises the question of a common *Cc*-eIF2 axis within the cell death signalosome—that is, an evolutionarily conserved core controlling PCD. These recent findings indicate that *Cc*, upon its release into the cytosol, activates PCD not only, as is generally accepted, by activating the caspase cascade responsible for PCD, but also, as has recently been proposed (Martinez-Fabregas et al., 2014b) by inhibiting survival pathways.

12.5 CROSSTALK BETWEEN TRANSIENT HOMEOSTATIC AND STABLE APOPTOTIC PATHWAYS

As shown in Figure 12.6, *Cc* delicately tilts the balance from cell life (respiration) to cell death (apoptosis). Whereas respiration is governed by interactions of *Cc* for ET within the mitochondria that are highly transient (with lifespans within the μs–ms range), the nucleocytoplasmic adducts of *Cc* that lead to apoptosis are amazingly stable (with lifespans ranging from minutes to days) (Table 12.1).

Here it is important to mention two points: that all interactions with *Cc* in the IMS result in a high turnover and that all are involved in ET reactions (e.g. binding of *Cc* to Complex III, Complex IV, *Cc*P, F*Cb*$_2$, SCO, GALDH and p66^{Shc}). To this previous observation, there appear to exist only two exceptions:, the highly stable complex formed between *Cc* and CL (a lipid located in the IMM) prior to CL peroxidation and *Cc* release in the early stages of apoptosis and the highly dynamic but non-redox reaction-led interaction between *Cc* and Bcl-x$_L$ in the OMM.

In contrast to the mitochondrial interactions of *Cc*, none of those by extra-mitochondrial *Cc* is redox. This is true for interactions both in the cytoplasm (e.g., with ALDOA, APAF1, BiP1/2, eIF2α, eIF2γ, GAPC1, HSPA5, InsP$_3$R, MCM6, RD21 and 14-3-3ε) and in the nucleus (e.g., ANP32B, GLY2, hnRNP C1/C2, NCL, NRP1, SET, SmD1, STRAP and TCL).

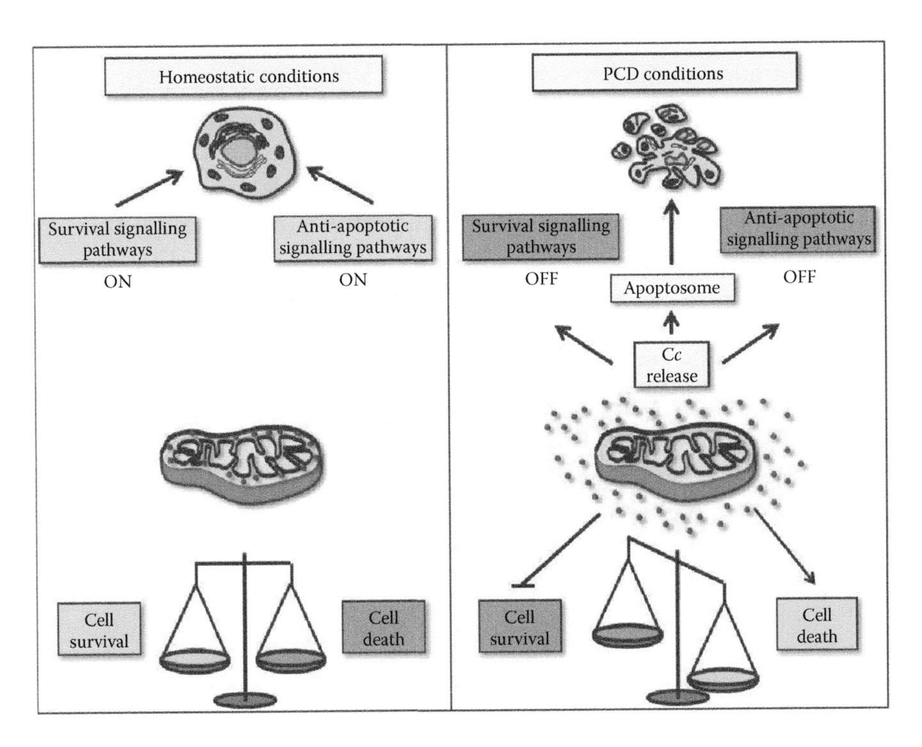

FIGURE 12.6 Cytochrome *c* tilts the balance between cell life and death. Under homeostatic conditions, both survival and anti-apoptotic signalling pathways are activated to keep cells living. However, under PCD conditions, extra-mitochondrial *Cc* not only activates the apoptosome-dependent pathway, but also blocks other survival and anti-apoptotic signalling pathways, effectively switching cell metabolism *off* and leading the cell to PCD.

TABLE 12.1 Lifespan of Bimolecular Complexes Formed by *Cc* with Other Protein Adducts

Complex Involving *Cc*	Lifespan (ms)	Reference
Cc–CcP	<0.8	Worrall et al., 2001
Cc–Complex IV	<1.5	Witt et al., 1998
Cc–Complex III	<6.5	Nyola and Hunte, 2008
Cc–14-3-3ε	537	Martínez-Fábregas et al., 2014a
Cc–SET	541	Martínez-Fábregas et al., 2014a
Cc–GLY2	602	Martínez-Fábregas et al., 2013
Cc–HSPA5	766	Martínez-Fábregas et al., 2014a
Cc–TCL	995	Martínez-Fábregas et al., 2013
Cc–eIF2α	1033	Martínez-Fábregas et al., 2014a
Cc–hnRNP C1/C2	1269	Martínez-Fábregas et al., 2014a
Cc–NRP1	1298	Martínez-Fábregas et al., 2013
Cc–APAF1	>3000	Acehan et al., 2002
Cc–CL	>3000	Sinibaldi et al., 2010

All these nucleo-cytoplasmic *Cc* ensembles are significantly stable, with small dissociation constants and long lifespans; whereas the complexes formed by *Cc* with its mitochondrial redox partners are rather transient, with rather short lifespans (Table 12.1).

Such a high affinity of *Cc* towards its nucleo-cytoplasmic targets, along with the high cytosolic concentration of the heme protein (0.4–0.5 mM) in the early stages of apoptosis, facilitates the role of *Cc* as an inter-organellar messenger. Upon apoptotic stimuli, the translocation of *Cc* into the nucleus and/or cytoplasm of the cell could both block survival pathways and unlock pro-apoptotic signals, thereby efficiently initiating PCD by preventing the spatial and temporal co-existence of opposing pro-survival and pro-apoptotic signals (Figure 12.6). Altogether, these findings suggest that *Cc* indeed plays a crucial role in controlling the fragile equilibrium between cell life and death.

ACKNOWLEDGEMENTS

We would like to thank all former and present members of the Biointeractomics group at the Institute for Plant Biochemistry and Photosynthesis (cicCartuja, University of Seville—CSIC). Financial support was provided by the Spanish Ministry of Economy and Competitiveness for several years (Grant No. BFU2003-00458/BMC, BFU2006-01361/BMC, BFU2009-07190/BMC and BFU2012-31670/BMC) and by the Andalusian Government (Grant PAI, BIO198, P07-CVI-02896 and P11-CVI-7216).

REFERENCES

Acehan, D., X. Jiang, D. G. Morgan, J. E. Heuser, X. Wang, and C. W. Akey. Three-dimensional structure of the apoptosome: Implications for assembly, procaspase-9 binding, and activation. *Molecular Cell* 9 (2002): 423–32.

Adams, J. M. Ways of dying: Multiple pathways to apoptosis. *Genes and Development* 17 (2003): 2481–95.

Alhagdow, M., F. Mounet, L. Gilbert et al. Silencing of the mitochondrial ascorbate synthesizing enzyme L-galactono-1,4-lactone dehydrogenase affects plant and fruit development in tomato. *Plant Physiology* 145 (2007): 1408–22.

Althoff, T., D. J. Mills, J.-L. Popot, and W. Kühlbrandt. Arrangement of electron transport chain components in bovine mitochondrial supercomplex $I_1III_2IV_1$. *The EMBO Journal* 30 (2011): 4652–64.

Arama, E., M. Bader, M. Srivastava, A. Bergmann, and H. Steller. The two *Drosophila* cytochrome *c* proteins can function in both respiration and caspase activation. *The EMBO Journal* 25 (2006): 232–43.

Balk, J., C. J. Leaver, and P. F. McCabe. Translocation of cytochrome *c* from the mitochondria to the cytosol occurs during heat-induced programmed cell death in cucumber plants. *FEBS Letters* 463 (1999): 151–4.

Banci, L., I. Bertini, G. Cavallaro, and A. Rosato. The functions of Sco proteins from genome-based analysis. *Journal of Proteome Research* 6 (2007): 1568–79.

Banci, L., I. Bertini, S. Ciofi-Baffoni, T. Kozyreva, M. Mori, and S. Wang. Sco proteins are involved in electron transfer processes. *Journal of Biological Inorganic Chemistry* 16 (2011): 391–403.

Bartoli, C. G., G. M. Pastori, and C. H. Foyer. Ascorbate biosynthesis in mitochondria is linked to the electron transport chain between complexes III and IV. *Plant Physiology* 123 (2000): 335–44.

Bashir, Q., A. N. Volkov, G. M. Ullmann, and M. Ubbink. Visualization of the encounter ensemble of the transient electron transfer complex of cytochrome *c* and cytochrome *c* peroxidase. *Journal of the American Chemical Society* 132 (2010): 241–7.

Bayir, H., B. Fadeel, M. J. Palladino et al. Apoptotic interactions of cytochrome *c*: Redox flirting with anionic phospholipids within and outside of mitochondria. *Biochimica et Biophysica Acta—Bioenergetics* 1757 (2006): 648–59.

Belikova, N. A., Y. A. Vladimirov, A. N. Osipov et al. Peroxidase activity and structural transitions of cytochrome *c* bound to cardiolipin-containing membranes. *Biochemistry* 45 (2006): 4998–5009.

Berry, E. A., M. Guergova-Kuras, L.-S. Huang, and A. R. Crofts. Structure and function of cytochrome *bc* complexes. *Annual Review of Biochemistry* 69 (2000): 1005–75.

Bertini, I., G. Cavallaro, and A. Rosato. A structural model for the adduct between cytochrome *c* and cytochrome *c* oxidase. *Journal of Biological Inorganic Chemistry* 10 (2005): 613–24.

Bertini, I., S. Chevance, R. Del Conte, D. Lalli, and P. Turano. The anti-apoptotic Bcl-x_L protein, a new piece in the puzzle of cytochrome *c* interactome. *PLoS One* 6 (2011): e18329.

Bihlmaier, K., N. Mesecke, N. Terziyska, M. Bien, K. Hell, and J. M. Herrmann. The disufilde relay system of mitochondria is connected to the respiratory chain. *The Journal of Cell Biology* 179 (2007): 389–95.

Boehning, D., R. L. Patterson, L. Sedaghat, N. O. Glebora, T. Kurosaki, and S. H. Snyder. Cytochrome *c* binds to inositol (1, 4, 5)-triphosphate receptor, amplifying calcium-dependent apoptosis. *Nature Cell Biology* 5 (2003): 1051–61.

Boehning, D., D. B. van Rossum, R. L. Patterson, and S. H. Snyder. A peptide inhibitor of cytochrome *c*/inositol 1,4,5-trisphosphate receptor binding blocks intrinsic and extrinsic cell death pathways. *Proceedings of the National Academy of Sciences of the United States of America* 102 (2005): 1466–71.

Bossy-Wetzel, E., D. D. Newmeyer, and D. R. Green. Mitochondrial cytochrome *c* release in apoptosis occurs upstream of DEVD-specific caspase activation and independently of mitochondrial transmembrane depolarization. *The EMBO Journal* 17 (1998): 37–49.

Capeillere-Blandin, C. Flavocytochrome b_2-cytochrome *c* interactions: The electron transfer reaction revisited. *Biochimie* 77 (1995): 516–30.

Carthy, C. M., B. Yanagawa, H. Luo et al. Bcl-2 and Bcl-x_L overexpression inhibits cytochrome *c* release, activation of multiple caspases, and virus release following coxsackievirus B3 infection. *Virology* 313 (2003): 147–57.

Casaus, J. L., J. A. Navarro, M. Hervás et al. *Anabaena sp.* PCC 7119 flavodoxin as electron carrier from photosystem I to Ferredoxin-NADP(+) reductase—Role of Trp57 and Tyr94. *Journal of Biological Chemistry* 277 (2002): 22338–44.

Chipuk, J. E., and D. R. Green. Do inducers of apoptosis trigger caspase-independent cell death?. *Nature Reviews Molecular Cell Biology* 6 (2005): 268–75.

Crowley, P. B., A. Díaz-Quintana, F. P. Molina-Heredia et al. The interactions of cyanobacterial cytochrome c_6 and cytochrome *f*, characterized by NMR. *Journal of Biological Chemistry* 277 (2002): 48685–9.

Díaz, A., M. Hervás, J. A. Navarro, M. A. De la Rosa, and G. Tollin. A thermodynamic study by laser-flash photolysis of plastocyanin and cytochrome c_6 oxidation by photosystem I from the green alga *Monoraphidium braunii*. *European Journal of Biochemistry* 222 (1994b): 1001–7.

Díaz, A., F. Navarro, M. Hervás et al. Cloning and correct expression in *Escherichia coli* of the petJ gene encoding cytochrome c_6 from *Synechocystis* 6803. *FEBS Letters* 347 (1994a): 173–7.

Díaz-Moreno, I., A. Díaz-Quintana, M. A. De la Rosa, P. B. Crowley, and M. Ubbink. Different modes of interaction in cyanobacterial complexes of plastocyanin and cytochrome *f*. *Biochemistry* 44 (2005a): 3176–83.

Díaz-Moreno, I., A. Díaz-Quintana, F. P. Molina-Heredia et al. NMR analysis of the transient complex between membrane photosystem I and soluble cytochrome c_6. *Journal of Biological Chemistry* 280 (2005b): 7925–31.

Díaz-Moreno, I., A. Díaz-Quintana, M. Ubbink, and M. A. De la Rosa. An NMR-based docking model for the physiological transient complex between cytochrome *f* and cytochrome c_6. *FEBS Letters* 579 (2005c): 2891–6.

Díaz-Moreno, I., J. M. García-Heredia, A. Díaz-Quintana, M. Teixeira, and M. A. De la Rosa. Nitration of tyrosines 46 and 48 induces the specific degradation of cytochrome *c* upon change of the heme iron state to high-spin. *Biochimica et Biophysica Acta—Bioenergetics* 1807 (2011): 1616–23.

Díaz-Moreno, I., J. M. García-Heredia, K. González-Arzola, A. Díaz-Quintana, and M. A. De la Rosa. Recent methodological advances in the analysis of protein tyrosine nitration. *ChemPhysChem* 14 (2013): 3095–102.

Díaz-Moreno, I., P. M. Nieto, R. Del Conte et al. A non-damaging method to analyze the configuration and dynamics of nitrotyrosines in proteins. *Chemistry—A European Journal* 18 (2012): 3872–8.

Engstrom, G., R. Rajagukguk, A. Saunders et al. Design of a ruthenium-labeled cytochrome *c* derivative to study electron transfer with the cytochrome bc_1 complex. *Biochemistry* 42 (2003): 2816–24.

Fraaije, M. W., W. J. H. van Berkel, J. A. E. Benen, J. Visser, and A. Mattevi. A novel oxidoreductase family sharing a conserved FAD-binding domain. *Trends in Biochemical Sciences* 23 (1998): 206–7.

Garber, E. A., and E. Margoliash. Interaction of cytochrome *c* with cytochrome *c* oxidase: An understanding of the high- to low-affinity transition. *Biochimica et Biophysica Acta—Bioenergetics* 1015 (1990): 279–87.

García-Heredia, J. M., I. Díaz-Moreno, A. Díaz-Quintana et al. Specific nitration of tyrosines 46 and 48 makes cytochrome *c* assemble a non-functional apoptosome. *FEBS Letters* 586 (2012): 154–8.

García-Heredia, J. M., I. Díaz-Moreno, P. M. Nieto et al. Nitration of tyrosine 74 prevents human cytochrome *c* to play a key role in apoptosis signaling by blocking caspase-9 activation. *Biochimica et Biophysica Acta—Bioenergetics* 1797 (2010): 981–3.

García-Heredia, J. M., A. Díaz-Quintana, M. Salzano et al. Tyrosine phosphorylation turns alkaline transition into a biologically relevant process and makes human cytochrome *c* behave as an anti-apoptotic switch. *Journal of Biological Inorganic Chemistry* 16 (2011): 1155–68.

García-Heredia, J. M., M. Hervás, M. A. De la Rosa, and J. A. Navarro. Acetylsalicylic acid induces programmed cell death in *Arabidopsis* cell cultures. *Planta* 228 (2008): 89–97.

Genova, M. L., and G. Lenaz. A critical appraisal of the role of respiratory supercomplexes in mitochondria. *Biological Chemistry* 394 (2013): 631–9.

Gertz, M., F. Fischer, D. Wolters, and C. Steegborn. Activation of the lifespan regulator $p66^{Shc}$ through reversible disulfide bond formation. *Proceedings of the National Academy of Sciences of the United States of America* 105 (2008): 5705–9.

Giannattasio, S., A. Atlante, L. Antonacci et al. Cytochrome *c* is released from coupled mitochondria of yeast en route to acetic acid-induced programmed cell death and can work as an electron donor and a ROS scavenger. *FEBS Letters* 582 (2008): 1519–25.

Giles, S. S., J. R. Perfect, and G. M. Cox. Cytochrome *c* peroxidase contributes to the antioxidant defense of *Cryptococcus neoformans*. *Fungal Genetics and Biology* 42 (2005): 20–9.

Giorgio, M., E. Migliaccio, F. Orsini et al. Electron transfer between cytochrome *c* and $p66^{Shc}$ generates reactive oxygen species that trigger mitochondrial apoptosis. *Cell* 122 (2005): 221–33.

Hackenbrock, C., B. Chazotte, and S. S. Gupte. The random collision model and a critical assessment of diffusion and collision in mitochondrial electron transport. *Journal of Bioenergetics and Biomembranes* 18 (1986): 331–68.

Haines, T. H., and N. A. Dencherb. Cardiolipin: A proton trap for oxidative phosphorylation. *FEBS Letters* 528 (2002): 35–9.

Heazlewood, J. L., K. A. Howell, and A. H. Millar. Mitochondrial complex I from *Arabidopsis* and rice. Orthologs of mammalian and fungal components coupled with plant-specific subunits. *Biochimica et Biophysica Acta—Bioenergetics* 1604 (2003): 159–69.

Heinemeyer, J., H.-P. Braun, E. J. Boekema, and R. Kouril. A structural model of the cytochrome *c* reductase/oxidase supercomplex from yeast mitochondria. *The Journal of Biological Chemistry* 282 (2007): 12240–8.

Hervás, M., Q. Bashir, N. G. H. Leferink et al. Communication between L–galactono–1,4–lactone dehydrogenase and cytochrome *c*. *The FEBS Journal* 280 (2013): 1830–40.

Hüttemann, M., P. Pecina, M. Rainbolt et al. The multiple functions of cytochrome *c* and their regulation in life and death decisions of the mammalian cell: From respiration to apoptosis. *Mitochondrion* 11 (2011): 369–81.

Janot, J.-M., C. Capeillere-Blandin, and F. Labeyrie. L-Lactate cytochrome *c* reductase: Rapid kinetic studies of electron transfers within the flavocytochrome b_2-cytochrome c assembly. *Biochimica et Biophysica Acta—Bioenergetics* 1016 (1990): 165–76.

Janzon, J., Q. Yuan, F. Malatesta et al. Probing the *Paracoccus denitrificans* cytochrome c_1-cytochrome c_{522} interaction by mutagenesis and fast kinetics. *Biochemistry* 47 (2008): 12974–84.

Jeng, W.-Y., C.-Y. Chen, H.-C. Chang, and W.-J. Chuang. Expression and characterization of recombinant human cytochrome *c* in *E. coli*. *Journal of Bioenergetics and Biomembranes* 34 (2002): 423–31.

Jiang, H., and A. M. English. Phenotypic analysis of the ccp1Δ and ccp1Δ -$ccp1^{W191F}$ mutant strains of *Saccharomyces cerevisiae* indicates that cytochrome *c* peroxidase functions in oxidative-stress signaling. *Journal of Inorganic Biochemistry* 100 (2006): 1996–2008.

Kagan, V. E., G. G. Borisenko, Y. Y. Tyurina et al. Oxidative lipidomics of apoptosis: Redox catalytic interactions of cytochrome *c* with cardiolipin and phosphatidylserine. *Free Radical Biology and Medicine* 37 (2004): 1963–85.

Kagan, V. E., V. A. Tyurin, J. Jiang et al. Cytochrome *c* acts as a cardiolipin oxygenase required for release of proapoptotic factors. *Nature Chemical Biology* 1 (2005): 223–32.

Kagan, V. E., Y. Y. Tyurina, H. Bayir et al. The "pro-apoptotic genies" get out of mitochondria: Oxidative lipidomics and redox activity of cytochrome *c*/cardiolipin complexes. *Chemico-Biological Interactions* 163 (2006): 15–28.

Kapralov, A. A., N. Yanamala, Y. Y. Tyurina et al. Topography of tyrosine residues and their involvement in peroxidation of polyunsaturated cardiolipin in cytochrome *c*/cardiolipin peroxidase complexes. *Biochimica et Biophysica Acta – Biomembranes* 1808 (2011): 2147–55.

Kluck, R. M., L. M. Ellerby, M. Ellerby et al. Determinants of cytochrome *c* pro-apoptotic activity: The role of Lysine 72 trimethylation. *The Journal of Biological Chemistry* 275 (2000): 16127–33.

Kokhan, O., C. Wraight, and E. Tajkhorshid. The binding interface of cytochrome *c* and cytochrome c_1 in the bc_1 complex: Rationalizing the role of key residues. *Biophysical Journal* 99 (2010): 2647–56.

Krammer, P. H. CD95's deadly mission in the immune system. *Nature* 407 (2000): 789–95.

Lange, C., and C. Hunte. Crystal structure of the yeast cytochrome bc_1 complex with its bound substrate cytochrome *c*. *Proceedings of the National Academy of Sciences of the United States of America* 99 (2002): 2800–5.

Leary, S. C., P. A. Cobine, B. A. Kaufman et al. The human cytochrome *c* oxidase assembly factors SCO1 and SCO2 have regulatory roles in the maintenance of cellular copper homeostasis. *Cell Metabolism* 5 (2007): 9–20.

Lee, I., A. R. Salomon, K. Yu, J. W. Doan, L. I. Grossman, and M. Hüttemann. New prospects for and old enzyme: Mammalian cytochrome *c* is tyrosine-phosphorylated *in vivo*. *Biochemistry* 45 (2006): 9121–8.

Leesch, V. W., J. Bujons, A. G. Mauk, and B. M. Hoffman. Cytochrome *c* peroxidase-cytochrome *c* complex: Locating the second binding domain on cytochrome *c* peroxidase with site-directed mutagenesis. *Biochemistry* 39 (2000): 10132–9.

Leferink, N. G. H., M. D. F. Jose, W. A. M. van den Berg, and W. J. H. van Berkel. Functional assignment of Glu386 and Arg388 in the active site of L-galactono-γ-lactone dehydrogenase. *FEBS Letters* 583 (2009): 3199–203.

Leferink, N. G. H., W. A. M. van den Berg, and W. J. H. van Berkel. L-Galactono-γ-lactone dehydrogenase from *Arabidopsis thaliana*, a flavoprotein involved in vitamin C biosynthesis. *The FEBS Journal* 275 (2008): 713–26.

Lenaz, G., and M. L. Genova. Structure and organization of mitochondrial respiratory complexes: A new understanding of an old subject. *Antioxidants and Redox Signaling* 12 (2010): 961–1008.

Li, P., D. Nijhawan, I. Budihardjo et al. Cytochrome *c* and dATP-dependent formation of Apaf-1/caspase-9 complex initiates an apoptotic protease cascade. *Cell* 91 (1997): 479–89.

Liu, R.-Q., L. Geren, P. Anderson et al. Design of ruthenium-cytochrome *c* derivatives to measure electron transfer to cytochrome *c* peroxidase. *Biochimie* 77 (1995): 549–61.

Locksley, R. M., N. Killeen, and M. J. Lenardo. The TNF and TNF receptors superfamilies: Integrating mammalian biology. *Cell* 104 (2001): 487–501.

Ly, H. K., T. Utesch, I. Díaz-Moreno, J. M. García-Heredia, M. A. De La Rosa, and P. Hildebrandt. Perturbation of the redox site structure of cytochrome *c* variants upon tyrosine nitration. *Journal of Physical Chemistry B* 116 (2012): 5694–702.

Marsden, V. S., L. O'Connor, L. A. O'Reilly et al. Apoptosis initiated by Bcl-2-regulated caspase activation independently of the cytochrome *c*/Apaf-1/caspase-9 apoptosome. *Nature* 419 (2002): 634–7.

Martínez-Fábregas, J., I. Díaz-Moreno, K. González-Arzola et al. New *Arabidopsis thaliana* cytochrome *c* partners: A look into the elusive role of cytochrome *c* in programmed cell death in plants. *Molecular and Cellular Proteomics* 12 (2013): 3666–76.

Martínez-Fábregas, J., I. Díaz-Moreno, K. González-Arzola et al. Structural and functional analysis of novel human cytochrome *c* targets in apoptosis. *Molecular and Cellular Proteomics* 13 (2014a): 1439–56.

Martínez-Fábregas, J., I. Díaz-Moreno, K. González-Arzola, A. Díaz-Quintana, and M. A. De la Rosa. A common signalosome for programmed cell death in humans and plants. *Cell Death and Disease* 5 (2014b): e1314.

Medina, M., M. Hervás, J. A. Navarro, M. A. De la Rosa, C. Gómez-Moreno, and G. Tollin. A laser flash absorption spectroscopy study of *Anabaena sp.* PCC 7119 flavodoxin photoreduction by photosystem I particles from spinach. *FEBS Letters* 313 (1992): 239–42.

Mei, H., L. Geren, M. A. Miller, B. Durham, and F. Millett. Role of the low-affinity binding site in electron transfer from cytochrome *c* to cytochrome *c* peroxidase. *Biochemistry* 41 (2002): 3968–76.

Meier, P., A. Finch, and G. Evan. Apoptosis in development. *Nature* 407 (2000): 769–801.

Mendes, C. S., E. Arama, S. Brown et al. Cytochrome *c-d* regulates developmental apoptosis in the *Drosophila* retina. *EMBO Reports* 7 (2006): 933–9.

Millar, A. H., V. Mittova, G. Kiddle et al. Control of ascorbate synthesis by respiration and its implications for stress responses. *Plant Physiology* 133 (2003): 443–47.

Millett, F., J. Havens, S. Rajagukguk, and B. Durham. Design and use of photoactive ruthenium complexes to study electron transfer within cytochrome bc_1 and from cytochrome bc_1 to cytochrome *c*. *Biochimica et Biophysica Acta—Bioenergetics* 1827 (2013): 1309–19.

Minina, E. A., A. P. Smertenko, and P. V. Bozhkov. Vacuolar cell death in plants: metacaspase releases the brakes on autophagy. *Autophagy* 10 (2014): 928–29.

Molina-Heredia, F. P., A. Díaz-Quintana, M. Hervás, J. A. Navarro, and M. A. De la Rosa. Site-directed mutagenesis of cytochrome c_6 from *Anabaena* species PCC 7119—Identification of surface residues of the hemeprotein involved in photosystem I reduction. *Journal of Biological Chemistry* 274 (1999): 33565–70.

Moreno-Beltrán, B., A. Díaz-Quintana, K. González-Arzola, A. Velázquez-Campoy, M. A. De la Rosa, and I. Díaz-Moreno. Cytochrome c_1 exhibits two binding sites for cytochrome *c* in plants. *Biochimica et Biophysica Acta—Bioenergetics* (2014) 1837: 1717–29.

Navarro, J. A., M. Hervás, B. De la Cerda, and M. A. De la Rosa. Purification and physicochemical properties of the low-potential cytochrome c_{549} from the cyanobacterium *Synechocystis sp.* PCC 6803. *Archives of Biochemistry and Biophysics* 318 (1995): 46–52.

Nuñez, G., M. A. Benedict, Y. Hu and N. Inohara. Caspases: The proteases of the apoptotic pathway. *Oncogene* 17 (1998): 3237–45.

Nur-E-Kamal, A., S. R. Gross, Z. Pan, Z. Balklava, J. Ma, and L. F. Liu. Nuclear translocation of cytochrome *c* during apoptosis. *The Journal of Biological Chemistry* 279 (2004): 24911–4.

Nyola, A., and C. Hunte. A structural analysis of the transient interactions between the cytochrome bc_1 complex and its substrate cytochrome *c*. *Biochemical Society Transactions* 36 (2008): 981–5.

Orsini, F., E. Migliaccio, M. Moroni et al. The life span determinant p66^{Shc} localizes to mitochondria where it associates with mitochondrial heat shock protein 70 and regulates trans-membrane potential. *The Journal of Biological Chemistry* 279 (2004): 25689–95.

Papa, S., N. Capitanio, and G. Capitanio. A cooperative model for proton pumping in cytochrome *c* oxidase. *Biochimica et Biophysica Acta—Bioenergetics* 1655 (2004): 353–64.

Pearl, N. M., T. Jacobson, M. Arisa, L. B. Vitello, and J. E. Erman. Effect of single-site charge-reversal mutations on the ccatalytic properties of yeast cytochrome *c* peroxidase: Mutations near the high-affinity cytochrome *c* binding site. *Biochemistry* 46 (2007): 8263–72.

Pearl, N. M., T. Jacobson, C. Meyen et al. Effect of single-site charge-reversal mutations on the catalytic properties of yeast cytochrome *c* peroxidase: Evidence for a single, catalytically active, cytochrome *c* binding domain. *Biochemistry* 47 (2008): 2766–75.

Pecina, P., G. Borisenko, N. Belikova et al. Phosphomimetic substitution of cytochrome *c*, tyrosine 48 decreases respiration and binding to cardiolipin and abolishes ability to trigger downstream caspase activation. *Biochemistry* 49 (2010): 6705–14.

Pelletier, H., and J. Kraut. Crystal structure of a complex between electron transfer partners, cytochrome *c* peroxidase and cytochrome *c*. *Science* 258 (1992): 1748–55.

Pietras, R., M. Sarewicz, and A. Osyczka. Molecular organization of cytochrome c_2 near the binding domain of cytochrome bc_1 studied by electron spin-lattice relaxation enhancement. *The Journal of Physical Chemistry B* 118 (2014): 6634–43.

Pineau, B., O. Layoune, A. Danon, and R. De Paepe. L-Galactono-1,4-lactone dehydrogenase is required for the accumulation of plant respiratory complex I. *The Journal of Biological Chemistry* 283 (2008): 32500–05.

Pinton, P., A. Rimessi, S. Marchi et al. Protein Kinase C β and prolyl isomerase 1 regulate mitochondrial effects of the life-span determinant p66^{Shc}. *Science* 315 (2007): 659–63.

Pollock, B. R., F. I. Rosell, M. B. Twitchett, M. E. Dumont, and A. G. Mauk. Bacterial expression of a mitochondrial cytochrome *c*. Trimethylation of Lys72 in yeast *iso*-1-cytochrome *c* and the alkaline conformational transition. *Biochemistry* 37 (1998): 6124–31.

Poulos, T. L., S. T. Freer, R. A. Alden et al. The crystal structure of cytochrome *c* peroxidase. *The Journal of Biological Chemistry* 255 (1980): 575–80.

Prudêncio, M., and M. Ubbink. Transient complexes of redox proteins: structural and dynamic details from NMR studies. *Journal of Molecular Recognition* 17 (2004): 524–39.

Qi, S., Y. Pang, Q. Hu et al. Crystal structure of the *Caenorhabditis elegans* apoptosome reveals an octameric assembly of CED-4. *Cell* 141 (2010): 446–57.

Roberts, V. A., and M. E. Pique. Definition of the interaction domain for cytochrome *c* on cytochrome *c* oxidase: III. Prediction of the docked complex by a complete, systematic search. *The Journal of Biological Chemistry* 274 (1999): 38051–60.

Rodríguez-Roldán, V., J. M. García-Heredia, J. A. Navarro, M. A. De la Rosa, and M. Hervás. Effect of nitration on the physicochemical and kinetic features of wild-type and monotyrosine mutants of human respiratory cytochrome *c*. *Biochemistry* 47 (2008): 12371–9.

Ruíz-Vela, A., G. González de Buitrago, and C. Martínez-A. Nuclear Apaf-1 and cytochrome *c* redistribution following stress-induced apoptosis. *FEBS Letters* 517 (2002): 133–8.

Rytömaa, M., P. Mustonen, and P. J. Kinnunen. Reversible, nonionic, and pH-dependent association of cytochrome *c* with cardiolipin-phosphatidylcholine liposomes. *The Journal of Biological Chemistry* 261 (1992): 22243–8.

Saenkham, P., P. Vattanaviboon, and S. Mongkolsuk. Mutation in *sco* affects cytochrome *c* assembly and alters oxidative stress resistance in *Agrobacterium tumefaciens*. *FEMS Microbiology Letters* 293 (2009): 122–9.

Sakamoto, K., M. Kamiya, M. Ima et al. NMR basis for interprotein electron transfer gating between cytochrome *c* and cytochrome *c* oxidase. *Proceedings of the National Academy of Sciences of the United States of America* 108 (2011): 12271–6.

Saraste, M. Oxidative phosphorylation at the *fin* de siècle. *Science* 283 (1999): 1488–93.

Sarewicz, M., A. Borek, F. Daldal, W. Froncisz, and A. Osyczka. Demonstration of short-lived complexes of cytochrome *c* with cytochrome bc_1 by EPR spectroscopy: Implications for the mechanism of interprotein electron transfer. *The Journal of Biological Chemistry* 283 (2008): 24826–36.

Savino, C., P. G. Pelicci, and M. Giorgio. The $p66^{Shc}$/mitochondrial permeability transition pore pathway determines neurodegeneration. *Oxidative Medicine and Cellular Longevity* 2013 (2013): 1–7.

Schertl, P., S. Sunderhaus, J. Klodmann, G. E. G. Grozeff, C. G. Bartoli, and H.-P. Braun. L-Galactono-1,4-lactone dehydrogenase (GLDH) forms part of three subcomplexes of mitochondrial complex I in *Arabidopsis thaliana*. *The Journal of Biological Chemistry* 287 (2012): 14412–9.

Schilder, J., F. Löhr, H. Schwalbe, and M. Ubbink. The cytochrome *c* peroxidase and cytochrome *c* encounter complex: The other side of the story. *FEBS Letters* 588 (2014): 1873–8.

Schug, Z. T., and E. Gottlieb. Cardiolipin acts as a mitochondrial signalling platform to launch apoptosis. *Biochimica et Biophysica Acta—Biomembranes* 1788 (2009): 2022–31.

Seib, K. L., M. P. Jennings, and A. G. McEwan. A Sco homologue plays a role in defence against oxidative stress in pathogenic *Neisseria*. *FEBS Letters* 546 (2003): 411–5.

Shawgo, M. E., S. N. Shelton, and J. D. Robertson. Caspase-9 activation by the apoptosome is not required for Fas-mediated apoptosis in type II *Jurkat* cells. *The Journal of Biological Chemistry* 284 (2009): 33447–55.

Short, D. M., M. D. Walkinshaw, P. Taylor, G. A. Reid, and S. K. Chapman. Location of a cytochrome *c* binding site on the surface of flavocytochrome b_2. *Journal of Biological Inorganic Chemistry* 3 (1998): 246–52.

Sinibaldi, F., L. Fiorucci, A. Patriarca et al. Insights into cytochrome *c*-cardiolipin interaction. Role played by ionic strength. *Biochemistry* 47 (2008): 6928–35.

Sinibaldi, F., B. D. Howes, M. C. Piro et al. Extended cardiolipin anchorage to cytochrome *c*: A model from protein-mitochondrial membrane binding. *Journal Biological Inorganic Chemistry* 15 (2010): 689–700.

Solmaz, S. R. N., and C. Hunte. Structure of complex III with bound cytochrome *c* in reduced state and definition of a minimal core interface for electron transfer. *The Journal of Biological Chemistry* 283 (2008): 17542–9.

Speck, S. H., D. Dye, and E. Margoliash. Single catalytic site model for the oxidation of ferrocytochrome *c* by mitochondrial cytochrome *c* oxidase. *Proceedings of the National Academy of Sciences of the United States of America* 81 (1984): 347–51.

Suen, D.-F., K. L. Norris, and R. J. Youle. Mitochondrial dynamics and apoptosis. *Genes and Development* 22 (2008): 1577–90.

Sun, J., W. Xu, M. Hervás, J. A. Navarro, M. A. De la Rosa, and P. R. Chitnis. Oxidizing side of the cyanobacterial photosystem I—Evidence for interaction between the electron donor proteins and a luminal surface helix of the PsaB subunit. *Journal of Biological Chemistry* 274 (1999): 19048–54.

Swierczek, M., E. Cieluch, M. Sarewicz et al. An electronic bus bar lies in the core of cytochrome bc_1. *Science* 329 (2010): 451–4.

Szado, T., V. Vanderheyden, J. B. Parys et al. Phosphorylation of inositol 1, 4, 5-triphosphate receptors by protein kinase B/Akt inhibits Ca2+ release and apoptosis. *Proceedings of the National Academy of Sciences of the United States of America* 105 (2008): 2427–32.

Tegoni, M., A. Mozzarelli, G. L. Rossi, and F. Labeyrie. Complex formation and intermolecular electron transfer between flavocytochrome b_2 in the crystal and cytochrome *c*. *The Journal of Biological Chemistry* 258 (1983): 5424–7.

Tegoni, M., S. A. White, A. Roussel, F. S. Mathews, and C. Cambillau. A hypothetical complex between crystalline flavocytochrome b_2 and cytochrome *c*. *Proteins* 16 (1993): 408–22.

Tian, H., R. Sadoski, L. Zhang et al. Definition of the interaction domain for cytochrome *c* on the cytochrome bc_1 complex: Steady-state and rapid kinetic analysis of electron transfer between cytochrome *c* and *Rhodobacter sphaeroides* cytochrome bc_1 surface mutants. *The Journal of Biological Chemistry* 275 (2000): 9587–95.

Trinei, M., M. Giorgio, A. Cicalese et al. A p53-p66^{Shc} signalling pathway controls intracellular redox status, levels of oxidation-damaged DNA and oxidative stress-induced apoptosis. *Oncogene* 21 (2002): 3872–8.

Trinei, M., E. Migliaccio, P. Bernardi, F. Paolucci, P. Pelicci, and M. Giorgio. p66^{shc}, mitochondria, and the generation of reactive oxygen species. *Methods in Enzymology* 528 (2013): 99–110.

Tsukihara, T., H. Aoyama, E. Yamashita et al. The whole structure of the 13-subunit oxidized cytochrome *c* oxidase at 2.8 Å. *Science* 272 (1996): 1136–44.

van Dijk, A. D. J., S. Ciofi-Baffoni, L. Banci, I. Bertini, R. Boelens, and A. M. J. J. Bonvin. Modeling protein-protein complexes involved in the cytochrome *c* oxidase copper-delivery pathway. *Journal of Proteome Research* 6 (2007): 1530–9.

van Nocker, S., and P. Ludwig. The WD-repeat protein superfamily in *Arabidopsis*: Conservation and divergence in structure and function. *BMC Genomics* 4 (2003): 50.

Vartapetian, A. B., A. I. Tuzhikov, N. V. Chichkova, M. Taliansky, and T. J. Wolpert. A plant alternative to animal caspases: subtilisin-like proteases. *Cell Death and Differentiation* 18 (2011): 1289–97.

Vempati, U. D., F. Diaz, A. Barrientos et al. Role of cytochrome *c* in apoptosis: Increased sensitivity to tumor necrosis factor alpha is associated with respiratory defects but not with lack of cytochrome *c* release. *Molecular and Cellular Biology* 27 (2007): 1771–83.

Vik, S. B., G. Georgevich, and R. A. Capaldi. Diphosphatidylglycerol is required for optimal activity of beef cytochrome *c* oxidase. *Proceedings of the National Academy of Sciences of the United States of America* 78 (1981): 1456–60.

Volkov, A. N., Q. Bashir, J. A. R. Worrall, and M. Ubbink. Binding hot spot in the weak protein complex of physiological redox partners yeast cytochrome *c* and cytochrome *c* peroxidase. *The Journal of Molecular Biology* 385 (2009): 1003–13.

Volkov, A. N., M. Ubbink, and N. A. J. van Nuland. Mapping the encounter state of a transient protein complex by PRE NMR spectroscopy. *Journal of Biomolecular NMR* 48 (2010): 225–36.

Volkov, A. N., and N. A. J. van Nuland. Solution NMR study of the yeast cytochrome *c* peroxidase: cytochrome *c* interaction. *Journal of Biomolecular NMR* 56 (2013): 255–63.

Volkov, A. N., J. A. R. Worrall, E. Holtzmann, and M. Ubbink. Solution structure and dynamics of the complex between cytochrome *c* and cytochrome *c* peroxidase determined by paramagnetic NMR. *Proceedings of the National Academy of Sciences of the United States of America* 103 (2006): 18945–50.

Wang, G., Z. Chen, F. Zhang et al. Blockade of PKCβ protects against remote organ injury induced by intestinal ischemia and reperfusion via a p66^{shc}-mediated mitochondrial apoptotic pathway. *Apoptosis* (2014). doi:10.1007/s10495-014-1008-x.

Wang, J., R. W. Larsen, S. J. Moench, J. D. Satterlee, D. L. Rousseau, and M. R. Ondrias. Cytochrome *c* peroxidase complexed with cytochrome *c* has an unperturbed heme moiety. *Biochemistry* 35 (1996): 453–63.

Williams, J. C., C. Sue, G. S. Banting et al. Crystal structure of human SCO1: Implications for redox signaling by a mitochondrial cytochrome *c* oxidase "assembly" protein. *The Journal of Biochemical Chemistry* 280 (2005): 15202–11.

Witt, H., F. Malatesta, F. Nicoletti, M. Brunori, and B. Ludwig. Tryptophan 121 of subunit II is the electron entry site to cytochrome *c* oxidase in *Paracoccus denitrificans*. Involvement of a hydrophobic patch in the docking reaction. *The Journal of Biological Chemistry* 273 (1998): 5132–6.

Worrall, J. A. R., U. Kolczak, G. W. Canters, and M. Ubbink. Interaction of yeast iso-1-cytochrome *c* with cytochrome *c* peroxidase investigated by [^{15}N, ^{1}H] heteronuclear NMR spectroscopy. *Biochemistry* 42 (2001): 7068–76.

Xia, Z.-X., and F. S. Mathews. Molecular structure of flavocytochrome b_2 at 2.4 Å resolution. *The Journal of Molecular Biology* 212 (1990): 837–63.

Yadaiah, M., P. N. Rao, P. Harish, and A. K. Bhuyan. High affinity binding of Bcl-x_L to cytochrome *c*: Possible relevance for interception of translocated cytochrome *c* in apoptosis. *Biochimica et Biophysica Acta—Proteins and Proteomics* 1774 (2007): 1370–9.

Yonetani, T., and T. Ohnishi. Cytochrome *c* peroxidase, a mitochondrial enzyme of yeast. *The Journal of Biological Chemistry* 241 (1966): 2983–4.

Yoshikawa, S., K. Shinzawa-Itoh, R. Nakashima *et al.* Redox-coupled crystal structural changes in bovine heart cytochrome *c* oxidase. *Science* 280 (1998): 1723–9.

Yu, C.-A., X. Wen, K. Xiao, D. Xia, and L. Yu. Inter- and intra-molecular electron transfer in the cytochrome bc_1 complex. *Biochimica et Biophysica Acta—Bioenergetics* 1555 (2002): 65–70.

Yu, H., I. Lee, A. R. Salomon, K. Yu, and M. Hüttemann. Mammalian liver cytochrome *c* is tyrosine-48 phosphorylated in vivo, inhibiting mitochondrial respiration. *Biochimica et Biophysica Acta—Bioenergetics* 1777 (2008): 1066–71.

Yu, X., D. Acehan, J.-F. Ménétret et al. A structure of the human apoptosome at 12.8 Å resolution provides insights into this cell death platform. *Structure* 13 (2005): 1725–35.

Yuan, S., X. Yu, M. Topf et al. Structure of the *Drosophila* apoptosome at 6.9 Å resolution. *Structure* 19 (2011): 128–40.

Zhao, X., I. R. Léon, S. Bak et al. Phosphoproteome analysis of functional mitochondria isolated from resting human muscle reveals extensive phosphorylation of inner membrane protein complexes and enzymes. *Molecular and Cellular Proteomics* 10 (2011): 1–14.

Zhou, J. S., and B. M. Hoffman. Stern-volmer in reverse: 2:1 stoichiometry of the cytochrome *c*-cytochrome *c* peroxidase electron-transfer complex. *Science* 265 (1994): 1693–6.

CHAPTER 13

Cell Membrane Raft Redox Signalosomes

Platform Responding to Danger Signals

Pin-Lan Li, Justine M. Abais, Jun-Xiang Bao and Yang Zhang

CONTENTS

13.1 INTRODUCTION

Since 1970s, the model of dynamic cell membrane microdomains, namely, the fluid mosaic blocks in the cell membrane, has been widely used to describe cell membrane structure and its function as well as related signalling or regulatory mechanisms (Israelachvili 1977; Jain and White 1977). Among these microdomains, a sphingolipid- and cholesterol-rich microdomain that was originally named as lipid raft (LR) (Simons and Ikonen 1997, 2000a; Simons and Toomre 2000b) and now as membrane raft (MR) (Pike 2006) has been implicated in numerous biological and physiological processes of many mammalian cells, determining their function or even their life (Mattjus et al. 1996; Rinia and de Kruijff 2001a; Rinia et al. 2001). MRs are functioning mainly through the formation of macrodomains or platforms (Alonso and Millan 2001; Sowa et al. 2001) and associated molecular activation in the platforms, which critically contributes to transmembrane signalling in a variety of mammalian cells. In particular, MRs clustering may result in aggregation of many cell surface receptors in plasma membranes (Grassme et al. 2002; Gulbins and Grassme 2002) such as T-cell receptor/CD3 complexes, B-cell receptors, CD2, CD38, CD40, CD44, L-selectin, insulin receptors and various death receptors (Jin et al. 2011; Lotocki et al. 2004; Simons and Toomre 2000b). These aggregated receptors and associated signalling molecules in MRs interact with each other to form signalosomes, which transduct transmembrane signals to intracellular effectors and produce cellular activity or functional changes (Brown and London 1998; Cheng et al. 1999). In this regard, our laboratories and others recently defined a novel MR signalosome, which was named as LR redox signalling signalosome or MR redox signalling platform (Jin et al. 2011; Li et al. 2007; Zhang et al. 2006). This MR signalosome is characterized by aggregation of cell surface receptors such as various death receptors, acid sphingomyelinase (ASMase) product, ceramide, sphingomyelin (SM), cholesterol, NADPH oxidase subunits and other signalling molecules. In some reports, this MR signalosome is called membrane ceramide-enriched redox signalling platform or signalosome (Jin et al. 2011; Li et al. 2007; Zhang et al. 2006). MRs clustering has been shown as a physical force that allows NADPH oxidase subunits such as gp91^{phox} and p47^{phox} to cluster and form an enzyme complex, where $O_2^{\bullet-}$ and reactive oxygen species (ROS) are produced to conduct transmembrane or intracellular signalling and regulate cell function under physiological and pathological conditions (Jin et al. 2011; Zhang and Li 2010). This chapter will briefly introduce the concept of MR redox signalosomes and related regulatory mechanisms, and then discuss several important functional events associated with this signalosome such as the cell response to danger signals including pathogen-associated molecular patterns (PAMPs) and damage-associated molecular patterns (DAMPs) of molecules and inflammasome activation to produce inflammatory responses and non-inflammatory injury.

13.2 MRs AND THEIR CLUSTERING

In early studies, it was proposed that MRs consist of cholesterol and lipids with saturated acyl chains such as sphingolipids and glycosphingolipids in the exoplasmic leaflets and phospholipids and cholesterol in the inner leaflets of the membrane bilayer (Oakley et al. 2009). Due to the long fatty acid of sphingolipids in the outer leaflets, these sphingolipids couple with the exoplasmic and cytoplasmic leaflets and interdigitation and transmembrane proteins further stabilized such coupling. This coupling and its stabilization result in a most important feature of MRs, namely, the formation of very stable and detergent resistant microdomains or blocks (Magee and Parmryd 2003; Nicolau et al. 2006). These membrane microdomains have different sizes depending upon different cell types, which are in the range of 50–200 nm in diameter and each may only carry no more than 10–30 proteins (Pralle et al. 2000). It was assumed that sphingolipid-associated rafts or microdomains constitute about 45% of the cell surface in fibroblasts and about 30% in lymphocytes (Gulbins 2008; Pralle et al. 2000). Although current studies using most currently advanced techniques have provided evidence showing the existence of this MR microdomain in living cells (Lingwood et al. 2009; Szanto et al. 2005; Vetrivel et al. 2004), many scientists failed to detect MRs on living cells and therefore they are not convinced that such MRs are present in living cell membranes. Due to this reason, at a *Keystone Symposium on LRs and Cell Functions* in 2006, with attendance of many leading scientists in the raft field, the term *lipid raft (LR)* was replaced with *membrane raft (MR)* (Pike 2006). Since this conference, the term *MR* is widely used in the literature; despite that *LRs* remain to be used in some publications.

13.2.1 Nature and Behaviour of MRs

To describe and explain the nature and behaviour of MRs, two molecular models are described in the literature. The first model is called sphingolipid-enriched model, which describes MRs as relatively small structures enriched in cholesterol and sphingolipids, where various proteins aggregate and assemble as a signalling unit that is regulated upon different stimuli (Simons and Ikonen 1997). SM composed of a highly hydrophobic ceramide moiety and a hydrophilic phosphorylcholine headgroup is considered a most prevalent sphingolipid in the MRs. Depending upon the tight interaction between the cholesterol-sterol-ring system and the ceramide moiety of the SM, a lateral association between sphingolipids and cholesterol is formed to produce distinct microdomains. In addition, the interactions of cholesterol and SM determine the existing status of MRs on the cell membrane, either in a liquid-ordered or gel-like phase. Without such interactions of cholesterol and SM, many other domains in cell membranes are primarily present in a more disordered fluid or liquid phase (Gulbins and Li 2006). Another model for interpretation of the nature and behaviour of MRs is known as the shell model. Based on this model, the generation of MRs is thought to depend on protein–lipid or protein–protein interactions, where MRs consist of lipid shells that are formed by proteins with their preferential types of lipids. It is the protein–protein interaction in such lipid shells to create larger functional units responsible for MR-related activities (Anderson and Jacobson 2002). Moreover, these proteins may be oligomerized to form larger MR platforms, again depending upon protein–lipid interactions or protein–protein interactions (Helms and Zurzolo 2004).

In mammalian cells, proteins associated with MRs are many and most of them are with specific posttranslational modifications including the glycolphosphotidylinosital (GPI)-anchoring proteins, Src family tyrosine kinase, caveolin and so on (Lingwood and Simons 2010; Lingwood et al. 2009; Simons and Ikonen 1997). These proteins can be present within MRs such as GPI-anchored proteins and some transmembrane proteins like G protein Gα subunits and nitric oxide synthase (NOS). However, there are some proteins present outside MRs or between or around LRs, which are certain proteins in low affinity with MR lipids under resting status. However, they may be oligomerized or transferred into MRs upon stimulation. Based on recent proteomic analysis, around 241 authentic proteins are detectable in MRs (Foster et al. 2003). Several types of post-translational modifications of proteins are featured for their high binding capacity to sphingolipids such as GPI-anchoring, palmitoylation and myristoylation (Munday and Lopez 2007; Patterson 2002). In particular, the palmitoylation of proteins enhances surface hydrophobicity, promoting the membrane affinity of protein substrates and having high propensity to associate with MRs (Draper et al. 2007; Linder and Deschenes 2007).

13.2.2 Caveolar and Non-Caveolar MRs

There are two types of MRs in the cell plasma membrane based on their structure and components, which are caveolar and non-caveolar rafts. In some cell types such as vascular endothelial cells, caveolin can bend to form caveolae. Many studies have revealed that caveolae is an important cell membrane structure, which was established even prior to emerging of the general MR concept (Frank et al. 2003; Schlegel et al. 1998; Stan 2002). In particular, numerous studies have shown that caveolae serves as an important platform for endothelial nitric oxide synthase (eNOS) to produce NO and thereby regulate vascular tone (Goligorsky et al. 2002). It has been demonstrated that eNOS activity is inhibited if it is bound to the caveolin scaffolding (Garcia-Cardena et al. 1996; Razani et al. 2001). To our knowledge, caveolin-1-mediated formation of caveolae in ECs is a constitutive mechanism of MR clustering and eNOS regulation. The activators of eNOS cannot change the location of the NOS in caveolae, which differs from non-caveolar MRs, where enzymes or proteins may change their position in the cell membrane, depending on clustering or de-clustering of MRs (Goligorsky et al. 2002; Grassme et al. 2001).

Non-caveolar MRs may have similar lipid profile compared to caveolar MRs, but they may be only seen in some cell types or exist with caveolar MRs in other cell types (Waugh et al. 2001). It is obvious that non-caveolar MRs could not form caveolae due to the lack of caveolins locally. These non-caveolar MRs are able to associate with a variety of proteins or signalling molecules. For example, even eNOS typically existing in caveolar MRs are also present in non-caveolar MRs. Non-caveolar MRs are often clustered upon stimuli. In cells with both caveolar and non-caveolar MRs, a temporal–spatial membrane signalling may be controlled by different molecules in both types of MRs when cells face different stimuli or agonists. For example, NADPH oxidase is found in both caveolar and non-caveolar rafts (Hilenski et al. 2004; Ushio-Fukai and Alexander 2004). Angiotensin II stimulates caveolae-associated NADPH oxidase to produce $O_2^{\bullet-}$, while other stimuli such as FasL or endostatin activate non-caveolar MRs clustering to produce $O_2^{\bullet-}$, conducting signalling or inducing redox injury (Zuo et al. 2005). NOS in endothelial caveolae produce NO, which is negatively regulated by caveolin-1 (Goligorsky et al. 2002; Stan 2002). However, non-caveolar eNOS activity also relates to the formation of caveolae under different conditions, where caveolae facilitate interfacing or juxtaposing of NOS with other signalling partners including dynamin-2, heat shock protein 90 and Akt (Pritchard et al. 2002).

13.2.3 Ceramide-Enriched Microdomains

In a broad sense, the ceramide-enriched microdomains are also called MRs. However, this ceramide-enriched MRs or macrodomain are usually formed in the absence of classically defined rafts, which are characterized with the small structures consisting of sphingolipids, cholesterol and associated proteins. Ceramide for this MR is derived from SM by hydrolysis via various sphingomyelinases (SMase) and in some cells via a de novo ceramide synthase-mediated synthesis. The pathway for production of ceramide via ASMase has been extensively studied since the 1990s. It was demonstrated to be a major enzyme that is responsible for ceramide production to form ceramide-enriched membrane platforms (Dumitru et al. 2007; Gulbins and Li 2006). In this ceramide-enriched MR, cholesterol serves as a spacer among the hydrocarbon chains of ceramide and as dynamic glue allowing the MRs to maintain appropriate assembly. In addition, cholesterol may separate the raft from non-raft phase of cell membranes, leading to a higher affinity to ceramide within MRs than that to unsaturated phospholipids. It has been confirmed that ceramide is produced by ASMase-mediated hydrolysis of SM and at the same time choline is released, but the hydrocarbon chains remain (Brown and London 2000; Simons and Ehehalt 2002). In summary, ceramide and cholesterol are two most important molecules for the formation and clustering of ceramide-enriched membrane rafts, which may largely contribute to the activity of MRs as a signalosomes complex.

13.2.4 Macrodomains as Signalling Platforms

It has been proposed that MRs usually conduct signalling by the formation of macrodomains or membrane platforms, which may be driven by the interactions of SM–cholesterol and ceramide–ceramide or by the protein–protein interactions in the shell protein MR model (Lingwood and Simons 2010). Such macrodomain formation may be an important function of MRs to form signalling platforms or signalosomes (Alonso and Millan 2001; Sowa et al. 2001). It has been demonstrated that aggregation of cell surface receptors is first activated through MR clustering in the plasma membrane (Grassme et al. 2002; Gulbins and Grassme 2002), which transfers signals accepted by receptors in the clustered MR to interact with other signalling molecules transferring signals to signalling proteins or proteins in the inner leaflets of the cell membrane, indicating completion of the transmembrane signalling process (Alonso and Millan 2001; Boniface et al. 1998; Grassme et al. 2001; Simons and Ikonen 2000a). In these MR clusters, aggregated receptors or other signalling molecules may be the constitutive elements in MRs or those translocated by transporters or trafficking during stimulations (Brown and London 1998; Cheng et al. 1999). This dynamic clustering of lipid microdomains may represent a critical common mechanism in transmembrane signal transduction. Ceramide-enriched membrane platforms are formed due to ceramide production or enrichment from SMase in response to agonists or stimuli (Gulbins and Kolesnick 2003; Hoekstra et al. 2003). The formation of these ceramide-enriched membrane platforms may occur without the presence of classically defined sphingolipid MRs, but they may be formed simply through a fusion of several ceramide molecules upon stimulations. The clustering of receptors within ceramide-enriched membrane platforms usually leads to aggregation of many receptor molecules in close proximity to facilitate transactivation of signalling molecules or amplification of the specific signal from activated receptors. However, in some cells, such as erythrocytes, MRs may mediate the scrambling of cell membrane but not assembling. This ceramide microdomain-associated scrambling of cell membrane leads to eryptosis during osmotic shock, which is another form of MR redox signalling characterized by scrambling of constitutively existing signalling complex (Lang et al. 2006; Lang et al. 2010).

13.3 FORMATION AND REGULATION OF MR REDOX SIGNALOSOMES

Recent studies have indicated that there may be three different types of MR redox signalosomes that efficiently and robustly conduct redox signalling and that they may be together or separately existing in different mammalian cells. These redox signalosomes include the following: (1) Non-caveolar MR redox signalosome. It is formed by clustering of non-caveolar MRs in plasma membrane. The signalling molecules within these signalosomes may be already located in the membrane, in particular, in the raft areas and they can aggregate upon stimulations. Some other signalling molecules may be recruited or traffic to the membrane during MR clustering. Among molecules within this signalosome, NADPH oxidase subunits and cofactors are essential to conduct redox signalling. (2) MR redox signalosome associated with caveolae endocytosis. Before formation of this signalosome, NADPH oxidase subunits and related cofactors are located in caveolar MRs, which may aggregate and be assembled into an enzyme complex during MRs clustering upon stimulations. At the same time, caveolae are endocytosed leading to the formation of MRs containing redoxosomes within cells, where $O_2^{\bullet -}$ and other ROS are produced for signal transduction. (3) MR redox signalosome associated with endosomes exocytosis. This MR redox signalosome is first formed within the cells and then moves to the cell plasma membrane to conduct redox signalling. In response to some stimuli, the MRs of endosomes, lysosome or related vesicles are clustered leading to aggregation or recruitment of NADPH oxidase subunits and associated cofactors. At the same time, these intracellular organelles or vesicles move to the cell plasma membrane and fuse there to form a signalosome, where $O_2^{\bullet -}$ and ROS are produced for signalling (Jin et al. 2011). The mechanisms mediating the formation and regulation of these MR signalosomes are discussed in the following section.

13.3.1 Production of Ceramide and Its Action as a Membrane Fusigen

Ceramide Production: As shown in Figure 13.1, there is considerable evidence that ceramide production in mammalian cells are mainly involved in two enzymatic pathways, namely, de novo synthesis pathway and SMase pathway. The de novo synthesis of ceramide within the endoplasmic reticulum begins with the action of ceramidase synthase to convert dihydro-sphingosine into dihydro-ceramide and then remove dihydro group via the catalytic action of dihydro-sphingosine reductase (Levy and Futerman 2010).

SMases hydrolyse phosphodiester bond in SM to ceramide and choline phosphate (Kitatani et al. 2008). In essence, SMases are a member of the phospholipase C (PLC) superfamily that acts on SM as a hydrolase (Cremesti et al. 2002). Importantly, SMases may act on SM with the MR area of cell membranes, where ceramide is produced to form ceramide-enriched platforms or signalosomes (Cremesti et al. 2002). Several types of SMases are expressed in different types of cells including ASMase, secretory SMase, neutral SMase (NSMase) (Mg^{2+}-dependent and Mg^{2+}-independent NSMase) and alkaline SMase (Goni and Alonso 2002). However, only ASMase and NSMase may participate in

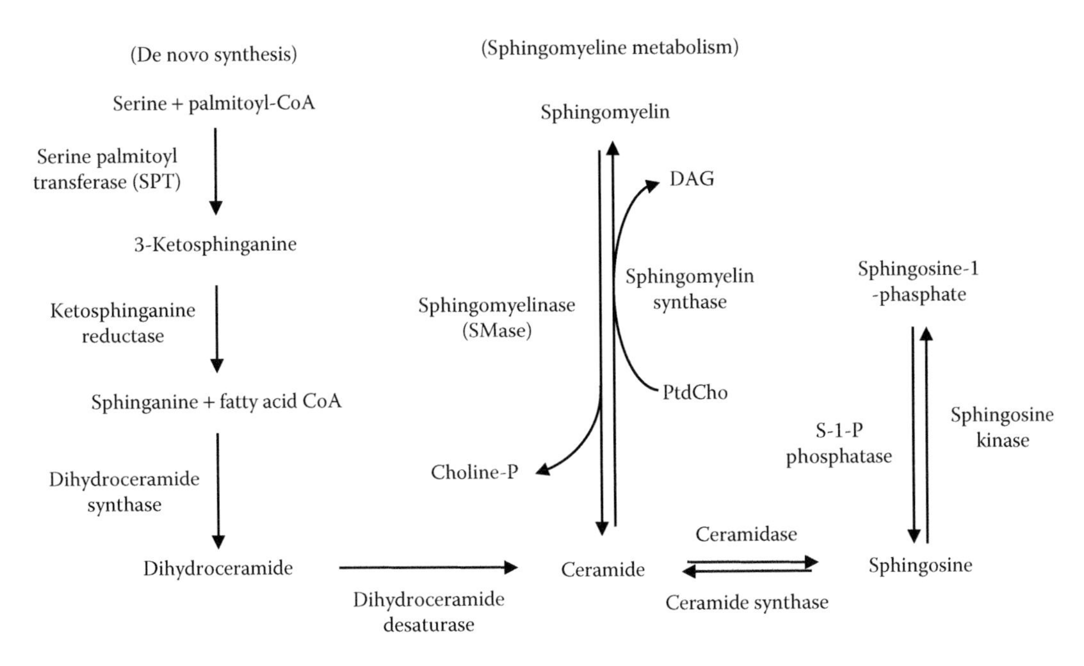

FIGURE 13.1 Ceramide synthesis and metabolism. Ceramide is generated from sphingomyelin via sphingomyelinase (SMase pathway) by hydrolysis of sphingomyelin or via a de novo synthesis pathway where cells synthesize ceramide from serine and palmitoyl-CoA. Ceramide is converted into other sphingolipids such as ceramide-1-phosphate, sphingosine and sphingosine-1-phosphate by different enzymes.

mammalian cell signalling, where ASMase accounts for 90% of total SMase activity (Goni and Alonso 2002). In these cells, ASMase is mainly found in lysosomes and it is important in the flip-flop of cell membranes (Murate et al. 2002). Its activity will be highest when it works in an environment with pH less than 5 and therefore ASMases in lysosomes work very actively given a low pH there. In response to various stimuli, ASMase is rapidly activated and released to the cell surface or even to extracellular space to hydrolyse SM, which can result in a rapid increase in cellular or local tissue ceramide levels within seconds to minutes. As observed in macrophages, fibroblasts and ECs, ASMase can be secreted and released to the cell plasma membrane, where ceramide is produced to form ceramide-enriched platforms for signalosomes (Murate et al. 2002).

Ceramide degradation: Two enzymatic pathways are responsible for ceramide metabolism or conversion. The first pathway is the synthesis of SM using ceramide as a substrate, which uses the SM synthase to catalyse the transfer of choline phosphate groups from phosphatidylcholine to ceramide. This enzymatic reaction generates both SM and diacylglycerol (DAG) (Cutler and Mattson 2001). Another pathway involved in its degradation is ceramidase. Ceramide is first phosphorylated by certain kinases and converted into ceramide-1-phosphate and then into sphingosine by the action of ceramidase. Sphingosine can also be converted back into sphingosine-1 phosphate (S1P) sphingosine kinase (Tani et al. 2007). The activity of ceramidase is also dependent on pH and therefore it also has several isoforms including acidic, neutral and alkaline ceramidase, which work in different cells and spatiotemporally to metabolize cellular ceramide (Tani et al. 2007).

Ceramide as Fusigen in MR Clustering: Ceramide is often limited in the membrane due to its hydrophobicity and therefore ceramide, in general, functions in its generating site. Since ceramide is a much less polar molecule than other sphingolipids, it can spontaneously aggregate and therefore some reports considered ceramide as a membrane fusigen. In this respect, ceramide is able to fuse MRs together and promote the formation of MR platforms, where various signalling molecules and intracellular vesicles are trafficked and aggregated, exerting the role as signalosomes (Jin and Zhou 2009; Utermohlen et al. 2008). When ASMase is activated, SM is hydrolysed to ceramide in the membrane as long as ASMase can reach it. Ceramide fuses to form platforms, leading to aggregation or recruitment of NADPH oxidase subunits and cofactors and consequently activating NADPH oxidase to produce $O_2^{\bullet-}$ for signalling. This ceramide-mediated signalosome formation is a response of cells to various different stimuli. Different cell types may respond to different agonists or stimuli. For example, in arterial ECs, individual MRs with attached death receptors including Fas and endostatin receptors are present in the plasma membrane under resting conditions (resting). These MRs are dynamic microdomains and carry several membrane-bound or attached proteins or enzymes such as G-proteins, protein kinases or the subunits of NADPH oxidase ($gp91^{phox}$) (shown as gp91). When the ligands (death factors) bind to their receptors on individual MRs, ASMase are activated to produce ceramide from SM, which induces MR clustering on the cell membrane to form a number of MR redox signalosomes, in which ASMase, NADPH oxidase subunits such as $gp91^{phox}$ and $p47^{phox}$ (p47) and other proteins are aggregated and activated, resulting in a prominent amplification of the transmembrane signal. The activation of NADPH oxidase and production of $O_2^{\bullet-}$ lead to decreases in NO bioavailability and consequent endothelial dysfunction (Becker et al. 2008; Bollinger et al. 2005; Li et al. 2010; Stancevic and Kolesnick 2010; Zhang et al. 2009). This represents a typical cell response of ECs to danger factors such as death receptor ligands (Figure 13.2).

13.3.2 Lysosome Trafficking and Fusion in MRs Clustering

Since ASMase is lysosomal hydrolase, a rapid movement and consequent fusion of lysosomes with the plasma membrane to supply ASMase is important in the formation of MR redox signalosomes (Jin et al. 2008a,b). Recent studies have indeed demonstrated that lysosomes can rapidly fuse with cell plasma membrane, leading to ASMase translocation to the surface of vascular ECs (Jaiswal et al. 2002; Jin et al. 2007, 2008a). It has been suggested that lysosomal vesicles translocated and fused via lysosome trafficking upon stimulations are vital contributors to the formation of MR-redox signalling platforms associated with NADPH oxidase. This rapid lysosome fusion into cell plasma membrane critically contributes to the formation of MR redox signalosomes, producing $O_2^{\bullet-}$ to conduct signalling process.

In addition to ASMase, sortilin, a glycoprotein responsible for transferring ASMase from the Golgi apparatus to lysosomes, is also transferred into the cell plasma membrane during stimulation such as death receptors action. This sortilin in lysosomes importantly initiates the movement of lysosomes and promotes their fusion (Bao et al. 2010a,b). This 95 kDa glycoprotein has a Vps10p domain in the luminal region, which serves as a binding site for the saposin-like motif of ASMase. Its cytoplasmic tail containing an acidic cluster-dileucine motif

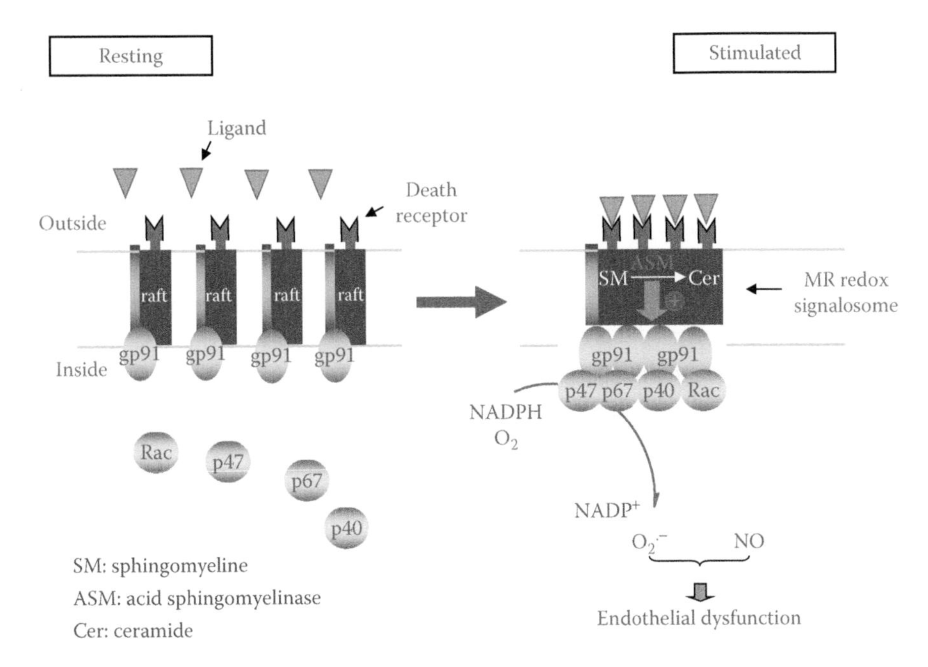

FIGURE 13.2 A working model of MR redox signalosomes. Under resting conditions, individual MR with attached death receptors including Fas and endostatin receptors are present in cell plasma membrane (resting). These MRs are dynamic microdomains and carry several membrane-bound or attached proteins or enzymes such as G-proteins, protein kinases or subunits of NADPH oxidase (gp91*phox*) (shown as gp91). When the ligands or death factors bind to their receptors on individual MR, ASMases are activated to produce ceramide from SM. Ceramide activates MRs clustering on the cell membrane to form a number of MR redox signalling platforms or signalosomes, in which ASMase, NADPH oxidase subunits such as gp91*phox* and p47*phox* (p47) and other proteins are aggregated and activated, resulting in a prominent amplification of the transmembrane signal. The activation of NADPH oxidase produces $O_2^{\bullet-}$, leading to NO decrease and consequent endothelial dysfunction.

is able to bind the monomeric adaptor protein GGA. These structural features determine sortilin to serve as an intracellular protein transporter, which mediates the sorting of soluble hydrolases such as ASMase to lysosomes. We have demonstrated in vascular ECs that sortilin functionally interacts with ASMase in addition to its action to target ASMase to lysosomes (Bao et al. 2010a), and this interaction of sortilin and ASMase promotes the movement of lysosomes towards the cell membrane, ultimately activating ASMase to produce ceramide leading to MR clustering and formation of redox signalosomes.

Recent studies have also shown how lysosomes fuse to cell membrane and transport ASMase into MR platforms. Among different factors regulating vesicle trafficking, SNARE (soluble N-ethylmaleimide-sensitive factor attachment protein receptor)-centred exocytic machinery has been reported to contribute to MR clustering to form redox signalosomes (Zhang and Li 2010). As a superfamily of small, mostly membrane-anchored proteins that mediate membrane fusion between organelles or from organelles to cell plasma membranes (Gerst 1999), SNARE mediates membrane fusion as a key contributor in the secretory pathway of various eukaryotic cells, which is named as SNARE or SNARE-centred exocytic machinery (Blank et al. 2002). Using different approaches such as confocal microscopic detection of FRET between molecules, gene silencing and co-immunoprecipitation, we have demonstrated that SNARE as a membrane fusion facilitator is also present in MR redox signalosomes, which may help recognize lysosomes containing ASMase and direct them to fuse with MRs (Zhang and Li 2010).

13.3.3 Role of Cytoskeleton in Molecular Trafficking and Recruitment

Given the enrichment of cytoskeleton in MRs, the relationship between cytoskeletal elements and MRs as well as the role of cytoskeleton in the MRs clustering or the formation of MR redox signalosomes has been recently addressed by some studies. Tubulin, a major component of cytoskeleton, is present in MRs co-localized with caveolin-1 in some tissues such as rat forebrain extracts (Dremina et al. 2005). It has been shown that the microtubules in vascular smooth muscle cells are stabilized by coexisting caveolins, which is due to interference with the interaction between the microtubule-destabilizing protein stathmin and tubulin. In addition, there are evidences that microtubules are linked to MRs in glias or cardiac myocytes, and treatment of these cells with microtubule-disrupting agents such as colchicines resulted in the loss of many signalling molecules from MRs, such as those associated with adrenergic receptor signalling (Donati and Rasenick 2005; Head et al. 2006). Another cytoskeleton component, actin, has also been reported to be involved in the MR clustering or related signalling. It is able to bind to phosphoinositide lipids such as PtdIns(4,5) P2 and PtdIns(3,4)P2 that are known to direct actin assembly into filaments, leading to the accumulation of these lipids in the MRs (Caroni 2001; Pollard and Borisy 2003). Moreover, actin also helps molecular trafficking and molecular recruitment into MRs to form signalosomes (Jaksits et al. 2004; Plowman et al. 2005).

13.3.4 Molecular Interactions and Signal Amplification in MR Redox Signalosomes

The molecular interactions are very common between aggregated molecules in the MR redox signalosomes. Such molecular interaction is responsible for signal transduction because the major function of MR redox signalling platforms is to transmit and amplify signals received by corresponding receptors by aggregation or recruitment of signalling molecules. One of such interaction mechanisms in MR redox signalosomes has feedforward amplifying action to enhance the formation of MR clustering and consequent novel signalosomes, which serves as a perpetual mechanism for redox signalling in the cell membrane. This mechanism is based on the production of ROS within MR redox signalosomes, which not only transduce signals to downstream molecules but also enhance the formation of more signalosomes, forming an endless ROS production and signalosomes formation, which may result in cell injury or dysfunction. This is because ROS activate ASMase due to dimer formation. This modification of ASMase by ROS relates to oxidation of the free C-terminal cysteine (Qiu et al. 2003). The ASMase activation has also been shown by exogenous administration of xanthine/xanthine oxidase, a $O_2^{\bullet-}$ generating system, resulting in a dramatic increase in MRs clustering in the cell plasma membrane

(Qiu et al. 2003; Zhang et al. 2007), suggesting that the oxidative action on ASMase is a critical mechanism of such perpetual MRs redox signalosomes. In addition, ROS may also produce lipid oxidation to form lipid peroxides, which promote the formation of MR clustering or signalosomes on the cell membrane (Ayuyan and Cohen 2006). It is concluded that ROS from MR signalosomes can act on the formation mechanism of signalosomes in a feedforward manner (Lu et al. 2007), thereby amplifying or enhancing cellular signalling and even resulting in multiple oxidative stress related diseases, if such signalosomes are excessively enhanced or amplified.

13.4 MR REDOX SIGNALOSOMES—PLATFORM RESPONDING TO DANGER SIGNALS

Although the MR redox signalosomes participate in the regulation of many cellular activities and tissue or organ functions, they may be more importantly involved in the response of different cells to a variety of danger signals (Gallucci and Matzinger 2001; Li et al. 2007, 2010). Both exogenous danger signals elaborated by pathogens and endogenous danger signals released by tissues undergoing stress, damage or abnormal death may stimulate the formation and activation of MR redox signalosomes, which produces necessary or appropriate reactions to protect cells from the injury or to maintain the structural and functional integrity of cells. Several cell responses or regulation associated with MR redox signalosomes are summarized in the following sections.

13.4.1 Response to Infection and PAMPs

In the innate immune or host defence response to microbial invasion, neutrophils play an essential role due to their phagocytosis of the pathogen and consequent release of free radicals, which kill a large number of the invading microorganisms and/or pathogens. It has been reported that the free radicals or $O_2^{\bullet-}$ are mainly generated by NADPH oxidase (Rosen et al. 1995), which are dependent on the formation of MR redox signalosomes centred by NADPH oxidase assembling or activation. In patients with myelodysplasia who always suffer from multiple types of infection, ROS production in their neutrophils is deficient because of the lack of Lyn, gp91*phox* and p22*phox* in MR fractions and down-regulation of plasma membrane expression of MR components (Fuhler et al. 2007). These results indicate that the host defence response to microbial invasion mediated by neutrophils or other cell types may be through MR clustering, MR redox signalosome formation and activation and consequent ROS production. Without this mechanism mediated by MR redox signalosomes, there will be no onset of phagocytic respiratory bursts in neutrophils or other phagocytes (Fuhler et al. 2007).

On the other hand, many studies on pathogen-receptor interactions have suggested that MR clustering and associated internalization of pathogens including bacteria, viruses, parasites and even fungi may facilitate their invasion into cells (Manes et al. 2003; Simons and Ehehalt 2002). In this regard, ceramide-enriched MRs or platforms have been intensively studied. The results have indicated that these MR platforms mediate the infection of mammalian cells with *Pseudomonas aeruginosa* (Grassme et al. 2003), *Staphylococcus aureus* (Esen et al. 2001), *Neisseriae gonorrhoeae* (Grassme et al. 1997; Hauck et al. 2000),

Rhinoviruses (Grassme et al. 2005) and Sindbis virus (Jan et al. 2000). Although the precise role of ROS from MR redox signalosomes in killing these microorganisms (Rikihisa 2010) remains unclear, the fate of microbial invasion under certain conditions is certainly dependent upon the local redox environment, which is provided by MR redox signalosomes (Gamalei et al. 2006).

13.4.2 Signalling of Death Receptor Activation

In many types of mammalian cells, a redox signalling response is usually produced to stimulation of death receptor ligands or activators, which are a group of endogenous danger signals. In many studies (Zhang et al. 2006), death receptor ligands were found to increase $O_2^{\bullet-}$ production mediated by NADPH oxidase activation (Jung et al. 2004). This NADPH oxidase-derived $O_2^{\bullet-}$ often contributes to more than 95% of $O_2^{\bullet-}$ production in ECs and some other cell types with cellular death factors (Gorlach et al. 2000; Jung et al. 2004). The clustering of MRs and consequent formation of MR redox signalosomes may be a primary mechanism to couple death receptor activation to NADPH oxidase-derived $O_2^{\bullet-}$ production. Many death receptor ligands or activators were found to stimulate MR redox signalosomes formation such as Fas ligand, TNF-α, endostatin, H_2O_2, homocysteine, 7-dehydrocholesterol, vasifatin, platelet aggregation factor and ASMase activators (Li et al. 2007; Zhang and Li 2010). Based on intensive studies by different groups of scientists, a general model of this MR redox signalosomes during death receptor activation is proposed. Under resting conditions, individual MRs or ceramide is diffusively present in the plasma membrane of different cells such as ECs, macrophages, kidney podocytes and others. These MRs are directly linked to some NADPH oxidase subunits such as gp91*phox* and p22*phox* or in proximity to them. When these cells are stimulated by different agonists or other danger signals with or without receptors binding, the MRs clustering is activated to form MR macrodomains or platforms. Many signalling molecules aggregate in these MR macrodomains or platforms including all NADPH oxidase subunits and other cofactors or activators such as Rac GTPase, ASMase, ceramide, sortilin, SNARE and others. Via molecular interactions, NADPH oxidase subunits are assembled into an enzyme complex and activated to produce $O_2^{\bullet-}$, forming a signalosome to transfer extracellular signals to intracellular effectors (Bao et al. 2010b; Jin et al. 2008a; Samhan-Arias et al. 2009; Zhang and Li 2010). $O_2^{\bullet-}$ may act with NO to increase ONOO- levels and thereby reduces the bioavailability of NO, leading to cell dysfunction. $O_2^{\bullet-}$ may also be converted into H_2O_2 by SOD, which provides redox signals to a wide range of cells or their extracellular or intracellular effectors influencing cell or organ functions extensively (Dikalova et al. 2010; Fukai et al. 2002; Gongora et al. 2008; Saitoh et al. 2006).

13.4.3 Triggering Inflammasomes

Recent studies have indicated that the nucleotide-binding oligomerization domain (Nod)-like receptor containing pyrin domain 3 (NLRP3, also known as NALP3) inflammasome is a key mechanism mediating the activation of the innate immune system to produce inflammatory response to different danger signals including PAMPs and DAMPs (Bakker et al. 2014; Martinon et al. 2004; O'Connor et al. 2003; Petrilli and Martinon 2007;

Yin et al. 2013). NLRP3 inflammasome activation may also directly trigger other cell injury responses through non-inflammatory actions such as pyroptosis, glycolysis, disturbance of lipid metabolism, cell survival and cell proliferation (Henao-Mejia et al. 2012; Lamkanfi 2011; Stienstra et al. 2010; Yang et al. 2014). Since NLRP3 inflammasome activation leads to cleavage of pro-caspase-1 to active form of caspase-1, which is shown to have more than 121 substrates, the role of NLRP3 inflammasome activation in the regulation of cell or organ function and related pathogenic relevance to different diseases may be far beyond the canonical inflammatory responses (Denes et al. 2012). It is now becoming a highly interesting topic to explore the mechanisms by which NLRP3 inflammasomes are activated, thereby producing both inflammatory and non-inflammatory responses in different cell type or organs, which may be relevant to the development of different human diseases and identification of new therapeutic targets for treatment of related diseases (Martinon and Tschopp 2005; Sutterwala et al. 2007).

With rapidly growing and extensive studies, several underlying mechanisms are proposed to mediate the activation of NLRP3 inflammasomes in response to diverse danger signals. There is considerable evidence showing that ROS, in particular, NADPH oxidase-derived ROS, play a critical role in the formation or activation of NLRP3 inflammasome in response to different stimuli such as ATP, asbestos and silica (Cruz et al. 2007; Dostert et al. 2008; Hewinson et al. 2008). Deletion of NADPH oxidase subunit $p22^{phox}$ or use of general ROS scavengers like *N*-acetylcysteine and antioxidant ammonium pyrrolidine dithiocarbamate was found to inhibit NLRP3 inflammasome activation and IL-1β release in monocyte THP-1 cells upon various stimulations (Dostert et al. 2008). More recently, we first demonstrated that NLRP3 inflammasome activation is importantly attributed to the pathogenesis of hyperhomocysteinemia (hHcys)-induced glomerular sclerosis (Abais et al. 2013; Zhang et al. 2012). Such role of NLRP3 inflammasome was confirmed by the results that locally *ASC* gene knocking down in the kidney significantly reduced NLRP3 inflammasome activation and IL-1β production, accompanied by decreased albuminuria, less foot process effacement of podocytes and reduced loss of podocyte slit diaphragm molecules in glomeruli of mice with hHcys (Abais et al. 2013; Zhang et al. 2012). NLRP3 inflammasomes were also shown to be activated in vascular ECs or carotid arterial wall by the adipokine visfatin, indicating that activation of NLRP3 inflammasomes turns on the inflammatory response in ECs and ultimately results in atherogenesis in mice during early stage obesity (Xia et al. 2014). This action of visfatin to activate NLRP3 inflammasome is due to activation of NADPH oxidase (Xia et al. 2011, 2014). Based on these results and previous reports on ceramide-enriched MR signalosomes in NADPH oxidase-dependent $O_2^{\bullet-}$ production (Abais et al. 2013; Boini et al. 2014; Xia et al. 2014; Zhang et al. 2012), a working model has been proposed to define the mechanisms mediating ROS production to instigate NLRP3 inflammasomes. It is believed that MR redox signalosomes is essential for the activation of NADPH oxidase providing $O_2^{\bullet-}$ to trigger NLRP3 inflammasome activation (Jin et al. 2011).

Further studies have shown that NLRP3 senses oxidative stress to activate the formation of inflammasome possibly via its interaction thioredoxin-interacting protein (TXNIP) and mitochondrial anti-viral signalling protein (MAVS). As shown by Zhou et al. (2010), TXNIP, the

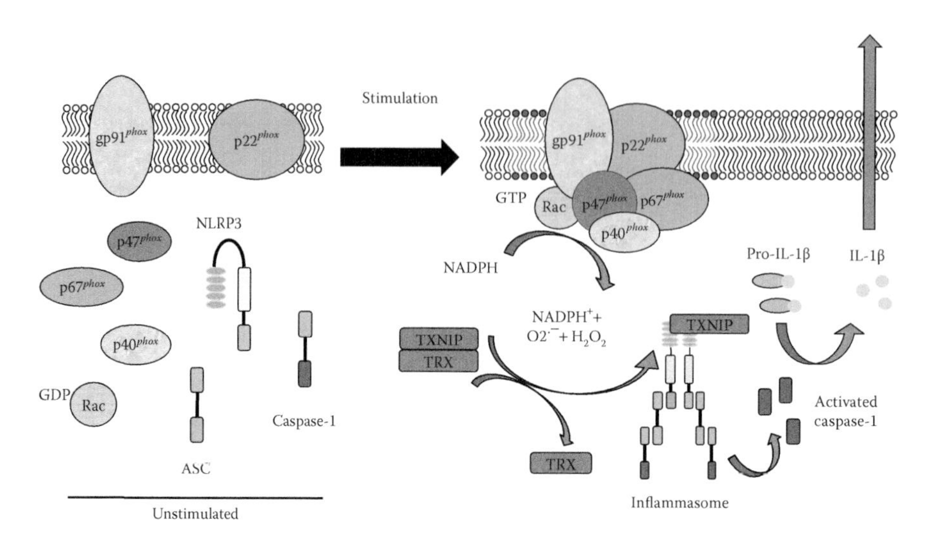

FIGURE 13.3 **(See colour insert.)** Activation of NLRP3 inflammasomes linking to MR redox signalosomes formation. NLRP3 inflammasomes are activated by NADPH oxidase-derived ROS through lipid raft clustering. ROS derived from MR redox signalosomes activate NLRP3 inflammasomes through the binding of thioredoxin-interacting protein (TXNIP) to NLRP3. TXNIP is a binding partner to NLRP3, which dissociates from TRX and then binds with NLRP3 when ROS is increased.

negative regulator of the anti-oxidant thioredoxin (TRX) is able to dissociate from TRX and then bind to NLRP3 upon stimulation, which leads to inflammasome formation and activation. This interaction of TXNIP and NLRP3 to activate inflammasomes has been confirmed in our studies (Abais et al. 2014) and by others (Chen et al. 2009; El-Azab et al. 2014; Mohamed et al. 2014; Wang et al. 2013). It is suggested that the formation and activation of MR redox signalosomes produce ROS, which dissociate TXNIP from TRX to bind with NLRP3, leading to the activation of inflammasomes and cell injury either through the inflammatory response or through direct injurious effects of inflammasome products. This model of NLRP3 inflammasome activation is summarized in Figure 13.3. With respect to the role of MAVS in mediating redox activation of NLRP3, this mitochondrial protein also associates with NLRP3 to facilitate its oligomerization with ASC and caspase-1 (Park et al. 2013) and recruitment of NLRP3 to mitochondria in response to mitochondrial ROS increase (Subramanian et al. 2013). This MAVS-mediated NLRP3 inflammasome action remains controversial.

13.5 CONCLUDING REMARKS

Recent studies have indicated that MR redox signalosomes can be formed in a variety of mammalian cells, which serve as platforms to mediate several important signalling pathways related to redox regulation of cell functions. It is the MRs clustering that provides the driving force to aggregate and assemble NADPH oxidase subunits into an enzymatic complex in the cell membrane, where $O_2^{\bullet-}$ is produced to conduct redox signalling with appropriate compartmentalization and amplifications upon stimulations. Different types of such NADPH oxidase-centred MR redox signalosomes may work in a spatiotemporal way to conduct redox signalling and thereby fine-regulate cell or organ function. Given the extensive studies on the downstream targets of MR redox signalosomes, a focus on their role in response to danger signals may link this important signalling mechanism to the activation of NRLP3 inflammasomes and consequent inflammatory response or action, which are considered as a common route of and root of many diseases. It is believed that understanding the MR redox signalosomes and their downstream targets will help clarify pathogenesis of diseases associated with inflammasome activation or inflammatory response. The knowledge of MR redox signalosomes may help develop more effective therapeutic strategies for intervention of early steps of different diseases.

ACKNOWLEDGEMENT

The authors thank the National Institutes of Health for providing grants to support most of works cited in this chapter (HL075316, HL057244, HL122937).

REFERENCES

Abais JM, Xia M, Li G et al. Nod-like Receptor Protein 3 (NLRP3) inflammasome activation and podocyte injury via Thioredoxin-Interacting Protein (TXNIP) during hyperhomocysteinemia. *The Journal of Biological Chemistry* **289**(2014): 27159–27168.

Abais JM, Zhang C, Xia M et al. NADPH oxidase-mediated triggering of inflammasome activation in mouse podocytes and glomeruli during hyperhomocysteinemia. *Antioxidants and Redox Signaling* **18**(2013): 1537–1548.

Alonso MA, Millan J. The role of lipid rafts in signalling and membrane trafficking in T lymphocytes. *Journal of Cell Science* **114**(2001): 3957–3965.

Anderson RG, Jacobson K. A role for lipid shells in targeting proteins to caveolae, rafts, and other lipid domains. *Science* **296**(2002): 1821–1825.

Ayuyan AG, Cohen FS. Lipid peroxides promote large rafts: Effects of excitation of probes in fluorescence microscopy and electrochemical reactions during vesicle formation. *Biophysical Journal* **91**(2006): 2172–2183.

Bakker PJ, Butter LM, Kors L et al. Nlrp3 is a key modulator of diet-induced nephropathy and renal cholesterol accumulation. *Kidney International* **85**(2014): 1112–1122.

Bao JX, Jin S, Zhang F, Wang ZC, Li N, Li PL. Activation of membrane NADPH oxidase associated with lysosome-targeted acid sphingomyelinase in coronary endothelial cells. *Antioxidants and Redox Signaling* **12**(2010a): 703–712.

Bao JX, Xia M, Poklis JL, Han WQ, Brimson C, Li PL. Triggering role of acid sphingomyelinase in endothelial lysosome-membrane fusion and dysfunction in coronary arteries. *The American Journal of Physiology—Heart and Circulatory Physiology* **298**(2010b): H992–H1002.

Becker KA, Gellhaus A, Winterhager E, Gulbins E. Ceramide-enriched membrane domains in infectious biology and development. *Subcellular Biochemistry* **49**(2008): 523–538.

Blank U, Cyprien B, Martin-Verdeaux S et al. SNAREs and associated regulators in the control of exocytosis in the RBL-2H3 mast cell line. *Molecular Immunology* **38**(2002): 1341–1345.

Boini KM, Xia M, Abais JM et al. Activation of inflammasomes in podocyte injury of mice on the high fat diet: Effects of ASC gene deletion and silencing. *Biochimica et Biophysica Acta* **1843**(2014): 836–845.

Bollinger CR, Teichgraber V, Gulbins E. Ceramide-enriched membrane domains. *Biochimica et Biophysica Acta* **1746**(2005): 284–294.

Boniface JJ, Rabinowitz JD, Wulfing C et al. Initiation of signal transduction through the T cell receptor requires the multivalent engagement of peptide/MHC ligands [corrected]. *Immunity* **9**(1998): 459–466.

Brown DA, London E. Functions of lipid rafts in biological membranes. *Annual Review of Cell and Developmental Biology* **14**(1998): 111–136.
Brown DA, London E. Structure and function of sphingolipid- and cholesterol-rich membrane rafts. *The Journal of Biological Chemistry* **275**(2000): 17221–17224.
Caroni P. New EMBO members' review: Actin cytoskeleton regulation through modulation of PI(4,5)P(2) rafts. *The EMBO Journal* **20**(2001): 4332–4336.
Chen J, Cha-Molstad H, Szabo A, Shalev A. Diabetes induces and calcium channel blockers prevent cardiac expression of proapoptotic thioredoxin-interacting protein. *The American Journal of Physiology—Endocrinology and Metabolism* **296**(2009): E1133–E1139.
Cheng PC, Dykstra ML, Mitchell RN, Pierce SK. A role for lipid rafts in B cell antigen receptor signaling and antigen targeting. *The Journal of Experimental Medicine* **190**(1999): 1549–1560.
Cremesti AE, Goni FM, Kolesnick R. Role of sphingomyelinase and ceramide in modulating rafts: Do biophysical properties determine biologic outcome? *FEBS Letters* **531**(2002): 47–53.
Cruz CM, Rinna A, Forman HJ, Ventura AL, Persechini PM, Ojcius DM. ATP activates a reactive oxygen species-dependent oxidative stress response and secretion of proinflammatory cytokines in macrophages. *The Journal of Biological Chemistry* **282**(2007): 2871–2879.
Cutler RG, Mattson MP. Sphingomyelin and ceramide as regulators of development and lifespan. *Mechanisms of Ageing and Development* **122**(2001): 895–908.
Denes A, Lopez-Castejon G, Brough D. Caspase-1: Is IL-1 just the tip of the ICEberg? *Cell Death and Disease* **3**(2012): e338.
Dikalova AE, Gongora MC, Harrison DG, Lambeth JD, Dikalov S, Griendling KK. Upregulation of Nox1 in vascular smooth muscle leads to impaired endothelium-dependent relaxation via enos uncoupling. *American Journal of Physiology—Heart and Circulatory* **299**(2010): H673–679.
Donati RJ, Rasenick MM. Chronic antidepressant treatment prevents accumulation of gsalpha in cholesterol-rich, cytoskeletal-associated, plasma membrane domains (lipid rafts). *Neuropsychopharmacology* **30**(2005): 1238–1245.
Dostert C, Petrilli V, Van Bruggen R, Steele C, Mossman BT, Tschopp J. Innate immune activation through Nalp3 inflammasome sensing of asbestos and silica. *Science* **320**(2008): 674–677.
Draper JM, Xia Z, Smith CD. Cellular palmitoylation and trafficking of lipidated peptides. *Journal of Lipid Research* **48**(2007): 1873–1884.
Dremina ES, Sharov VS, Schoneich C. Protein tyrosine nitration in rat brain is associated with raft proteins, flotillin-1 and alpha-tubulin: Effect of biological aging. *Journal of Neurochemistry* **93**(2005): 1262–1271.
Dumitru CA, Zhang Y, Li X, Gulbins E. Ceramide: A novel player in reactive oxygen species-induced signaling? *Antioxidants and Redox Signaling* **9**(2007): 1535–1540.
El-Azab MF, Baldowski BR, Mysona BA et al. Deletion of thioredoxin-interacting protein preserves retinal neuronal function by preventing inflammation and vascular injury. *British Journal of Pharmacology* **171**(2014): 1299–1313.
Esen M, Schreiner B, Jendrossek V et al. Mechanisms of Staphylococcus aureus induced apoptosis of human endothelial cells. *Apoptosis* **6**(2001): 431–439.
Foster LJ, De Hoog CL, Mann M. Unbiased quantitative proteomics of lipid rafts reveals high specificity for signaling factors. *Proceedings of the National Academy of Sciences of the United States of America* **100**(2003): 5813–5818.
Frank PG, Woodman SE, Park DS, Lisanti MP. Caveolin, caveolae, and endothelial cell function. *Arteriosclerosis, Thrombosis, and Vascular Biology* **23**(2003): 1161–1168.
Fuhler GM, Blom NR, Coffer PJ, Drayer AL, Vellenga E. The reduced GM-CSF priming of ROS production in granulocytes from patients with myelodysplasia is associated with an impaired lipid raft formation. *Journal of Leukocyte Biology* **81**(2007): 449–457.
Fukai T, Folz RJ, Landmesser U, Harrison DG. Extracellular superoxide dismutase and cardiovascular disease. *Cardiovascular Research* **55**(2002): 239–249.

Gallucci S, Matzinger P. Danger signals: SOS to the immune system. *Current Opinion in Immunology* **13**(2001): 114–119.

Gamalei IA, Efremova TN, Kirpichnikova KM et al. Decreased sensitivity of transformed 3T3-SV40 cells treated with N-acetylcysteine to bacterial invasion. *Bulletin of Experimental Biology and Medicine* **142**(2006): 90–93.

Garcia-Cardena G, Oh P, Liu J, Schnitzer JE, Sessa WC. Targeting of nitric oxide synthase to endothelial cell caveolae via palmitoylation: Implications for nitric oxide signaling. *Proceedings of the National Academy of Sciences of the United States of America* **93**(1996): 6448–6453.

Gerst JE. SNAREs and SNARE regulators in membrane fusion and exocytosis. *Cellular and Molecular Life Sciences* **55**(1999): 707–734.

Goligorsky MS, Li H, Brodsky S, Chen J. Relationships between caveolae and eNOS: Everything in proximity and the proximity of everything. *American Journal of Physiology* **283**(2002): F1–F10.

Gongora MC, Lob HE, Landmesser U et al. Loss of extracellular superoxide dismutase leads to acute lung damage in the presence of ambient air: A potential mechanism underlying adult respiratory distress syndrome. *The American Journal of Pathology* **173**(2008): 915–926.

Goni FM, Alonso A. Sphingomyelinases: Enzymology and membrane activity. *FEBS Letters* **531**(2002): 38–46.

Gorlach A, Brandes RP, Nguyen K, Amidi M, Dehghani F, Busse R. A gp91phox containing NADPH oxidase selectively expressed in endothelial cells is a major source of oxygen radical generation in the arterial wall. *Circulation Research* **87**(2000): 26–32.

Grassme H, Gulbins E, Brenner B et al. Acidic sphingomyelinase mediates entry of N. gonorrhoeae into nonphagocytic cells. *Cell* **91**(1997): 605–615.

Grassme H, Jekle A, Riehle A et al. CD95 signaling via ceramide-rich membrane rafts. *The Journal of Biological Chemistry* **276**(2001): 20589–20596.

Grassme H, Jendrossek V, Bock J, Riehle A, Gulbins E. Ceramide-rich membrane rafts mediate CD40 clustering. *Journal of Immunology* **168**(2002): 298–307.

Grassme H, Jendrossek V, Riehle A et al. Host defense against Pseudomonas aeruginosa requires ceramide-rich membrane rafts. *Nature Medicine* **9**(2003): 322–330.

Grassme H, Riehle A, Wilker B, Gulbins E. Rhinoviruses infect human epithelial cells via ceramide-enriched membrane platforms. *The Journal of Biological Chemistry* **280**(2005): 26256–26262.

Gulbins E. Highlight: sphingolipids—signals and disease. *Biological Chemistry* **389**(2008): 1347–1348.

Gulbins E, Grassme H. Ceramide and cell death receptor clustering. *Biochimica et Biophysica Acta* **1585**(2002): 139–145.

Gulbins E, Kolesnick R. Raft ceramide in molecular medicine. *Oncogene* **22**(2003): 7070–7077.

Gulbins E, Li PL. Physiological and pathophysiological aspects of ceramide. *American Journal of Physiology. Regulatory, Integrative and Comparative Physiology* **290**(2006): R11–R26.

Hauck CR, Grassme H, Bock J et al. Acid sphingomyelinase is involved in CEACAM receptor-mediated phagocytosis of Neisseria gonorrhoeae. *FEBS Letters* **478**(2000): 260–266.

Head BP, Patel HH, Roth DM et al. Microtubules and actin microfilaments regulate lipid raft/caveolae localization of adenylyl cyclase signaling components. *The Journal of Biological Chemistry* **281**(2006): 26391–26399.

Helms JB, Zurzolo C. Lipids as targeting signals: Lipid rafts and intracellular trafficking. *Traffic* **5**(2004): 247–254.

Henao-Mejia J, Elinav E, Strowig T, Flavell RA. Inflammasomes: Far beyond inflammation. *Nature Immunology* **13**(2012): 321–324.

Hewinson J, Moore SF, Glover C, Watts AG, MacKenzie AB. A key role for redox signaling in rapid P2X7 receptor-induced IL-1 beta processing in human monocytes. *Journal of Immunology* **180**(2008): 8410–8420.

Hilenski LL, Clempus RE, Quinn MT, Lambeth JD, Griendling KK. Distinct subcellular localizations of Nox1 and Nox4 in vascular smooth muscle cells. *Arteriosclerosis, Thrombosis, and Vascular Biology* **24**(2004): 677–683.

Hoekstra D, Maier O, van der Wouden JM, Slimane TA, van ISC. Membrane dynamics and cell polarity: The role of sphingolipids. *Journal of Lipid Research* **44**(2003): 869–877.

Israelachvili JN. Refinement of the fluid-mosaic model of membrane structure. *Biochimica et Biophysica Acta* **469**(1977): 221–225.

Jain MK, White HB, 3rd. Long-range order in biomembranes. *Advances in Lipid Research* **15**(1977): 1–60.

Jaiswal JK, Andrews NW, Simon SM. Membrane proximal lysosomes are the major vesicles responsible for calcium-dependent exocytosis in nonsecretory cells. *The Journal of Cell Biology* **159**(2002): 625–635.

Jaksits S, Bauer W, Kriehuber E et al. Lipid raft-associated GTPase signaling controls morphology and CD8 | T cell stimulatory capacity of human dendritic cells. *Journal of Immunology* **173**(2004): 1628–1639.

Jan JT, Chatterjee S, Griffin DE. Sindbis virus entry into cells triggers apoptosis by activating sphingomyelinase, leading to the release of ceramide. *Journal of Virology* **74**(2000): 6425–6432.

Jin S, Yi F, Li PL. Contribution of lysosomal vesicles to the formation of lipid raft redox signaling platforms in endothelial cells. *Antioxidants and Redox Signaling* **9**(2007): 1417–1426.

Jin S, Yi F, Zhang F, Poklis JL, Li PL. Lysosomal targeting and trafficking of acid sphingomyelinase to lipid raft platforms in coronary endothelial cells. *Arteriosclerosis, Thrombosis, and Vascular Biology* **28**(2008a): 2056–2062.

Jin S, Zhang Y, Yi F, Li PL. Critical role of lipid raft redox signaling platforms in endostatin-induced coronary endothelial dysfunction. *Arteriosclerosis, Thrombosis, and Vascular Biology* **28**(2008b): 485–490.

Jin S, Zhou F. Lipid raft redox signaling platforms in vascular dysfunction: Features and mechanisms. *Current Atherosclerosis Reports* **11**(2009): 220–226.

Jin S, Zhou F, Katirai F, Li PL. Lipid raft redox signaling: Molecular mechanisms in health and disease. *Antioxidants and Redox Signaling* **15**(2011): 1043–1083.

Jung O, Schreiber JG, Geiger H, Pedrazzini T, Busse R, Brandes RP. gp91phox-containing NADPH oxidase mediates endothelial dysfunction in renovascular hypertension. *Circulation* **109**(2004): 1795–1801.

Kitatani K, Idkowiak-Baldys J, Hannun YA. The sphingolipid salvage pathway in ceramide metabolism and signaling. *Cell Signaling* **20**(2008): 1010–1018.

Lamkanfi M. Emerging inflammasome effector mechanisms. *Nature Reviews Immunology* **11**(2011): 213–220.

Lang F, Gulbins E, Lang PA, Zappulla D, Foller M. Ceramide in suicidal death of erythrocytes. *Cellular Physiology and Biochemistry* **26**(2010): 21–28.

Lang F, Lang KS, Lang PA, Huber SM, Wieder T. Mechanisms and significance of eryptosis. *Antioxidants and Redox Signaling* **8**(2006): 1183–1192.

Levy M, Futerman AH. Mammalian ceramide synthases. *IUBMB Life* **62**(2010): 347–356.

Li PL, Zhang Y, Yi F. Lipid raft redox signaling platforms in endothelial dysfunction. *Antioxidants and Redox Signaling* **9**(2007): 1457–1470.

Li X, Becker KA, Zhang Y. Ceramide in redox signaling and cardiovascular diseases. *Cellular Physiology and Biochemistry* **26**(2010): 41–48.

Linder ME, Deschenes RJ. Palmitoylation: Policing protein stability and traffic. *Nature Reviews Molecular Cell Biology* **8**(2007): 74–84.

Lingwood D, Kaiser HJ, Levental I, Simons K. Lipid rafts as functional heterogeneity in cell membranes. *Biochemical Society Transactions* **37**(2009): 955–960.

Lingwood D, Simons K. Lipid rafts as a membrane-organizing principle. *Science* **327**(2010): 46–50.

Lotocki G, Alonso OF, Dietrich WD, Keane RW. Tumor necrosis factor receptor 1 and its signaling intermediates are recruited to lipid rafts in the traumatized brain. *Journal of Neuroscience* **24**(2004): 11010–11016.

Lu SP, Lin Feng MH, Huang HL, Huang YC, Tsou WI, Lai MZ. Reactive oxygen species promote raft formation in T lymphocytes. *Free Radical Biology and Medicine* **42**(2007): 936–944.

Magee AI, Parmryd I. Detergent-resistant membranes and the protein composition of lipid rafts. *Genome Biology* **4**(2003): 234.

Manes S, del Real G, Martinez AC. Pathogens: Raft hijackers. *Nature Reviews Immunology* **3**(2003): 557–568.

Martinon F, Agostini L, Meylan E, Tschopp J. Identification of bacterial muramyl dipeptide as activator of the NALP3/cryopyrin inflammasome. *Current Biology* **14**(2004): 1929–1934.

Martinon F, Tschopp J. NLRs join TLRs as innate sensors of pathogens. *Trends in Immunology* **26**(2005): 447–454.

Mattjus P, Slotte JP. Does cholesterol discriminate between sphingomyelin and phosphatidylcholine in mixed monolayers containing both phospholipids? *Chemistry and Physics of Lipids* **81**(1996): 69–80.

Mohamed IN, Hafez SS, Fairaq A, Ergul A, Imig JD, El-Remessy AB. Thioredoxin-interacting protein is required for endothelial NLRP3 inflammasome activation and cell death in a rat model of high-fat diet. *Diabetologia* **57**(2014): 413–423.

Munday AD, Lopez JA. Posttranslational protein palmitoylation: promoting platelet purpose. *Arteriosclerosis, Thrombosis, and Vascular Biology* **27**(2007): 1496–1499.

Murate T, Suzuki M, Hattori M et al. Up-regulation of acid sphingomyelinase during retinoic acid-induced myeloid differentiation of NB4, a human acute promyelocytic leukemia cell line. *The Journal of Biological Chemistry* **277**(2002): 9936–9943.

Nicolau DV, Jr., Burrage K, Parton RG, Hancock JF. Identifying optimal lipid raft characteristics required to promote nanoscale protein-protein interactions on the plasma membrane. *Molecular and Cellular Biology* **26**(2006): 313–323.

Oakley FD, Abbott D, Li Q, Engelhardt JF. Signaling components of redox active endosomes: the redoxosomes. *Antioxidants and Redox Signaling* **11**(2009): 1313–1333.

O'Connor W, Jr., Harton JA, Zhu X, Linhoff MW, Ting JP. Cutting edge: CIAS1/cryopyrin/PYPAF1/NALP3/CATERPILLER 1.1 is an inducible inflammatory mediator with NF-kappa B suppressive properties. *Journal of Immunology* **171**(2003): 6329–6333.

Park S, Juliana C, Hong S et al. The mitochondrial antiviral protein MAVS associates with NLRP3 and regulates its inflammasome activity. *Journal of Immunology* **191**(2013): 4358–4366.

Patterson SI. Posttranslational protein S-palmitoylation and the compartmentalization of signaling molecules in neurons. *Biological Research* **35**(2002): 139–150.

Petrilli V, Martinon F. The inflammasome, autoinflammatory diseases, and gout. *Joint Bone Spine* **74**(2007): 571–576.

Pike LJ. Rafts defined: a report on the keystone symposium on lipid rafts and cell function. *Journal of Lipid Research* **47**(2006): 1597–1598.

Plowman SJ, Muncke C, Parton RG, Hancock JF. H-ras, K-ras, and inner plasma membrane raft proteins operate in nanoclusters with differential dependence on the actin cytoskeleton. *Proceedings of the National Academy of Sciences of the United States of America* **102**(2005): 15500–15505.

Pollard TD, Borisy GG. Cellular motility driven by assembly and disassembly of actin filaments. *Cell* **112**(2003): 453–465.

Pralle A, Keller P, Florin EL, Simons K, Horber JK. Sphingolipid-cholesterol rafts diffuse as small entities in the plasma membrane of mammalian cells. *The Journal of Cell Biology* **148**(2000): 997–1008.

Pritchard KA, Ackerman AW, Ou J et al. Native low-density lipoprotein induces endothelial nitric oxide synthase dysfunction: role of heat shock protein 90 and caveolin-1. *Free Radical Biology and Medicine* **33**(2002): 52–62.

Qiu H, Edmunds T, Baker-Malcolm J et al. Activation of human acid sphingomyelinase through modification or deletion of C-terminal cysteine. *The Journal of Biological Chemistry* **278**(2003): 32744–32752.

Razani B, Engelman JA, Wang XB et al. Caveolin-1 null mice are viable but show evidence of hyperproliferative and vascular abnormalities. *The Journal of Biological Chemistry* **276**(2001): 38121–38138.

Rikihisa Y. Molecular events involved in cellular invasion by Ehrlichia chaffeensis and Anaplasma phagocytophilum. *Veterinary Parasitology* **167**(2010): 155–166.

Rinia HA, de Kruijff B. Imaging domains in model membranes with atomic force microscopy. *FEBS Letters* **504**(2001a): 194–199.

Rinia HA, Snel MM, van der Eerden JP, de Kruijff B. Visualizing detergent resistant domains in model membranes with atomic force microscopy. *FEBS Letters* **501**(2001): 92–96.

Rosen GM, Pou S, Ramos CL, Cohen MS, Britigan BE. Free radicals and phagocytic cells. *The FASEB Journal* **9**(1995): 200–209.

Saitoh S, Zhang C, Tune JD et al. Hydrogen peroxide: A feed-forward dilator that couples myocardial metabolism to coronary blood flow. *Arteriosclerosis, Thrombosis, and Vascular Biology* **26**(2006): 2614–2621.

Samhan-Arias AK, Garcia-Bereguiain MA, Martin-Romero FJ, Gutierrez-Merino C. Clustering of plasma membrane-bound cytochrome b5 reductase within "lipid raft" microdomains of the neuronal plasma membrane. *Molecular and Cellular Neurosciences* **40**(2009): 14–26.

Schlegel A, Volonte D, Engelman JA et al. Crowded little caves: structure and function of caveolae. *Cell Signaling* **10**(1998): 457–463.

Simons K, Ehehalt R. Cholesterol, lipid rafts, and disease. *The Journal of Clinical Investigation* **110**(2002): 597–603.

Simons K, Ikonen E. Functional rafts in cell membranes. *Nature* **387**(1997): 569–572.

Simons K, Ikonen E. How cells handle cholesterol. *Science* **290**(2000a): 1721–1726.

Simons K, Toomre D. Lipid rafts and signal transduction. *Nature Reviews Molecular Cell Biology* **1**(2000b): 31–39.

Sowa G, Pypaert M, Sessa WC. Distinction between signaling mechanisms in lipid rafts vs. caveolae. *Proceedings of the National Academy of Sciences of the United States of America* **98**(2001): 14072–14077.

Stan RV. Structure and function of endothelial caveolae. *Microscopy Research and Technique* **57**(2002): 350–364.

Stancevic B, Kolesnick R. Ceramide-rich platforms in transmembrane signaling. *FEBS Letters* **584**(2010): 1728–1740.

Stienstra R, Joosten LA, Koenen T et al. The inflammasome-mediated caspase-1 activation controls adipocyte differentiation and insulin sensitivity. *Cell Metabolism* **12**(2010): 593–605.

Subramanian N, Natarajan K, Clatworthy MR, Wang Z, Germain RN. The adaptor MAVS promotes NLRP3 mitochondrial localization and inflammasome activation. *Cell* **153**(2013): 348–361.

Sutterwala FS, Ogura Y, Flavell RA. The inflammasome in pathogen recognition and inflammation. *Journal of Leukocyte Biology* **82**(2007): 259–264.

Szanto I, Rubbia-Brandt L, Kiss P et al. Expression of NOX1, a superoxide-generating NADPH oxidase, in colon cancer and inflammatory bowel disease. *The Journal of Pathology* **207**(2005): 164–176.

Tani M, Ito M, Igarashi Y. Ceramide/sphingosine/sphingosine 1-phosphate metabolism on the cell surface and in the extracellular space. *Cell Signaling* **19**(2007): 229–237.

Ushio-Fukai M, Alexander RW. Reactive oxygen species as mediators of angiogenesis signaling: Role of NAD(P)H oxidase. *Molecular and Cellular Biochemistry* **264**(2004): 85–97.

Utermohlen O, Herz J, Schramm M, Kronke M. Fusogenicity of membranes: The impact of acid sphingomyelinase on innate immune responses. *Immunobiology* **213**(2008): 307–314.

Vetrivel KS, Cheng H, Lin W et al. Association of gamma-secretase with lipid rafts in post-Golgi and endosome membranes. *The Journal of Biological Chemistry* **279**(2004): 44945–44954.

Wang W, Wang C, Ding XQ et al. Quercetin and allopurinol reduce liver thioredoxin-interacting protein to alleviate inflammation and lipid accumulation in diabetic rats. *British Journal of Pharmacology* **169**(2013): 1352–1371.

Waugh MG, Minogue S, Anderson JS, dos Santos M, Hsuan JJ. Signalling and non-caveolar rafts. *Biochemical Society Transactions* **29**(2001): 509–511.

Xia M, Boini KM, Abais JM, Xu M, Zhang Y, Li PL. Endothelial NLRP3 inflammasome activation and enhanced neointima formation in mice by adipokine visfatin. *The American Journal of Pathology* **184**(2014): 1617–1628.

Xia M, Zhang C, Boini KM, Thacker AM, Li PL. Membrane raft-lysosome redox signalling platforms in coronary endothelial dysfunction induced by adipokine visfatin. *Cardiovascular Research* **89**(2011): 401–409.

Yang F, Wang Z, Wei X et al. NLRP3 deficiency ameliorates neurovascular damage in experimental ischemic stroke. *Journal of Cerebral Blood Flow and Metabolism* **34**(2014): 660–667.

Yin Y, Pastrana JL, Li X et al. Inflammasomes: Sensors of metabolic stresses for vascular inflammation. *Fronters in Bioscience (Landmark Ed)* **18**(2013): 638–649.

Zhang AY, Yi F, Jin S et al. Acid sphingomyelinase and its redox amplification in formation of lipid raft redox signaling platforms in endothelial cells. *Antioxidants and Redox Signaling* **9**(2007): 817–828.

Zhang AY, Yi F, Zhang G, Gulbins E, Li PL. Lipid raft clustering and redox signaling platform formation in coronary arterial endothelial cells. *Hypertension* **47**(2006): 74–80.

Zhang C, Boini KM, Xia M et al. Activation of Nod-like receptor protein 3 inflammasomes turns on podocyte injury and glomerular sclerosis in hyperhomocysteinemia. *Hypertension* **60**(2012): 154–162.

Zhang C, Li PL. Membrane raft redox signalosomes in endothelial cells. *Free Radical Research* **44**(2010): 831–842.

Zhang Y, Li X, Becker KA, Gulbins E. Ceramide-enriched membrane domains–structure and function. *Biochimica et Biophysica Acta* **1788**(2009): 178–183.

Zhou R, Tardivel A, Thorens B, Choi I, Tschopp J. Thioredoxin-interacting protein links oxidative stress to inflammasome activation. *Nature Immunology* **11**(2010): 136–140.

Zuo L, Ushio-Fukai M, Ikeda S, Hilenski L, Patrushev N, Alexander RW. Caveolin-1 is essential for activation of Rac1 and NAD(P)H oxidase after angiotensin II type 1 receptor stimulation in vascular smooth muscle cells: Role in redox signaling and vascular hypertrophy. *Arteriosclerosis, Thrombosis, and Vascular Biology* **25**(2005): 1824–1830.

CHAPTER 14

Conclusions

Irene Díaz-Moreno and Ricardo O. Louro

Paradigms, like tides, ebb and flood. The original proposal that bioenergetic complexes are organized in solid-like supercomplexes with restricted diffusion of redox shuttles to ensure efficient electron transfer was clearly ahead of its time (Keilin and Hartree, 1947). The overwhelming power of reductionist biology to identify and characterize individual functional elements, and the inability of contemporary structural biology methods to characterize targets that are dynamic and heterogeneous in space and time, played against that idea. Those factors established the opportunity for the emergence of the model of random collision of the complexes (Gupte et al., 1984), based on the fluid mosaic model for the organization of biological membranes (Nicolson, 2014). These seemingly opposing views were since reconciled in the plasticity model (Acín-Pérez and Enriquez, 2014). This model appeared following the systems biology revolution supported by methodological and technological developments at the level of gentle conditions for the purification of membrane proteins and advanced high-resolution imaging. In the frame of the plasticity model, the solid and the fluid models represent extreme manifestations of a dynamic network of multiple association arrangements of the various bioenergetic complexes displaying variable stoichiometry.

Bioenergetic supercomplexes are associated with cell membranes. In this context, it cannot be overlooked that the organization of the cell envelope in Gram-negative bacteria with two membranes, as also found in mitochondria and chloroplasts, is different from that of Gram-positive bacteria with a membrane and a cell wall dominated by peptidoglican, and from that of archaea with different lipid and cell wall composition and chemistry. Furthermore, the individual complexes in Eukarya, Bacteria and Archaea have a different composition and number of subunits. Despite these dramatic differences in the structure and composition of the cell envelope, bioenergetic supercomplexes are found in all domains of life, which is a strong indication that living organisms extract a powerful evolutionary advantage from the existence of supercomplexes. The advantage results from changes in the kinetic and thermodynamic properties of redox proteins when integrated in supercomplexes. Theoretical and experimental procedures to functionally characterize these proteins and their interactions were developed over the years. These allowed the establishment of detailed mechanistic descriptions even in the absence of high-resolution structures, which in the majority of the cases were only obtained much later. Substrate channelling between the various components is the core characteristic of supercomplexes, and is provided for example by confined pools of quinones that do not readily exchange with the bulk quinone pool. Supercomplexes also facilitate fast intermolecular electron transfer, making them independent of the much slower diffusional rate of the individual redox complexes. The common theme in both cases is the establishment of conditions that restrict the mobility of redox components. This prevents dissipation of reducing power to neighbouring redox pools or to off-path redox complexes that would lead to the formation of R(N)OS.

One dramatic example of the consequences of release of redox components is provided by the role of mitochondrial cytochrome *c* in initiating apoptosis. Cytochrome *c* provides a clear example of the connections that exist between supercomplexes of redox proteins that may be called *respirasomes* (Chance and Williams, 1955) and those that form the so-called *signalosomes*. For example, cytochrome *c* mediates peroxidation of cardiolipin, which leads to disruption of supercomplexes and consequent increase in R(N)OS formation. The formation of membrane rafts is also correlated with oxidative modification of membrane lipids. This provides yet another mechanism for linking the activity of bioenergetic supercomplexes and signalosomes. Indeed, the capability of redox proteins to associate in bioenergetic supercomplexes or signalosomes and the dynamic characteristics of these associations are very much contingent on the properties of the membranes to which they are associated. Lipid-protein interactions contribute significantly to the stability, stoichiometry and association-dissociation dynamics of supercomplexes. Therefore, it is not surprising that perturbations of the availability of particular lipids such as ceramide, cardiolipin or cholesterol, or their chemical modification, lead to pathological manifestations with clear phenotypes.

It is becoming more apparent that, particularly at the level of mitochondria, the distinction between bioenergetic supercomplexes and signalosomes is becoming blurred. The original view of mitochondria as the cellular powerhouses gave way to a more nuanced description derived from the myriad of services that they provide to eukaryotic cells. Mitochondria modulate cellular phenotypes through diverse signalling pathways. The mitochondrial membrane potential as well as the phosphorylation/dephosphorylation of the respiratory complexes is modulated by the energetic state of the cell. Under these stimuli, modifications of amino acid sidechains by phosphorylation or oxidation take place. These modifications in turn modulate the degree of supercomplex association, with the ensuing changes in the release of R(N)OS subsequently modulating signalling pathways.

R(N)OS management in the cells and mitochondria is therefore essential for survival and metabolic regulation. Enzymes that participate in this process via the synthesis of antioxidant molecules or by directly scavenging R(N)OS such as thioredoxin become essential. These proteins are key elements of signalosomes due to their role in maintaining redox homeostasis. Disturbances of supercomplex assembly with ROS formation and the consequent accumulation of damage in mtDNA are at the basis of the mitochondrial theory of ageing. But even after damage is inflicted, transient associations between redox proteins and

DNA participate in the process of error detection and correction. Disturbances of charge transfer along the stacked base pairs in double-stranded DNA, resulting from mismatched or chemically modified bases, are used to guide the cooperative search and correction of errors in the DNA pairing (Boal et al., 2009).

The bioenergetic evolutionary pressure on Archaea and Bacteria appears to be different from that imposed on multi-cellular Eukaryotes. The microbial metabolic versatility arises from the fact that homeostasis around the cells can hardly be assured in an open environment. Even microbial biofilms often disassemble in the face of different redox environments (Thormann et al., 2005). This leads to the evolution of enzymatic machinery capable of opportunistically using *exotic* respiratory terminal electron acceptors. These are often not gases, and sometimes not even soluble. These organisms make use of unique strategies to link the intracellular metabolism with extracellular acceptors and control the organization of multi-centre redox proteins at the surface of cells. By contrast, both mitochondria and chloroplasts appear to have been subjected to similar evolutionary pressures that lead to an increase in the membrane surface area relative to the volume in comparison to what is found in Prokaryotes. This enables these organelles to produce more energy because ATP production scales with surface area (Lane, 2005).

The complex topology of the inner membranes of mitochondria and chloroplasts provides hotspots for protein interaction, with a consequent inhomogeneous distribution of the complexes on the membrane surface leading to functionally specialized areas. The vast majority of the respiratory chain complexes are located in the cristae, and therefore, the topology of the inner membrane and its remodelling by proteins that induce curvature in the membrane provide an added mechanism of controlling supercomplex assembly and disassembly. This topological control of the respiratory chain allows a fine regulation of the level of pool behaviour versus substrate channelling of mobile redox carriers. Dynamic rearrangement of photosystems and antennae are essential for the response of light-harvesting organisms to fluctuating conditions of illumination. Also, different oligomerization states of bioenergetic complexes lead to different yields of ATP per NADH, providing a way of balancing the redox status of the cell in face of variable nutrient availability, and energy demand. The localization of the ATP synthases at the cristae apex with the other complexes located along the sides is proposed to drive the proton flow from the source complexes to the ATP synthase. This would minimize their dissipation to the immediate surrounding environment, similar to what happens in substrate channelling and direct intermolecular electron transfer.

In conclusion, bioenergetic supercomplexes of redox proteins assemble and disassemble with variable rates and stoichiometries on the basis of multiple environmental cues. They appear to be designed to provide a tight grip on the length scale of electron and proton transfer, providing speed due to proximity and control due to restricted diffusion. Relaxation of this tight control leads to release of components that initiate signalling cascades, showing the interconnectedness between bioenergetic supercomplexes and signalosomes. In no place better than mitochondria are these interconnectedness more apparent given the added attention that OXPHOS supramolecular organization have attracted due to their relevance for pathological dysfunctions.

REFERENCES

Acín-Pérez, R., and Enriquez, J. (2014). The function of the respiratory supercomplexes: The plasticity model. *Biochim Biophys Acta, 1837*, 444–450.

Boal, A., Genereux, J., Sontz, P., Gralnick, J., Newman, D., and Barton, J. (2009). Redox signaling between DNA repair proteins for efficient lesion detection. *Proc Natl Acad Sci U S A, 106*, 15237–15242.

Chance, B., and Williams, G. (1955). A method for the localization of sites for oxidative phosphorylation. *Nature, 176*, 250–254.

Gupte, S., Wu, E., Hoechli, L., Hoechli, M., Jacobson, K., Sowers, A., and Hackenbrock, C. (1984). Relationship between lateral diffusion, collision frequency, and electron transfer of mitochondrial inner membrane oxidation-reduction components. *Proc Natl Acad Sci U S A, 81*, 2606–2610.

Keilin, D., and Hartree, E. (1947). Activity of the cytochrome system in heart muscle preparations. *Biochem J, 41*, 500–502.

Lane, N. (2005). *Power, Sex, Suicide: Mitochondria and the Meaning of Life*. Oxford University Press.

Nicolson, G. L. (2014). The fluid-mosaic model of membrane structure: Still relevant to understanding the structure, function and dynamics of biological membranes after more than 40 years. *Biochim Biophys Acta, 1838*, 1451–1466.

Thormann, K., Saville, R., Shukla, S., and Spormann, A. (2005). Induction of rapid detachment in Shewanella oneidensis MR-1 biofilms. *J Bacteriol, 187*, 1014–1021.

Index

Note: Locator followed by '*f*' and '*t*' denotes figure and table in the text

D

E

M

N

T

U

V

W